Design and Analysis
of Clinical Trials

WILEY SERIES IN PROBABILITY AND STATISTICS
APPLIED PROBABILITY AND STATISTICS SECTION

Established by WALTER A. SHEWHART and SAMUEL S. WILKS

Editors: *Vic Barnett, Ralph A. Bradley, Noel A. C. Cressie, Nicholas I. Fisher, Iain M. Johnstone, J. B. Kadane, David G. Kendall, David W. Scott, Bernard W. Silverman, Adrian F. M. Smith, Jozef L. Teugels; J. Stuart Hunter, Emeritus*

A complete list of the titles in this series appears at the end of this volume.

Design and Analysis of Clinical Trials

Concept and Methodologies

SHEIN-CHUNG CHOW
Biostatistics and Data Management
Covance, Inc.
Princeton, New Jersey

JEN-PEI LIU
Department of Statistics
National Cheng-Kung University
Tainan, Taiwan

A Wiley-Interscience Publication
JOHN WILEY & SONS, INC.
New York • Chichester • Weinheim • Brisbane • Singapore • Toronto

This book is printed on acid-free paper. ∞

Copyright © 1998 by John Wiley & Sons, Inc. All rights reserved. 0

Published simultaneously in Canada.

No part of this publication may be reproduced, stored in a retrieval system or transmitted in any form or by any means, electronic, mechanical, photocopying, recording, scanning or otherwise, except as permitted under Sections 107 or 108 of the 1976 United States Copyright Act, without either the prior written permission of the Publisher, or authorization through payment of the appropriate per-copy fee to the Copyright Clearance Center, 222 Rosewood Drive, Danvers, MA 01923, (978) 750-8400, fax (978) 750-4744. Requests to the Publisher for permission should be addressed to the Permissions Department, John Wiley & Sons, Inc., 605 Third Avenue, New York, NY 10158-0012, (212) 850-6011, fax (212) 850-6008, E-Mail: PERMREQ @ WILEY.COM.

Library of Congress Cataloging in Publication Data:

Chow, Shein-Chung, 1955–
   Design and analysis of clinical trials : concepts and methodologies / Shein-Chung Chow and Jen-pei Liu.
      p.   cm. — (Wiley series in probability and statistics. Applied probability and statistics)
   Includes bibliographical references and index.
   ISBN 0-471-13404-X (cloth : alk. paper)
   1. Clinical trials—Methodology. 2. Clinical trials–Statistical methods. I. Liu, Jen-Pei, 1952– II. Title. III. Series.
R853.S7C48 1998
615'.19'072—dc21   97-28675

Printed in the United States of America

10 9 8 7 6 5 4 3

Contents

Preface   ix

1 Introduction   1
- 1.1 What Are Clinical Trials? 1
- 1.2 History of Clinical Trials, 3
- 1.3 Regulatory Process and Requirements, 7
- 1.4 Investigational New Drug Application, 16
- 1.5 New Drug Application, 25
- 1.6 Clinical Development and Practice, 34
- 1.7 Aims and Structure of the Book, 43

2 Basic Statistical Concepts   47
- 2.1 Introduction, 47
- 2.2 Uncertainty and Probability, 48
- 2.3 Bias and Variability, 52
- 2.4 Confounding and Interaction, 63
- 2.5 Descriptive and Inferential Statistics, 74
- 2.6 Hypotheses Testing and p-Values, 81
- 2.7 Significant Difference and Clinical Equivalence, 89

3 Basic Design Consideration   93
- 3.1 Introduction, 93
- 3.2 Patient Selection, 94
- 3.3 Selection of Controls, 103
- 3.4 Statistical Consideration, 109
- 3.5 Other Issues, 117
- 3.6 Discussion, 120

4 Randomization and Blinding   121

  4.1 Introduction, 121
  4.2 Randomization Models, 123
  4.3 Randomization Methods, 129
  4.4 Implementation of Randomization, 154
  4.5 Generalization of Controlled Randomized Trials, 160
  4.6 Blinding, 165
  4.7 Discussion, 174

5 Designs for Clinical Trials   176

  5.1 Introduction, 176
  5.2 Parallel Designs, 178
  5.3 Crossover Designs, 183
  5.4 Titration Designs, 191
  5.5 Enrichment Designs, 203
  5.6 Discussion, 210

6 Classification of Clinical Trials   212

  6.1 Introduction, 212
  6.2 Multicenter Trial, 213
  6.3 Active Control Trial, 221
  6.4 Combination Trial, 226
  6.5 Equivalence Trial, 241
  6.6 Discussion, 254

7 Analysis of Continuous Data   259

  7.1 Introduction, 259
  7.2 Estimation, 260
  7.3 Test Statistics, 265
  7.4 Analysis of Variance, 272
  7.5 Analysis of Covariance, 281
  7.6 Nonparametrics, 284
  7.7 Repeated Measures, 291
  7.8 Discussion, 304

8 Analysis of Categorical Data   306

  8.1 Introduction, 306
  8.2 Statistical Inference for One Sample, 312

CONTENTS vii

    8.3 Inference of Independent Samples, 326
    8.4 Ordered Categorical Data, 334
    8.5 Combining Categorical Data, 339
    8.6 Model-Based Methods, 346
    8.7 Repeated Categorical Data, 354
    8.8 Discussion, 360

9 Censored Data and Interim Analysis   363

    9.1 Introduction, 363
    9.2 Estimation of the Survival Function, 365
    9.3 Comparison between Survival Functions, 373
    9.4 Cox's Proportional Hazard Model, 382
    9.5 Calendar Time and Information Time, 400
    9.6 Group Sequential Methods, 407
    9.7 Discussion, 420

10 Sample Size Determination   424

    10.1 Introduction, 424
    10.2 Basic Concept, 425
    10.3 Two Samples, 431
    10.4 Multiple Samples, 441
    10.5 Survival Analysis, 453
    10.6 Dose-Response Studies, 457
    10.7 Crossover Designs, 467
    10.8 Multiple-Stage Design in Cancer Trials, 475
    10.9 Discussion, 481

11 Issues in Efficacy Evaluation   483

    11.1 Introduction, 483
    11.2 Baseline Comparison, 485
    11.3 Intention-to-Treat and Efficacy Analysis, 491
    11.4 Adjustment for Covariates, 497
    11.5 Multicenter Trials, 505
    11.6 Multiplicity, 514
    11.7 Data Monitoring, 525
    11.8 Discussion, 531

12 Safety Assessment — **535**

12.1 Introduction, 535
12.2 Extent of Exposure, 537
12.3 Coding of Adverse Events, 542
12.4 Analysis of Adverse Events, 551
12.5 Analysis of Laboratory Data, 560
12.6 Discussion, 570

References — **572**

Appendix A — **599**

Appendix B — **627**

Index — **635**

Preface

Clinical trials are scientific investigations that examine and evaluate safety and efficacy of drug therapies in human subjects. Biostatistics has been recognized and extensively employed as an indispensable tool for planning, conduct, and interpretation of clinical trials. In clinical research and development, the biostatistician plays an important role that contributes toward the success of the trial. An open and effective communication among clinician, biostatistician, and other related clinical scientists will result in a successful clinical trial. The mutual communication, however, is a two-way street: not only (1) the biostatistician must effectively deliver statistical concepts and methodologies to his/her colleagues but also (2) the clinician must communicate thoroughly clinical and scientific principles embedded in clinical research to the biostatistician. The biostatistician can then formulate these clinical and scientific principles into valid statistical hypotheses, models, and methodologies for data analyses. The integrity, quality, and success of a clinical trial depend on the interaction, mutual respect, and understanding among the clinician, the biostatistician, and other clinical scientists.

There are many books on clinical trials already on market. These books, however, emphasize either statistical or clinical aspects. None of these books provides a balanced view of statistical concepts and clinical issues. Therefore the purpose of this book is not only to fill the gap between clinical and statistical disciplines but also to provide a comprehensive and unified presentation of clinical and scientific issues, statistical concepts, and methodologies. Moreover this book focuses on the interactions among clinicians, biostatisticians, and other clinical scientists that often occur during the various phases of clinical research and development. This book is intended to give a well-balanced overview of current and emerging clinical issues and newly developed statistical methodologies. Although this book is written from a viewpoint of pharmaceutical research and development, the principles and concepts presented in this book can be applied to nonbiopharmaceutical settings.

It is our goal to provide a concise and comprehensive reference book for physicians, clinical researchers, pharmaceutical scientists, clinical or medical research associates, clinical programmers or data coordinators, and biostatisticians in the areas of clinical research and development, regulatory agencies, and academe. Hence this book is written for readers with minimal mathematical and statistical backgrounds. Although it is not required, an introductory statistics course that covers the concepts of probability, sampling distribution, estimation, and hypothesis testing would be helpful. This book can also serve as a textbook for graduate courses in the areas of clinical and pharmaceutical research and development. Readers are encouraged to pay attention to clinical issues and their statistical interpretations as illustrated through real examples from various phases of clinical research and development.

The issues covered in this book may occur during the various phases of clinical trials in pharmaceutical research and development, and their corresponding statistical interpretations, concepts, designs and analyses. All the important clinical issues are addressed in terms of the concepts and methodologies of the design and analysis of clinical trials. For this reason this book is composed of clinical concepts and methodologies. Each chapter with different topics is self-contained.

Chapter 1 provides an overview of clinical development for pharmaceutical entities, the process of drug research and development in pharmaceutical industry, and regulatory processes and requirements. The aim and structure of the book is also discussed in this chapter. The concepts of design and analysis of clinical trials are covered from Chapters 2 through 6. Basic statistical concepts such as uncertainty, bias, variability, confounding, interaction, statistical versus clinical significance are introduced in Chapter 2. Fundamental considerations for the selection of a suitable design in achieving certain objectives of a particular trial under various circumstances are provided in Chapter 3. Chapter 4 illustrates the concepts and different methods of randomization and blinding that are indispensable to the success and integrity of a clinical trial. Chapter 5 introduces different types of statistical designs for clinical trials such as parallel, crossover, titration, and enrichment designs and discusses their relative advantages and drawbacks. Various types of clinical trials, which include multicenter, active control, combination, and equivalence trials, are the subject of Chapter 6.

Methodologies and the issues for clinical data analysis are addressed in Chapters 7 through 12. Since clinical endpoints can generally be classified into three types: continuous, categorical, and censored data, various statistical methods for analyses of these three types of clinical data and their advantages and limitations are provided in Chapters 7, 8, and 9, respectively. In adition, group sequential procedures for interim analysis are given in Chapter 9. Different procedures for sample size determination are provided in Chapter 10 for data under different designs. Statistical issues in analyzing efficacy data are discussed in Chapter 11. These issues include baseline comparisons, intention-to-treat analyses versus evaluable or per-protocol analyses, adjustment of covariates, multiplicity issues, and data monitoring. Chapter 12 focuses on the issues of analysis of safety data, including the extent of exposure, coding and analysis of adverse events, and analysis of laboratory data.

For each chapter, whenever possible, real examples from clinical trials are included to demonstrate the clinical and statistical concepts, interpretations, and their relationships and interactions. Comparisons of the relative merits and disadvantages of statistical methodology for addressing different clinical issues in various therapeutic areas are discussed in appropriate chapters. In addition, if applicable, topics for future development are provided. All computations in this book were performed using SAS. Other statistical packages such as SPSS, BMDP, or MINTAB may also be applied.

At John Wiley, we would like to thank Acquisition Editor Steve Quigley for providing us with the opportunity to work on this book and for his outstanding effort in preparing this book for publication. We greatly indebted to the Bristol-Myers Squibb Company and Covance, Inc. for their support, in particular, to S. A. Henry, L. Meinert, and H. Koffer. We are grateful for A. P. Pong, C. C. Hsieh, and G. Y. Han for their assistance in preparing the many charts, figures, graphs, and tables in this book. We are grateful to Y. C. Chi, F. Ki, and C. S. Lin for many helpful discussions and for reviewing the manuscript. We also wish to thank A. P. Pong, M. L. Lee, and E. Nordbrock for their constant support and encouragement. The first author also wishes to express his appreciation to his wife Yueh-Ji and their daughters, Emily and Lilly, for their patience and understanding during the preparation of this book.

Finally, we are fully responsible for any errors remaining in the book. The views expressed are those of the authors and are not necessarily those of Covance, Inc. and the National Cheng-Kung University.

SHEIN-CHUNG CHOW
JEN-PEI LIU

Princeton, New Jersey
Tainan, Taiwan
October 1997

Design and Analysis
of Clinical Trials

CHAPTER 1

Introduction

1.1 WHAT ARE CLINICAL TRIALS?

Clinical trials are clinical investigations. They have evolved with different meanings by different individuals and organizations at different times. For example, Meinert (1986) indicates that a clinical trial is a research activity that involves administration of a test *treatment* to some *experimental unit* in order to *evaluate* the treatment. Meinert (1986) also defines a clinical trial as a planned experiment designed to assess the efficacy of a treatment in humans by comparing the outcomes in a group of patients treated with the test treatment with those observed in a comparable group of patients receiving a control treatment, where patients in both groups are enrolled, treated, and followed over the same time period. This definition indicates that a clinical trial is used to evaluate the effectiveness of a treatment. Piantadosi (1997) simply defined a clinical trial as an experimental testing medical treatment on human subject. On the other hand, Spilker (1991) considers clinical trials as a subset of clinical studies that evaluate investigational medicines in phases I, II, and III, the clinical studies being the class of all scientific approaches to evaluate medical disease preventions, diagnostic techniques, and treatments. This definition is somewhat narrow in the sense that it restricts to the clinical investigation conducted by pharmaceutical companies during various stages of clinical development of pharmaceutical entities which are intended for marketing approval. The Code of Federal Regulations (CFR) defines a clinical trial as the clinical investigation of a drug that is administered or dispensed to, or used involving one or more human subjects (21 CFR 312.3, April, 1994). Three important key words in these definitions of clinical trials are *experimental unit*, *treatment*, and *evaluation* of the treatment.

Experimental Unit

An *experimental unit* is usually referred to as a subject from a targeted population under study. Therefore the experimental unit is usually used to specify the intended study population to which the results of the study are inferenced.

For example, the intended population could be patients with certain diseases at certain stages or healthy human subjects. In practice, although a majority of clinical trials are usually conducted in patients to evaluate certain test treatments, it is not uncommon that some clinical trials may involve healthy human subjects. For example, at very early phase trials of clinical development, initial investigation of a new pharmaceutical entity may only involve a small number of healthy subjects, say fewer than 30. Large primary prevention trials are often conducted with healthy human subjects with size in tens of thousand subjects. See, for example, *Physician's Health Study* (PHSRG, 1988), *Helsinki Health Study* (Frick et al., 1987), and *Women Health Trial* (Self et al., 1988).

Treatment

In clinical trials a *treatment* can be a placebo or any combinations of a new pharmaceutical identity (e.g., a compound or drug), a new diet, a surgical procedure, a diagnostic test, a medical device, or no treatment. For example, in the *Physician's Health Study*, one treatment arm is a combination of low-dose aspirin and beta carotene. Other examples include lumpectomy, radiotherapy, and chemotherapy as a combination of surgical procedure and drug therapy for breast cancer; magnetic resonance imaging (MRI) with a contrast imaging agent as a combination of diagnostic test and a drug for enhancement of diagnostic enhancement; or a class III antiarrhythmic agent and an implanted cardioverter defibrillator as a combination of a drug and a medical device for treatment of patients with ventricular arrhythmia. As a result a *treatment* is any intervention to be evaluated in human subjects regardless that it is a new intervention to be tested or serves as a referenced control group for comparison.

Evaluation

In his definition of clinical trials, Meinert (1986) emphasizes the *evaluation* of efficacy of a test treatment. It, however, should be noted that the assessment of safety of an intervention such as adverse experiences, elevation of certain laboratory parameters, or change in findings of physical examination after administration of the treatment is at least as important as that of efficacy. Recently, in addition to the traditional evaluation of effectiveness and safety of a test treatment, clinical trials are also designed to assess quality of life and pharmacoeconomics such as cost-minimization, cost-effectiveness, and cost-benefit analyses to human subjects associated with the treatment under study. It is therefore recommended that clinical trials should not only evaluate the effectiveness and safety of the treatment but also assess quality of life, pharmacoeconomics, and outcomes research associated with the treatment.

Throughout this book we will define a clinical trial as a clinical investigation in which treatments are administered, dispensed, or used involving one or more human subjects for evaluation of the treatment. By this definition, the experimental units are human subjects either with a pre-existing disease under study or healthy. Unless otherwise specified, clinical trials in this book are referred to as all clinical investigations in human subjects that may be conducted by pharmaceutical companies, clinical research organizations such as the U.S. National Institutes of Health (NIH), university hospitals, or any other medical research centers.

1.2 HISTORY OF CLINICAL TRIALS

We humans since our early days on earth have been seeking or trying to identify some interventions, whether they be a procedure or a drug, to remedy ailments that inflict ourselves and our loved ones. In this century the explosion of modern and advanced science and technology has led to many successful discoveries of promising treatments such as new medicines. Over the years there has been a tremendous need for clinical investigations of these newly discovered and promising medicines. In parallel, different laws have been enacted and regulations imposed at different times to ensure that the discovered treatments are effective and safe. The purpose for imposing regulations on the evaluation and approval of treatments is to minimize potential risks that they may have for human subjects, especially for those treatments whose efficacy and safety are unknown or are still under investigation.

In 1906 the United States Congress passed the *Pure Food and Drug Act*. The purpose of this act is to prevent misbranding and adulteration of food and drugs. However, the scope of this act is rather limited. No preclearance of drugs is required. Moreover the act does not give the government any authority to inspect food and drugs. Since the act does not regulate the claims made for a product, the Sherley Amendment to the act was passed in 1912 to prohibit labeling medicines with false and fraudulent claims. In 1931 the U.S. Food and Drug Administration (FDA) was formed. The provisions of the FDA are intended to ensure that (1) food is safe and wholesome, (2) drugs, biological products, and medical devices are safe and effective, (3) cosmetics are unadulterated, (4) the use of radiological products does not result in unnecessary exposure to radiation, and (5) all of these products are honestly and informatively labeled (Fairweather, 1994).

The concept of testing marketed drugs in human subjects did not become a public issue until the Elixir Sulfanilamide disaster occurred in the late 1930s. The disaster was a safety concern of a liquid formulation of a sulfa drug that caused more than 100 deaths. This drug had never been tested in humans before its marketing. This safety concern led to the pass of the *Federal Food, Drug and Cosmetic Act* (FD&C Act) in 1938. The FD&C Act extended its coverage to cosmetics and therapeutic devices. More important, the FD&C Act requires the pharmaceutical companies to submit full reports of investigations regarding the safety of new drugs. In 1962 a significant Kefauver-Harris Drug Amendment to the FD&C Act was passed. The Kefauver-Harris Amendment not only strengthened the safety requirements for new drugs but also established an efficacy requirement for new drugs for the first time. In 1984 the Congress passed the *Price Competition and Patent Term Restoration Act* to provide for increased patent protection to compensate for patent life lost during the approval process. Based on this act, the FDA was also authorized to approve generic drugs only based on bioavailability and bioequivalence trials on healthy male subjects. It should be noted that the FDA also has the authority for designation of prescription drugs or over-the-counter drugs (Durham-Humphrey, 1952).

The concept of randomization in clinical trials was not adopted until the early 1920s (Fisher and Mackenzie, 1923). Amerson et al. (1931) first considered randomization of patients to treatments in clinical trials to reduce potential bias and consequently to increase statistical power for detection of a clinically important difference. At the same time a Committee on Clinical Trials was formed by the Medical Research Council of the Great Britain (Medical Research Council, 1931) to promulgate good clinical practice by developing guidelines governing the conduct of clinical studies from which data will be used to support application for marketing approval. In 1937 the NIH awarded its first research grant in clinical trial. At the same time the U.S. National Cancer Institute (NCI) was also formed to enhance clinical research in the area of cancer. In 1944 the first publication of results from a multicenter trial appeared in *Lancet* (Patulin Clinical Trials Committee, 1944). Table 1.2.1 provides a chronic accounts of historical events for both clinical trials and the associated regulations for treatments intended for marketing approval. Table 1.2.1 reveals that the advance of clinical trials goes hand in hand with the development of regulations.

Oklin (1995) indicated that there are at least 8,000 randomized controlled clinical trials conducted each year whose size can include as many as 100,000 subjects. These trials are usually sponsored by the pharmaceutical industry, government agencies, clinical research institutions, or more recently a third party such as health maintenance organizations (HMO) or insurance companies. In recent years clinical trials conducted by the pharmaceutical industry for marketing approval have become more extensive. However, the sizes of clinical trials funded by other organizations are even larger. The trials conducted by the pharmaceutical industry are mainly for the purpose of registration for marketing approval. Therefore they usually follow a rigorously clinical development plan which is usually carried out in phases (e.g., phases I, II, and III trials, which will be discussed later in this chapter) that progress from very tightly controlled dosing of a small number of normal subjects to less tightly controlled studies involving large number of patients.

According to *USA Today* (Feb. 3, 1993), the average time that a pharmaceutical company spends getting a drug to market is 12 years and 8 months. Of this figure, six years and 8 months are spent in clinical trials to obtain the required information for market registration. The FDA review takes 2 years and 6 months. This lengthy clinical development process is necessary to assure the efficacy and safety of the drug product. As a result this lengthy development period sometimes does not allow the access of promising drugs or therapies to subjects with serious or life-threatening illnesses. Kessler and Feiden (1995) point out that the FDA may permit promising drugs or therapies

Table 1.2.1 Significant Historical Events in Clinical Trials and Regulations

Year	Clinical Trials	Regulations
1906		Pure Food and Drug Act (Dr. Harvey Wiley)
1912		Sherley Amendment
1923	First randomization to experiments (Fisher and Mackenzie, 1923)	
1931	First randomization of patients to treatments in clinical trials (Amberson, et al., 1931) Committee on clinical trials by the Medical Research Council of Great Britain (Medical Research Council, 1931)	Formation of U.S. Food and Drug Administration
1937	Formation of National Cancer Institute and First Research Grant by National Institutes of Health (National Institutes of Health, 1981)	
1938		U.S. Federal Food, Drug and Cosmetic Act (Dr. R. Tugwell)
1944	First publication of results from a multicenter trial (Patulin Clinical Trial Committee, 1944)	
1952	Publication of *Elementary Medical Statistics* (Mainland, 1952)	FDA makes designation of Prescription Drug or OTC (Durham-Humphrey, 1952)
1962	Publication of *Statistical Methods in Clinical and Preventive Medicine* (Hill, 1962)	Amendment to the U.S. Food, Drug, and Cosmetic Act (Kefauver-Harris, 1962)
1966		Mandated creation of the local boards (IRB) for Funding by U.S. Public Health Service
1976		Medical Device Amendment to the U.S. Food, Drug Cosmetic Act (1976)
1977		Publications of *General Considerations for Clinical Evaluatin of Drugs* (HEW (FDA), 1977)

Table 1.2.1 *(Continued)*

Year	Clinical Trials	Regulations
1984		Drug Price Competition and Patent Term Restoration Act (Waxman and Hatch, 1984)
1985		NDA rewrite
1988		Publication of *Guidelines for the Format and Content of the Clinical and Statistical Section of an Application* (FDA, 1988)
1990		Publication of *Good Clinical Practice for Trials on Medicinal Products in the European Community* (EC Commission, 1990)
1987		Treatment IND (FDA, 1987)
1992		Parallel track and accelerated approval (FDA, 1992)
1997		Publication of *Good Clinical Practice: Consolidated Guidelines* (ICH, 1996)

currently under investigation to be available to patients with serious or life-threatening diseases under the so-called *treatment IND* in 1987. The *Parallel Track Regulations* in 1992 allow promising therapies for serious or life-threatening diseases to become available with considerably fewer data than required for approval. In the same year, the FDA published the regulations for the *Accelerated Approval* based only on surrogate endpoints to accelerate the approval process for promising drugs or therapies indicated for life-threatening diseases.

The size of trials conducted by the pharmaceutical industry can be as small as a dozen subjects for the phase I trial in human, or it can be as large as a few thousands for support of approval of ticlopidine for stroke prevention (Temple, 1993). The design of the trial can be very simple as the single-arm trial with no control group, or it can be very complicated as a 12-group factorial design for the evaluation of the dose responses of combination drugs. Temple (1993) points out that information accumulated from previous experience in the database of preapproval New Drug Application (NDA) or Product License Application (PLA) can range from a few hundred subjects (e.g., contrast imaging agents) to four or five thousand subjects (antidepressants or antihypertensives, antibiotics, etc.).

When the safety profile and mechanism of action for the efficacy of a new drug or therapy are well established, probably after its approval, a simple but large confirmatory trial is usually conducted to validate the safety and effectiveness of the new drug or therapy. This kind of trial is large in the sense that there are relaxed the entrance criteria to enroll a large number of subjects (e.g., tens of thousands) with various characteristics and care settings. The purpose of this kind of trial is to increase the exposure of a new drug or therapy to more subjects with the indicated diseases. For example, the *Global Utilization of Streptokinase and Tissue Plasminogen Activator for Occluded Coronary Arteries Trial* (GUSTO, 1993) enrolled over 41,000 subjects in 1,081 hospitals from 15 countries while in the *Physician's Health Study* funded by the NIH over 22,000 physicians were randomized to one of four arms in the trial. In addition these trials usually follow subjects for a much longer period of time than most trials for marketing approval. For example, *Helsinki Heart Study* followed a cohort over 4,000 middle-aged men with dyslipidemia for five years (Frick et al., 1987). The recent *Prostate Cancer Prevention Trial* (PCPT) plans to follow 18,000 healthy men over age 55 for 7 years (Feigl et al., 1995). Such trials are simple in the sense that only few important data are collected from each subject. Because the sizes of these trials are considerably large, they can detect a relatively small yet important and valuable treatment effects that previous smaller studies failed to detect. Sometimes, public funded clinical trials can also be used as a basis for approval of certain indications. An example is the combined therapy of leuprolide with flutamide for patients with disseminated, previously untreated D_2 stage prostate cancer. Approval of flutamide was based on a study funded by NCI.

In the very near future, health care providers such as HMO or insurance companies will be more interested in providing funding for rigorous clinical trials to evaluate not only efficacy and safety of therapies but also quality of life, pharmacoeconomics, and outcomes. The purpose of this kind of clinical trial is to study the cost associated with the health care provided. The concept is to minimize the cost with the optimal therapeutic effect under the same quality of health care. Temple (1993) points out that from the results of the study of *Systolic Hypertension in the Elderly* (SHEP), a potential savings of six billion dollars per year can be provided by the treatment regimen of chlorthalidone with a beta blocker backup such as atenol as compared to the combined treatment of an angiotensin converting enzyme (ACE) inhibitor with a calcium channel blocker backup. Temple (1993) also indicates that a multicooperative group study supported by health care providers is already under way to evaluate the effects of bone marrow transplant with aggressive chemotherapy for breast cancer.

1.3 REGULATORY PROCESS AND REQUIREMENTS

Chow and Liu (1995a) indicated that the development of a pharmaceutical entity is a lengthy process involving drug discovery, laboratory development, animal studies, clinical trials, and regulatory registration. The drug development can be classified into nonclinical, preclinical, and clinical development phases. As indicated by the *USA Today* (Feb. 3, 1993), approximately 75% of drug development is devoted to clinical development and regulatory registration. In this section we will focus on regulatory process and requirements for clinical development of a pharmaceutical entity.

For marketing approval of pharmaceutical entities, the regulatory process and requirements may vary from country (or region) to country (or region). For example, the European Community (EC), Japan, and the United States have similar but different requirements as to the conduct of clinical trials and the submission, review, and approval of clinical results for pharmaceutical entities. In this section, for simplicity, we will focus on the regulatory process and requirements for the conduct, submission, review, and approval of clinical trials currently adopted in the United States. As was indicated earlier, the FDA was formed in 1931 to enforce the FD&C Act for marketing approval of drugs, biological products, and medical devices. With very few exceptions, since the enactment of the FD&C Act, treatment interventions such as drugs, biological products, and medical devices either currently on the market or still under investigation are the results of a joint effort between the pharmaceutical industry and the FDA. To introduce regulatory process and requirements for marketing approval of drugs, biological products, and medical devices, it is helpful to be familiar with the functional structure of the FDA.

The Food and Drug Administration

The FDA is a subcabinet organization within the Department of Health and Human Services (HHS) which is one of the major cabinets in the United States government. The FDA is headed by a commissioner with several deputy or associate commissioners to assist him or her in various issues such as regulatory affairs, management and operations, health affairs, science, legislative affairs, public affairs, planning and evaluation, and consumer affairs. Under the office of commissioner, there are currently six different *centers* of various functions for evaluation of food, drugs, and cosmetics. They are Center for Drug Evaluation and Research (CDER), Center for Biologics Evaluation and Research (CBER), Center for Devices and Radiological Health (CDRH), National Center for Toxicological Research (NCTR), Center for Veterinary Medicine (CVM), and Center for Food Safety and Applied Nutrition (CFSAN).

Recently, in the interest of shortening the review process, the sponsors are required to provide the so-called user's fee for review of submission of applications to the FDA. In October 1995 CDER was reorganized to reflect the challenge of improving efficiency and shortening the review and approval process as demanded by the United States Congress and the pharmaceutical industry. Figure 1.3.1 provides the current structure of CDER. Under the current CDER organization, there are two deputy directors for review management and pharmaceutical science, and two associate directors for policy and medical policy assist the center director to ensure the function of CDER. Figure 1.3.1 indicates that CDER is currently composed of six major *offices*. These offices include Office of Management, Office of Training and Communications, Office of Compliance, Office of Information Technology, Office of Review Management, and

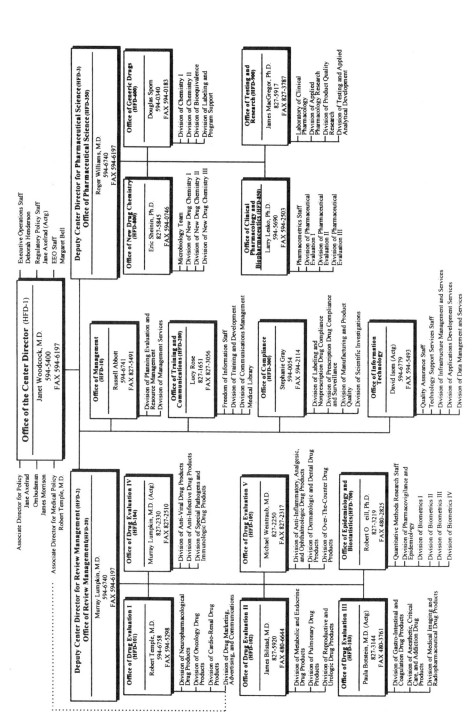

Figure 1.3.1 Center for Drug Evaluation and Research.

Office of Pharmaceutical Science. The Office of Review Management is responsible for drug evaluation and consists of six offices including Offices of Drug Evaluation I–V and the Office of Epidemiology and Biostatistics. On the other hand, the Office of Pharmaceutical Science consists of four offices, which are Office of New Drug Chemistry, Office of Generic Drugs, Office of Clinical Pharmacology and Biopharmaceutics, and Office of Testing and Research. Note that each of these offices are in turn composed of several divisions. Figures 1.3.2, 1.3.3, and 1.3.4 provide respective organizations of Offices of the Center Directors, Pharmaceutical Science, and Review Management. Note that CBER has a similar functional structure though it has fewer offices than CDER.

FDA Regulations for Clinical Trials

For evaluation and marketing approval of drugs, biological products, and medical devices, the sponsors are required to submit substantial evidence of effectiveness and safety accumulated from adequate and well-controlled clinical trials to CDER, CBER, or CDRH of the FDA, respectively. The current regulations for conducting clinical trials and the submission, review and approval of clinical results for pharmaceutical entities in the United States can be found in CFR (e.g., see 21 CFR Parts 50, 56, 312, and 314). These regulations are developed based on the FD&C Act passed in 1938. Table 1.3.1 summarizes the most relevant regulations with respect to clinical trials. These regulations cover not only pharmaceutical entities such as drugs, biological products, and medical devices under investigation but also the welfare of participating subjects and the labeling and advertising of pharmaceutical products. It can be seen from Table 1.3.1 that pharmaceutical entities can be roughly divided into three categories based on the FD&C Act and hence the CFR. These categories include drug products, biological products, and medical devices. For the first category, a drug is as defined in the FD&C Act (21 U.S.C. 321) as an article that is (1) recognized in the U.S. Pharmacopeia, official Homeopathic Pharmacopeia of the United States, or official National Formulary, or a supplement to any of them; (2) intended for use in the diagnosis, cure, mitigation, treatment, or prevention of disease in humans or other animals, or (3) intended to affect the structure or function of the body of humans or other animals. For the second category, a biological product is defined in the 1944 *Biologics Act* (46 U.S.C. 262) as a virus, therapeutic serum, toxin, antitoxin, bacterial or viral vaccine, blood, blood component or derivative, allergenic product, or analogous product, applicable to the prevention, treatment, or cure of disease or injuries in humans. Finally, a medical device is defined as an instrument, apparatus, implement, machine contrivance, implant, *in vitro* reagent, or other similar or related article, including any component, part, or accessory that—similar to a drug—is (1) recognized in the official National Formulary or the U.S. Pharmacopeia or any supplement in them; (2) intended for use in the diagnosis in humans or other animals; or (3) intended to affect the structure or function of the body of humans or other animals.

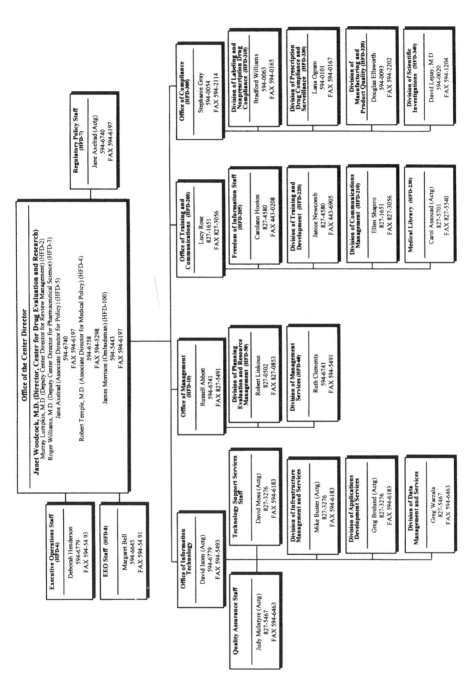

Figure 1.3.2 Office of the Center Director.

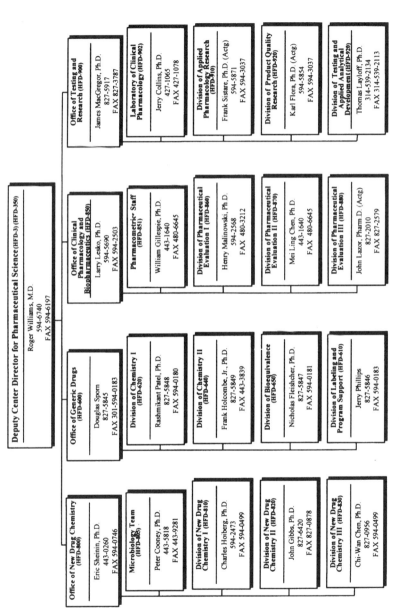

Figure 1.3.3 Office of Pharmaceutical Science.

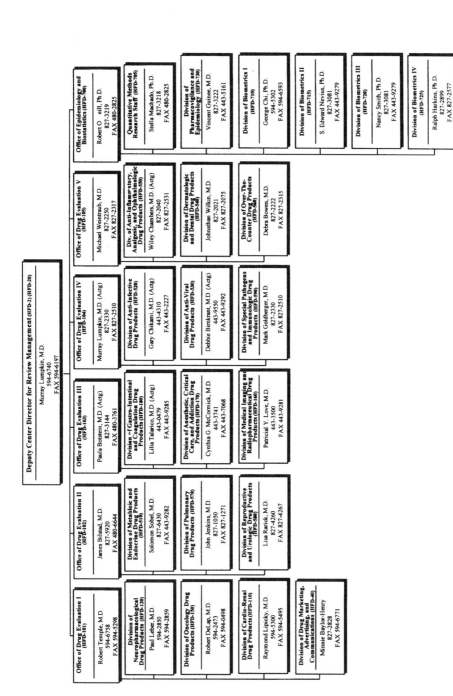

Figure 1.3.4 Office of Review Management.

Table 1.3.1 U.S. Codes of Federal Regulation (CFR) for Clinical Trials Used to Approve Pharmaceutical Entities

CFR Number	Regulations
21 CFR 50	Protection of human subjects
21 CFR 56	Institutional review boards (IRB)
21 CFR 312	Investigational new drug application (IND)
Subpart E	Treatment IND
21 CFR 314	New drug application (NDA)
Subpart C	Abbreviated applications
Subpart H	Accelerated approval
21 CFR 601	Establishment license and product license applications (ELA and PLA)
Subpart E	Accelerated approval
21 CFR 316	Orphan drugs
21 CFR 320	Bioavailability and bioequivalence requirements
21 CFR 330	Over-the-counter (OTC) human drugs
21 CFR 812	Investigational device exemptions (IDE)
21 CFR 814	Premarket approval of medical devices (PMA)
21 CFR 60	Patent term restoration
21 CFR 201	Labeling
21 CFR 202	Prescription drug advertising

The CDER of the FDA has jurisdiction over administration of regulation and approval of pharmaceutical products classified as *drug*. These regulations include Investigational New Drug Application (IND) and New Drug Application (NDA) for new drugs, orphan drugs, and over-the-counter (OTC) human drugs and Abbreviated New Drug Application (ANDA) for generic drugs. On the other hand, the CBER is responsible for enforcing the regulations of biological products through processes such an Establishment License Application (ELA) or Product License Application (PLA). Administration of the regulations for medical devices belongs to the jurisdiction of the CDRH through Investigational Device Exemptions (IDE) and Premarket Approval of Medical Devices (PMA) and other means.

A treatment for a single illness might consist of a combination of drugs, biological products, and/or medical devices. If a treatment consists of a number of drugs, then it is called a combined therapy. For example, leuprolide and flutamide for disseminated, previously untreated D_2 stage prostate cancer or an ACE inhibitor and a calcium channel blocker as a backup for essential hypertension. However, if a treatment consists of a combination of drugs, biologics, and/or devices such as drug with device, biologic with device, drug with biologic, drug with biologic in conjunction with device, then it is defined as a combined product. For a combined product consisting of different pharmaceutical entities, FDA requires that each of entities should be reviewed separately by appropriate centers at the FDA. In order to avoid confusion of jurisdiction over a combination product and to improve efficiency of approval process, the principle of primary mode of action of a combination product was established in the *Safe Medical Devices Act* (SMDA) in 1990 (21 U.S.C. 353). In 1992, based on this principle, three intercenter agreements were signed between CDER and CBER, between CDER and CDRH, and between CBER and CDRH to establish the ground rules for assignment of a combined product and intercenter consultation (Margolies, 1994).

Phases of Clinical Development

In a set of new regulations promulgated in 1987 and known as the *IND Rewrite*, the phases of clinical investigation adopted by the FDA since the late 1970s is generally divided into three phases (21 CFR 312.21). These phases of clinical investigation are usually conducted sequentially but may overlap.

Phase I clinical investigation provides an initial introduction of an investigational new drug to humans. The primary objectives of phase I clinical investigation are twofold. First, it is to determine the metabolism and pharmacologic activities of the drug in humans, the side effects associated with increasing doses, and early evidence on effectiveness. In addition it is to obtain sufficient information about the drug's pharmacokinetics and pharmacological effects to permit the design of well-controlled and scientifically valid phase II clinical studies. Thus phase I clinical investigation includes studies of drug metabolism, bioavailability, dose ranging, and multiple doses. Phase I clinical investigation usually involves 20 to 80 normal volunteer subjects or patients. In general, protocols for phase I studies are less detailed and more flexible than for subsequent phases, but they must provide an outline of the investigation and also specify in detail those elements that are critical to safety. For phase I investigation, FDA's review will focus on the assessment of safety. Therefore extensive safety information such as detailed laboratory evaluations are usually collected at very intensive schedules.

Phase II studies are the first controlled clinical studies of the drug, and they involve no more than several hundred patients. The primary objectives of phase II studies are not only to first evaluate the effectiveness of a drug based on clinical endpoints for a particular indication or indications in patients with the disease or condition under study but also to determine the dosing ranges and doses for phase III studies and the common short-term side effects and risks associated with the drug. Although the clinical investigation usually involves no more than several hundred patients, expanded phase II clinical studies may involve up to several thousand patients. Note that some pharmaceutical companies further differentiate this phase into phases IIA and IIB. Clinical studies designed to evaluate dosing are referred to as phase IIA studies, and studies designed to determine the effectiveness of the drug are called phase IIB.

Phase III studies are expanded controlled and uncontrolled trials. The primary objectives of phase III studies are not only to gather the additional information about effectiveness and safety needed to evaluate the overall benefit-risk relationship of the drug but also to provide an adequate basis for physician labeling. Phase III studies, which can involve from several hundred to several thousand patients, are performed after preliminary evidence regarding the effectiveness of the drug has been demonstrated. Note that studies performed after submission before approval are generally referred to as phase IIIB studies.

In drug development, phase I studies refer to an early stage of clinical pharmacology, and phase II and III studies correspond to a later stage of clinical development. For different phases of clinical studies, the investigational processes are regulated differently, for example, the FDA review of submissions in phase I ensures that subjects are not exposed to unreasonable risks, while the review of submissions in phases II and III also ensures that the scientific design of the study is likely to produce data capable of meeting statutory standards for marketing approval.

Phase IV trials generally refer to studies performed after a drug is approved for marketing. The purpose for conducting phase IV studies is to elucidate further the incidence of adverse reactions and determine the effect of a drug on morbidity or mortality. In addition a phase IV trial is also conducted to study a patient population not previously studied such as children. In practice, phase IV studies are usually considered useful market-oriented comparison studies against competitor products.

Note that there is considerable variation within the pharmaceutical industry in categorizing clinical studies into phases. For example, in addition to phases I through IV described above, some pharmaceutical companies consider clinical studies conducted for new indications and/or new formulations (or dosage forms) as phase V studies.

1.4 INVESTIGATIONAL NEW DRUG APPLICATION

As indicated in the previous section, different regulations exist for different products, such as IND and NDA for drug products, ELA and PLA for biological products, IDE and PMA for medical devices. However, the spirit and principles for the conduct, submission, review, and approval of clinical trials are the same. Therefore, for the purpose of illustration, we will only give a detailed discussion on IND and NDA for drug products.

Before a drug can be studied in humans, its sponsor must submit an IND to the FDA. Unless notified otherwise, the sponsor may begin to investigate the drug 30 days after the FDA has received the application. The IND requirements extend throughout the period during which a drug is under study. As mentioned in Sections 312.1 and 312.3 of 21 CFR, an IND is synonymous with *Notice of Claimed Investigational Exemption for a New Drug*. Therefore an IND is, legally speaking, an exemption to the law that prevents the shipment of a new drug for interstate commerce. Consequently the drug companies that file an IND have flexibility of conducting clinical investigations of products across the United States. However, it should be noted that different states might have different laws that may require the sponsors to file separate IND to the state governments. As indicated by Kessler (1989), there are two types of INDs, commercial and noncommercial. A commercial IND permits the sponsor to gather the data on the clinical safety and effectiveness needed for an NDA. If the drug is approved by the FDA, the sponsor is allowed to market the drug for specific uses. A noncommercial IND allows the sponsor to use the drug in research or early clinical investigation to obtain advanced scientific knowledge of the drug. Note that the FDA itself does not investigate new drugs or conduct clinical trials. Pharmaceutical manufacturers, physicians, and other research organizations such as NIH may sponsor INDs. If a commercial IND proves successful, the sponsor ordinarily submits an NDA. During this period the sponsor and the FDA usually negotiate over the adequacy of the clinical data and the wording proposed for the label accompanying the drug, which sets out description, clinical pharmacology, indications and usage, contraindications, warnings, precautions, adverse reactions, and dosage and administration.

By the time an IND is filed, the sponsor should have enough information about the chemistry, manufacturing, and controls of the drug substance and drug product to ensure the identity, strength, quality, and purity of the investigational drug covered by the IND. In addition the sponsor should provide adequate information about pharmacological studies for absorption, distribution, metabolism, and excretion (ADME) and acute, subacute, and chronic toxicological studies and reproductive tests in various animal species to support that the investigational drug is reasonably safe to be evaluated in clinical trials of various durations in humans.

A very important component of an IND is the general investigational plan, which is in fact an abbreviated version of the clinical development plan for the particular pharmaceutical entity covered by the IND. However, the investigational plan should identify the phases of clinical investigation to be conducted that depend on the previous human experience with the investigational drug. Usually if a new investigational drug is developed in the United States, it is very likely that at the time of filing the IND no clinical trial on human has evern been conducted. Consequently the investigational plan might consist of all clinical trials planned for each stage of phases I, II, and III during the entire development period. On the other hand, some investigational pharmaceutical entities may be developed outside the United States. In this case sufficient human experiences may have already been accumulated. For example, for an investigational drug, suppose that the clinical development plan outside the United States has already completed phase II stage. Then the initial safety and pharmacological ADME information can be obtained from phase I clinical trials. In addition phase II dose response (ranging) studies may provide adequate dose information for the doses to be employed in the planned phase III studies. Consequently the investigational plan may only include the plan for phase III trials and some trials for specific subject population such as renal or hepatic impaired subjects. However, all information and results from phases I and II studies should be adequately documented in the section of previous human experience with the investigational drug in the IND. A general investigational plan may consist of more than one protocol depending on the stage of the clinical investigational plan to be conducted.

An IND plays an important role in the clinical development of a pharmaceutical entity. An IND should include all information about the drug product available to the company up to the time point of filing. Table 1.4.1 lists the contents of an IND provided in Section 312.23 (a) (6) of 21 CFR that a sponsor must follow and submit. A cover sheet usually refers to the form of FDA-1571. The form reinforces the sponsor's commitment to conduct the investigation in accordance with applicable regulatory requirements. A table of contents should also be included to indicate the information attached in the IND submission. The investigational plan should clearly state the rationale for the study of the drug, the indication(s) to be studied, the approach for the evaluation of the drug, the kinds of clinical trials to be conducted, the estimated number of patients, and any risks of particular severity or seriousness anticipated. For completeness, an investigator's brochure should also be provided. As mentioned earlier, the central focus of the initial IND submission should be on the general investigational plan and protocols for specific human studies. Therefore a copy of protocol(s) which includes study objectives, investigators, criteria for inclusion and exclusion, study design, dosing schedule, end-point measurements, and clinical procedure should be submitted along with the investigational plan and other information such as chemistry, manufacturing, and controls, pharmacology and toxicology, previous human experiences with the investigational drug, and any additional information relevant to the investigational drug. Note that the FDA requires that all sponsors should submit an original and two copies of all submissions to the IND file, including the original submission and all amendments and reports.

Clinical Trial Protocol

To ensure the success of an IND, a well-designed protocol is essential when conducting a clinical trial. A protocol is a plan that details how a clinical trial

Table 1.4.1 Documents to Accompany an IND Submission

A cover sheet
A table of contents
The investigational plan
The investigator's brochure
Protocol
Chemistry, manufacturing, and controls information
Pharmacology and toxicology information
Previous human experiences with the investigational drug
Additional information
Relevant information

is to be carried out and how the data are to be collected and analyzed. It is an extremely critical and the most important document, since it ensures the quality and integrity of the clinical investigation in terms of its planning, execution, and conduct of the trial as well as the analysis of the data. Section 312.23 of 21 CFR provides minimum requirements for the protocol of a clinical trial. In addition the *Guideline for the Format and Content of the Clinical and Statistical Sections of an Application* was issued by CDER of the FDA in October 1988. Appendix C of this guideline describes key elements for a well-designed protocol. All of these requirements and elements are centered around experimental units, treatments, and evaluations of the treatments as discussed previously in Section 1.1.

Table 1.4.2 gives an example for format and contents of a well-controlled protocol for a majority of clinical trials. A well-designed protocol should always have a protocol cover sheet to provide a synopsis of the protocol. A proposed protocol cover sheet can be found in Appendix C of the FDA guideline. The objective of the study should be clearly stated at the beginning of any protocols. The study objectives are concise and precise statements of prespecified hypotheses based on clinical responses for evaluation of the drug product under study. The objectives usually consist of the primary objective, secondary objectives, and sometimes the subgroup analyses. In addition these objectives should be such that can be translated into statistical hypotheses. The subject inclusion and exclusion criteria should also be stated unambiguously in the protocol to define the targeted population to which the study results are inferred. The experimental design then employed should be able to address the study objectives with certain statistical inference. A valid experimental design should include any initial baseline or run-in periods, the treatments to be compared, the study configuration such as parallel, crossover, or forced titration, and duration of the treatment. It is extremely important to provide a description of the control groups with the rationale as to why the particular control groups are chosen for comparison.

The methods of blinding used in the study to minimize any potential known biases should be described in detail in the protocol. Likewise the protocol should provide the methods of assignment for subjects to the treatment groups. The methods of assignment are usually different randomization procedures to prevent any systematic selection bias and to ensure comparability of the treatment groups with respect to pertinent variables. Only the randomization of subjects can provide the foundation of a valid statistical inference. A well-designed protocol should describe the efficacy and safety variables to be recorded, the time that they will be evaluated, and the methods to measure them. In addition the methods for measuring the efficacy endpoints such as symptom scores for benign prostatic hyperplasis or some safety endpoints such as some important laboratory assay should be validated and results of validation need to be adequately documented in the protocol. The FDA guideline also call for designation of primary efficacy endpoints. From the primary objective based on

Table 1.4.2 Format and Contents of a Protocol

1. Protocol cover sheet
2. Background
3. Objectives
   Primary
   Secondary
4. Study plan
   Study design
   Subject inclusion criteria
   Subject exclusion criteria
   Treatment plan
5. Study drugs
   Dose and route
   Method of dispensing
   Method and time of administration
   Description of controls
   Methods of randomization and blinding
   Package and labeling
   Duration of treatment
   Concomitant medications
   Concomitant procedures
5. Measurements and observations
   Efficacy endpoints
   Safety endpoints
   Validity of measurements
   Time and events schedules
   Screening, baseline, treatment periods, and
      post-treatment follow-up
6. Statistical methods
   Database management procedures
   Methods to minimize bias
   Sample size determination
   Statistical general considerations
      Randomization and blinding
      Dropouts, premature termination, and missing data
      Baseline, statistical parameters, and covariates
      Multicenter studies
      Multiple testing
      Subgroup analysis
      Interim analysis
   Statistical analysis of demography and baseline characteristics
   Statistical analysis of efficacy data
   Statistical analysis of safety data
7. Adverse events
   Serious adverse events
   Adverse events attributions
   Adverse event intensity

Table 1.4.2 (*Continued*)

7. Adverse events (*Continued*)
    Adverse event reporting
    Laboratory test abnormalities
8. Warning and precautions
9. Subject withdrawal and discontinuation
    Subject withdrawal
    End of treatment
    End of study
10. Protocol changes and protocol deviations
    Protocol changes
    Protocol deviation
    Study termination
11. Institutional review and consent requirements
    Institutional review board (IRB)
    Informed consent
12. Obligations of investigators and administrative aspects
    Study drug accountability
    Case report forms
    Laboratory and other reports
    Study monitoring
    Study registry
    Record retention
    Form FDA 1572
    Signatures of investigators
    Confidentiality
    Publication of results
13. Flow chart of studies activities
14. References
15. Appendixes

the primary efficacy endpoint, the statistical hypothesis for sample size determination can be formulated and stated in the protocol. The treatment effects assumed in both null and alternative hypotheses with respect to the experimental design employed in the protocol and the variability assumed for sample size determination should be described in full detail in the protocol as should the procedures for accurate, consistent, and reliable data. The statistical method section of any protocols should address general statistical issues often encountered in the study. These issues include randomization and blinding, handling of dropouts, premature termination of subjects, and missing data, defining the baseline and calculation of statistical parameters such as percent change from baseline and use of covariates such as age or gender in the analysis, the issues of multicenter studies, and multiple comparisons and subgroup analysis.

If interim analyses or administrative looks are expected, the protocol needs to describe any planned interim analyses or administrative looks of the data and the composition, function, and responsibilities of a possible outside data-monitoring committee. The description of interim analyses consists of monitoring procedures, the variables to be analyzed, the frequency of the interim analyses, adjustment of nominal level of significance, and decision rules for termination of the study. In addition the statistical methods for analyses of demography and baseline characteristics together with the various efficacy and safety endpoints should be described fully in the protocol. The protocol must define adverse events, serious adverse events, and attributions and intensity of adverse events and describe how the adverse events are reported. Other ethical and administration issues should also be addressed in the protocol. They are warnings and precautions, subject withdrawal and discontinuation, protocol changes and deviations, institutional review board and consent form, obligation of investigators, case report form, and others.

It should be noted that once an IND is in effect, the sponsor is required to submit a protocol amendment if there are any changes in protocol that significantly affect the subjects' safety. Under 21 CFR 312.30(b) several examples of changes requiring an amendment are given. These examples include (1) any increase in drug dosage, duration, and number of subjects, (2) any significant change in the study design, (3) the addition of a new test or procedure that is intended for monitoring side effects or an adverse event. In addition the FDA also requires an amendment be submitted if the sponsor intends to conduct a study that is not covered by the protocol. As stated in 21 CFR 312.30(a) the sponsor may begin such study provided that a new protocol is submitted to the FDA for review and is approved by the institutional review board. Furthermore, when a new investigator is added to the study, the sponsor must submit a protocol amendment and notify FDA of the new investigator within 30 days of the investigator being added. Note that modifications of the design for phase I studies that do not affect critical safety assessment are required to be reported to FDA only in the annual report.

Institutional Review Board

Since 1971 the FDA has required that all proposed clinical studies be reviewed both by the FDA and an institutional review board (IRB). The responsibility of an IRB is not only to evaluate the ethical acceptability of the proposed clinical research but also to examine the scientific validity of the study to the extent needed to be confident that the study does not expose its subjects to unreasonable risk (Petricciani, 1981). This IRB is formally designated by a public or private institution in which research is conducted to review, approve, and monitor research involving human subjects. Each participating clinical investigator is required to submit all protocols to an IRB. An IRB must formally grant approval before an investigation may proceed, which is in contrast to the 30-day notification that the sponsors must give the FDA. To ensure that the investigators are included in the review process, the FDA requires that the clinical investigators communicate with the IRB. The IRB must monitor activities within their institutions.

The composition and function of an IRB are subject to FDA requirements. Section 56.107 in Part 56 of 21 CFR states that each IRB should have at least five members with varying backgrounds to promote a complete review of research activities commonly conducted by the institution. In order to avoid conflict of interest and to provide an unbiased and objective evaluation of scientific merits, ethical conduct of clinical trials, and protection of human subjects, the CFR enforces a very strict requirement for the composition of members of an IRB. The research institution should make every effort to ensure that no IRB is entirely composed of one gender. In addition no IRB may consist entirely of members of one profession. In particular, each IRB should include at least one member whose primary concerns are in the scientific area and at least one member whose primary concerns are in nonscientific areas. On the other hand, each IRB should include at least one member who is not affiliated with the institution and who is not part of the immediate family of a person who is affiliated with the institution. Furthermore no IRB should have a member participate in the IRB's initial or continuous review of any project in which the member has a conflicting interest, except to provide information requested by the IRB.

Safety Report

The sponsor of an IND is required to notify FDA and all participating investigators in a written IND safety report of any adverse experience associated with use of the drug. Adverse experiences need to be reported include serious and unexpected adverse experiences. A serious adverse experience is defined as any experience that is fatal, life-threatening, permanently disabling requiring inpatient hospitalization, or congenital anomaly, cancer, or overdose. An unexpected adverse experience is referred to as any adverse experience that is not identified in nature, severity, or frequency in the current investigator brochure or the general investigational plan or elsewhere in the current application, as amended.

The FDA requires that any serious and unexpected adverse experience associated with use of the drug in the clinical studies conducted under the IND in writing to the agency and all participating investigators within 10 working days. The sponsor is required to fill out the FDA-1639 form to report an adverse experience. Fatal or immediately life-threatening experience require a telephone report to the agency within three working days after receipt of the information. A follow-up of the investigation of all safety information is also expected.

Treatment IND

During the clinical investigation of the drug under an IND, it may be necessary and ethical to make the drug available to those patients who are not in the clinical trials. Since 1987 the FDA permits an investigational drug to be used under a treatment protocol or treatment IND if the drug is intended to treat a serious or immediately life-threatening disease, especially when there is no comparable or satisfactory alternative drug or other therapy available to treat that stage of the disease in the intended patient population. FDA, however, may deny a request for treatment use of an investigational drug under a treatment protocol or treatment IND if the sponsor fails to show that the drug may be effective for its intended use in its intended patient population or that the drug may expose the patients to an unreasonable and significant additional risk of illness or injury.

Withdraw and Termination of an IND

At any time a sponsor may withdraw an effective IND without prejudice. However, if an IND is withdrawn, FDA must be notified and all clinical investigations conducted under the IND shall be ended. If an IND is withdrawn because of a safety reason, the sponsor has to promptly inform FDA, all investigators, and all reviewing IRBs with the reasons for such withdrawal.

If there are any deficiencies in the IND or in the conduct of an investigation under an IND, the FDA may terminate an IND. If an IND is terminated, the sponsor must end all clinical investigations conducted under the IND and recall or dispose all unused supplies of the drug. Some examples of deficiencies in an IND are discussed under 21 CFR 312.44. For example, FDA may propose to terminate IND if it finds that human subjects would be exposed to an unreasonable and significant risk of illness or injury. In such a case the FDA will notify the sponsor in writing and invite correction or explanation within a period of 30 days. A terminated IND is subject to reinstatement based on additional submissions that eliminate such risk. In this case a regulatory hearing on the question of whether the IND should be reinstated will be held.

Communication with the FDA

FDA encourages open communication regarding any scientific or medical question that may be raised during the clinical investigation. Basically it is suggested that such communication be arranged at the end of the phase II study and prior to a marketing application. The purpose of an end-of-phase II meeting is to review the safety of the drug proceeding to phase III. This meeting is helpful not only in that it evaluates the phase III plan and protocols but also in that it identifies any additional information necessary to support a marketing application for the uses under investigation. Note that a similar meeting may be held at the end of phase I in order to review results of tolerance/safety studies and the adequacy of the remaining development program. At the end of phase I, a meeting would be requested by a sponsor when the drug or biologic product is being developed for a life-threatening disease and the sponsor wishes to file under the expedited registration regulations. The purpose of pre-NDA meetings is not only to uncover any major unresolved problems but also to identify those studies that are needed for establishment of drug effectiveness. In addition the communication enables the sponsor to acquaint FDA reviewers with the general information to be submitted in the marketing application. More important, the communication provides the opportunity to discuss (1) appropriate methods for statistical analysis of the data and (2) the best approach to the presentation and formatting of the data.

1.5 NEW DRUG APPLICATION

For approval of a new drug, the FDA requires at least two adequate well-controlled clinical studies be conducted in humans to demonstrate substantial evidence of the effectiveness and safety of the drug. The *substantial evidence* as required in the Kefaurer-Harris amendments to the FD&C Act in 1962 is defined as the evidence consisting of adequate and well-controlled investigations, including clinical investigations, by experts qualified by scientific training and experience to evaluate the effectiveness of the drug involved, on the basis of which it could fairly and responsibly be concluded by such experts that the drug will have the effect it purports to is represented to have under the conditions of use prescribed, recommended, or suggested in the labeling or proposed labeling thereof. Based on this amendment, the FDA requests that reports of adequate and well-controlled investigations provide the primary basis for determining whether there is *substantial evidence* to support the claims of new drugs and antibiotics. Section 314.126 of 21 CFR provides the definition of an adequate and well-controlled study, which is summarized in Table 1.5.1. It can be seen from Table 1.5.1 that an adequate and well-controlled study is judged by eight criteria specified in the CFR. These criteria are objectives, method of analy-

Table 1.5.1 Characteristics of an Adequate and Well-Controlled Study

Criteria	Characteristics
Objectives	Clear statement of investigation's purpose
Methods of analysis	Summary of proposed or actual methods of analysis
Design	Valid comparison with a control to provide a quantitative assessment of drug effect
Selection of subjects	Adequate assurance of the disease or conditions under study
Assignment of subjects	Minimization of bias and assurance of comparability of groups
Participants of studies	Minimization of bias on the part of subjects, observers, and analysts
Assessment of responses	Well-defined and reliable
Assessment of the effect	Requirements of appropriate statistical methods

sis, design of studies, selection of subjects, assignment of subjects, participants of studies, assessment of responses, and effect. First, each study should have a very clear statement of objectives for clinical investigation such that they can be reformulated into statistical hypotheses and estimation procedures. In addition proposed methods of analyses should be described in the protocol and actual statistical methods used for analyses of data should be described in detail in the report. Second, each clinical study should employ a design that allows a valid comparison with a control for an unbiased assessment of drug effect. Therefore selection of a suitable control is one of keys to integrity and quality of an adequate and well-controlled study. The CFR recognizes the following controls: placebo concurrent control, dose-comparison concurrent control, no treatment control, active concurrent control, and historical control. Next, the subjects in the study should have the disease or condition under study. Furthermore subjects should be randomly assigned to different groups in the study to minimize potential bias and ensure comparability of the groups with respect to pertinent variables such as age, gender, race, and other important prognostic factors. All statistical inferences are based on such randomization and possibly stratification to achieve these goals. However, bias will still occur if no adequate measures are taken on the part of subjects, investigator, and analysts of the study. Therefore blinding is extremely crucial to eliminate the potential bias from this source. Usually an adequate and well-controlled study is at least double blinded whereby investigators and subjects are blinded to the treatments during the study. However, currently a triple-blind study in which the sponsor (i.e., clinical monitor) of the study is also blinded to the treatment is not uncommon. Another critical criterion is the validity and reliability of assessment of responses. For example, the methods for measurement of responses such as symptom scores for benign prostate hyperplasia should be validated before their usage in the study (Barry et al., 1992). Finally, appropriate statistical methods should be used for assessment of comparability among treatment groups with respect to pertinent variables mentioned above and for unbiased evaluation of drug effects.

Section 314.50 of 21 CFR specifies the format and content of an NDA, which is summarized in Table 1.5.2. The FDA requests that the applicant should submit a complete archival copy of the new drug application form (A) to (F) with a cover letter. In addition, the sponsor needs to submit a review copy for each of the six technical sections with the cover letter, application form (356H) of (A), index of (B), and summary of (C) as given in Table 1.5.2 to each of six reviewing disciplines. The reviewing disciplines include chemistry reviewers for the chemistry, manufacturing, and controls; pharmacology reviewers for nonclinical pharmacology and toxicology; medical reviewers for clinical data section; and statisticians for statistical technical section. The outline of review copies for clinical reviewing divisions include (1) cover letter, (2) application form (356H), (3) index, (4) summary, and (5) clinical section. The outline of review copies for statistical reviewing division consists of (1) cover letter, (2) application form (356H), (3) index, (4) summary, and (5) statistical section.

Table 1.5.2 A Summary of Contents and Format of a New Drug Application (NDA)

Cover letter
A. Application form (365H)
B. Index
C. Summary
D. Technical sections
    1. Chemistry, manufacturing, and controls
    2. Nonclinical pharmacology and toxicology
    3. Human pharmacology and bioavailability
    4. Microbiology (for anti-infective drugs)
    5. Clinical data
    6. Statistical
E. Samples and labeling
F. Case report forms and tabulations
    1. Case report tabulations
    2. Case report forms
    3. Additional data

Note: Based on Section 314.50 of Part 21 of Codes of Federal Regulation (4-1-94 edition).

Table 1.5.3 provides a summary of the format and content of a registration dossier for the European Economic Community (EEC). A comparison of Table 1.5.2 and Table 1.5.3 reveals that the information required by the FDA and ECC for marketing approval of a drug is essentially the same. However, no statistical technical section is required in the ECC registration. In October 1988, to assist an applicant in presenting the clinical and statistical data required as part of an NDA submission, the CDER of the FDA issued the *Guideline for the Format and Content of the Clinical and Statistical Sections of an Application* under 21 CFR 314.50, which is summarized in Table 1.5.4. The guideline indicates the preference of having an integrated clinical and statistical report rather than two separate reports. A complete submission should include clinical section [21 CFR 314.50(d)(5)], statistical section [21 CFR 314.50(d)(6)], and case report forms and tabulations [21 CFR 314.50(f)]. The same guideline also provides the content and format of the fully integrated clinical and statistical report of a controlled clinical study in an NDA. A summary of it is given in Table 1.5.5. Based on the content and format of the fully integrated and statistical report of a controlled study required by the FDA, the *Structure and Content of Clinical Study Reports* was also issued by the European Community in May 1993. A summary is given in Table 1.5.6. In addition the European Community also published a guideline entitled *Biostatistical Methodology in Clinical Trials in Applications for Marketing Authorizations for Medicinal Products* in March 1993.

Table 1.5.3 Format and Contents of a Registration Dossier for the European Economic Community (EEC)

Flyleaf	
Annex I:	General information
Annex II:	Information and documents on physicochemical, biological, or microbiological tests
Annex II.A:	Complete qualitative and quantitative composition
Annex II.B:	Method of preparation
Annex II.C:	Controls of starting materials
Annex II.D:	Control tests on intermediate products (if necessary)
Annex II.E:	Control tests for the finished product
Annex II.F:	Stability tests
Annex II.G:	Conclusions
Annex III:	Toxicological and pharmacological tests
Annex III.A:	Acute toxicity
Annex III.B:	Toxicity with repeated administration
Annex III.C:	Fetal toxicity
Annex III.D:	Fertility studies
Annex III.E:	Carcinogenicity and mutagenicity
Annex III.F:	Pharmacodynamics
Annex III.G:	Pharmacokinetics
Annex IV:	Clinical trials
Annex IV.A:	Human pharmacology
Annex IV.B:	Clinical data
Annex IV.C:	Side effects and interactions
Annex V:	Special particulars
Annex V.A:	Dosage forms
Annex V.B:	Samples
Annex V.C:	Manufacturing authorization
Annex V.D:	Marketing authorization

Expanded Access

As mentioned earlier, a standard clinical development program of phases I, II, and III clinical trials and traditional approval of a new pharmaceutical entity through IND and NDA processes by the FDA will generally take between 8 to 10 years with an average cost around $300 million. Kessler and Feiden (1995) indicated that on average, the FDA receives around 100 original NDAs each year. For each NDA submission, FDA requires substantial evidence of efficacy and safety be provided with fully matured and complete data generated from at least two adequate and well-controlled studies before it can be considered for approval. This requirement is necessary for drugs with marginal clinical advantages and for treatment of conditions or diseases that are not life-threatening. However, if the diseases are life-threatening or severely debilitating, then the

Table 1.5.4 Summary of the Clinical and Statistical Section of an NDA

A. List of investigators; list of INDs and NDAs
B. Background/overview of clinical investigations
C. Clinical pharmacology
D. Control clinical studies
E. Uncontrolled clinical studies
F. Other studies and information
G. Integrated summary of effectiveness data
H. Integrated summary of safety data
I. Drug abuse and overdosage
J. Integrated summary of benefits and risks of the drug

Source: Based on *Guideline for the Format and Content of the Clinical and Statistical Sections of an Application* (July, 1988, Center for Drug Evaluation and Research, FDA)

traditional clinical development and approval process might not be soon enough for the subjects whose life may be saved by the promising drugs. According to Section 312.81 in 21 CFR, life-threatening diseases are defined as (1) the diseases or conditions where the likelihood of death is high unless the course of the disease is interrupted and (2) diseases or conditions with potentially fatal

Table 1.5.5 Summary of Format and Contents of a Fully Integrated Clinical and Statistical Report for a Controlled Study in an NDA

A. Introduction
B. Fully integrated clinical and statistical report of a controlled clinical study
   1. Title page
   2. Table of contents for the study
   3. Identity of the test materials, lot numbers, etc.
   4. Introduction
   5. Study objectives
   6. Investigational plan
   7. Statistical methods planned in the protocol
   8. Disposition of patients entered
   9. Effectiveness results
   10. Safety results
   11. Summary and conclusion
   12. References
   13. Appendixes

Source: Based on *Guideline for the Format and Content of the Clinical and Statistical Sections of an Application* (July, 1988, Center for Drug Evaluation and Research, FDA).

Table 1.5.6 Summary of Format and Contents of Clinical Study Reports for the European Economic Community (EEC)

1. Title page
2. Table of contents for the study
3. Synopsis
4. Investigators
5. Introduction
6. Study objectives
7. Investigational plan
8. Study subjects
9. Effectiveness evaluation
10. Safety evaluation
11. Discussion
12. Overall conclusions
13. Summary tables, figures, and graphs cited in text
14. Reference list
15. Appendixes

Source: Based on Structure and Content of Clinical Study Reports Joint EFPIA/CPMP Document—May 13, 1993.

outcomes, where the endpoint of clinical trial analysis is survival. On the other hand, severely debilitating diseases are those that cause major irreversible morbidity. Since 1987 regulations have been established for early access to promising experimental drugs and for accelerated approval of drugs for treatment of life-threatening or severely debilitating diseases.

Expanded access is devised through treatment IND (Section 312.34 of 21 CFR) and parallel track regulations. For a serious or immediately life-threatening disease with no satisfactory therapy available, as mentioned before, a treatment IND allows promising new drugs to be widely distributed even when data and experience are not sufficient enough for a full marketing approval. On the other hand, for example, for the patients infected with human immunodeficiency virus (HIV) who are not qualified for clinical trials and have no other alternative treatment, parallel track regulations issued in 1992 provide a means for these patients to obtain experimental therapy very early in the development stage through their private physicians. In 1992 the FDA also established the regulations for accelerated approval of the drug for serious or life-threatening diseases based on a surrogate clinical endpoint other than survival or irreversible morbidity (Subpart H of Section 314 in 21 CFR). A new concept for approval called *Telescoping Trials* has also emerged (Kessler and Feiden, 1995). Under this concept, phase III clinical trials might be totally eliminated. For example, the FDA might consider approval of a drug for a serious disease which, during phase II clinical trials, demonstrates a positive impact on survival or irreversible morbidity. The time table for drug evaluation and approval is illustrated in Figure 1.5.1 which is adopted from Kessler and Feiden (1995).

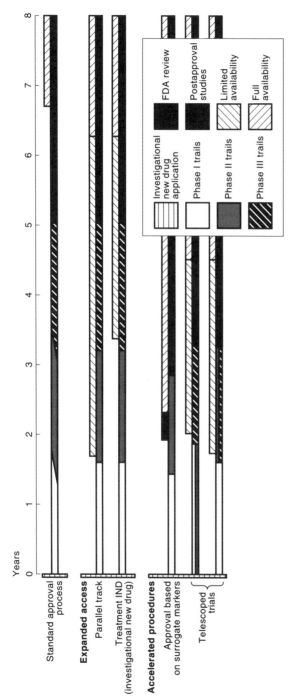

Figure 1.5.1 Time table for drug evaluation and approval. Drug development can take many years of successively larger clinical studies (*top*). The FDA's expanded-access rules now make available to patients serious illness drugs that are still under investigation (*middle*). The agency has also reduced the time it takes to approve new drugs for sale, and it may issue provisional approval for widespread marketing of a compound on the basis of significantly fewer data than it once required (*bottom*). A drug may then be removed from the market if later evaluations show that it is not beneficial. (Source: Kessler and Faiden, 1995.)

A successful example of expanded access and accelerated approval provided by these regulations is the review and approval of dideoxyinosine (ddI) of Bristol-Myers Squibb Company for patients with HIV. An expanded access to ddI was initiated in September 1989. The new drug application based on the data of phase I clinical trials with no control group was filed in April 1991. The FDA granted conditional approval of the drug in October 1991 based on a clinical surrogate end point called a CD4+ lymphocyte count. With the data from phases II and III clinical trials submitted in April 1992, the approval of ddI was broadened in September 1992. The history of ddI case is illustrated in Figure 1.5.2 (also adopted from Kessler and Feiden, 1995). Another example for fast-track development and accelerated approval is the case of fludarabine phosphate (fludara) for treatment of refractory chronic lymphocytic leukemia (CLL) (Tessman, Gipson, and Levins, 1994). Fludara is the first new drug approved for this common form of adult leukemia in the United States over 50 years. The NDA, filed in November 1989 and approved in April 1991, was in fact based on retrospective analyses of phase II clinical trials conducted by NCI through cooperative groups including Southwest Oncology Group (SWOG) and M.D. Anderson Cancer Center in Houston, Texas. In addition an early excess to the drug was provided in 1989 through NCI's Group C protocol, which is equivalent to NCI's version of treatment IND.

Abbreviated New Drug Application

An abbreviated NDA (ANDA) is usually reserved for drug products (e.g., generics) that duplicate products previously approved under a full NDA. For an ANDA, reports of nonclinical laboratory studies and clinical investigations except for those pertaining to *in vivo* bioavailability of the drug product are not

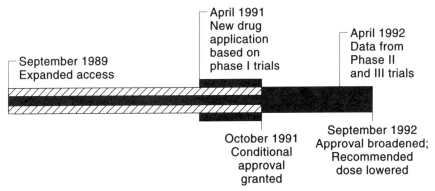

Figure 1.5.2 Case history of dideoxyinosine (ddI) shows how efforts to streamline the regulatory process have paid off. In September 1989 the drug was made available to many AIDs patients on an expanded-access basis. The FDA approved the drug for sale after reviewing preliminary results from ongoing studies and then expanded the approval once final results came in. (Source: Kessler and Faiden, 1995.)

required. The information may be omitted when the FDA has determined that the information already available to it is adequate to establish that a particular dosage form of a drug meets the statutory standards for safety and effectiveness. The duplicate products are usually referred to as products with the same active ingredient(s), route of administration, dosage form, strength, or condition of use that may be made by different manufacturers.

As mentioned earlier, under the *Drug Price Competition and Patient Term Restoration Act* passed in 1984, the FDA may approve generic drug products if the generic drug companies can provide evidence that the rates and extents of absorption of their drug products do not show a significant difference from those of the innovator drug products when administered at the same molar dose of the therapeutic moiety under similar experimental conditions (21 CFR 320). The *Drug Price Competition and Patent Term Restoration* Act states FDA's authority for all generic drug approvals through an ANDA submission for bioequivalence review. An ANDA submission should include product information, pharmacokinetic data and analysis, statistical analysis, analytical methodology and validation, and clinical data. In the ANDA submission the FDA requires the sponsor to provide necessary information regarding the drug product such as formulation, potency, expiration dating period (or shelf life), and dissolution data. For example, the dissolution profile of the generic drug product should be comparable with that of the innovator drug product for drug release. Before the conduct of a bioavailability and bioequivalence study, the FDA also requires the sponsor to provide validation data for the analytical method used in the study. The analytic method should be validated according to standards specified in the U.S. Pharmacopeia and National Formulary (USP/NF, 1995). For example, the analytical method needs to be validated in terms of its accuracy, precision, selectivity, limit of detection, limit of quantitation, range, linearity, and ruggedness (Chow and Liu, 1995). For pharmacokinetic data, descriptive statistics should be given by the sampling time point and for each pharmacokinetic responses. To ensure the validity of bioequivalence assessment, the Division of Bioequivalence, Office of Generic Drugs of CDER at the FDA issued a *Guidance on Statistical Procedures for Bioequivalence Studies Using a Standard Two-Treatment Crossover Design* in July 1992. The guidance sets forth regulations for valid statistical analysis for bioequivalence assessment. Note that detailed information regarding statistical design and analysis of bioavailability and bioequivalence studies can be found in Chow and Liu (1992a). In addition any relevant clinical findings, adverse reactions and deviation from the protocol need to be included in the ANDA submission. Note that a new guidance (issued in October, 1997) *In Vivo Bioequivalence Studies Based on Population and Individual Bioequivalence Approaches* is currently being distributed for comment.

Supplemental New Drug Application

A supplemental NDA (SNDA) is referred to as documentation submitted to FDA on a drug substance or product that is already the subject of an approved NDA. Supplements may be submitted for a variety of reasons such as labeling changes, a new or expanded clinical indication, or a new dosage form. For example, for labeling changes, the sponsor may want to add a new specification or test method or changes in the methods, facility, or controls to provide increased assurance that the drug will have the characteristics of identity, strength, quality, and purity that it purports to possess. For drug substance and/or drug product, the sponsor may want to relax the limits for a specification, establish a new regulatory analytical method, or delete a specification or regulatory analytical method. In addition the sponsor may want to extend the expiration date of the drug product based on data obtained under a new or revised stability testing protocol that has not been approved in the application or to establish a new procedure for reprocessing a batch of the drug product that fails to meet specification. It, however, should be noted that in an SNDA, the sponsor is required to fully describe the change in each condition established in an approved application beyond the variation already provided for in the application.

Advisory Committee

The FDA has established advisory committees each consisting of clinical, pharmacological, and statistical experts and one consumer advocate (not employed by the FDA) in designated drug classes and subspecialities. The responsibilities of the committees are to review data presented in NDA's and to advise FDA as to whether there exists substantial evidence of safety and effectiveness based on adequate and well-controlled clinical studies. In addition the committee may also be asked at times to review certain INDs, protocols, or important issues relating to marketed drugs and biologics. The advisory committees not only supplement the FDA's expertise but also allow an independent peer review during the regulatory process. Note that the FDA usually prepares a set of questions for the advisory committee to address at the meeting. The following is a list of some typical questions:

1. Are there two or more adequate and well-controlled trials?
2. Have the patient populations been well enough characterized?
3. Has the dose-response relationship been sufficiently characterized?
4. Do you recommend the use of the drug for the indication sought by the sponsor for the intended patient population?

The FDA usually will follow the recommendations made by the Advisory Committee for marketing approval, though they do not have to legally.

1.6 CLINICAL DEVELOPMENT AND PRACTICE

Clinical research and development in pharmaceutical environment is to scientifically evaluate the benefits and risks of promising pharmaceutical entities at a minimal cost and within a very short timeframe. To ensure the success of the development of the pharmaceutical entity, a clinical development plan is necessary.

Clinical Development Plan

A clinical development plan (CDP) is a description of clinical studies that will be carried out in order to assess the safety and effectiveness of the drug. A clinical development plan typically includes a development rationale, listing of trial characteristics, timeline, cost, and resource requirements. A good and flexible clinical development plan hence is extremely crucial and important to the success and unbiased assessment of a potential pharmaceutical entity. Although a typical CDP is based primarily on the validity of medical and scientific considerations, other factors that involve issues such as biostatistics, regulatory, marketing, and management are equally important. For a successful CDP, we first need to define a product profile for the promising pharmaceutical entity before any clinical development. Table 1.6.1 lists essential components of a product

Table 1.6.1 Components of a Pharmaceutical Product Profile

Target population
Innovation protentials
  Therapeutic concepts
  Innovative elements
  Technological advances
  Patent status
Route of administrations
  Doses
  Formulations
  Regimens
Duration of dosing
Status of market
Current Competitors
  On market
  Under development
  Advantages
  Disadvantages
Minimum requirements
  Efficacy
  Safety
Termination criteria
  Efficacy
  Safety
Time frames
  Milestones

profile. These components set the goals and objectives for the clinical development program of a pharmaceutical entity. A clinical development program is referred to as the set of different clinical trials plans at different stages with milestones for assessment and decision making to evaluate the goals and objectives stated in the product profile. For example, if the drug product under development is for an indication intended for a particular population, the relative merits and disadvantages of the product as compared to other products either on the market or still under development should objectively be assessed. In order to evaluate the relative merits, minimum requirements and termination criteria on the effectiveness and safety of the product are usually set. These requirements and criteria are evaluated through statistical analysis of data collected from a series of clinical trials. The deadlines for milestones and decision making should also be scheduled in CDP according to the time when certain clinical trials to evaluate the requirements and criteria are completed and the data are adequately analyzed. Since a hugh investment is usually necessarily committed to develop a new pharmaceutical entity, information based on efficacy and safety alone may not be enough to evaluate a potential product. It is therefore recommended that cost-effectiveness and quality of life be evaluated, especially for the me-too products in a saturated market. In this case requirements and criteria for cost-effectiveness and quality of life need to be included at milestones and/or decision-making points. As indicated earlier, although many factors such as statistics, marketing, regulatory, and management need to be considered in a CDP, the scientific validity of clinical investigations is the key to the success of a clinical development program.

In the pharmaceutical industry clinical development of a pharmaceutical entity starts with seeking alternatives or new drug therapies for an existing health problem (e.g., hypertension) or a newly identified health problem (e.g., AIDS). The health problem of interest may be related to virus, cardiovascular, cancer diseases, or other diseases. Once the health problem is selected or identified, whether it is worth developing an alternative or a new pharmaceutical entity for this particular disease is a critical development decision point. A clear decision point can increase the success of the project and consequently reduce the risk and cost. Suppose that it is decided to proceed with the development of a pharmaceutical entity (e.g., enzymes or receptors), a number of chemical modifications and ADME tests in animals may be necessary before it can be tested on humans. ADME studies are used to determine how a drug is taken up by the body, where it goes in the body, the chemical changes it undergoes in the body, and how it is eliminated from the body. ADME studies describe the pharmacokinetics and bioavailability of a drug. If the drug shows promising effectiveness and safety in animals, the sponsor normally will make a decision to go for an IND. As indicated in Section 1.4, an IND is a synthesis process that includes formulation, analytical method development and validation, stability, animal toxicity, pharmacokinetic/pharmacology, previous human experience, and clinical development. The sponsor will then prepare a registration document that combines all the relevant data to allow the FDA to review and decide whether to approve marketing of the new drug. As discussed in Section 1.5, an NDA submission should include chemistry, pharmacology, toxicology, metabolism, manufacturing, quality controls, and clinical data along with the proposed labeling.

Good Clinical Practices

Good clinical practices (GCP) is usually referred to as a set of standards for clinical studies to achieve and maintain high-quality clinical research in a sensible and responsible manner. The FDA, the Committee for Proprietary Medicinal Products (CPMP) for the European Community, the Ministry of Health and Welfare of Japan, and other countries worldwide have each issued guidelines on good clinical practices. For example, the FDA promulgated a number of regulations and guidelines governing the conduct of clinical studies from which data will be used to support applications for marketing approval of drug products. The FDA regulations refer to those regulations specified in 21 CFR Parts 50, 56, 312, and 314, while the FDA guidelines are guidelines issued for different drug products such as *Guidelines for the Clinical Evaluation of Anti-Anginal Drugs and Guidelines for the Clinical Evaluation of Bronchodilator Drugs.* On the other hand, the European Community established the principles for their own GCP standard in all four phases of clinical investigation of medicinal products in July 1990. Basically these guidelines define the responsibilities of sponsors, monitors, and investigators in the initiation, conduct, documentation, and verification of clinical studies to establish the credibility of data and to protect the rights and integrity of study participants.

In essence GCP concerns patient protection and the quality of data used to prove the efficacy and safety of a drug product. GCP ensures that all data, information, and documents relating to a clinical study can be confirmed as being properly generated, recorded, and reported through the institution by independent audits. Therefore the basic GCP concerns are not only the protection of study subjects through informed consent and consultation by ethics committees such as IRB but also the responsibilities of the sponsors and monitors to establish written procedures for study monitoring and conduct and to ensure that such procedures are followed. In addition GCP emphasizes the responsibilities of investigator to conduct the study according to the protocol and joint responsibilities for data reporting, recording, analysis, and archiving as well as prompt reporting of serious adverse events. Moreover GCP calls for the most appropriate design for a valid statistical evaluation of the hypotheses of the clinical trials. The chosen design must suit the purpose with the best possible fit. Incorporating the concerns of GCP in the protocol will ensure a protocol of high standard, which in turn will help generate high-quality data.

Study conduct according to GCP standards requires regular visits to investigating center to monitor study progress. The activities of the sponsor's monitors that will affect the investigator and support staff should be stated in the protocol. Not only this is courteous, it prevents misunderstanding, facilitates cooperation, and aids the speedy acquisition of completed case report form. The activities include frequency of monitoring visits, activities while on site (e.g., auditing CRFs), and departments to be visited (e.g., pharmacy). The practical effects of adopting GCP are that the investigator is audited by the sponsor's monitors (to confirm data on CRFs are a true transcript of original records), by a sponsor administratively separate from the clinical function and in some countries, by the national regulatory agency. The sponsor's monitors are audited by a compliance staff and by national regulatory agencies to confirm the accuracy of data recorded and the implementation of all written procedures such as standard operating procedure (SOP) and protocol.

Most of pharmaceutical companies and research institutions have a protocol review committee (PRC) to evaluate the quality and integrity of the protocol and hence to approve or disapprove the protocol. Some companies also ask the principal study medical monitor and statistician to submit a case report form (CRF) and a statistical analysis plan with mock tables and listing for presentation of the results to PRC at the same time when the protocol is submitted for review.

Lisook (1992) has assembled a GCP packet to assist the sponsors in the planning, execution, data analysis, and submission of results to the FDA. A summary of this GCP packet is given in Table 1.6.2. Most of these regulations have been discussed in the previous sections of this chapter.

Table 1.6.2 References to Keep at Hand for Good Clinical Practice

1. Information on FDA regulations
2. Center for Drug Evaluation and Research publications
3. Clinical Investigations (excerpt from the *Federal Register*, 9-27-77)
4. Protection of Human Subjects, Informed Consent Forms
5. New Drug, Antibiotic, and Biologic Drug Product Regulations; Final Rule (excerpt from the *Federal Register*, 3-19-87)
6. Investigational New Drug, Antibiotic, and Biologic Drug Product Regulations; Treatment Use and Sale; Final Rule (excerpt from the *Federal Register*, 5-22-87)
7. *Guideline for the Monitoring of Clinical Investigations*
8. Investigational New Drug, Antibiotic, and Biologic Drug Product Regulations; Procedure Intended to Treat Life-Threatening and Severely Debilitating Illness; Interim Rule (excerpt from the *Federal Register*, 10-22-88)
9. FDA IRB (Institution Review Board) Information Sheets
10. FDA Clinical Investigator Sheet
11. Reprint of Alan B. Lisook, M.D. FDA audits of clinical studies: Policy and procedure, *Journal of Clinical Pharmacology*, **30** (April 1990) 296–302.
12. Federal Policy for the Protection of Human Subjects; Notices and Rules (excerpt from the *Federal Register*, 6-18-91)
13. *FDA Compliance Program Guidance Manual-Clinical Investigators* (9-1-91)
14. *FDA Compliance Program Guidance Manual-Sponsors, Contract Research Organization and Monitors* (12-3-91)

In the past, as demonstrated in Tables 1.5.2 and 1.5.3, Tables 1.5.5 and 1.5.6, health regulatory authorities in different countries have different requirements for approval of commercial use of the drug products. As a result, lots of resource had been spent by the pharmaceutical industry in the preparation of different documents for applications of the same pharmaceutical product to meet different regulatory requirements requested by different countries or regions. However, because of globalization of the pharmaceutical industry, arbitrary differences in regulations, increase of health care costs, need for reduction of time for patients to access new drugs, and of experimental use of humans and animals without compromising safety, the necessity to standardize these similar yet different regulatory requirements has been recognized by both regulatory authorities and pharmaceutical industry. Hence, The International Conference on Harmonization of Technical Requirements for the Registration of Pharmaceuticals for Human Use (ICH) was organized to provide an opportunity for important initiatives to be developed by regulatory authorities as well as industry association for the promotion of international harmonization of regulatory requirements.

Currently, ICH, however, is only concerned with tripartite harmonization of technical requirements for the registration of pharmaceutical products among three regions: The European Union, Japan, and the United States. As a result, the organization of the ICH consists of six parties of these three regions which include the European Commission of the European Union, the European Federation of Pharmaceutical Industries' Associations (EFPIA), the Japanese Ministry of Health and Welfare (MHW), the Japanese Pharmaceutical Manufacturers Association (JPMA), the Centers for Drug Evaluation and Research and Biologics Evaluation and Research of the US FDA, and the Pharmaceutical Research and Manufacturers of America (PhRMA). The ICH steering committee was established in April, 1990 to (1) determine policies and procedures, (2) to select topics, (3) to monitor progress, and (4) to oversee preparation of biannual conferences. Each of the six parties has two seats on the ICH steering committee. The ICH steering committee also includes observers from the World Health Organization, the Canadian Health Protection Branch, and the European Free Trade Area which have one seat each on the committee. In addition, two seats of the ICH Steering Committee are given to the International Federation of Pharmaceutical Manufacturers Association (IFPMA), which represents the research-based pharmaceutical industry from 56 countries outside ICH regions. IFPMA also runs the ICH Secretariat at Geneva, Switzerland which coordinates the preparation of documentation.

In order to harmonize technical procedures the ICH has issued a number of guidelines and draft guidelines. After the ICH steering committee selected the topics, the ICH guidelines initiated by a concept paper and went through a 5-step review process given in Table 1.6.3. The number of ICH guidelines and draft guidelines at various stages of review process is given in Table 1.6.4. Table 1.6.5 provides a list of currently available ICH guidelines or draft guidelines pertaining to clinical trials while Table 1.6.6 gives the table of contents for the ICH draft guideline on general considerations for clinical trials. In addition,

Table 1.6.3 Review Steps for the ICH Guidelines

Step 1
1. Harmonize topic identified
2. Expert working group (EWG) formed
3. Each party has a topic leader and a deputy
4. Rapporteur for EWR selected
5. Other parties represented on EWG as appropriate
6. Produce a guideline, policy statement, "points to consider"
7. Agreement on scientific issues
8. Sign-off and submit to the ICH steering committee

Step 2
1. Review of ICH document by steering committee
2. Sign-off by all six parties
3. Formal consultation in accord with regional requirements

Step 3
1. Regulatory rapporteur appointed
2. Collection and review of comments across all three regions
3. Step 2 draft revised
4. Sign-off by EWR regulatory members

Step 4
1. Forward to steering committee
2. Review and sign-off by three regulatory members of ICH
3. Recommend for adoption to regulatory bodies

Step 5
1. Recommendations are adopted by regulatory agencies
2. Incorporation into domestic regulations and guidelines

the table of contents of the ICH guidelines for good clinical practices: consolidated guidelines, for structure and content of clinical study reports, and for statistical principles for clinical trials are given, respectively in Tables 1.6.7, 1.6.8, and 1.6.9. From these tables, it can be seen that these guidelines are not only for harmonization of design, conduct, analysis, and report for a single clinical trial but also for consensus in protecting and maintaining the scientific integrity of the entire clinical development plan of a pharmaceutical entity. Along this line, Chow (1997) introduced the concept of *good statistics practice* (GSP) in drug development and regulatory approval process as the foundation of ICH GCP.

Table 1.6.4 Summary of the Number of ICH Guidelines or Draft Guidelines

	Step 1	Step 2	Step 3	Step 4	Step 5
Efficacy	1	3	1	2	5
Safety	1	3	1	0	7
Quality	2	2	0	5	5

Table 1.6.5 The ICH Clinical Guidelines or Draft Guidelines

Draft Guidelines

Timing of nonclinical studies for the conduct of human clinical trials for pharmaceuticals
Standardization of medical terminology for regulatory purposes
Clinical safety data management: Data elements for transmission of individual case report forms
Ethnic factors in the acceptability of foreign clinical data
General considerations for clinical trials
Statistical principles for clinical trials

Guidelines

The extent of population exposure to assess clinical safety for drugs intended for long-term treatment of nonlife-threatening conditions
Clinical safety data management: Definitions and standards for expedited reporting
Clinical safety data management: Periodic safety update reports for marketed drugs
Structure and content of clinical studies
Dose-response information to support drug registration
Good clinical practice: Consolidated guideline
Studies in support of special populations: Geriatrics
Choice of control group in clinical trials
Electronic standards for the transfer of regulatory information

Table 1.6.6 The Table of Contents for the Draft Guideline on General Considerations for Clinical Trials

1. Objectives of this document
2. General principles
   2.1 Protection of clinical trial subjects
   2.2 Scientific approach in design and analysis
3. Development methodology
   3.1 Considerations for development
      3.1.1 Nonclinical studies
      3.1.2 Quality of investigational medicinal products
      3.1.3 Phases of clinical development
      3.1.4 Special considerations
   3.2 Considerations for individual clinical trials
      3.2.1 Objectives
      3.2.2 Design
      3.2.3 Conduct
      3.2.4 Analysis
      3.2.5 Reporting

Table 1.6.7 Table of Contents for the ICH Guideline on Good Clinical Practice: Consolidated Guideline

Introduction
1. Glossary
2. The Principles of ICH GCP
3. The Institutional Review Board/Independent Ethnic Committee (IRB/IEC)
    3.1 Responsibilities
    3.2 Composition, functions, and operations
    3.3 Procedures
    3.4 Records
4. Investigators
    4.1 Investigator's qualifications and agreements
    4.2 Adequate resources
    4.3 Medical care of trial subjects
    4.4 Communication with IRB/IEC
    4.5 Compliance with protocol
    4.6 Investigational products
    4.7 Randomization procedures and unblinding
    4.8 Informed consent of trial subjects
    4.9 Records and reports
    4.10 Progress reports
    4.11 Safety reporting
    4.12 Premature termination or suspension of a trial
    4.13 Final report(s) by investigator/institution
5. Sponsor
    5.1 Quality assurance and quality control
    5.2 Contract research organization
    5.3 Medical expertise
    5.4 Trial design
    5.5 Trial management, data handling, recordingkeeping, and independent data monitoring committee
    5.6 Investigator selection
    5.7 Allocation of duties and functions
    5.8 Compensation to subjects and investigators
    5.9 Financing
    5.10 Notification/submission to regulatory authority(ies)
    5.11 Confirmation review by IRE/IEC
    5.12 Information on investigational product(s)
    5.13 Manufacturing, packaging, labeling, coding investigation product(s)
    5.14 Supplying and handling, investigational product(s)
    5.15 Record access
    5.16 Safety information
    5.17 Adverse drug reaction reporting
    5.18 Monitoring
    5.19 Audit

Table 1.6.7 *(Continued)*

5. Sponsor (*Continued*)
    5.20   Noncompliance
    5.21   Premature termination or suspension of a trial
    5.22   Clinical trial/study reports
    5.23   Multicenter trials
6. Clinical trial protocol and protocol amendment(s)
    6.1   General information
    6.2   Background information
    6.3   Trial objectives and purpose
    6.4   Trial design
    6.5   Selection and withdrawal of subjects
    6.6   Treatment of subjects
    6.7   Assessment of efficacy
    6.8   Assessment of safety
    6.9   Statistics
    6.10   Direct assess to source data/documents
    6.11   Quality control and quality assurance
    6.12   Ethics
    6.13   Data handling and recordkeeping
    6.14   Financing and insurance
    6.15   Publication
    6.16   Supplements
7. Investigator's Brochure
    7.1   Introduction
    7.2   General considerations
    7.3   Contents of the investigator's brochure
    7.4   Appendix 1
    7.5   Appendix 2
8. Essential documents for the conduct of a clinical trial
    8.1   Introduction
    8.2   Before and clinical phase of the trial commences
    8.3   During the clinical conduct of the trial
    8.4   After completion or termination of the trial

The concepts and principles stated in the ICH clinical guidelines will be introduced, addressed, and discussed in the subsequent chapters of this book.

1.7 AIMS AND STRUCTURE OF THE BOOK

As indicated earlier, clinical trials are scientific investigations that examine and evaluate drug therapies in human subjects. Biostatistics has been recognized and extensively employed as an indispensable tool for planning, conduct, and interpretation of clinical trials. In clinical research and development the biostatistician plays an important role that contributes toward the success of clinical trials. Well-prepared and open communication among clinicians, biostatisticians, and other

Table 1.6.8 Table of Contents for the ICH Guideline on Structure and Contents of Clinical Study Reports

Introduction to the guideline
1. Title page
2. Synopsis
3. Table of contents for the individual clinical study report
4. List of abbreviations and definition of terms
5. Ethics
6. Investigators and study administrative structure
7. Introduction
8. Study objectives
9. Investigational plan
10. Study patients
11. Efficacy evaluation
12. Safety evaluation
13. Discussion and overall conclusions
14. Tables, figures, graphs referred to but not included in the text
15. Reference list
16. Appendices

related clinical research scientists will result in a successful clinical trial. Communication, however, is a two-way street: Not only (1) must the biostatistician effectively deliver statistical concepts and methodologies to his or her clinical colleagues but also (2) the clinicians must communicate thoroughly clinical and scientific principles embedded in clinical research to the biostatisticians. The biostatisticians can then formulate these clinical and scientific principles into valid statistical hypotheses under an appropriate statistical model. Overall, the integrity, quality, and success of a clinical trial depends on the interaction, mutual respect, and understanding between the clinicians and the biostatisticians.

The aim of this book is not only to fill the gap between clinical and statistical disciplines but also to provide a comprehensive and unified presentation of clinical and scientific issues, statistical concepts, and methodology. Moreover the book will focus on the interactions between clinicians and biostatisticians that often occur during various phases of clinical research and development. This book is also intended to give a well-balanced summarization of current and emerging clinical issues and recently developed corresponding statistical methodologies. Although this book is written from the viewpoint of pharmaceutical research and development, the principles and concepts presented in this book can also be applied to a nonbiopharmaceutical setting.

It is our goal to provide a comprehensive reference book for physicians, clinical researchers, pharmaceutical scientists, clinical or medical research associates, clinical programmers or data coordinators, and biostatisticians or statisticians in the areas of clinical research and development, regulatory agencies, and academe. The scope of this book covers clinical issues, such as may occur during various phases of clinical trials in clinical and pharmaceutical research and develop-

Table 1.6.9 Table of Contents for the ICH Guideline on Statistical Principles for Clinical Trials

I. Introduction
    1.1 Background and purposes
    1.2 Scope and direction
II. Considerations for overall clinical development
    2.1 Study context
    2.2 Study scope
    2.3 Design techniques to avoid bias
III. Study design considerations
    3.1 Study configuration
    3.2 Multicenter trials
    3.3 Type of comparisons
    3.4 Group sequential designs
    3.5 Sample size
    3.6 Data capture and processing
IV. Study conduct
    4.1 Study monitoring
    4.2 Changes in inclusion and exclusion criteria
    4.3 Accrual rates
    4.4 Sample size adjustment
    4.5 Interim analysis and early stopping
    4.6 Role of independent data monitoring committee (IDMC)
V. Data analysis
    5.1 Prespecified analysis plan
    5.2 Analysis sets
    5.3 Missing values and outliers
    5.4 Data transformation/modification
    5.5 Estimation, confidence interval and hypothesis testing
    5.6 Adjustment of type I error rate and confidence levels
    5.7 Subgroup, interactions, and covariates
    5.8 Integrity of data and computer software
VI. Evaluation of safety and tolerability
    6.1 Scope of evaluation
    6.2 Choice of variables and data collection
    6.3 Set of subjects to be evaluated and presentation of data
    6.4 Statistical evaluation
    6.5 Single study versus integrated summary
VII. Reporting
    7.1 Evaluation and reporting
    7.2 Summarizing the clinical database
Annex 1 Glossary

ment, their corresponding statistical interpretations, concepts, designs, and analyses, such as are adopted to address these important clinical issues.

The scope of this book covers clinical issues, which may occur during various phases of clinical trials in pharmaceutical research and development, their corresponding statistical interpretations, concepts, designs and analyses, which are adopted to address these important clinical issues. Basically, this book is devoted to the concepts and methodologies of design and analysis of clinical trials. As a result, this book can be divded into two parts: concepts and methodologies. Each part consists of several chapters with different topics. Each part and each chapter are self-contained. But, at the same time, parts and chapters are arranged in a sensible manner such that there is a smooth transition between parts and from chapter to chapter within each part.

Chapter 1 provides an overview of clinical development for pharmaceutical entities, the process of drug research and development in the pharmaceutical industry, and regulatory processes and requirements. The aim and structure of the book are also given in this chapter. The concepts of design and analysis of clinical trials are covered from Chapter 2 through Chapter 6. Basic statistical concepts such as uncertainty, bias, variability, confounding, interaction, and statistical versus clinical significance are introduced in Chapter 2. Fundamental considerations for selection of a suitable design for achieving certain objectives of a particular trial under various circumstances are provided in Chapter 3. Chapter 4 illustrates the concepts and different methods of randomization and blinding, which are critically indispensable for the success and integrity of a clinical trial. Chapter 5 introduces different types of statistical designs for clinical trials, such as, parallel, crossover, titration, and enrichment designs, as well as their relative advantages and drawbacks. Various types of clinical trials, which include multicenter, active control, combination, and equivalence trials will be discussed in Chapter 6.

Methodologies and various issues for the analysis of clinical data are addressed from Chapter 7 to Chapter 12. Since clinical endpoints can generally be classified into three types: continuous, categorical, and censored data, various statistical methods for analyses of these three types of clinical data and their advantages and limitations are provided in Chapters 7, 8, and 9, respectively. In addition, group sequential procedures for interim analysis are also given in Chapter 9. Different procedures for sample size determination are provided in Chapter 10 for various types of data under different designs. Statistical issues in analyzing efficacy data are discussed in Chapter 11. These issues include baseline comparison, intention-to-treat analysis versus evaluable or per-protocol analysis, adjustment of covariates, multiplicity issues, and data monitoring. Chapter 12 focuses on the issues of analysis of safety data, including extent of exposure, coding and analysis of adverse events, and analysis of laboratory data.

For each chapter, whenever possible, real examples from clinical trials are included to demonstrate the clinical and statistical concepts, interpretations, and their relationships and interactions. Comparisons regarding the relative merits and disadvantages of the statistical methodology for addressing different clinical issues in various therapeutic areas are discussed wherever deemed appropriate. In addition, if applicable, topics for future research development are provided. All computations in this book are performed using version 6.10 of SAS. Other statistical packages such as BMDP or MINITAB can also be applied.

CHAPTER 2

Basic Statistical Concepts

2.1 INTRODUCTION

As was indicated in the preceding chapter, the FDA requires that two adequate well-controlled clinical trials be conducted to demonstrate the effectiveness and safety of a drug product. The success of an adequate well-controlled clinical trial depends on a well-designed protocol. A well-designed protocol describes how the clinical trial is to be carried out, which ensures the quality of clinical data collected from the trial. Based on the high-quality clinical data, appropriate statistical methods can then be applied to provide a valid and unbiased assessment of the efficacy and safety of the drug product. Spilker (1991) indicated that the greater the attention paid to the planning phase of a clinical trial, the greater the likelihood that the clinical trial will be conducted as desired. In this chapter we will describe several basic statistical concepts and issues that have a great impact on the success of a clinical trial during its planning, design, execution, analysis, and reporting phases. These basic statistical concepts include uncertainty and probability, bias and variability, confounding and interaction, and descriptive and inferential statistics using hypotheses testing and p-values, for example.

In the medical community there are many unknowns remaining in the clinical research of certain diseases such as AIDS. These unknowns or uncertainties are often scientific questions of particular interest to clinical scientists. Once a scientific question regarding the uncertainty of interest is clearly stated, clinical trials are necessarily conducted to provide scientific or clinical evidence to statistically address the uncertainty. Under some underlying probability distribution assumption, a statistical inference can then be derived based on clinical data collected from a representative sample of the targeted patient population. To provide a valid statistical assessment of the uncertainty with a desired accuracy and reliability, statistical and/or estimation procedures should possess the properties of unbiasedness and least variability whenever possible. In practice, well-planned statistical designs can generally serve to avoid unnecessary bias and minimize the potential variability that can occur during the conduct of the clinical trials. In some clinical trials design factors such as race and gender may have an impact on the statistical inference of clinical evaluation of the study medication. In this case it is suggested that possible confounding and/or interaction effects be carefully identified and separated from the treatment effect in order to have a valid and unbiased assessment of the clinical evaluation of the study medication. After the completion of a clinical trial, the collected clinical data can either be summarized descriptively to provide a quick overview of clinical results or be analyzed to provide statistical inference on clinical endpoints of interest. Descriptive statistics usually provide useful information regarding a potential treatment effect. This information can be confirmed by a valid statistical inference with a certain assurance that can be obtained through an appropriate statistical analysis such as hypotheses testing and p-values.

These basic statistical concepts play an important role in the success of clinical trials. They are helpful not only at the very early stage of a study's concept statement development but also at the stage of protocol development. In addition to these basic statistical concepts, there are many statistical/medical issues that can affect the success of a clinical trial. For example, clinical scientists may be interested in establishing clinical efficacy using a one-sided test procedure rather that a two-sided test procedure based on prior experience of the study medication. It should be noted that different test procedures require different sample sizes and address different kinds of uncertainty regarding the study medication. In addition clinical scientists always focus on clinical difference rather than statistical difference. It should be noted that the statistical test is meant to detect a statistical difference with a desired power. The discrepancy between a clinical difference and a statistical difference has an impact on the establishment of clinical equivalence between treatments. As a result these issues are also critical in the success of clinical trials which often involve considerations from different perspectives such as political, medical, marketing, regulatory, and statistical.

2.2 UNCERTAINTY AND PROBABILITY

For a medication under investigation, there are usually many questions regarding the properties of the medication that are of particular interest to clinical scientists. For example, the clinical scientists are interested to know whether the study medication works for the intended indication and patient population. In addition the clinical scientists may be interested in knowing whether the study medication can be used as a substitute for other medications currently available on the market. To address these questions (or uncertainties) regarding the study medication, clinical trials are necessarily conducted to provide scientific/clinical evidence for a fair scientific/clinical evaluation/justification. In order to address the uncertainty regarding the study medication for the targeted patient population, a representative sample is typically drawn for clinical evaluation according to a well-designed protocol. Based on clinical results from the study, statistical inference on the uncertainty can then be made under some underlying probability distribution assumption. As a result the concept of uncertainty and probability plays an important role for clinical evaluation of a study medication.

Uncertainty

Bailar (1992) indicated that uncertainties of interest to clinical scientists include uncertainty from confounders, uncertainty regarding scientific or medical assumptions (or hypotheses), and uncertainty about the generalization of the results from animals to humans. These uncertainties are to be verified through clinical trials. A recent example of uncertainty on medical assumptions is the cardiac arrhythmia suppression trial (CAST). Two antiarrhythmic agents, namely, encainide and flecainide were approved for the indication of ventricular arrhythmia by the FDA based on indisputable evidence on the suppression of objective endpoint premature ventricular beats (PVB) per hours as documented by ambulatory 24-hour Holter monitor. This is because the occurrence of premature ventricular depolarization is considered a risk factor in the survivor of myocardial infarction. The approval of encainide and flecainide in treating patients who survived myocardial infarction is based on the fundamental assumption that the suppression of PVB will reduce the chance of subsequent sudden death. This crucial assumption of treating patients with asymptomatic or symptomatic ventricular arrhythmia after myocardial infarction was never challenged until the CAST trial initiated by the U.S. National Heart, Lung, and Blood Institute. After an average of 10 months of follow-up, an interim analysis conducted by the investigators of the CAST discovered that the relative risk of deaths from arrhythmia and nonfatal cardiac arrest of the patients receiving encainide or flecainide (33 of 725) as compared to the placebo (9 of 725) is 3.5. In other words, the chance of death or suffering cardiac arrest for patients who took either encainide or flecainide were three times as high as those who took placebo. The study was terminated shortly after this finding. As a result the investigators recommended that neither encainide nor flecainide be used to treat patients with asymptomatic or mildly symptomatic ventricular arrhythmia after myocardial infarction. The saga of CAST demonstrates that the assumption of the suppression of PVB as a surrogate clinical endpoint for survival of this patient population is inadequate. In clinical trials we usually make scientific or medical assumptions that are based on previous animal or human experiences. It is suggested that these critical assumptions be precisely stated in the protocol for clinical test, evaluation, and interpretation.

Uncertainty regarding confounders and scientific or medical assumptions can not usually be quantified until the confounders are controlled or assumptions are properly investigated. Note that the concept of confounders will be introduced later in this chapter. In practice, the uncertainty caused by known variations can be statistically quantified. For example, in a recent GUSTO (the Global Use of Strategies to Open Occluded Coronary Arteries) study it was suggested that for the patients with evolving myocardial infarction, on the average, the intravenous administration of t-PA (accelerated) over a period of one and a half hour produced a 14.5% reduction in 30-day mortality as compared with the streptokinase therapy. An interesting question is, How likely is it that the same mortality reduction will be observed in patients with similar characteristics as the 41,021 patients enrolled in the GUSTO study? It depends on underlying source of variations. The accelerated t-PA provided a 21% mortality reduction for patients who are younger than 75 years old, while only a 9% reduction in the patients older than 75. This illustrates the uncertainty due to biological variation. Thus, due to the various sources of variation, before the administration of the accelerated t-PA, cardiologists can only expect that on the average, a 93.7% myocardial infarction will be saved. Although the 30-day survival of a patient can be realized by the accelerated t-PA, until it is administrated and the patient is observed over a period of 30 days. Even though the relationship between the accelerated t-PA and 30-day mortality was deterministic due to the variation from different causes, clinicians have to think probabilistically in the application of any clinical data, such as the results of the GUSTO study.

Probability

The purpose of clinical trials is not only to investigate or verify some scientific or medical hypotheses of certain interventions in a group of patients but also to be able to apply the results to the targeted patient population with similar characteristics. This process is called (statistical) inference. It is the process by which clinicians can draw conclusions based on the results observed from the targeted patient population. Suppose that a clinical trial is planned to study the effectiveness of a newly developed cholesterol-lowering agent in patients with hypercholesterolemia as defined by nonfasting plasma total cholesterol level being greater than 250 mg/dL. Furthermore assume that this new agent is extremely promising as shown by previous small studies and that the elevation of the cholesterol level is a critical factor for reduction of the incidence of coronary heart disease. For this reason the government is willing to provide unlimited resources so that every patient with hypercholesterolemia in the country has the opportunity to be treated with this promising agent. One of the primary clinical responses is the mean reduction in total cholesterol level after six months of treatment from the baseline. In this hypothetical trial the targeted patient population is patients with hypercholesterolemia. The mean reduction of total cholesterol level after six months of treatment is a characteristic regarding the targeted patient population that we are interested in this study. If we measure mean reduction in total cholesterol level for each patient, the mean reductions in total cholesterol level from baseline would form a population distribution. Statistically a distribution can be characterized by its location, spread and skewness. The location of a distribution is also referred to as the central tendency of the distribution. The most commonly used measures for the central tendency are arithmetic mean, median, and mode. The *arithmetic mean* is defined as the sum of the reductions divided by the number of patients in the patient population. The most frequently occurring reductions are called the *mode*, while the median is the middle value of the reductions among all patients. In other words, half of the patients have their reductions above the median and the other half have theirs below it. *Spread* is the variation or dispersion among the patients. The commonly used measures for variation are range, variance, and standard deviation (SD). The *range* is simply the difference between the largest and smallest reductions in the patient population. The *variance*, however, is the sum of the squares of the deviations from the mean divided by the number of the patients. The most commonly employed measure for dispersion is the *standard deviation*, which is defined as the positive square root of the variance. These measures for description of the population distribution are called parameters. Statistically we can impose some probability laws to describe the population distribution. The most important and frequently used probability law is probably the bell-shaped normal distribution which serves an adequate and satisfactory model for description of many responses in clinical research such as height, weight, total cholesterol level, blood pressure, and many others. For this hypothetical study, suppose that mean and standard deviation of the reduction in total cholesterol level from baseline are 50 and 10 mg/dL, respectively. Then about 68% patients are expected to have reductions between 40 and 60 mg/dL. The chance that a reduction exceeds 70 mg/dL is only about 2.5%. Under the normal probability law, the uncertainty of reduction in total cholesterol level can be completely quantified because the scope of this trial is the entire population of patients with hypercholesterolemia. In reality, however, the government will have a limited budget/resource. We therefore could not conduct such intensive study. Alternatively, we randomly select a representative sample from the patient population. For this sample, similar descriptive measures for central tendency and variation are computed. They are called *statistics*. These statistics are estimates for the corresponding parameters of the patient population. Since it is almost impossible to conduct a clinical trial on the entire patient population and the parameters are always unknown, statistics, in turn, can be used to approximate its corresponding population parameters. This process of making a definite conclusion about the patient population based on the results of randomly selected samples is referred to as *statistical inference*. Statistical inference provides an approximation to the parameters of the patient population with certain assurance. It therefore involves uncertainty too. The closeness of the approximation to the unknown population parameters by the sample statistics can also be quantified by the probability laws in statistics. These probability laws are called the sampling distribution of sample statistics. The sampling distributions are the basis of statistical inference. To provide a better understanding, we continue the above example concerning the study of reduction in total cholesterol level in patients with hypercholesterolemia. Suppose we randomly select a sample of 100 patients from the patient population and calculate the mean reduction after six months of treatment. In practice, we can select another sample of 100 patients from the same patient population with replacement and compute its mean reduction. We can repeat this sampling process indefinitely and compute mean reduction for each sample drawn from the same population. The sampling distribution for the sample mean reductions in total cholesterol level for the sample size of 100 patients can then be determined as the frequency distribution of these mean reductions. It can be verified that the mean of the sampling distribution of the sample mean reductions in total cholesterol level is equal to its unknown population parameter. This desirable statistical property is called unbiasedness. In practice, we only draw one sample from the patient population, and hence we only have one sample mean cholesterol reduction. It is then of interest to know how we would judge the closeness of the sample mean cholesterol reduction to its corresponding population parameter. This can be determined by the precision of the sample mean cholesterol reduction. With the sample size of 100, therefore, the variance of the sample mean is simply equal to the population variance divided by 100. The square root of the variance of the sample mean is called the *standard error* (SE) of the sample mean. The approximation of the population mean by the sample mean can be quantified by its standard error. The smaller the standard error is, the closer to the unknown population mean it is. When the sample size is sufficiently large, the sampling distribution will behave like a normal distribution regardless of its corresponding population distribution. This important property is called the *central limit theorem*. As a result we can quantify the standard error of the sample mean in conjunction with the central limit theorem in terms of probability, that is, the closeness of the approximation of the sample mean to the population mean in the total cholesterol reduction provided that the sample size is at least of moderate size.

2.3 BIAS AND VARIABILITY

As was indicated earlier, the FDA requires that the results from clinical trials be accurate and reliable in order to provide a valid and unbiased assessment of true efficacy and safety of the study medication. The accuracy and reliability are usually referred to as the closeness and the degree of the closeness of the clinical results to the true value regarding the targeted patient population. The accuracy and reliability can be assessed by the bias and variability of the primary clinical endpoint used for clinical assessment of the study medication. In what follows we will provide more insight regarding bias and variability separately.

Bias

Since the accuracy of the clinical results is referred to as closeness to the true value, we measure any deviation from the true value. The deviation from the true value is considered as a *bias*. In clinical trials, clinical scientists would make any attempt to avoid bias in order to ensure that the collected clinical results are accurate. It, however, should be noted that most biases are probably caused by human errors. *Webster's II New Riverside University Dictionary* (1984) defines bias as an inclination or preference, namely one that interferes with impartial judgment. Along this line, Minert (1986) considers bias as a preconceived personal preference or inclination that influences the way in which a measurement, analysis, assessment, or procedure is performed or reported. Spilker (1991), on the other hand, views bias as a systematic error that enters a clinical trial and distorts the data obtained, as opposed to a random error that might enter a clinical trial. Yet another definition, by Sackett (1979), describes bias as any process at any stage of inference that tends to produce results or conclusions that differ systematically from the truth. As a summary of these different definitions, we define bias as a systematic error that deviates data from the truth caused by the partial judgment or personal preference that can occur at any stage of a clinical trial. Clinical trials are usually planned, designed, executed, analyzed, and reported by a team that consists of clinical scientists from different disciplines to evaluate the effects of the treatments in a targeted population of human subjects. When there are such nonnegligible differences in human background, education, training, and opinions, it is extremely difficult to remain totally impartial to every aspect at all stages of a clinical trial. Bias inevitably occurs. Where bias occurs, the true effects of the treatment cannot be accurately estimated from the collected data. Since it is almost impossible for a clinical trial to be free of any biases, it is crucial to identify any potential bias that may occur at every stage of a clinical trial. Once the potential bias are identified, one can then implement some procedures such as blinding or randomization to minimize or eliminate the bias.

Sackett (1979) partitions a clinical trial (or research) into seven stages at which bias can occur. These seven stages are (1) in reading up on the field, (2) in specifying and selecting the study sample, (3) in executing the experimental maneuver, (4) in measuring exposures and outcomes, (5) in analyzing the data, (6) in interpreting the analysis result, and (7) in publishing the results. Sackett also provides a detailed catalog of biases for each stage of a case-control trial (see Table 2.3.1). Although the purpose of this catalog is to demonstrate that more biases can occur in case-control studies than in randomized control trials, it still can be applied to the different phases of clinical trials. In addition to the 57 types of bias listed by Sackett (1979), Spilker (1991) describes the six types of biases summarized in Table 2.3.2. Here we will classify all the possible biases into three groups: bias due to selection, observation, and statistical procedures.

Selection bias is probably the most common source of bias that can occur in clinical trials. For example, at the planning stage of a clinical trial, selection bias can occur if clinical scientists review only a partial existing literature on current treatments for a certain disease. A review that does not provide a full spectrum of all possible positive and negative results of certain treatments in all possible demographic subpopulations will in turn bias the thinking of the clinical scientists who are planning the study. As a result a serious selection bias will be introduced in the selection of patients and the corresponding treatment assignment.

Table 2.3.1 Catalog of Biases

1. In Reading the Literature

1.1 *Bias of rhetoric.* Any of several techniques used to convince the reader without appealing to reasons.
1.2 *All's well literature bias.* Scientific or professional societies may publish reports or editorials that omit or play down controversies or disparate results.
1.3 *One-sided reference bias.* Authors may restrict their references to only those works that support their position; a literature review with a single starting point risks confinement to a single side of the issue.
1.4 *Positive results bias.* Authors are more likely to submit, and editors accept, positive than null results.
1.5 *Hot stuff bias.* When a topic is hot, neither investigators or editors may be able to resist the temptation to publish additional results, no matter how preliminary or shaky.

2. In Specifying and Selecting the Study Sample

2.1 *Popularity bias.* The admission of patients to some practices, institutions, or procedures (surgery, autopsy) is influenced by the interest stirred by the presenting and its certain causes.
2.2 *Centripetal bias.* Reputations of certain clinicians and institutions cause individuals with certain disorders or exposures to gravitate toward them.
2.3 *Referral filter bias.* As a group of ill persons referred from primary to secondary to tertiary care, the number of rare causes, multiple diagnoses, and hopeless cases may increase, such as secondary hypertension at the Cleveland Clinic (Gifford, 1969).
2.4 *Diagnostic access bias.* Individuals differ in their geographic, temporal, and economic access to the diagnostic procedures which label them as having a given disease.
2.5 *Diagnostic suspicion bias.* Knowledge of the subject's prior exposure to a putative cause (ethnicity, taking a certain drug, having a second disorder, being exposed in an epidemic) may influence both the intensity and the outcome of the diagnostic procedure.
2.6 *Unmasking (detection signal) bias.* An innocent exposure may become suspect if, rather than causing a disease, it causes a sign or symptom that precipitates a search for the disease, such as the current controversy over postmenopausal estrogens and cancer of the endometrium.
2.7 *Mimicry bias.* An innocent exposure may become suspect if, rather than causing a disease, it causes a (benign) disorder that resembles the disease.
2.8 *Previous opinion bias.* The tactics and results of a previous diagnostic process on a patient, if known, may affect the tactics and results of a subsequent diagnostic process on the same patient, such as multiple referrals among hypertensive patients.
2.9 *Wrong sample size bias.* Samples that are too small can prove nothing; samples that are too large can prove anything.

Table 2.3.1 (*Continued*)

2.10 *Admission rate (Berkson) bias.* If hospitalization rates differ for different exposure/disease groups, the relation between exposure and disease will become distorted in hospital-based studies.
2.11 *Prevalence-incidence (Neyman) bias.* A late look at those exposed (or affected) early will miss fatal and other short episodes, plus mild or "silent" cases and cases where evidence of exposure disappears with disease onset.
2.12 *Diagnostic vogue bias.* The same illness may receive different diagnostic labels at different points in space or time.
2.13 *Diagnostic purity bias.* "Pure" diagnostic groups that exclude comorbidity may become nonrepresentative.
2.14 *Procedure selection bias.* Certain clinical procedures may be preferentially offered to those who are poor risk, such as selection of patients for "medical" versus "surgical" therapy.
2.15 *Missing clinical data bias.* Clinical data may be missing if they are normal, negative, never measured, or measured but never recorded.
2.16 *Noncontemporaneous control bias.* Secular changes in definitions, exposures, diagnoses, diseases, and treatments may render noncontemporaneous controls noncomparable.
2.17 *Starting time bias.* Failure to identify a common starting time for exposure or illness may lead to systematic misclassification.
2.18 *Unacceptable disease bias.* Socially unacceptable disorders (V.D., suicide, insanity) tend to be under-reported.
2.19 *Migrator bias.* Migrants may differ systematically from those who stay home.
2.20 *Membership bias.* Membership in a group (the employed, joggers, etc.) may imply a degree of health that differs systematically from the general population, such as more exercise and recurrent myocardial infarction.
2.21 *Nonrespondent bias.* Nonrespondents (or "latecomers") from a specified sample may exhibit exposures or outcomes that differ from those of early respondents such as in cigarette smoking.
2.22 *Volunteer bias.* Volunteers in a study sample may exhibit exposures or outcomes (they tend to be healthier) that differ from those of nonvolunteers or latecomers, such as in the screening selection.

3. In Executing the Experiment Maneuver (or Exposure)

3.1 *Contamination bias.* In an experiment when members of the control group inadvertently receive the experiment maneuver, the difference in outcomes between experimental and control patients may be systematically reduced.
3.2 *Withdrawal bias.* Patients who are withdrawn from an experiment may be differ systematically from those who remain, such as in a neurosurgical trial of surgical versus medical therapy of cerebrovascular disease, patients who died during surgery were withdrawn as "unavailable for follow-up" and excluded from early analyses.
3.3 *Compliance bias.* In requiring patient adherence to therapy, issues of efficacy become confounded with those of compliance.

Table 2.3.1 (*Continued*)

3.5 *Bogus control bias.* When patients allocated to an experimental maneuver die or sicken before or during its administration and are omitted or reallocated to the control group, the experimental maneuver will appear spuriously superior.

4. In Measuring Exposure and Outcomes

4.1 *Insensitive exposures and outcomes.* When outcome measures are incapable of detecting clinically significant changes or difference, type II errors occur.
4.2 *Underlying causing bias (rumination bias).* Patients may ruminate about possible causes for their illness and exhibit different recall or prior exposure than controls.
4.3 *End-digit preference bias.* In converting analog to digital data, observers may record some terminal digits with an unusual frequency.
4.4 *Apprehension bias.* Certain measures (pulse, blood pressure) may alter systematically from their usual levels when the subject is apprehensive.
4.5 *Unacceptability bias.* Measurements that hurt, embarrass, or invade privacy may be systematically refused or evaded.
4.6 *Obsequiousness bias.* Subjects may systematically alter questionnaire responses in the direction they perceive desired by the investigator.
4.7 *Expectation bias.* Observers may systematically err in measuring and recording observations so that they concur with prior expectations.
4.8 *Substitution game.* The substitution of a risk factor that has not been established as causal for its associated outcome.
4.9 *Family information bias.* The flow of family information about exposure and illness is stimulated by, and directed to, a newly arising case.
4.10 *Exposure suspicion bias.* A knowledge of the subject's disease status may influence both the intensity and outcome of a search for exposure to the putative cause.
4.11 *Recall bias.* Questions about specific exposures may be asked several times of cases but only once of controls.
4.12 *Attention bias.* Study subjects may systematically alter their behavior when they know they are being observed.
4.13 *Instrument bias.* Defects in the calibration or maintenance of measurement instruments may lead to systematic deviations from true value.

5. In Analyzing Data

5.1 *Post hoc significant bias.* When decision levels or "tails" for α and β are selected after the data have been examined, conclusions may be biased.
5.2 *Data dredging bias (looking for the pony).* When data are reviewed for all possible associations without prior hypothesis, the results are suitable for hypothesis-forming activities only.
5.3 *Scale degradation bias.* The degradation and collapsing of measurement scales tend to obscure differences between groups under comparison
5.4 *Tidying-up bias.* The exclusion of outliers or other untidy results cannot be justified on statistical grounds and may lead to bias.
5.5 *Reported peek bias.* Repeated peeks at accumulating data in a randomized trial are not independent and may lead to inappropriate termination.

Table 2.3.1 *(Continued)*

6. In Interpreting the Analysis

6.1 *Mistaken identity bias.* In compliance trials, strategies directed toward improving the patients compliance may, instead or in addition, cause the treating clinician to prescribe more vigorously such that the effect on achievement of the treatment goal may be misinterpreted.

6.2 *Cognitive dissonance bias.* The belief in a given mechanism may increase rather than decrease in the face of contradictory evidence.

6.3 *Magnitude bias.* In interpreting a finding, the selection of a scale of measurement may markedly affect the interpretation; for example, 1,000,000 may be also be 0.0003% of the national budget.

6.4 *Significance bias.* The confusion of statistical significance with biologic or clinical or health care significance can lead to fruitless and useless conclusion.

6.5 *Correlation bias.* Equating correlation with causation leads to errors of both kinds.

6.6 *Underexhaustion bias.* The failure to exhaust the hypothesis space may lead to authoritarian rather than authoritative interpretation.

As indicated in the previous chapter, one of the most critical questions often asked at the FDA advisory committee meeting is whether the patient population has been sufficiently characterized. It should be recognized that the patients in a clinical trial are only a sample with certain characterizations of demography and disease status defined by the inclusion and exclusion criteria of the protocol. The real question is whether the patients are a true representative sample of the targeted patient population. Ideally the patients in a clinical trial should be a representative sample randomly selected from the targeted patient popu-

Table 2.3.2 Other Types of Bias

1. *Selection bias.* Physicians may recruit patients for clinical trials in ways that abuse the data.
2. *Information bias.* The information patients provide to physicians (or others) is heavily tainted by their own beliefs and values.
3. *Observer bias.* The objectivity of physicians or others who measure the magnitude of patient responses varies greatly, even in tests with objective endpoints.
4. *Interviewer bias.* Interviewer bias is a well-known and obvious source of bias in clinical trials, particularly when interviews are used in measuring endpoints that determine the clinical trial's outcome.
5. *Use of nonvalidated instruments.* The use of nonvalidated instruments is widespread in clinical trials.
6. *Active control bias.* Biased higher cure rates for a new antifungal medicine compared with an active medicine rather than compared with a placebo.

lation to which the results of the trial can be inferenced. However, in practice, most of clinical trials enrolled patients sequentially as long as they meet the inclusion and exclusion criteria. This practice is quite different from that of survey sampling or political poll for which samples can easily be selected at random using a telephone book or voter's registration list. For clinical trials it is almost impossible to achieve the ideal goal of random selection of representative sample from the targeted population. If the size of the study is quite large, for example, the GUSTO trial (1993), valid inference and bias may not be issues for major efficacy and safety evaluation in a trial size of 41,000 patients. On the other hand, most clinical trials conducted by the pharmaceutical industry for registration are of small to moderate size. Bias will occur if care is not exercised in the selection of random samples from the targeted population. This bias is particularly crucial for a clinical development program because it will accumulate from phase I to phase III when adequate and well-controlled studies must produce substantial evidence for the approval of the drug.

In many clinical trials the enrollment is slow due to a seasoning of the disease or the geographical location. In such cases the sponsors will open more study sites or enroll more patients at existing sites, whenever possible, in order to reach the required number of evaluable patients. For example, let us consider a phase II dose-ranging trial on basal cell carcinoma (a common skin cancer) conducted by a major pharmaceutical company. This study consisted of a placebo group and three treatment groups of low, medium, and high doses. The study called for a total of 200 patients with 50 patients per dose group. This study was a multicenter study with four study sites in the United States. Three sites were located in the south: San Diego, California, Phoenix, Arizona, and Houston, Texas, which are known to have high prevalence and incidence of skin cancers. The other site was Minneapolis, Minnesota. It turned out that the three sites in the south had no problem of enrollment but that the site at Minneapolis only enrolled a total of three patients. In order to finish the study on time, a decision was made not to open new sites but to enroll as many patients as possible at the three other sites. In order to meet the required sample size of 200 patients, this study ended up with one site of only three patients and three sites with 65 to 70 patients. As a result the validity of statistical inference based on the results from the study is doubtful due to the possible bias caused by the fact that one site in the north only enrolled three patients and the other three sites in the south enrolled most patients which may not constitute a representative sample of the targeted patient population.

Variability

As mentioned earlier, the reliability is referred to as the degree of the closeness (or precision) of the clinical results to the true value regarding the targeted patient population. The reliability of a clinical trial is an assessment of the precision of the clinical trial which measures the degree of the closeness of the clinical results to the true value. Therefore the reliability of a clinical trial reflects the ability to repeat or reproduce similar clinical outcomes in the targeted patient population to which the clinical trial is inferred. The higher precision a clinical trial has, the more likely the results will be reproducible. The precision of a clinical trial can be characterized by the variability of an estimated treatment effect based on some clinical endpoints used for clinical evaluation of the trial. In practice, sample size and the variability of the primary clinical endpoint play an important role in determining the precision and reproducibility of the clinical trial. The larger the sample size of the clinical trial is, the higher the precision and the more reliable the result will be. In clinical trials, however, the sample size is usually not large, and it cannot be increased indefinitely due to limited budget, resources, and often difficulty in patients recruitment. Indeed, the cost of achieving a desired precision can be extremely prohibitive. As an alternative, we can carefully define patient inclusion and exclusion criteria to reduce the variability of primary clinical endpoints and consequently reduce the cost. It is therefore critical that detailed procedures be implemented in the protocol to ensure that all the participating investigators keep to a clinical evaluation that is as homogeneous (less variable) as possible. To draw a reliable statistical inference on the efficacy and safety of the study medication, it is equally imperative to identify all sources of variations that may occur during the conduct of the trial. After these known sources of variations are identified, an appropriate statistical methodology based on the study design can be used to separate the known variabilities from any naturally inherent variability (or random error) for clinical evaluation of the study medication.

Note that for a clinical response consisting of several components, the variation of the clinical response may involve some known and unknown sources of variations. For example, a clinical response may include some continuous measurements such as systolic and diastolic blood pressures (mm Hg) or direct bilirubin (mg/dL), ordinal categorical data such as NIH stroke scale (NIHSS) for quantification of neurologic deficit in the patients with acute ischemic stroke, or some binary data such as the cure of a patient infected with a certain bacteria by some antibiotic. All these clinical responses may be viewed as a sum of several components. One of the components may constitute the true unknown variation, and the others may consist of known sources of variations. These variations cause the variation of the clinical response.

Colton (1974) classifies sources of variation of quantitative clinical responses into three types: true biological and temporal variation and variability due to the measurement error. True biological variation is caused by the difference between subjects. In other words, factors such as age, gender, race, genotype, education status, smoking habit, sexual orientation, study center, and underlying disease characteristics at the baseline possibly can cause variation among subjects. True biological variation explains the differences among individuals. This source of variation is classified as a source of variation for the intersubject variability. The second type of the variation is the variability of clinical responses from the same subject measured at different time points at which the status of the subject may change and vary. A well-known phenomenon of this type of variation is the circadian rhythms. For example, the systolic and diastolic blood pressures measured every 30 minutes by a 24-hour ambulatory device, premature ventricular contraction (PVC) as recorded on a 24-hour Holter monitor, or iron level measured every 8 hours over a 7-day period by either ICP or colorimetric methods all demonstrate the circadian rhythms of these clinical measurements. As a result, if the clinical responses exhibit temporal variation such as circadian rhythms, it is very important to eliminate this type of variation from comparison between treatments. Since the temporal variation reflects the fluctuation of clinical responses within the same subject, it is also known as a source of the intrasubject variability. Note that since the status for the cause of the temporal variation may be different treatments received by the same subject at different time points, the information may be used at the planning stage of clinical trials. Dose titration studies and crossover designs are typical examples utilizing temporal variation for comparison of treatments. As compared to parallel designs that involve both intersubject and intrasubject variabilities, dose titration studies and crossover designs utilizing intrasubject variabilities will provide better precision.

For assessment of efficacy and safety of study medication, a question of particular interest to clinical scientists is, Can similar clinical results be observed if the trial is to be repeatedly carried out under the same conditions? This concerns the variation of a clinical response due to the so-called measurement error. *Measurement error* is probably the most important variation that is difficult to detect and/or control. Measurement error induces the variation among the repeated measurements for the same clinical endpoint obtained from the same subject under the same environment. The possible causes of measurement error include observers such as clinicians or study nurses and laboratory errors caused by an instrument, the technician, or others. If a clinical endpoint produces an unacceptable measurement error, then the results cannot be reproduced under the same conditions. In this case the data have very little value in the clinical evaluation because they do not provide reliably the intended measurement. Since the variation due to measurement error is the variation of replication from the same subject, it can also be classified as a source of intrasubject variability.

Both true biological and temporal variations can be controlled by appropriate statistical designs, blocking, or stratification. Their impact on the precision of the estimates for treatment effects may be eliminated through adequate statistical analyses. In practice, the variation of a primary clinical endpoint due to measurement error is often used for the determination of sample size. Although variation due to measurement error cannot be eliminated completely, it can, however, be reduced tremendously by specifying standard procedures for measuring clinical endpoints in the protocol. Training can then take place, at the investigators' initiatives of the appropriate personnel who will be involved in the trial. Increasingly the training of clinicians, study nurses, and laboratory technicians is becoming important, since the data are gathered by sophisticated machinery, computers, or questionnaires. The clinical personnel must not only understand the rationale and newly developed technology but also be able to perform consistently throughout the study according to the procedures specified in the protocol. For example, the bone mineral density (BMD, in g/cm^3) of the spine is one of the primary and objective clinical responses for clinical evaluation of osteoporosis by certain interventions. A densitometer, which is a dual-energy X-ray absorptiometer, and its accompanying computer algorithm are used to determine bone mineral density. To ensure the reproducibility and consistency of BMD measurements, technicians are trained on the densitometer using an anthropomorphic spine phantom with a User's Manual provided by the manufacturer. In addition the spine phantom be used daily for calibration to lessen the drift effect of the instrument. The regular quality control evaluation should be maintained by the manufacturer. Another source of variation for BMD measurements is the position of the patients. In order to reduce this variation, the positioning of the patient at the various visits should be as close as possible to that at his/her baseline visit. Since the measurement error associated with BMD determination is mainly with the performance of technician, it is preferable that one technician carry out the measurements through the entire study.

Another example for possible reduction of measurement error by training is the NIH Stroke Scale (NIHSS) which was developed by the U.S. National Institute of Neurologic Disorder and Stroke (NINDS) from the original scale devised at the University of Cincinnati. Technically it is a rating scale for quantifying a neurological deficit by a total of 42 points in 11 categories such as given in Table 2.3.3. In practice, it is a rating scale based on the subjective judgment of qualified neurologists and emergency physicians. This scale has recently been used in several clinical trials as the primary outcome measure for efficacy assessment in various thrombolytic agents in patients with acute ischemic stroke (The National Institute of Neurological Disorders and Stroke rt-PA Stroke Study Group, 1995, TOAST, 1994). In these multicenter clinical trials, in order to maintain double-blinding, different clinicians were asked to perform the baseline and subsequent NIHSS. Standardization of the procedure for rating NIHSS in the protocol by the clinicians within the same center as well as by investigators at different centers is considered crucial in order to reduce measurement error and produce reliable, consistent, and reproducible results. As a result, video training was selected to train and certify investigators who perform NIHSS so that the measurement error in those clinical trials could be reduced.

In summary, to have an accurate and reliable assessment of the true efficacy and safety of a study medication, it is important to avoid bias and to minimize the variability of the primary clinical endpoint whenever possible. Figure 2.3.1 shows the impact of bias and variability in tackling the truth. As can be seen from the figure, the ideal situation is to have an unbiased estimate with no variability. If the estimate has a nonnegligible bias, the truth is compromised

Table 2.3.3 Summary of the National Institutes of Health Stroke Scale

Item	Name	Response
1A	Level of consciousness	0 = Alert
		2 = Not alert, obtunded
		3 = Unresponsive
1B	Questions	0 = Answers both correctly
		1 = Answers one correctly
		2 = Answers neither correctly
1C	Commands	0 = Performs both tasks correctly
		1 = Performs one task correctly
		2 = Performs neither correctly
2	Gaze	0 = Normal
		1 = Partial gaze palsy
		2 = Total gaze palsy
3	Visual fields	0 = No visual loss
		1 = Partial hemianopsia
		2 = Complete hemianopsia
		3 = Bilateral hemianopsia
4	Facial palsy	0 = Normal
		1 = Minor paralysis
		2 = Partial paralysis
		3 = Complete paralysis
5	Motor arm	0 = No drift
	a. Left	1 = Drift before 10 seconds
	b. Right	2 = Fall before 10 seconds
		3 = No effort against gravity
		4 = No movement
6	Motor leg	0 = No drift
	a. Left	1 = Drift before 5 seconds
	b. Right	2 = Fall before 5 seconds
		3 = No effort against gravity
		4 = No movement
7	Ataxia	0 = Absent
		1 = One limb
		2 = Two limbs
8	Sensory	0 = Normal
		1 = Mild loss
		2 = Severe loss
9	Language	0 = Normal
		1 = Mild aphasia
		2 = Severe aphasia
		3 = Mute or global aphasia
10	Dysarthria	0 = Normal
		1 = Mild
		2 = Severe
11	Extinction/inattention	0 = Normal
		1 = Mild
		2 = Severe

Source: Lyden et al. (1994).

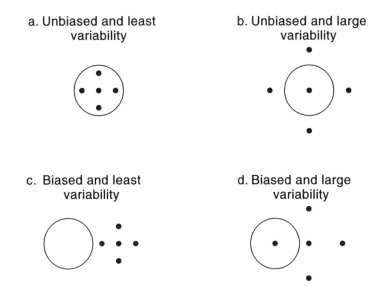

Figure 2.3.1 Accuracy and precision for assessment of treatment effect. (*a*) Unbiased and least variability, (*b*) unbiased and large variability, (*c*) biased and least variability, and (*d*) biased and large variability.

even for the smallest variability. If the estimate is biased with a lot of variability, the assessment for the treatment effect must be discarded.

2.4 CONFOUNDING AND INTERACTION

In clinical trials, confounding and interaction effects are the most common distortions in the evaluation of medication. *Confounding effects* are contributed by various factors such as race and gender that cannot be separated by the design under study; an *interaction effect* between factors is a joint effect with one or more contributing factors (Chow and Liu, 1995a). Confounding and interaction are important considerations in clinical trials. For example, when confounding effects are observed, we cannot assess the treatment effect because it is contaminated by other effects. On the other hand, when interactions among factors are observed, the treatment must be carefully evaluated for those effects.

Confounding

In clinical trials there are many sources of variation that have an impact on the primary clinical endpoints for evaluation relating to a certain new regimen or intervention. If some of these variations are not identified and properly controlled, they can become mixed in with the treatment effect that the trial is designed to demonstrate. Then the treatment effect is said to be confounded by effects due to these variations. To provide a better understanding, consider the following example. Suppose that last winter Dr. Smith noticed that the temperature in the emergency room was relatively low and caused some discomfort among medical personnel and patients. Dr. Smith suspected that the heating system might not be functioning properly and called on the hospital to improve it. As a result the temperature of the emergency room is at a comfortable level this winter. However, this winter is not as cold as last winter. Therefore it is not clear whether the improvement in the emergency room temperature was due to the improvement in the heating system or the effect of a warmer winter. In fact the effect due to the improvement of the heating system and that due to a warmer winter are confounded and cannot be separated from each other. In clinical trials there are many subtle, unrecognizable, and seemingly innocent confounding factors that can cause ruinous results of clinical trials. Moses (1992) gives the example of the devastating result in the confounder being the personal choice of a patient. The example concerns a polio-vaccine trial that was conducted on two million children worldwide to investigate the effect of Salk poliomyelitis vaccine. This trial reported that the incidence rate of polio was lower in the children whose parents refused injection than whose who received placebo after their parent gave permission (Meier, 1989). After an exhaustive examination of the data, it was found that susceptibility to poliomyelitis was related to the differences between the families who gave the permission and those who did not.

Sometimes confounding factors are inherent in the design of the studies. For example, dose titration studies in escalating levels are often used to investigate the dose-response relationship of the antihypertensive agents during phase II stage of clinical development. For a typical dose titration study, after a washout period during which previous medication stops and the placebo is prescribed, N subjects start at the lowest dose for a prespecified time interval. At the end of the interval, each patient is evaluated as a responder to the treatment or a nonresponder according to some criteria prespecified in the protocol. In a titration study a subject will continue to receive the next higher dose if he or she fails, at the current level, to meet some objective physiological criteria such as reduction of diastolic blood pressure by a prespecified amount and has not experienced any unacceptable adverse experience. Figure 2.4.1 provides a graphical presentation of a typical titration study (Shih, Gould, and Hwang, 1989). Dose titration studies are quite popular among clinicians because they mimic real clinical practice in the care of patients. The major problem with this typical design for a dose titration study is that the dose-response relationship is often confounded with time course and the unavoidable carryover effects from the previous dose levels which cannot be estimated and eliminated. One can always argue that the relationship found in a dose titration study is not due to the dose but to the time. Statistical methods for binary data from dose titration studies have been suggested under some rather strong assumptions (e.g., see Chuang, 1987; Shih, Gould, and Hwang, 1989). Due to the fact that the dose level is confounded with time, estimation of the dose-response relationship based on continuous data

CONFOUNDING AND INTERACTION

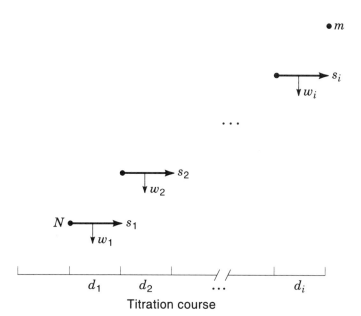

Figure 2.4.1 Graphical display of a titration trial. d_i, the ith dose level; s_i, the number of subjects who responded at the ith dose; w_i, the number of subjects who withdraw at the ith dose; and m, the number of subjects who completed the study without a response. (Source: Shih, Gould, and Hwang, 1989.)

has not yet been resolved. Another type of design that can induce confounding problems when it is conducted inappropriately is the crossover design. For a standard 2×2 crossover design, each subject is randomly assigned to one of the two sequences. In sequence 1, subjects receive the reference (or control) treatment at the first dosing period and the test treatment at the second dosing period after a washout period of sufficient length. The order of treatments is reversed for the subjects in sequence 2. The issues in analysis of the data from a 2×2 crossover design is twofold. First, unbiased estimates of treatment effect cannot be obtained from the data of both periods in the presence of a nonzero carryover effect. The second problem is that the carryover effect is confounded with sequence effect and treatment-by-period interaction. In the presence of a significant sequence effect, however, the treatment effect can be estimated unbiasedly from the data of both periods. In practice, it is not clear whether an observed statistically significant sequence effect (or carryover effect) is a true sequence effect (or carryover effect). As a result this remains a major drawback of the standard 2×2 crossover design, since the primary interest is to estimate a treatment effect that is still an issue in the presence of a significant nuisance parameter. The sequence and carryover effects, however, are not confounded to each other in higher-order crossover designs that compare two treatments and can provide unbiased estimation of treatment effect in the presence of a significant carryover effect (Chow and Liu, 1992a, 1992b).

Bailar (1992) provided another example of subtle and unrecognizable confounding factors. In the same issue of *New England Journal of Medicine*, Wilson et al. (1985) and Stampfer et al. (1985) both reported the results on the incidence of cardiovascular diseases in postmenopausal women who had been taking hormones compared to those who had not. Their conclusions, however, were quite different. One reported that the incidence rate of cardiovascular disease among the women taking hormones was twice that in the control group, while the other reported a totally opposite conclusion in which the incidence of the experimental group was only half that of women who were not taking hormones. Although these trials were not randomized studies, both studies were well planned and conducted. Both studies had carefully considered the differences in known risk factors between the two groups in each study. As a result the puzzling difference in the two studies may be due to some subtle confounding factors such as the dose of hormones, study populations, research methods, and other related causes. This example indicates that it is imperative to identify and take into account all confounding factors for the two adequate, well-controlled studies that are required for demonstration of effectiveness and safety of the study medication.

In clinical trials it is not uncommon for some subjects not to follow instructions in taking the prescribed does at the scheduled time as specified in the protocol. If the treatment effect is related to (or confounded with) patients' compliance, any estimates of the treatment effect are biased unless there is a placebo group in which the differences in treatment effects between subjects with good compliance and poor compliance can be estimated. As a result interpretation and extrapolation of the findings are inappropriate. In practice, it is very difficult to identify compliers and noncompliers and to quantify the relationship between treatment and compliance. On the other hand, subject withdrawals or dropouts from clinical trials are the ultimate examples of noncompliance. There are several possible reasons for dropouts. For example, a subject with severe disease did not improve and hence dropped out from the study. The estimate of treatment effect will be biased in favor of a false positive efficacy if the subjects with mild disease remain and improve. On the other hand, subjects will withdraw from a study if their conditions improve, and those who did not improve will remain until the scheduled termination of a study. The estimation of efficacy will then be biased and hence indicate a false negative efficacy. Noncompliance and subject dropouts are only two of the many confounding factors that can occur in many aspects of clinical trials. If there is an unequal proportion of the subjects who withdraw from the study or comply to the dosing regimen among different treatment groups, it is very important to perform an analysis on these two groups of subjects to determine whether confounded factors exist and the direction of possible bias. In addition every effort must be made to continue subsequent evaluation of withdrawals in primary clinical endpoints such as survival or any serious adverse events. For analyses of data with noncompliance or withdrawals, it is suggested that an "intention-to-treat" analysis be performed. An intention-to-treat analysis includes all available data based on all randomized subjects with the degree of compliance or reasons for withdrawal as possible covariates.

Interaction

The objective of a statistical interaction investigation is to conclude whether the joint contribution of two or more factors is the same as the sum of the contributions from each factor when considered alone. The factors may be different drugs, different doses of two drugs, or some stratification variables such as severity of underlying disease, gender, or other important covariates. To illustrate the concept of statistical interaction, we consider the Second International Study of Infarct Survival (ISIS-2, 1988). This study employed a 2×2 factorial design (two factor with two levels at each factor) to study the effect of streptokinase and aspirin in the reduction of vascular mortality in patients with suspected acute myocardial infarction. The two factors are one-hour intravenous infusion of 1.5 MU of streptokinase and one month of 150 mg per day enteric-coated aspirin. The two levels for each factor are either active treatment and their respective placebo infusion or tablets. A total of 17,187 patients were enrolled in this study. The numbers of the patients randomized to each arm is illustrated in Table 2.4.1. The key efficacy endpoint is the cumulative vascular mortality within 35 days after randomization. Table 2.4.2 provides the cumulative vascular mortality for each of the four arms as well as those for streptokinase and aspirin alone. From Table 2.4.2 the mortality of streptokinase group is about 9.2%, with the corresponding placebo mortality being 12.0%. The improvement in mortality rate attributed to streptokinase is 2.8% (12.0%–9.2%). This is referred to as the main effect of streptokinase. Similarly the main effect of aspirin tablets can also be estimated from Table 2.4.2 as 2.4% (11.8%–9.4%). The left two panels of Figure 2.4.2 give the cumulative vascular moralities of main effects for both streptokinase and placebo. The right panel of Figure 2.4.2 provides mortality for combination of streptokinase and aspirin against that of both placebos. From either Table 2.4.2 or Figure 2.4.2, the joint

Table 2.4.1 Treatment of ISIS-2 with Number of Patients Randomized

Aspirin	IV Infusion of Streptokinase		
	Active	Placebo	Total
Active	4292	4295	8,587
Placebo	4300	4300	8,600
Total	8292	8595	17,187

Source: ISIS-2 (1988).

Table 2.4.2 Cumulative Vascular Mortality in Days 0–35 of ISIS-2

Aspirin	IV Infusion of Streptokinase		
	Active	Placebo	Total
Active	8.0%	10.7%	9.4%
Placebo	10.4%	13.2%	11.8%
Total	9.2%	12.0%	

Source: ISIS-2 (1988).

contribution of both streptokinase and aspirin in improvement in mortality is 5.2% (13.2%–8.0%) which is exactly equal to the contribution in mortality by streptokinase (2.8%) plus that by aspirin (2.4%). This is a typical example that no interaction exists between streptokinase and aspirin because the reduction in mortality by joint administration of both streptokinase and aspirin can be expected as the sum of reduction in mortality attributed to each antithrombolytic agent when administrated alone. In other words, the difference between the two levels in one factor does not depend on the level of the other factor. For example, the difference in vascular mortality between streptokinase and placebo for the patients taking aspirin tablets is 2.7% (10.7%–8.0%). A similar difference of 2.8% is observed between streptokinase (10.4%) and placebo (13.2%) for the patients taking placebo tablets. Therefore the reduction in mortality attributed to streptokinase is homogeneous for the two levels of aspirin tablets. As a result there is no interaction between streptokinase infusion and aspirin tablets. This phenomenon is also observed in Figure 2.4.3.

The ISIS-2 trial provides an example of an investigation of interaction between two treatments. However, in the clinical trial it is common to check interaction between treatment and other important prognostic and stratification factors. For example, almost all adequate well-controlled studies for the establishment of effectiveness and safety for approval of pharmaceutical agents are multicenter studies. For multicenter trials the FDA requires that the treatment-by-center interaction be examined to evaluate whether the treatment is consistent across all centers.

One of the objectives of the National Institute of Neurological Disorders and Stroke rt-PA Stroke Study is to investigate whether the improvement of neurological deficit upon administration of intravenous recombinant t-PA over the placebo group is consistent over the time to treatment after the onset of stroke as stratified from 0 to 90 minutes and from 90 to 180 minutes. The results based on NIHSS are reproduced is Table 2.4.3. It can be seen that the difference in NIHSS between t-PA and placebo (i.e., treatment effect) is homogeneous between the two time intervals. Consequently no interaction exists between the treatment and time to treatment.

When the difference among levels of one factor is not the same at different

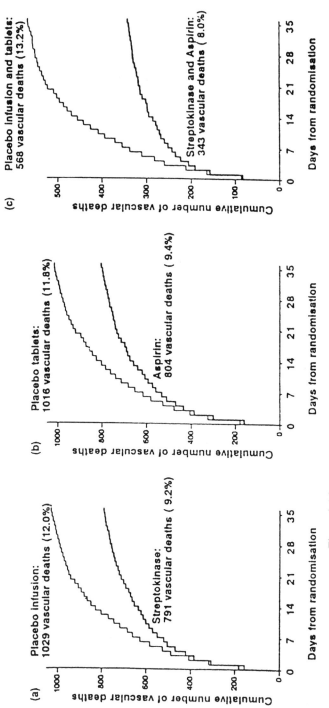

Figure 2.4.2 Cumulative vascular mortality in days 0–35. (Source: ISIS-2, 1988.)

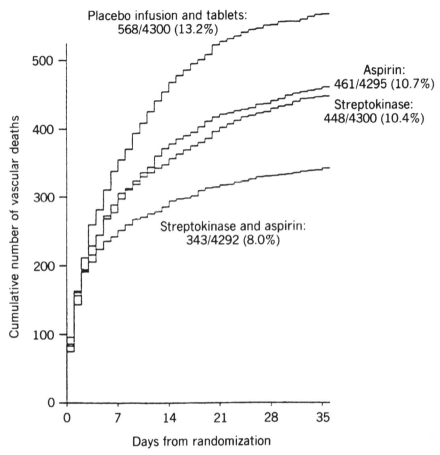

Figure 2.4.3 Cumulative vascular mortality in days 0–35 by treatment groups. (Source: ISIS-2, 1988.)

Table 2.4.3 Mean NIHSS Score of the National Institute of Neurological Disorders and Stroke rt-PA Stroke Study Group

Time to Treatment after Stroke Onset	t-PA	Placebo
0–90 minutes	$N = 157$ 9(2–17)	$N = 145$ 12(6–18)
0–180 minutes	$N = 155$ 8(3–19)	$N = 167$ 13(7–19)

Source: National Institute of Neurological Disorders and Stroke rt-PA Stroke Study Group (1995).
Note: Numbers in the parentheses are ranges.

levels of other factors, then it is said that interaction exist between these two factors. In general, interactions can be classified as quantitative or qualitative (Gail and Simon, 1985). A quantitative interaction is the one for which the magnitude of the treatment effect is not the same across the levels of other factors but the direction of the treatment remains the same for all levels of other factors. A qualitative interaction is the interaction in which the direction of the treatment effect changes in some levels of other factors. To provide a better understanding, we consider the following hypothetical example of the treatment of an irreversible inhibitor of steroid aromatase in the patients with benign prostatic hyperplasia. Suppose that one of the objectives of the trial is to investigate whether improvement of peak urinary flow rate (mL/sec) of the treatment over placebo is the same for patients with an American Urinary Association (AUA) symptom score between 8 and 19 inclusively and those with AUA score greater than 19. There are two factors, each with two levels. One can display the four treatment-by-symptom means in a figure for visual inspection of possible interaction. The vertical axis is the mean change from baseline in peak urinary flow rate (mL/sec). The two levels of treatment can be represented on the horizontal axis. The mean peak urinary flow rate of all levels of AUA symptom score (other factor) then can be plotted at each level of treatment on the horizontal axis, and the means of the same levels of AUA symptom score are connected over the two levels of treatment. Panel A of Figure 2.4.4 exhibits a pattern of no interaction in which the two lines are parallel and the distance between the two lines is the same at all levels of treatment. Panels B and C demonstrate a possible quantitative interaction between the treatment and AUA symptom score because the two lines do not cross and treatment effect does not change its direction, though the distance between the lines is not the same. Both panels B and C indicate that the treatment is more effective than placebo for increasing peak flow rate. Panel B shows that treatment induces a better improvement in peak urinary flow rate in patients with an AUA symptom score greater than 19 than those between 8 and 19, while the opposite observation is seen in panel C. In panel D a positive difference in peak urinary flow rate between treatment and placebo is observed in patients with an AUA score greater than 19, while the difference is negative in patients with an AUA symptom score between 8 and 19. Panel D shows a possible qualitative interaction between treatment and symptom score. Since the two lines cross each other in panel D, a qualitative interaction of this kind is also called a crossover interaction.

Interaction can be used to investigate whether the effectiveness of treatment is homogeneous across groups of patients with different characteristics. It therefore is important in the interpretation and inference of the trial results. Gail and Simon (1985) provided an graphical illustration of different interactions for two subgroups of patients, which is reproduced in Figure 2.4.5. The true differences in efficacy between two treatments in subgroups 1 and 2 are δ_1 and δ_2. The 45° line is the line where $\delta_1 = \delta_2$. It therefore represents the line of no interaction. In the unshaded areas (the first and third quadrants) δ_1 and δ_2 are either both

Panel A

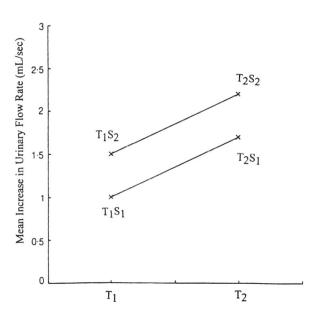

Panel B

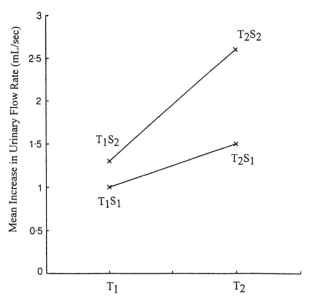

Figure 2.4.4 Graphical presentation of two-factor interaction.

Panel C

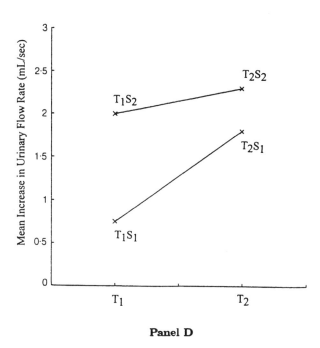

Panel D

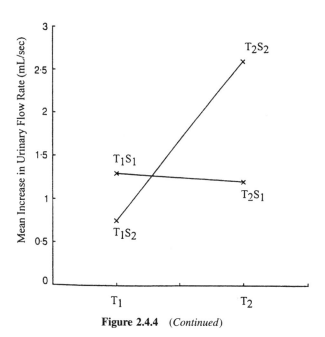

Figure 2.4.4 (*Continued*)

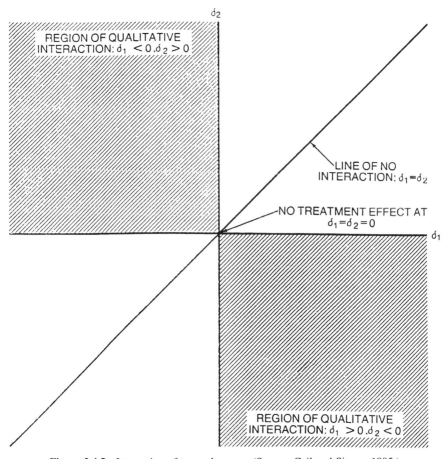

Figure 2.4.5 Interaction of two subgroups. (Source: Gail and Simon, 1985.)

positive or both negative. Any point in the unshaded area except for those on the 45° line represents a quantitative interaction. The two shaded areas (the second and fourth quadrants) consist of points for qualitative (or crossover) interaction.

2.5 DESCRIPTIVE AND INFERENTIAL STATISTICS

Dietrich and Kearns (1986) divided statistics into two broad areas, namely descriptive and inferential statistics. Descriptive statistics is the science of summarizing or describing data, while inferential statistics is the science of inter-

DESCRIPTIVE AND INFERENTIAL STATISTICS

preting data in order to make estimates, hypotheses testing, predictions, or decisions from the samples to the targeted population.

In clinical trials, data are usually collected through case report forms which are designed to capture clinical information from the studies. The information on the case report forms is then entered into the database. The raw database is always messy, though it does contain valuable clinical information from the study. In practice, it is often of interest to summarize the raw database by a graphical presentation (e.g., a data plot) or by descriptive (or summary) statistics. Descriptive statistics are simple sample statistics such as means and standard deviations (or standard errors) of clinical variables or endpoints. Note that the standard deviation describes the variability of a distribution, either a population distribution or a sample distribution, whereas the standard error is the variability of a sample statistic (e.g., sample mean or sample variance). Descriptive statistics are often used to describe the targeted population before and after the study. For example, at baseline, descriptive statistics are often employed to describe the comparability between treatment groups. After the completion of the study, descriptive statistics are useful tools to reveal possible clinical differences (or effects) or trends of study drugs. As an example, Table 2.5.1 provides a partial listing of individual patient demographics and baseline characteristics from a study comparing the effects of captopril and enalapril on quality of life in the older hypertensive patients (Testa et al., 1993). As can be seen from Table 2.5.1, although as a whole, the patient listing gives a detailed description of the characteristics for individual patients, it does not provide much summary information regarding the study population. In addition descriptive statistics for demographic and baseline information describe not only the characteristics of the study population but also the comparability between treatment groups (see Table 2.5.2). In addition, for descriptive purposes, Table 2.5.3 groups patients into low, medium, and high categories according to the ranking of their scores on the baseline quality of life scale. It can be seen that there is a potential difference in treatment effect among the three groups with regard to the change from baseline on the quality of life. These differences were confirmed to be statistically significant by valid statistical tests. Therefore a preliminary investigation of descriptive statistics of primary clinical endpoints may reveal a potential drug effect.

When we observe some potential differences (effects) or trends, it is necessary to further confirm with certain assurance that the differences (effects) or trends indeed exist and are not due to chance alone. For this purpose it is necessary to provide inferential statistics for the observed differences (effects) or trends. Inferential statistics such as confidence intervals and hypotheses testing are often performed to provide statistical inference on the possible differences (effects) or trends that can be detected based on descriptive statistics. For the rest of this section, we will focus on confidence intervals (or interval estimates). Hypotheses testing will be discussed in more detail in the following section.

Clinical endpoints are often used to assess the efficacy and safety of drug

Table 2.5.1 Partial Data of Capoten Quality of Life Study

Patient	Race	Age (years)	Height (inches)	Weight (pounds)	Heart Rate (per minute)	Systolic BP (mmHg)	Diastolic BP (mmHg)	Alcohol Consumption	Tobacco Consumption
C01-003	Caucasian	69	70	195	72	146	94	No	No
C01-004	Caucasian	69	70	188	72	170	94	Yes	Yes
C01-006	Caucasian	62	76	231.5	84	158	91	No	No
C01-008	Caucasian	56	72	244	76	140	97	No	No
C01-010	Caucasian	55	75	258	60	139	100	Yes	No
C01-012	Caucasian	58	74	191	86	138	95	Yes	No
C02-004	Black	66	70	220	64	141	99	No	No
C02-006	Black	55	65	244.5	64	171	109	No	No
C02-008	Black	61	69	281	76	173	104	No	No
C02-009	Black	61	68	190.5	60	150	91	Yes	No
C02-013	Black	71	74	193	88	140	97	No	Yes
C03-001	Caucasian	55	74	303	76	151	103	Yes	No
C03-004	Caucasian	65	71	243	78	156	91	No	No
C03-005	Caucasian	55	69	178	100	150	91	Yes	No
C03-006	Caucasian	59	65	174	64	157	101	Yes	Yes
C03-010	Caucasian	74	67	171	80	188	109	No	No
C03-012	Caucasian	65	65	150	58	169	99	Yes	No
C04-003	Caucasian	64	72	194	96	161	99	Yes	No
C04-005	Black	59	69	201	76	179	96	No	No
C04-009	Caucasian	58	78	334	80	159	114	No	No

Table 2.5.2 Demographic, Clinical, and Quality of Life Variables at the Baseline

Variable	Captopril ($N = 192$)	Enalapril ($N = 187$)
Demographic		
Age (yr)	64.2 ± 5.5	64.6 ± 6.4
Education (%)		
No high school	9	7
Some high school	36	30
Some college	49	58
Postgraduate degree	6	5
Income (%)		
<$15,000	15	12
$15,000–40,999	41	47
$41,000–80,000	33	35
>$80,000	11	6
Percent married	87	88
Occupational status (%)		
Employed full-time	40	36
Employed part-time	14	15
Retired	43	49
Unemployed	3	1
Race (%)		
White	84	82
Black	14	18
Other	3	1
Clinical		
Weight (lb)	197.8 ± 36.4	198.7 ± 37.9
Body-mass index	28.8 ± 4.7	28.6 ± 5.0
Blood pressure (mm Hg)		
Systolic	155.0 ± 14.8	154.6 ± 15.8
Diastolic	97.3 ± 5.8	97.3 ± 5.9
Previous antihypertensive therapy (%)	89	87
Quality-of-life scales		
Psychological well-being	462 ± 78	452 ± 80
Psychological distress	526 ± 64	521 ± 59
General perceived health	493 ± 77	495 ± 67
Well-being at work or in daily routine	485 ± 61	480 ± 62
Sexual-symptom distress	518 ± 144	503 ± 162
Distress and stress indexes		
Side effects and symptoms distress	24 ± 40	25 ± 35
Life events	30 ± 40	28 ± 40
Stress	274 ± 144	296 ± 142

Source: Testa et al. (1993).

Table 2.5.3 Changes from the Baseline to End Point in Quality of Life According to Scores on the Quality of Life Scale at the Baseline

Scale	Baseline Scores for Randomized Patients			Captopril (N = 184)			Enalapril (N = 178)		
	Low	Medium	High	Low (N = 53)	Medium (N = 60)	High (N = 71)	Low (N = 60)	Medium (N = 65)	High (N = 53)
General perceived health	420	506	552	+21.0 ± 6.6	−2.7 ± 6.3	−1.6 ± 3.2	+1.7 ± 7.3	−10.4 ± 5.3	−15.8 ± 6.1
Psychological well-being	374	469	524	+19.3 ± 8.8	−1.4 ± 7.3	+8.2 ± 3.0	+17.7 ± 9.0	+1.3 ± 6.1	−7.9 ± 6.0
Psychological distress	457	533	577	+19.7 ± 6.8	−6.2 ± 5.9	−3.5 ± 2.6	+7.2 ± 6.5	−0.9 ± 5.1	−10.8 ± 4.5
Overall quality of life	427	502	545	+18.1 ± 5.3	−6.8 ± 5.4	−0.5 ± 2.4	+5.9 ± 6.1	−4.3 ± 4.5	−10.7 ± 4.6

Source: Testa et al (1993).

Note: Mean ±SD change from baseline score.

products. For example, diastolic blood pressure is one of the primary clinical endpoints for the study of ACE inhibitor agents in the treatment of hypertensive patients. The purpose of the diastolic blood pressure for hypertensive patients is to compare their average diastolic blood pressure with the norm for ordinary health subjects. However, the average diastolic blood pressure for the hypertensive patients is unknown. We will need to estimate the average diastolic blood pressure based on the observed diastolic pressures obtained from the hypertensive patients. The observed diastolic blood pressures and the average of these diastolic blood pressures are the sample and sample mean of the study. The sample mean is an estimate of the unknown population average diastolic blood pressure. Point estimates may not be of practical use. For example, suppose that the sample mean is 98 mm Hg. It is then important to know whether the population average for the hypertensive patients could reasonably be 90 mm Hg given that the sample average turned out to be 98 mm Hg. This kind of information depends on the knowledge of the standard error, not merely of the point estimate itself.

The observed diastolic blood pressures are usually scattered around the sample mean. Based on these observed diastolic blood pressures, the standard error of the sample mean of the observed diastolic blood pressures can be obtained. If the distribution of the diastolic blood pressure appears to be a bell shaped and the sample size is of moderate size, then there is about 95% chance that the unknown average diastolic blood pressure of the targeted population will fall within the area between approximate two (i.e., 1.96) standard errors below and above the sample mean. The lower and upper limits of the area constitute an interval estimate for the unknown population average diastolic blood pressure. An interval estimate is usually referred to as a confidence interval with a desired confidence level, such as 95%. Unlike a point estimate, a confidence interval provides a whole interval as an estimate for a population parameter instead of just a single value. A 95% confidence interval is an interval that is calculated according to a certain procedure that would produce a different interval for each sample upon repeated sampling from the population, and 95% of these intervals would contain the unknown fixed population parameter. A 95% confidence interval is not an interval that will contain 95% of the sample averages that would be obtained on repeating the sampling procedure, nor is a particular 95% confidence interval in which the population average will fall 95% of the times. It should be noted that the population average is an unknown constant and does not vary while a confidence interval is random. It is either in the confidence interval or not.

A classical confidence interval for the population average of a clinical variable is symmetric about the observed sample mean of the observed responses of the clinical variable. This classical confidence interval is sometimes called the shortest confidence interval because its width is the shortest among all of the confidence intervals of the same confidence level by other statistical procedures. In some situations it may be of interest to obtain a symmetric confidence with respect to a fixed number. For example, in bioequivalence trials it is of interest to obtain a confidence interval for the difference in a pharmacokinetic parameter such as area under the blood or plasma concentration time curve (AUC) between the test and reference drug product. If the 90% confidence interval falls within ±20% of the average of the reference product, then we conclude that the test product is bioequivalent to the reference product (e.g., Chow and Liu, 1992a). Since the limits are ±20%, which is symmetric about 0%, Westlake (1976) proposed the idea to consider a symmetric confidence interval with respect to 0 rather than the shortest confidence interval symmetric about the observed difference in the sample means. Note that as indicated in Chow and Liu (1992a), the most common criticisms of Westlake's symmetric confidence interval are that it has shifted away from the direction in which the sample difference was observed and that the tail probabilities associated with Westlake's symmetric confidence interval are not symmetric. As a result Westlake's symmetric confidence interval moves from a two-sided to a one-sided approach as the true difference and the random error increase.

The confidence level is the degree of certainty that the interval actually contains the unknown population parameter value. It provides the degree of assurance or confidence that the statement regarding the population parameter is correct. The more certainty we want, the wider the interval will have to be. A very wide interval estimate may not be of practical use because it fails to identify the population parameter closely. In practice, it is more usual to use 90%, 95%, or 99% as confidence levels. Table 2.5.4 summarizes the multiple of standard errors that are needed for confidence levels of 68, 95, and 99. Therefore we will have 68%, 95%, and more than 99% confidence that the population parameter will fall within one, two, and three standard errors of the observed value, respectively.

When we claim that a drug product is effective and safe with 95% assurance, it is expected that we will observe consistent significant results 95% of times if the clinical trial were repeatedly carried out with the same protocol. However, current FDA regulation only requires two adequate well-controlled clinical trials

Table 2.5.4 Confidence Levels with Various Standard Errors

Standard Errors	Confidence Level
0.5	0.3830
0.675	0.5000
1.0	0.6826
1.5	0.8664
1.96	0.9500
2.0	0.9544
2.5	0.9876
3.0	0.9974
4.0	1.0000

be conducted to provide substantial evidence for efficacy and safety. It is therefore of interest to estimate the probability that the drug is effective and safe based on clinical results obtained from the two adequate well-controlled trials.

2.6 HYPOTHESES TESTING AND *p*-VALUES

In clinical trials a hypothesis is a postulation, assumption, or statement that is made about the population regarding the efficacy, safety, and other pharmacoeconomics outcomes (e.g., quality of life) of a drug product under study. This statement or hypothesis is usually a scientific question that needs to be investigated. A clinical trial is often designed to address the question by translating it into specific study objective(s). Once the study objective(s) has been carefully selected and defined, a random sample can be drawn through an appropriate study design to evaluate the hypothesis about the drug product. For example, a scientific question regarding a drug product, say drug A, of interest could be either (1) Is the mortality reduced by drug A? or (2) Is drug A superior to drug B in treating hypertension? The hypothesis to be questioned is usually referred to as the *null hypothesis*, denoted by H_0. The hypothesis that the investigator wishes to establish is called the *alternative hypothesis*, denoted by H_a. In practice, we attempt to gain support for the alternative hypothesis by producing evidence to show that the null hypothesis is false. For the questions regarding drug A described above, the null hypotheses are that (1) there is no difference between drug A and the placebo in the reduction of mortality and (2) there is no difference between drug A and drug B in treating hypertension, respectively. The alternative hypotheses are that (1) drug A reduces the mortality and (2) drug A is superior to drug B in treating hypertension, respectively. These scientific questions or hypotheses to be tested can then be translated into specific study objectives as to compare (1) the efficacy of drug A with no therapy in the prevention of reinfarction or (2) the efficacy of drug A with that of drug B in reducing blood pressure in elderly patients, respectively.

Chow and Liu (1992a) recommended the following steps be taken to perform a hypothesis testing:

1. Choose the null hypothesis that is to be questioned.
2. Choose an alternative hypothesis that is of particular interest to the investigators.
3. Select a test statistic, and define the rejection region (or a rule) for decision making about when to reject the null hypothesis and when not to reject it.
4. Draw a random sample by conducting a clinical trial.
5. Calculate the test statistic and its corresponding *p-value*.
6. Make conclusion according to the predetermined rule specified in step 3.

When performing a hypotheses testing, basically two kinds of errors occur. If the null hypothesis is rejected when it is true, then a type I error has occurred. For example, a type I error has occurred if we claim that drug A reduces the mortality when in fact there is no difference between drug A and the placebo in the reduction of mortality. The probability of committing type I error is known as the level of significance. It is usually denoted by α. In practice, α represents the consumer's risk which is often chosen to be 5%. On the other hand, if the null hypothesis is not rejected when it is false, then a type II error has been made. For example, we have made a type II error if we claim that there is no difference between drug A and the placebo in the reduction of mortality when in fact drug A does reduce the mortality. The probability of committing type II error, denoted by β, is sometimes referred to as the producer's risk. In practice, $1 - \beta$ is known as the power of the test, which represents the probability of correctly rejecting the null hypothesis when it is false.

Table 2.6.1 summarizes the relationship between type I and type II errors when testing hypotheses. Furthermore a plot based on the null hypothesis of no difference is presented in Figure 2.6.1 to illustrate the relationship between α and β (or power) for various β's under H_0 for various alternatives at $\alpha =$ 5% and 10%. It can be seen that α decreases as β increases or α increases as β decreases. The only way of decreasing both α and β is to increase the sample size. In clinical trials a typical approach is to first choose a significant level α and then select a sample size to achieve a desired test power. In other words, a sample size is chosen to reduce type II error such that β is within an acceptable range at a prespecified significant level of α. From Table 2.6.1 and Figure 2.6.1 it can be seen that α and β depend on the selected null and alternative hypotheses. As indicated earlier, the hypothesis to be questioned is usually chosen as the null hypothesis. The alternative hypothesis is usually of particular interest to the investigators. In practice, the choice of the null hypothesis and the alternative hypothesis has an impact on the parameter to be tested. Chow and Liu (1992a) indicate that the null hypothesis may be selected based on the importance of the type I error. In either case, however, it should be noted that we will never be able to prove that H_0 is true even though the data fail to reject it.

Table 2.6.1 Relationship Between Type I and Type II Errors

When	If H_0 is True	If H_0 is False
Fail to reject	No error	Type II error
Reject	Type I error	No error

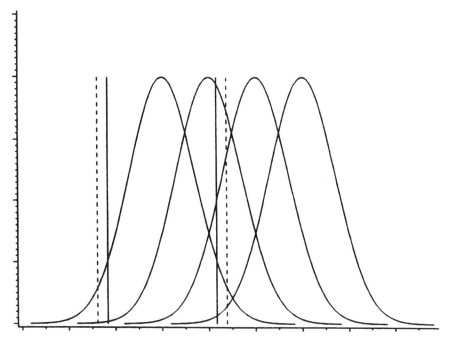

Figure 2.6.1 Relationship between probabilities of type I and type II errors.

p-Values

In medical literature *p*-values are often used to summarize results of clinical trials in a probabilistic way. For example, in a study of 10 patients with congestive heart failure, Davis et al. (1986) report that at single daily doses of captopril of 25 to 150 mg, the cardiac index rose from 1.75 ± 0.18 to 2.77 ± 0.39 (mean ± SD) liters per minute per square meter ($p < 0.001$). Powderly et al. (1995) confirmed that fluconazole was effective in preventing esophageal candidiasis (adjusted relative hazard, 5.8; 95% confidence interval 1.7 to 20.0; $p = 0.004$) in patients with advance human immunodeficiency virus (HIV) infection. In a multicenter trial Coniff et al. (1995) indicate that all active treatments (acarbose, tolbutamide, and acarbose plus tolbutamide) were superior ($p < 0.05$) to placebo in reducing postprandial hyperglycemia and HbA_{1c} levels in noninsulin-dependent diabetes mellitus (IDDM) patients. In a study evaluating the rate of bacteriologic failure of amoxicillin-clavulanate in the treatment of acute otitis media, Patel et al. (1995) reveal that the bacteriologic failure was higher in nonwhite boys ($p = 0.026$) and in subjects with a history of three or more previous episodes of acute otitis media ($p = 0.008$). These statements indicated that a difference at least as great as the observed would occur in less than 1 in 100 trials if a 1% level of significance were chosen or in less than 1 in 20 trials if a 5% level of significance were selected provided that the null hypothesis of no difference between treatments is true and the assumed statistical model is correct.

In practice, the smaller the p-value shows, the stronger the result is. However, the meaning of a p-value may not be well understood. The *p-value* is a measure of the chance that the difference at least as great as the observed difference would occur if the null hypothesis is true. Therefore, if the p-value is small, then the null hypothesis is unlikely to be true by chance, and the observed difference is unlikely to occur due to chance alone. The p-value is usually derived from a statistical test that depends on the size and direction of the effect (a null hypothesis and an alternative hypothesis). To show this, consider testing the following hypotheses at the 5% level of significance:

$$H_0: \text{There is no difference;}$$
$$\text{vs.} \quad H_a: \text{There is a difference.} \quad (2.6.1)$$

The statistical test for the above hypotheses is usually referred to as a *two-sided test*. If the null hypothesis (i.e., H_0) of no difference is rejected at the 5% level of significance, then we conclude there is a significant difference between the drug product and the placebo. In this case we may further evaluate whether the trial size is enough to effectively detect a clinically important difference (i.e., a difference that will lead the investigators to believe the drug is of clinical benefit and hence of effectiveness) when such difference exists. Typically the FDA requires at least 80% power for detecting such difference. In other words, the FDA requires there be at least 80% chance of correctly detecting such difference when the difference indeed exists.

Figure 2.6.2 displays the sampling distribution of a two-sided test under the null hypothesis in (2.6.1). It can be seen from Figure 2.6.2 that a two-sided test has equal chance to show that the drug is either effective in one side or ineffective in the other side. In Figure 2.6.2, C and $-C$ are critical values. The area under the probability curve between $-C$ and C constitutes the so-called acceptance region for the null hypothesis. In other words, any observed difference in means in this region is a piece of supportive information of the null hypothesis. The area under the probability curve below $-C$ and beyond C is known as the rejection region. An observed difference in means in this region is a doubt of the null hypothesis. Based on this concept, we can statistically evaluate whether the null hypothesis is a true statement. Let μ_D and μ_P be the means of the primary efficacy variable of the drug product and the placebo, respectively. Under the null hypothesis of no difference (i.e., $\mu_D = \mu_P$), a statistical test, say T can be derived. Suppose that t, the observed difference in means of the drug product and the placebo, is a realization of T. Under the null hypothesis we can expect that the majority of t will fall around the center, $\mu_D - \mu_P = 0$. There is a 2.5% chance that we would see t will fall in each tail. That is, there is a 2.5% chance that t will be either below the critical value $-C$ or beyond the critical value C. If t falls below $-C$, then the drug is worse than

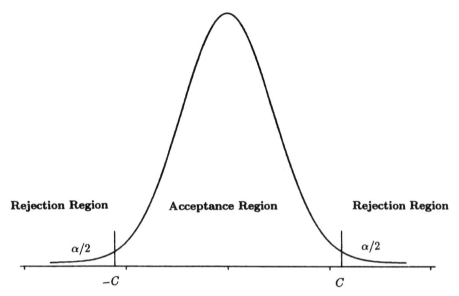

Figure 2.6.2 Sampling distribution of two-sided test.

the placebo. On the other hand, if t falls beyond C, then the drug is superior to the placebo. In both cases we would suspect the validity of the statement under the null hypothesis. Therefore we would reject the null hypothesis of no difference if

$$t > C \quad \text{or} \quad t < -C.$$

Furthermore we may want to evaluate how strong the evidence is. In this case, we calculate the area under the probability curve beyond the point t. This area is known as the *observed p-value*. Therefore the p-value is the probability that a result at least as extreme as that observed would occur by chance if the null hypothesis is true. It can be seen from Figure 2.6.2 that

$$p-\text{value} < 0.025 \quad \text{if and only if} \quad t < -C \quad \text{or} \quad t > C.$$

A smaller p-value indicates that t is further away from the center (i.e., $\mu_D - \mu_P = 0$) and consequently provides stronger evidence that supports the alternative hypothesis of a difference. In practice, we can construct a confidence interval for $\mu_D - \mu_P = 0$ and compare it with $(-C, C)$. If the constructed confidence interval is not within $(-C, C)$, then we reject the null hypothesis of no difference at the 5% level of significance. It should be noted that the above evaluations for the null hypothesis are operationally equivalent. In other words, they will reach the same conclusion regarding the rejection of the null hypothesis. However, a typical approach is to present the observed p-value. If the observed p-value is less than the level of significance, then the investigators would reject the null hypothesis in favor of the alternative hypothesis.

Although p-values measure the strength of evidence by indicating the probability that a result at least as extreme as that observed would occur due to random variation alone under the null hypothesis, they do not reflect sample size and the direction of treatment effect. Ware, Mosteller, and Ingelfinger (1986) indicate that p-values are a way of reporting the results of statistical analyses. It may be misleading to equate p-values with decisions. Therefore, in addition to p-values, Ware, Mosteller, and Ingelfinger (1986) recommend that the investigators also report summary statistics, confidence intervals, and the power of the tests used. Furthermore the effects of selection or multiplicity should also be reported.

Note that when a p-value is between 0.05 and 0.01, the result is usually called statistically significant; when it is less than 0.01, the result is often called highly statistically significant.

One-Sided versus Two-Sided Hypotheses

For marketing approval of a drug product, current FDA regulations require that substantial evidence of effectiveness and safety of the drug product be provided. Substantial evidence can be obtained through the conduct of two adequate well-controlled clinical trials. The evidence is considered substantial if the results from the two adequate well-controlled studies are consistent in the positive direction. In other words, both trials show that the drug product is significantly different from the placebo in the positive direction. In this case the alternative hypothesis of interest to the investigators will be that the drug is superior to the placebo. However, the hypotheses given in (2.6.1) do not specify the direction once the null hypothesis is rejected. As an alternative, the following hypotheses are proposed:

$$H_0: \text{There is no difference;}$$
$$\text{vs.} \quad H_a: \text{The drug is better than placebo.} \qquad (2.6.2)$$

The statistical test for the above hypotheses is known as *one-sided test*. If the null hypothesis of no difference is rejected at the 5% level of significance, then we conclude that the drug product is better than the placebo and hence is effective. Figure 2.6.3 gives the rejection region of a one-sided test. To further compare a one-sided and a two-sided test, let's consider the level of proof required for marketing approval of a drug product at the 5% level of significance. For a given clinical trial, if a two-sided test is employed, the level of proof required is one out of 40. In other words, at the 5% level of significance, there is 2.5% chance (or one out of 40) that we may reject the null hypothesis of no difference in the positive direction and conclude the drug is effective at one side.

HYPOTHESES TESTING AND p-VALUES

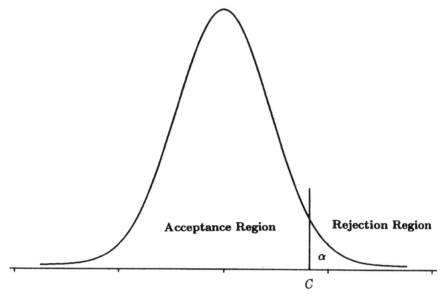

Figure 2.6.3 Sampling distribution of one-sided test.

On the other hand, if a one-sided test is used, the level of proof required is one out of 20. It turns out that the one-sided test allows more ineffective drugs to be approved because of chance as compared to the two-sided test. As indicated earlier, to demonstrate the effectiveness and safety of a drug product, FDA requires two adequate well-controlled clinical trials be conducted. Then the level of proof required should be squared regardless of which test is used. Table 2.6.2 summarizes the levels of proof required for the marketing approval of a drug product. As Table 2.6.2 indicates, the levels of proof required for one-sided and two-sided tests are one out of 400 and one out of 1600, respectively. Fisher (1991) argues that the level of proof of one out of 400 is a strong proof and is sufficient to be considered as substantial evidence for marketing approval, so the one-sided test is appropriate. However, there is no universal agreement among the regulatory agency (e.g., FDA), academia, and the pharma-

Table 2.6.2 Level of Proof Required for Clinical Investigation

	Type of Tests	
Number of Trials	One-Sided	Two-Sided
One trial	1/20	1/40
Two trials	1/400	1/1600

ceutical industry as to whether a one-sided test or a two-sided test should be used. The concern raised is based on the following two reasons:

1. Investigators would not run a trial if they thought the drug would be worse than the placebo. They would study the drug only if they believe that it might be of benefit.
2. When testing at the 0.05 significance level with 80% power, the sample size required is increased by 27% for the two-sided test as opposed to the one-sided test. As a result there is a substantial impact on cost when a one-sided test is used.

It should be noted that although investigators may believe that a drug is better than the placebo, it is never impossible that such belief might be unexpected (Fleiss, 1987). Ellenburg (1990) indicates that the use of a one-sided test is usually a *signal* that the trial has too small a sample size and that the investigators are attempting to squeeze out a significant result by a *statistical maneuver*. These observations certainly argue against the use of one-sided test for the evaluation of effectiveness in clinical trials. Cochran and Cox (1966) suggest that a one-sided test is used when *it is known* that the drug must be at least as good as the placebo, while a two-sided test is used when it is *not known* which treatment is better.

As indicated by Dubey (1991), the FDA tends to oppose the use of a one-sided test. However, this position has been challenged by several drug sponsors on the Drug Efficacy Study Implementation (DESI) drugs at the administrative hearings. As an example, Dubey (1991) points out that several views that favor the use of one-sided test were discussed in an administrative hearing. Some drug sponsors argued that the one-sided test is appropriate in the following situations: (1) where there is truly only concern with outcomes in one tail and (2) where it is completely inconceivable that the results can go in the opposite direction. In this hearing the sponsors inferred that the prophylactic value of the combination drug is greater than that posted by the null hypothesis of equal incidence, and therefore the risk of finding an effect when none in fact exists is located only in the upper tail. As a result a one-sided test is called for. However, the FDA feels that a two-sided test should be applied to account for not only the possibility that the combination drugs are better than the single agent alone at preventing candidiasis but also the possibility that they are worse at doing so.

Dubey's opinion is that one-sided tests may be justified in some situations such as toxicity studies, safety evaluation, analysis of occurrences of adverse drug reactions data, risk evaluation, and laboratory research data. Fisher (1991) argues that one-sided tests are appropriate for drugs that are tested against placebos at the 0.05 level of significance for two well-controlled trials. If, on the other hand, only one clinical trial rather than two is conducted, a one-sided test should be applied at the 0.025 level of significance. However, Fisher agrees that two-sided tests are more appropriate for active control trials.

It is critical to specify hypotheses to be tested in the protocol. A one-sided test or two-sided test can then be justified based on the hypotheses. It should be noted that the FDA is against a post hoc decision to create significance or near significance on any parameters when significance did not previously exist. This critical switch cannot be adequately explained and hence is considered an invalid practice by the FDA. More discussion regarding the use of one-sided test versus two-sided test from the perspectives of the pharmaceutical industry, academe, an FDA Advisory Committee member, and the FDA can be found in Peace (1991), Koch (1991), Fisher (1991), and Dubey (1991), respectively.

2.7 SIGNIFICANT DIFFERENCE AND CLINICAL EQUIVALENCE

In clinical trials a significant difference may be referred to as a statistically significant difference or a clinically significant difference. In hypotheses (2.6.1), if we reject the null hypothesis of no difference at the α level of significance, then we say that there is a *statistically significant difference* between treatments. Therefore a difference that is unlikely to occur by chance alone is considered as a statistically significant difference. Note that a statistically significant difference may be relatively small compared to the treatment mean. A large difference may not be statistically significant if the sample size is too small. A *clinically significant difference* is a difference that is considered important to the investigators. For example, for antidepressant agents (e.g., Serzone), a change from baseline of 8 in the Hamilton depression (Ham-D) scale may be considered of clinical importance. For antibiotics agents (e.g., Cefzil), a 15% reduction in bacteriologic eradication rate could be a clinically significant improvement. Similarly we could consider a reduction of 10 mm Hg in blood pressure as clinically significant for ACE inhibitor agents in treating hypertensive patients.

Basically there are four different outcomes for significant differences in a clinical trial. The result may show that (1) the difference is both statistically and clinically significant, (2) there is a statistically significant difference yet the difference is not clinically significant, (3) the difference is of clinical significance yet not statistically significant, and (4) the difference is neither statistically significant nor clinically significant. If the difference is both clinically and statistically significant or if it is neither clinically nor statistically significant, then there is no confusion. The conclusion can be drawn based on the results from the clinical data. However, in many cases a statistically significant difference does not agree with the clinically significant difference. For example, a statistical test may reveal that there is a statistically significant difference. However, if the difference is too small (it may be due to a unusually small variability or a relatively large sample size) to be of any clinical importance, then it is not clinically significant. In this case a small p-value may be instrumental in concluding the effectiveness of the treatment. On the other hand, the result may indicate that there is a clinically significant difference but the sample size is too small (or variability is too large) to claim a statistically significant difference. In this case the evidence of effectiveness is not substantial due to a large p-value. This inconsistency has created confusion/arguments among clinicians and biostatisticians in assessment of the efficacy and safety of clinical trials.

As indicated earlier, for the assessment of efficacy and safety of a drug product, a typical approach is to first demonstrate that there is a statistically significant difference between the drug products in terms of some clinical endpoints by testing hypotheses (2.6.1) repeated below:

$$H_0: \text{There is no difference;}$$
$$\text{vs.} \quad H_a: \text{There is a difference.}$$

Equivalently

$$H_0: \mu_D = \mu_P$$
$$\text{vs.} \quad H_a: \mu_D \neq \mu_P,$$

where μ_D and μ_P are the means of the primary clinical endpoint for the drug product and the placebo, respectively. If we reject the null hypothesis of no difference at the α level of significance, then there is a statistically significant difference between the drug product and the placebo in terms of the primary clinical endpoint. We then further evaluate whether there is sufficient power to correctly detect a clinically significant difference. If it does, then we can conclude that the drug product is effective and safe. Note that the above hypotheses are known as *point hypotheses*. In practice, it is recognized that no two treatments will have exactly the same mean responses. Therefore, if the mean responses of the two treatments differ by less than a meaningful limit (i.e., a clinically important difference), the two treatments can be considered clinically equivalent. Based on this idea, Schuirmann (1987) first introduces the use of *interval hypotheses* for assessing bioequivalence. The interval hypotheses for clinical equivalence can be formulated as

$$H_0: \text{The two drugs are not equivalent;}$$
$$\text{vs.} \quad H_a: \text{The two drugs are equivalent.} \qquad (2.6.3)$$

Or put differently,

$$H_0: \mu_A - \mu_B \leq L \quad \text{or} \quad \mu_A - \mu_B \geq U,$$
$$\text{vs.} \quad H_a: L < \mu_A - \mu_B < U,$$

where μ_A and μ_B are the means of the primary clinical endpoint for drugs A and B, respectively, and L and U are some clinically meaningful limits. The concept and interval hypotheses (2.6.3) is to show equivalence by rejecting the null hypothesis of inequivalence. The above hypotheses can be decomposed into two sets of one-sided hypotheses

SIGNIFICANT DIFFERENCE AND CLINICAL EQUIVALENCE 91

H_{01}: Drug A is superior to drug B (i.e., $\mu_A - \mu_B \geq U$);
vs. H_{a1}: Drug A is not superior to drug B;

and

H_{02}: Drug A is inferior to drug B (i.e., $\mu_A - \mu_B \leq L$);
vs. H_{a2}: Drug A is not inferior to drug B

The first set of hypotheses is to verify that drug A is not superior to drug B, while the second set of hypotheses is to verify that drug A is not worse than drug B. A relatively large or small observed difference may refer to the concern of the comparability between the two drug products. Therefore the rejection of H_{01} and H_{02} will lead to the conclusion of clinical equivalence. This is equivalent to rejecting H_0 in (2.6.3). In practice, if L is chosen to be $-U$, then we can conclude clinical equivalent if

$$|\mu_A - \mu_B| < \Delta,$$

where $\Delta = U = -L$ is the clinically significant difference. For example, for the assessment of bioequivalence between a generic drug product and an innovator drug product (or reference drug product), the bioequivalence limit Δ is often chosen to be 20% of the bioavailability of the reference product. In other words, in terms of the ratio of means μ_A/μ_B, the limits become $L = 80\%$ and $U = 120\%$. When log-transformed data are analyzed, the FDA suggests using $L = 80\%$ and $U = 125\%$. More detail on the assessment of bioequivalence between drug products can be found in Chow and Liu (1992a).

When two drugs are shown to be clinically equivalent, they are comparable to each other. Consequently they can be used as substitutes for each other. It should be noted that there is difference between the assessment of equality and equivalence. Hypotheses (2.6.1) are set for assessment of the equality between treatments, while hypotheses (2.6.3) are for the assessment of equivalence. The demonstration of equality does not necessarily imply equivalence. This is because the selected sample size for testing equality may not be sufficient for assessing the equivalence. Besides, when we fail to reject the null hypothesis of equality, it does not imply that the two treatments are equivalent, even if there is sufficient power for the detection of a clinically significant difference.

Note that current FDA regulations do not allow the sponsors to establish clinical equivalence based on clinical trials designed for the assessment of equality. Clinical equivalence between two drug products must be established based on the interval hypotheses as described in (2.6.3). For example, for a positive control study intended to show that a new therapy is at least as effective as the standard therapy, the *Guideline for the Format and Content of the Clinical and Statistical Sections of an Application* indicates that the sample size determination should specify a value for Δ, which is the difference between treatments that would be considered clinically meaningful. A difference smaller than this Δ would therefore indicate that the new therapy is clinically equivalent to the standard therapy. The power to detect a treatment difference of magnitude Δ or greater should be given. For testing interval hypotheses, several test statistics have been proposed. See, for example, Blackwelder (1982), Wellek (1993), Jennison and Turnball (1993), and Liu (1995a).

CHAPTER 3

Basic Design Considerations

3.1 INTRODUCTION

In the clinical development of a drug product, clinical trials are often conducted to address scientific and/or medical questions regarding the drug product in treatment of a specific patient population with certain diseases. At the planning stage of a clinical trial, it is therefore important to define "What is the question?" Defining "What is the question?" helps determine the study objective(s) and consequently helps set up appropriate hypotheses for scientific evaluation. The next question is "How to answer the question?" It is important that the intended clinical trial provide an unbiased and valid scientific evaluation of the question. Temple (1982) indicates that two kinds of difficulties are often encountered when the clinical scientists attempt to identify/answer pertinent scientific questions by conducting well-controlled clinical trials. The first difficulty is that individual studies may be designed without careful attention to the questions they really are capable of answering. Consequently the trial is either a useless trial that answers no question at all or it is a trial that answers some other question (but not the one intended) or only part of the intended question. Second, the total package of studies may be designed without a thoughtful consideration of all the questions that are pertinent. There are practical limitations on the number of studies that can reasonably be expected; nevertheless, it seems possible that more of the pertinent questions can be answered without any increase in the total number of patients exposed in clinical trials.

To best answer scientific and/or medical questions through clinical trials, the FDA suggests that an overall study plan and design be briefly but clearly described in the protocol of the intended clinical trial. A thoughtful and well-organized protocol includes study objective(s), study design, patient selection criteria, dosing schedules, statistical methods, and other medical related details. As a result, "How to choose an appropriate study design?" and "How to analyze the collected clinical data using valid statistical methods?" have become two serious aspects of a clinical trial plan. These two aspects are closely related to each other since statistical methods for data analysis depend on the design employed. Generally speaking, meaningful conclusions can only be drawn based on data collected from a valid scientific design using appropriate statistical methods. Therefore the selection of an appropriate study design is important in order to provide an unbiased and scientific evaluation of the scientific and/or medical questions regarding the study drug. Before a study design is chosen, some basic design considerations such as patient selection, randomization and blinding, the selection of control(s), and some statistical issues must be considered to justify the use of statistical analyses. In this chapter our efforts will be directed to the selection of patients for clinical trials, the selection of control(s), statistical considerations, and some other related issues. Randomization and blinding will be discussed in detail in the next chapter. Several commonly employed designs in clinical trials are reviewed in Chapter 5.

In the next section issues in selecting patients for a clinical trial are discussed. In Section 3.3 we discuss the selection of control(s) in clinical trials. Some statistical considerations regarding clinical evaluation of efficacy and safety, sample size estimation, interim analysis and data monitoring, and statistical and clinical inference are given in Section 3.4. Section 3.5 contains some specific issues related to designing a clinical trial such as single site versus multi-sites, treatment duration, patient compliance, and missing value and dropout. A brief concluding discussion is given in the last section.

3.2 PATIENT SELECTION

As was indicated earlier, one of the primary objectives of a clinical trial is to provide an accurate and reliable clinical evaluation of a study drug for an intended patient population with certain diseases. In practice, statistical and clinical inference are usually drawn based on a representative sample (a group of patients to be enrolled in the trial) selected from the intended patient population of the clinical trial. A representative sample provides the clinician with the ability to generalize the findings of the study. Therefore selecting patients for a clinical trial plays an important role to best answer the scientific and/or medical questions of interest regarding the study drug. Basically selecting patients for a clinical trial involves two steps. First, we need to define the intended patient population. Patients are then selected from the intended patient population for the clinical trial. For a given disease, the intended patient population is often rather heterogeneous with respect to patient characteristics and the severity of the disease. The heterogeneity of the intended patient population can certainly decrease the accuracy, reliability, and the generalization of the findings of the study. In clinical trials the intended patient population usually involves various sources of expected and unexpected biases and variabilities. For example, bias and variability due to differences in patient demographic characteristics such as age, sex, height, weight, and functional status are expected. Bias and variability caused by changes in disease status and concomitant therapies are unexpected. These sources of biases and variabilities will not only decrease the accuracy and reliability of the observed clinical results but also limit the clinician's ability to generalize the findings of the study. For good clinical practice it is therefore desirable to define the intended patient population in such a way that it is a homogeneous as possible with respect to these patient characteristics in order to reduce bias and to minimize variability. For this purpose Section 314.166 of CFR also requires that the method for selection of patients in clinical trials provide adequate assurance that the selected patients have the disease and condition being studied.

For patient selection Weintraub and Calimlim (1994) classify patients into two categories. These two categories are inpatients for short-term hospital studies and outpatients for chronic conditions. Different concerns/considerations may be raised depending on which type of patients are intended for the clinical trials. In this section we will focus on a general concept for selecting patients for a clinical trial which includes the development of eligibility criteria, selection process, and ethical considerations.

Eligibility Criteria

In clinical trials a set of eligibility criteria is usually developed to define the intended patient population from which qualified (or eligible) patients can be recruited to enroll the studies. Typically a set of eligibility criteria consists of a set of inclusion criteria and a set of exclusion criteria. The set of inclusion criteria is used to roughly outline the intended patient population, while the set of exclusion criteria is used to fine-tune the intended patient population by removing the expected sources of variabilities. To be eligible for the intended study, patients must meet *all* the inclusion criteria. Patients meeting *any* of the exclusion criteria will be excluded from the study. Eligibility criteria should be developed based on patient characteristics, diagnostic criteria, treatment duration, and the severity of the disease.

Before a set of well-defined eligibility criteria can be developed, it is necessary to have a clear understanding of the study medicine and the indication it is intended for. For example, some medicines are intended for specific patient population (e.g., female, children, or elderly) with a certain disease. The inclusion criteria usually describe the intended patient population based on the diagnosed symptoms or history of the intended disease. Patients who have history of hypersensitivity to the study medicine, treatment-resistance, disease changes, and/or concurrent diseases requiring treatments are usually excluded from the study. Different eligibility criteria will result in different study patient populations. These differences decrease the ability to apply the study results to any other patient population. In what follows, we provide three examples for the development of eligibility criteria for clinical trials from three major therapeutic areas: anti-infectives, cardiovascular, and central nervous system.

For the first example, consider a clinical trial comparing the clinical and microbiologic efficacy and safety of an antibiotic agent in the treatment of febrile episodes in neutropenic cancer patients. As indicated in the *Guidelines for the Use of Antimicrobial Agents in Neutropenic Patients with Unexplained Fever* (IDSA, 1990), anti-infective drugs have become a standard of medical practice whenever a neutropenic patient becomes febrile. For example, ceftazidime which is a marketed third-generation cephalosporin is indicated for the treatment of febrile episodes in neutropenic cancer patients caused by *Streptococcus spp.*, *Escherichia coli*, *Klebsiella spp.*, *Pseudomonas aeruginosa*, and *Proteus mirabilis* (PDR, 1992). With the more prompt and routine initiation of anti-infective therapy, the microbiologic confirmation of infection has declined such that as many as 50% to 70% of the febrile neutropenic episodes do not have a defined microbial etiology. These patients are categorized as having unexplained fever in which the infection may have been masked by the early introduction of antimicrobial therapy. Since unexplained fever constitutes the majority of febrile neutropenic events, the evaluation of empiric therapy has become more difficult. Consequently the question raised is how therapy can be adequately assessed when fever is the only evaluable parameter (IHS, 1990). For this reason the primary clinical endpoint being evaluated in this study is fever. It is suggested an oral temperature greater than $38.5°C$ once or greater than $38°C$ on two or more occasions during a 12-hour period be considered as an inclusion criterion for the study. Note that it may be a concern that the weak antistrephylococcal activity of ceftazidime in patients whose infections are frequently caused by gram-positive bacteria. As a result many practitioners have routinely added vancomycin to ceftazidime as initial coverage for the febrile neutropenic patient. Therefore no other antibacterial agents except intravenous vancomycin will be administered during the study. Patients who require other systemic antibacterial drugs concomitantly are then excluded from the study. Other considerations regarding the inclusion and exclusion criteria are summarized in Table 3.2.1. For example, patients who have history of hypersensitivity to a cephalosporin or penicillin are excluded from the study.

The second example concerns the evaluation of the efficacy of an oral agent for the treatment of patients with noninsulin-dependent diabetes mellitus (NIDDM). As indicated by Cooppan (1994), NIDDM is the most common form of diabetes seen in clinical practice. The prevalence in the United States is about 6.6% and rises to 18% in the elderly. The incidence is about 500,000 new patients every year. NIDDM often have hypertension and hyperlipidemia. In early stages patients are hyperinsulinemic. Most patients are overweight and have upper body or truncal obesity. The onset of NIDDM is usually above 40 years of age. The disease has a strong genetic basis. It is more frequently seen in native Americans, Mexican Americans, and blacks. The pathophysiology of the disease is due to changes in insulin production and secretion, insulin resistance in liver, muscle, and adipose tissue. A high glucose level could further reduce panneatic insulin secretion. The treatment for NIDDM patients normally includes (1) diet alone, (2) diet plus oral hypoglycemic drug, and (3) weight control. Note that a mild to moderate weight loss (e.g., 5 to 10 kg) has been shown to improve diabetes control and a moderate calorie restriction (e.g., 250 to 500 calories less than average daily intake) is recommended for

Table 3.2.1 Eligibility Criteria for Anti-Infectives Agents

A. Inclusion Criteria

1. Hospitalized patients aged 18 years or older.
2. An oral temperature greater than 38.5°C once or greater than 38°C on two or more occasions during a 12-hour period.
3. Fewer than 500 absolute neutrophils (polymorphonuclear and segmented) per mm^3, or patients presenting with between 500 and 1000 absolute neutrophils per mm^3, whose counts are anticipated to fall below 500 per mm^3 within 48 hours because of antecedent therapy.

B. Exclusion Criteria

1. History of hypersensitivity to a cephalosporin or penicillin.
2. Pregnant or breast-feeding.
3. Requiring other systemic antibacterial drugs concomitantly except for intravenous vancomycin.
4. Creatinine clearance ≤15 mL/min or requiring hemodialysis or peritoneal dialysis.
5. History of positive antibody test for HIV.
6. A severe underlying disease such as meningitis, osteomyelitis, or endocarditis.
7. Patients undergoing bone marrow transplantation or stem cell harvesting and infusion.
8. Any other condition that in the opinion of the investigator(s) would make the patient unsuitable for enrollment.

---

weight control. Based on the above considerations, Table 3.2.2 provides a sample eligibility criteria for the NIDDM study. As can be seen, patients aged 40 years or older—who have a previously established diagnosis of NIDDM and are currently controlled on diet alone or were previously managed on an oral sulfonylurea with a fasting plasma glucose greater than 180 mg/dL but without symptomatic diabetes—meet the inclusion criteria for entry. However, exclusion criteria exclude patients who are known to have a history of hypersensitivity of biguanides and significant cardiovascular diseases from the study. Significant cardiovascular diseases may include acute myocardial infarction, unstable angina, congestive heart failure, and arrhythmia.

For the third example, consider a clinical trial comparing the effects of an antidepressant compound to sertralin (Zoloft) on sexual function in patients with previously demonstrated sexual dysfunction with sertraline during treatment for major depression. Segraves (1988, 1992) indicated that patients treated with many psychotropic medications including antidepressants have sexual adverse effects. However, the mechanism by which antidepressants produce sexual dysfunction have not been clearly established. Symptoms of sexual dysfunction may include one or more of the followings: delayed or absent ejaculatory response, partial or total anorgasmia, inadequate lubrication or swelling. The incidence rate for sexual dysfunction for sertraline-treated male patients is 15.5% compared to 2.2% of placebo-treated male patients

Table 3.2.2 Eligibility Criteria for a NIDDM Study

A. Inclusion Criteria

1. Males or females aged ≥40 years old.
2. Females who are not postmenopausal; they must be nonlactating, incapable of becoming pregnant or of childbearing potential, practicing an effective method of of contraception, or have a negative serum pregnancy test documented at screening.
3. Currently suboptimally controlled on diet alone or previously managed on an oral sulfonylurea with an fasting plasma glucose ≥ 180 mg/dL but without symptomatic diabetes.
4. Detectable fasting serum insulin and c-peptide at screening.
5. Normal renal function as defined by serum creatinine of <1.5 mg/dL for men and <1.4 mg/dL for women, and ≤1 proteinuria on routine urinalysis.
6. Acceptable liver function as defined by SGOT/AST ≤ 62 U/L and SGPT/ALT ≤ 58 U/L for females and ≤90 U/L for males.

B. Exclusion Criteria

1. Markedly symptomatic diabetes, marked polyuria and weight loss > 10%.
2. History of hypersensitivity to biguanides.
3. Prior insulin therapy except for acute illness or sugery.
4. Significant cardiovascular disease.
5. Significant renal disease or renal functional impairment as evidenced by a serum creatinine ≥1.5 mg/dL for males and ≥1.4 mg/dL for females.
6. Significant hepatic disease as evidenced by abnormal liver function as defined as by SGOT/AST > 62 U/L and SGPT/ALT > 58 U/L for females and >90 U/L for males.
7. Active infectious process such as gangrene and pneumonia.
8. Pulmonary insufficiency.
9. Metabolic acidosis and acute/chronic diabetic ketoacidosis.
10. Any patient for any other condition which, in the investigator's opinion, would make the patient unsuitable for the study or would interfere with the evaluation of the study medication.

(Zoloft, 1992). It is believed that both potentiation of peripheral nervous system adrenergic/noradrenergic activity and increasing brain serotonin (5-HT) level by blocking the neuronal 5-HT reuptake process may induce sexual dysfunction. In order to be eligible for this study, patients must be experiencing sexual dysfunction while being treated with sertraline at a daily dose of 100 mg during their current depressive episode. In addition sertraline must have been prescribed for the *Diagnostic and Statistical Manual*, Third Edition–Revised (DSM-III-R), diagnosis of major depression based on documented patient history. Other considerations for excluding patients from the study including sexual dysfunction due to any organic condition, treatment-resistant depression, or some significant and/or uncontrolled medical conditions. Table 3.2.3 gives a sample list of eligibility criteria for the study. Note that significant and/or

PATIENT SELECTION 99

Table 3.2.3 Eligibility Criteria for Central Nervous System Agents

A. Inclusion Criteria

1. Males or females 18 to 65 years of age. Female patients of childbearing potential must be nonlactating, have a confirmed negative serum pregnancy test prior to enrollment, and be employing an acceptable method of birth control. employing an acceptable method of birth control.
2. Patients who are experiencing sexual dysfunction in response to sertraline at a to daily dose of 100 mg during their current depressive episode.
3. Treatment with sertraline must have been diagnosed of major depression.

B. Exclusion Criteria

1. Patient having a diagnosis of treatment-resistant depression.
2. History of sexual dysfunction due to any organic condition.
3. Patients who cannot discontinue their current psychotropic medications and/or are likely to require treatment with any prohibited concomitant therapy.
4. History of hypersensitivity to trazodone, etoperidone, or sertraline.
5. Patients receiving any concomitant medication that can produce sexual dysfunction.
6. Patients who have met DSM-III-R criteria for any significant psychoactive substance use disorder within the 12 months prior to screening.
7. Patients who exhibit a significant risk of committing suicide or have a score ≤ 3 or item 3 "suicide" of the HAM-D scale.
8. Patients who have a significant and/or uncontrolled medical condition.
9. Patients with any clinically significant deviation from normal in the physical or electrocardiographic examinations or medically significant values outside the normal range in clinical laboratory tests.
10. Patients with a positive urine drug screen.
11. Patients with implanted prosthetic devices.
12. Patients who have any other medical condition(s) that can confound the interpretation of the safety and the efficacy data.

uncontrolled conditions may include symptomatic paroxysmal, chronic cardiac arrhythmias, history of stroke, transient ischemic attacks, or history of a positive test for the HIV antibody or antigen.

Patient Selection Process

As discussed above, a set of well-developed eligibility criteria for patient selection can not only best describe the intended patient population but also provide a homogeneous sample. The criteria help in reducing bias and variability and consequently increase statistical power of the study. Therefore, in practice, it may be desirable to impose more inclusion and exclusion criteria to further eliminate bias and variability. However, it should be noted that the more criteria that are imposed, the smaller the intended patient population will be. Although a smaller patient population may be more homogeneous, it may result in difficulties in patient recruitment and limitations in the generalization of the findings of the study. Therefore it is suggested that the considerations not be too restrict to decrease patient enrollment and lose the generality of the intended patient population.

In clinical trials, however, the number of patients is usually called for by the study protocol to ensure that the clinical trials can provide valid clinical evaluation of the study medicines with the desired accuracy and precision. It is important then in patient selection to achieve enough patients for the intended trials. In practice, a single study site may not be feasible for an intended clinical trial due to its limited capacity and resources. Besides, there may not be sufficient patients with the disease available in the area within the intended time period of the study. To recruit enough number of patients and to complete the study within the time frame, as an alternative, a multicenter trial is usually considered. If a multicenter trial is to be conducted, the following two questions should be considered:

1. How many study sites should be used?
2. How to select these study sites?

As a rule of thumb, the number of sites should not be greater than the number of patients within each selected study site. This is because statistical comparison between treatments is usually made based on patients (i.e., experimental units) within study sites. It is therefore not desirable to have too many study sites, though it may speed up the enrollment and consequently shorten the completion time of the study. The selection of study sites depends primarily on the following criteria:

1. Individual investigator's qualification and experience for disease.
2. Feasibility of the investigator's site for conducting the proposed trial.
3. Dedication, education, training, and experience of the personnel at the investigator's site.
4. Availability of certain equipments (e.g., magnetic resonance imaging [MRI] or densitometer).
5. Geographic location.

Considerations 1 to 4 ensure that the intended study will be appropriately carried out in such a way that the differences among investigators is minimized. The geographic location guarantees that the patients enrolled into the study constitute a representative sample from the intended patient population. Another important consideration for selection of investigators or study sites is probably their ability to enroll patients and to complete the study within the planned time frame.

For a selected study center, the selection process for patients involves the following concerns:

PATIENT SELECTION 101

1. Initial guess of how many patients will meet the eligibility criteria.
2. Screening based on diagnostic criteria.
3. Patient's disease changes.
4. Concurrent diseases/medications.
5. Psychological factors.
6. Informed consent

At the selected study site the investigator is often concerned with whether he or she can enroll enough patients for the intended trial. The investigator usually provides an estimate of how many patients will meet the eligibility criteria based on how many patients he or she has seen at the study site. In practice, such an estimate often overestimates the actual number of patients who will participate the study. Bloomfield (1969) recommends that investigator check the availability and suitability of the patient population at hand through their records or perhaps a formal pilot study.

As indicated by Weintraub and Calimlim (1994), a majority of patients may be excluded at the screening stage of patient selection process due to some administrative reason and the rigor of the diagnostic criteria. Weintraub and Calimlim (1994) point out that administrative reasons such as nonavailability during screening can be as high as 40% in a study intended to evaluate the efficacy of three analgesic treatments and a placebo administered in single doses in double-blind fashion for postoperative pain. Furthermore in this case 86% were eliminated due to the rigor of diagnostic criteria such as insufficient severity of pain after the operation. Thus the diagnostic strictures imposed by the clinical trial can decrease the number of available patients even further. It should be noted that small changes in the criteria can make vast differences in patient availability without materially influencing the clinical outcome and its extrapolatability.

It is also recommended that the patient selection process be able to address the issue of specific disease requirements such as disease of a particular severity or duration. For example, a moderate disease status may be preferred because (1) it is realized that patients must be sick enough to get better and (2) patients may be too severe to study. At screening, many patients may be excluded from the study due to disease changes and concurrent diseases requiring concomitant medications. This is true especially for very sick patients who frequently have disease changes and/or concurrent diseases. If we exclude patients who have disease changes and/or concurrent diseases, the patient population under study will become much smaller. Consequently we may not be able to recruit enough patients for the study. In addition seasonal factor for some diseases must be taken into account.

Weintraub and Calimlim (1994) point out that ideal participants for clinical trials are patients who will carry through the clinical trials and actively interact with the investigators rather than be passive experimental subjects. It is suggested that psychological factors (e.g., fear of toxicity) be carefully analyzed to enable a patient to make a reasonable judgment about participation in a clinical trial.

At screening prior to the entry of the study, signed and dated written informed consent must be obtained by the investigator from the patient after full disclosure of the potential risks and their nature. Consent must be obtained before a prospective study candidate participates in any study-related procedure, including any change in current therapy required for entry into the study. The fact that such consent was obtained must be recorded in the case report form. In practice, it is not uncommon to allow the investigator to exclude any patient for any condition which in his/her opinion, would make the patient unsuitable for the study or would interfere with the evaluation of the study medication. For example, the investigator may decide to exclude the patient whose white blood cell count is less than $3500/\text{mm}^3$ or neutrophil count is less than $1500/\text{mm}^3$ from the study of an antidepressant agent comparing with sertraline on sexual function in patients with previously demonstrated sexual dysfunction with sertraline during treatment for major depression as described above. Note that many times the abnormalities that are observed in laboratory tests are due to illnesses unrelated to the disease under study or to other necessary therapeutic interventions.

Ethical Considerations

For many severely destructive diseases such as AIDS, Alzheimer's disease, and cancer, it is unethical to include placebo concurrent control in a clinical trial where an effective alternative remedy is available. It, however, should be noted that the effectiveness and safety of a test agent can only be established by inclusion of a placebo concurrent control. Ethical considerations will definitely affect the patient selection process. In such cases it is suggested that different numbers of patients be allocated to the treatment arm and the placebo arm in order to reduce the percentage of patients being assigned to the placebo arm. For example, we may consider a two-to-one ratio for a study. In other words, two-thirds of the patients who participate in the study will receive the active treatment and only one-third will receive the placebo. In some comparative clinical trials, if patients are too sick, a certain amount of standard therapy must be permitted for ethic reasons. The use of active control will be discussed further in the next section. Note that it is suggested that placebos be used for trivial, nondangerous, or self-limiting disorders provided that consent is obtained from the patient.

In recent years ethical considerations for the use of females, children, and the elderly have attracted much attention. For example, in its 1993 revised guidelines for clinical trials, the FDA suggests to include in clinical trials women of childbearing potential who are usually excluded in early drug studies of non-life-threatening disease. Since neonates, infants, and children respond to certain medicines differently than adults, trials of medicines to be used in this age group are always necessary. Therefore special consideration must be given to the conduct of trials in children. For this purpose the CPMP (Committee for Proprietary Medicinal Products) of the EC has adopted guidelines on clinical investigation in children. For the geriatric population it is also important to evaluate any medicine likely to be used in that age group due to the reasons that there is an increasing incidence of adverse events in the elderly and that there are altered pharmacokinetic profiles of some medicines and impaired homeostasis in the elderly.

Note that in the pharmaceutical industry a copy of the final Institution Review Board (IRB) approved informed consent form must be provided to the sponsor before drug supplies will be shipped or enrollment in the study can begin.

3.3 SELECTION OF CONTROLS

In clinical trials, bias and variability can occur in many ways depending on the experimental conditions. These bias and variability will have an impact on the accuracy and reliability of statistical and clinical inference of the trials. Uncontrolled (or noncomparative) studies are rarely of value in clinical research, since definitive efficacy data are unobtainable and data on adverse events can be difficult to interpret. For example, an increase in the incidence of hepatitis during an uncontrolled study may be attributed to the medicine under investigation. Therefore the FDA requires that adequate well-controlled clinical trials be conducted to provide an unbiased and valid evaluation of the effectiveness and safety of study medicines. The purpose of a well-controlled study is not only to eliminate bias but also to minimize the variability, and consequently to improve the accuracy and reliability of the statistical and clinical inference of the study.

In early 1970s it was not uncommon for clinical scientists to conduct an uncontrolled clinical trial for scientific evaluation of a therapeutic intervention. Table 3.3.1 summarizes a comparison of positive findings between uncontrolled and controlled clinical trials in selected therapeutic areas. As can be seen from

Table 3.3.1 Comparison of the Results Between Uncontrolled and Controlled Trials

Therapeutic Areas	Percent of Positive Findings	
	Uncontrolled	Controlled
Psychiatric (Foulds, 1958)	83%	25%
Antidepressant (Wechsler et al., 1965)	57%	29%
Antidepressant (Smith, et al., 1969)	58%	33%
Respiratory distress syndrome (Sinclair, 1966)	89%	50%
Rheumatoid arthritis (O'Brien, 1968)	62%	25%

Source: Summarized and tabulated from Spilker (1991).

Table 3.3.1, the positive findings of uncontrolled trials were obviously exaggerated. Since estimates of the treatment effects are usually extremely biased in the positive direction, the FDA requires that adequate well-controlled studies use a design that permits a valid comparison with a control to provide a quantitative assessment of drug effect (Section 314.126 in Part 21 of CFR). A well-controlled trial is referred to as a trial that is conducted under the experimental conditions such that patient characteristics between treatment groups are homogeneous. For a controlled clinical trial comparing a therapeutic intervention with a control, it is also desirable to avoid any possible confounding factor when evaluating the treatment effect. In other words, any difference in clinical endpoints between treatment groups is only due to the difference between treatments rather than other confounding factors. In general, there are two types of controls commonly employed in comparative clinical trials. They are concurrent control and historical control, which depend on whether the experimental treatments are investigated concurrently with the controlled groups and internally within the same clinical trial. Consequently the use of a concurrent control provides internal validity of the conclusions obtained from clinical trials because it provides a valid comparison for assessment of the experimental treatments with respect to a control within the same study. According to Section 314.126 in Part 21 of CFR, the concurrent control includes placebo concurrent control, dose-comparison control, no treatment control, and active concurrent control, which are described below.

Placebo Concurrent Control

In the past almost two centuries, there has been many heated debates over the use of an inactive placebo group as a reference for evaluation of a test therapy in treatment of patients under aliment or medication conditions. For example, see Brody (1981, 1982), Lundh (1987), Levine (1987), Stanley (1988), and Sanford (1994). Brody (1982) defines a placebo as a form of medical therapy, or an intervention designed to simulate medical therapy. A placebo is believed to be without specificity for the condition being treated. A placebo is used either for its symbolic effect or to eliminate observer bias in a controlled experiment. Brody (1982) also indicated that a placebo effect is the change in the patient's condition that is attributable to the symbolic import of the healing intervention rather than to the intervention's specific pharmacologic or physiologic effects. In clinical trials it is not uncommon to observe the placebo effect. Brody (1982) points out that the placebo effect can be as important as many treatments. Sackett (1989) also indicated that the placebo effect is the most probable cause for symptomatic relief following internal mammary artery ligation experienced by many angia patients. In addition the influence of the placebo effect is not restricted only to subjective psychological or psychiatric measurements. Placebo also alters objective clinical endpoints such as cholesterol level (Coronary Drug Project Research Group, 1980), laboratory values, and measures of physiologic change (Wolf, 1950), for even the pattern of placebo response resembles the pharmacologic response of the active treatment (Lasagna et al., 1958). For a better understanding of the placebo effect, consider a clinical trial with a 4-week single-blind placebo run-in phase and a 24-week double-blind randomized phase for evaluation of a new agent in three doses with a placebo in treatment of the patients with benign prostatic hyperplasia. The primary clinical endpoints are the proportions of the patients with an at least 3 improvement of total symptom score or an increase of maximum urinary flow rate greater than 3 mL/s at the end of 24 weeks of the double-blind phase (Boyarsky et al., 1977) which is given in Table 3.3.2. The placebo response rate based on subjective total symptom score is not statistically significantly different from the rates of three doses. Although the placebo effect for the more objective maximum urinary flow rate is less than that based on the symptom score, it is still about 23% and is not statistically significantly different from those of the three doses. The causes of the placebo effect have been speculated for a long time; the effect is now believed to be due to a combination of interactions among patients, physicians, and experimental conditions surrounding the clinical trials (Brody, 1981, 1982; Lundh, 1987; Sanford, 1994). In any cases a response observed from a patient receiving an active treatment is a function of three major components: the true pharmacological activity of the active ingredients, the symptomatic relief provided by the placebo, and the natural reversible healing process provided by the body. The effect contributed by the last two components cannot be unbiasedly estimated unless there is a placebo group in the trials. Therefore the inclusion of a placebo concurrent control in clinical trials is necessary to provide unequivocally and unbiasedly an assessment of the effectiveness and safety of the therapeutic intervention under study.

As indicated earlier, it is unethical to use placebo concurrent controls where symptoms are severe or hazardous and where there exists an alternative therapy with established effectiveness and safety. In practice, it is also unethical to expose patients with severe diseases to study medicines under investigation that may have unknown yet potentially serious even deadly adverse events. The

Table 3.3.2 Response Rates by Total Symptom Score and Maximum Urinary Flow Rate (mL/s) at the End of 24 Weeks of the Double-Blind Phase

	Percent of Patients with Improvement of	
Dose	Symptom Score > 3	Maximum Urinary Flow Rate >3 mL/s
Placebo	41/92(45%)	21/92(23%)
10 mg	36/89(40%)	25/88(28%)
30 mg	38/85(45%)	17/82(21%)
60 mg	36/85(42%)	20/80(25%)

saga of Cardiac Arrhythmia Suppression Trial (CAST, 1991) provides a vivid but sad example. If a placebo concurrent group had not been included, neither the excessive risk of death for flecainide and encainide could have been demonstrated nor the assumption of the use of surrogate endpoint ventricular premature contraction (VPC) for mortality could have been proved wrong. Kessler and Feiden (1995) also indicate that the AIDS activists now made an extraordinary plea to the top FDA officials not to approve drugs to treat the disease caused by the human immunodeficiency virus (HIV) too quickly. The reason is that they and their physicians must study in detail the approved antiviral AIDS drugs and examine the efficacy and safety of experimental therapy before they can make an optimal use of these treatments, since many of new experimental drugs were tested without a placebo concurrent group. Spilker (1991) lists the conditions for the ethical use of a placebo concurrent groups in clinical trials. These conditions are summarized in Table 3.3.3. As a result placebo concurrent control should not be selected as the internally controlled group for evaluation of a new treatment if there exists a treatment whose efficacy has already been established for the intended disease. It is not ethical to use placebo for the care of severe or life-threatening diseases. In all other cases, however, placebo concurrent control should be employed as the standard concurrent control, whenever operationally feasible, for evaluation of the effectiveness and safety of a new therapeutic intervention.

Dose-Comparison Concurrent Control

As discussed in Chapter 1, the primary objectives for phase II studies are (1) to establish the efficacy, (2) to characterize its dose-response relationship, and (3) to identify the minimum effective and maximum tolerable doses of the therapeutic agent under development. The dose proportionality studies for the

Table 3.3.3 Conditions for the Ethical Inclusion of a Placebo Concurrent Control

1. No Standard treatment exists.
2. Standard treatment is ineffective or unproved to be effective.
3. Standard treatment is appropriate for the particular clinical trials.
4. The placebo has been reported to be relatively effective in treating the disease or condition.
5. The disease is mild and lack of treatment is not considered to be medically important.
6. The placebo is given as an add-on treatment to an already existing regimen that is not sufficient to treat patients.
7. Allowing concomitant medicine is one measure of efficacy in these clinical trials.
8. The disease process is characterized by frequent spontaneous exacerbations and remission (e.g., peptic ulcer).
9. "Escape clauses" or points are designed into the protocol.

Source: Spilker (1991).

assessment of the assumption of linear pharmacokinetics of the test drug usually include at least three doses. Therefore a clinical trial with dose-comparison concurrent control includes at least two doses of the same test agent. Since the dose-comparison studies are usually conducted in the phase II stage where the efficacy of the test agent has not yet definitely been established, it is imperative to include a placebo concurrent control to provide an estimate of the absolute efficacy for each dose in addition to the dose-response relationship. The exclusion of placebo concurrent control in a dose-response study could be disastrous and costly. For example, in a major pharmaceutical company, a randomized, double-blind phase II study was conducted to establish the dose-response relationship for a new contrast enhancement agent in conjunction with MRI in diagnosis of malignant liver tumors in patients with known focal liver lesions. Despite suggestions of changes by the project statistician because of (1) the exclusion of a placebo concurrent control, (2) the use of an invalidated scale for diagnostic confidence, and (3) visual evaluation of the pre- and postcontrast films as the primary endpoint, the trial was conducted as planned without any of the modifications. Table 3.3.4 provides the proportion of the patients with good or excellent improvement for diagnostic confidence or visual evaluation. As can be seen from Table 3.3.4, there was no dose-response at all. It should be noted that without a placebo concurrent control, it is impossible to assess whether a response rate between 60 to 65% observed in this trial really demonstrates the true efficacy of the test agent because there were no trials with a placebo concurrent control ever conducted.

Active Treatment Concurrent Control

During the development of a new test agent, it may be of interest to establish a superior efficacy than the standard agent or to show therapeutic equivalence in efficacy to the standard therapy but with a better safety profile. For these purposes clinical trials are usually conducted with active agents concurrently. In many cases active treatments are employed for ethical reasons. If the trials are designed to serve as adequate well-controlled trials for providing substantial evidence of efficacy and safety for drug approval, the active treatment concurrent control must unequivocally demonstrate its superior effi-

Table 3.3.4 Percent of Patients with Good or Excellent Improvement in Diagnostic Confidence and Visual Evaluation

Dose	Percent of Patients with Good or Excellent Improvement	
	Diagnostic Confidence	Visual Evaluation
12.5 μmol	45/77 (58%)	51/79 (65%)
25.0 μmol	47/78 (60%)	52/82 (63%)
50.0 μmol	45/76 (59%)	50/78 (64%)

cacy in pivotal trials with a placebo concurrent control. Otherwise the trials must include a placebo concurrent placebo in addition to the active treatment concurrent control. In some cases clinical trials are conducted to establish therapeutic equivalence to a standard therapy because of no systematic absorption of a different route of administration such as the metered dose inhaler (MDI) for asthma and retin-A for acne, or because of inadequacy of pharmacokinetic measures for chemicals such as sucralfate for acute duodenal ulcer (Liu and Chow, 1993; Liu, 1995a, 1996). However, as indicated by Temple (1982) and Huque and Dubey (1990), equivalence between two active agents demonstrated in an active control trial can imply that both agents are efficacious or both are inefficacious. Therefore it is important to always include a placebo concurrent control in the active control trials unless a superior efficacy has been established and accepted by the regulatory authority. Note that active control trials will be discussed further in Chapter 5.

No Treatment Concurrent Control

For certain diseases, under the assumptions that (1) the objective measurements for effectiveness are available and can be obtained in a very short period of time and (2) the placebo effect is negligible, the test agent can be compared concurrently with no treatment. In these cases the FDA requires that patients be randomized to receive either the test agent or the no treatment concurrent control (Section 314.122 in Part 21 of CFR). In practice, however, it is recommended that no treatment concurrent controls should be avoided if possible during clinical development of phases I–III trials of new agents due to the reasons that it is not good clinical practice and that it fails to simulate the psychological effect of the placebo on efficacy.

Historical Control

In clinical research sometimes it is of interest to compare the results of the test treatment with those of other active treatments or the historical experience of a disease or condition that is adequately documented. Basically historical data are obtained in two ways. One is from the same group of patients who received no treatment, the same treatment, or different treatments at different times. The other is from different patients who received no treatment, the same treatment, or different treatments at different times. In either case the data of historical control are not obtained concurrently. Therefore the experimental conditions of the trials are not controlled concurrently for both the test and control groups. Hence Section 314.122 of Part 21 of CFR indicates that the historical control are reserved for the special diseases with high and predictable mortality such as certain malignant cancers or for the agents in which the effect of the drug is self-evident such as general anesthetics.

In summary, for clinical development of phases I–III trials of a new test agent, the principle of good clinical practice for regulatory approval is to dictate the placebo concurrent control as the fundamental referenced control for unbiased evaluation of effectiveness and safety unless unequivocal evidence proves that it is unnecessary.

3.4 STATISTICAL CONSIDERATIONS

At the planning stage some statistical considerations regarding the manner in which the data will be tabulated and analyzed at the end of the study should be carefully considered. These considerations include the primary and secondary response variables, the criteria for efficacy and safety assessment, sample size estimation, possible interim analysis and data monitoring, and statistical and clinical inference. We will now describe these considerations.

Efficacy and Safety Assessment

For a clinical trial, it is recognized that it is impossible to address all questions with one trial. Therefore it is important to identify the primary and secondary response variables that will be used to address the scientific and/or medical questions of interest. The response variables (or clinical endpoints) are usually chosen at the outset, since they are needed to fulfill the study objectives. Once the response variables are chosen, the possible outcomes of treatment are defined, and those showing efficacy and safety are clearly indicated. In practice, it is suggested that the selected clinical endpoints be validated (reliable and reproducible), widely available, understandable, and accepted. For example, in an antibiotic trial the outcome might be defined as cure, cure with relapse, or treatment failure, and the response variables may be pyrexia, dysuria, and frequency of urination. The criteria for the evaluation of a cure could be that all signs or symptoms of urinary tract infection are resolved during the study period. For another example, in an antihypertensive trial the outcome of treatment might be defined as normalization, partial response, or failure, and the response variable would be change in blood pressure. The criteria for normalization and partial response could be that diastolic pressure is less that 90 mm Hg and that diastolic blood pressure is reduced by more than 10% from baseline, respectively.

For efficacy assessment, once the primary efficacy variable is identified, the criteria for the evaluability of the patients should be precisely defined. For example, we may conduct an analysis based on all patients with any effectiveness observation or with a certain minimum number of observations. In some cases clinical scientists may be interested in analyzing patients who complete the trial (or completer analysis) or all patients with an observation during a particular time window. To provide a fair assessment of efficacy, sometimes it may be of interest to analyze only patients with a specified degree of compliance, such as patients who took 80% to 120% of the doses during the course of the trial. It should be noted that the evaluability criteria should be clearly defined in the study protocol. As indicated in the FDA guidelines, although a reduced subset of the patients is usually preferred for data analysis, it is recommended that an additional intent-to-treat analysis using all randomized patients be performed.

For safety evaluation the FDA requires that all patients entered into treatment who received at least one dose of the treatment must be included in the safety analysis. Safety evaluation is usually performed based on clinical and laboratory tests. To provide an effective evaluation, it is suggested that the following should be provided:

1. Parameters to be measured.
2. Timing and frequency.
3. Normal values for laboratory parameters.
4. Definition of test abnormalities.

The primary safety variable is the incidence of adverse event, which is defined as any illness, sign, or symptom that has appeared or worsened during the course of the clinical study regardless of causal relationship to the medicine under study. The FDA suggests that basic display of adverse event rates be used to compare rates in treatment and control groups. In addition, if the study size permits, the more common adverse events that seem to be drug related should be examined for their relationship to dosage and to mg/kg dose, to dose regimen, to duration of treatment, to total dose, to demographic characteristics, or to other baseline features if data are available. However, the FDA also points out that it is not intended that every adverse be subjected to rigorous statistical evaluation.

Sample Size Estimation

For assessment of the effectiveness and safety of a study drug, a typical approach is first to show that the study drug is statistically significant from a placebo control. If there is a statistically significant difference, we then demonstrate that the trial has a high probability of correctly detecting a clinically meaningful difference. The probability of correctly detecting a clinically meaningful difference is known as the (statistical) power of the trial. In clinical trials, for a given significance level, we can increase the statistical power by increasing the sample size. In practice, a pre-study power analysis for sample size estimation is usually performed to ensure that the intended trials have a desired power (e.g., 80%) for addressing the scientific/medical questions of interest. In clinical trials, we can classify sample size estimation as either sample size determination or sample size justification. The purpose of a sample size determination is to find an appropriate sample size based on the information (the desired power, variability and clinically meaningful differences, etc.) provided by clinical scientists. If the sample size has been chosen based on medical/marketing considerations, then it is necessary to provide a sample size justification for the

STATISTICAL CONSIDERATIONS

chosen sample size such as "What difference can be detected with the desired power for the chosen sample size?" It should be noted that a larger sample size will allow us to detect a smaller difference if the difference indeed exists.

For sample size determination, Table 3.4.1 provides some examples of required sample sizes for achieving desired power to detect some clinically meaningful differences under the assumption of various standard deviations. The estimated sample sizes were obtained based on a two-sided test for two independent samples with the 5% level of significance. For example, a total of 32 patients is needed to have a 80% power for detection of one standard deviation difference. Additional 12 patients are required to increase the power from 80% to 90%. As can be seen from Table 3.4.1, the sample size increases as the standard deviation increases. In addition a larger sample size is required to detect a smaller difference. On the other hand, Table 3.4.2 gives statistical justifications for differences that can be detected for some chosen sample sizes. For example, the selected sample size of 100 will have a 80% power for detection of an approximately half standard deviation difference. Note that the differences that can be detected based on the selected sample size may not be of clinically meaningful difference.

It should be noted that the sample size determination/justification should be carried out based on appropriate statistics under the selected design. Different study designs and testing hypotheses may result in different sample size requirements for achieving a desired power. Therefore it is recommended that the following be considered when performing a pre-study power analysis for sample size estimation:

1. What design is to be used?
2. What hypotheses are to be tested?
3. What statistic is to be performed?

Table 3.4.1 Sample Size Determination

Power	Standard Deviation (%)	Sample Size[a] Clinical 10%	Difference 15%
80%	10	32	14
	20	126	56
	30	284	126
90%	10	44	20
	20	170	76
	30	380	170

[a]Sample sizes were obtained based on a two-sided test for two independent samples at the $\alpha = 5\%$ level of significance.

Table 3.4.2 Sample Size Justification

		Detected Difference[a]	
	Standard	Power	
Sample Size	Deviation (%)	80%	90%
100	10	5.6	6.5
	20	11.2	13.0
	30	16.8	19.5
200	10	4.0	4.6
	20	8.0	9.2
	30	12.0	13.8

[a]The numbers were obtained based on a two-sided test for two independent samples at the $\alpha = 5\%$ level of significance.

If the selected design is a parallel design and/or a two-sided test is used, it may require more patients to reach the desired power. Under a selected design, sample size requirements are different for testing point hypotheses and interval hypotheses. As discussed in the previous chapter, an interval hypotheses is intended for establishment of clinical equivalence. The FDA indicates that for a positive control study intended to show that a new therapy is at least as effective as the standard therapy, the sample size determination should specify a clinically meaningful difference indicating that the new therapy is clinically equivalent if the difference is smaller than such a difference. The power to detect a treatment difference should be given.

Interim Analysis and Data Monitoring

Interim analysis and data monitoring is a process of examining and/or analyzing data accumulating in a clinical trial, either formally or informally, during the conduct of the clinical trial. The nature and intent of data monitoring and interim analysis in the pharmaceutical industry are often misunderstood. As indicated by the Biostatistics and Medical Ad Hoc Committee (BMAHC) on Interim Analysis of the Pharmaceutical Manufacturing Association (PMA), the following three issues should be addressed when planning an interim analysis (PMA, 1989):

- Protection of the overall type I error rate in formal confirmatory clinical trials designed to establish efficacy.
- Safeguarding of the blinding of a study.
- Use of interim analyses for administrative or planning purposes to generate hypotheses for future studies or to assess safety.

For the protection of the overall type I error, there are many methods available in the literature. For example, see Pocock (1977), O'Brien and Fleming

(1979), Peto et al. (1976), Slud and Wei (1982), Lan and DeMets (1983), and Lan and Wittes (1988). The protection of the overall type I error can usually be achieved through a carefully planned study protocol. The safeguarding of the blind is a critical issue that has a great impact on the credibility of the study. Therefore it is suggested that the sponsors to develop formal procedures to ensure that the dissemination of the results of interim analyses is controlled in such a way as to minimize the potential bias. The third issue considers an interim analysis as a study management tool for addressing some important questions during the conduct of the study. Such analyses, which are known as administrative look, are usually performed on an unblinded basis and without adjustment of p-values.

Interim analysis and/or data monitoring provides an administrative tool for terminating a trial during which is observed either a superior efficacy or an excessive safety risk in the treatments presented to patients. Currently all clinical trials sponsored by NIH are required to perform interim analyses. If no interim analyses is intended, the reasons why an interim analysis is not necessary for the study are to be clearly stated in the study protocol. Since the late 1980s pharmaceutical companies have begun to recognize the need of interim analyses and consequently have started to perform interim analyses and administrative data monitoring for their sponsored trials in an expeditious manner. Since blinding plays an important role in protecting the integrity of clinical trials, the BMAHC was formed to examine the impact of interim analysis on blindness (PMA, 1989, 1993). In their position paper (PMA, 1993), the BMAHC emphasized that blinding (masking) is an important issue, since the interim analysis requires that the study be unblinded. Knowledge of early trends or lack thereof can bias the remainder of the study and result in changes in the patient recruitment. Knowledge of treatment assigned to individual patients can also introduce bias and serious dropouts on the validity of results. The committee suggested that an SOP be developed to describe

- Who will have access to the randomization code?
- How the blinding will be broken?
- Who will have access to the interim results?
- Whether ongoing patients will be included in the analysis?

Williams et al. (1993) point out that major pharmaceutical companies such as Merck have developed their own internal SOPs for triple-blind policy for all phase III and IV studies and most phase II trials. Moreover they state that interim evaluation should be performed by a party that is not involved in the actual conduct of the study. The results of the interim analysis should not be provided in any form to those individuals involved in the conduct of the trial in order to avoid any temptation to alter the study design and to introduce any potential bias. They suggest that the following procedures be imposed to ensure the blindness when performing an interim analysis.

- Merge of randomization codes with patient identification numbers for the use of interim analysis must be performed by a low-level statistician who is not directly involved in the study.
- Identity of treatments received by individuals must not be known.
- Only the minimum information required to meet the objectives of the interim analysis can be presented and only to the few individuals who are responsible for decision making on the drug's development.
- Detailed procedures for the implementation of interim analysis such as unblindedness, decision making, and the frequency of interim analyses must be fully documented and available for external review.

Since interim analyses require not only that the randomization codes be unblinded during the clinical trials but also that the results be disseminated to either the external or internal data-monitoring board, blindness is in fact compromised to some extent and a bias will always be introduced. The FDA has expressed some concerns regarding the issue of only an internal data-monitoring board. O'Neill (1993) indicates that: FDA is primarily concerned that a study can be biased by monitoring practices and procedures, and since the monitoring group has a vested interest in the product being evaluated, the study might be compromised to the extent that it will not support the scientific regulatory standards for drug approval.

Therefore, in addition to the procedures recommended by Merck, we also suggest the following:

- The external data-monitoring board should include clinical and statistical experts from academic institutions in the therapeutic areas under investigation.
- The interim analysis should be performed by the statistical members of the data-monitoring board, using the database merged by the low-level statistician, who is not involved with the trial, from the randomization codes and patient identification numbers with pre-specified efficacy and safety endpoints for the interim analysis in the protocol.
- The external data-monitoring board should have the authority to make decisions regarding when, how, what, and to whom (including those in the top management who make the decision on the drug's development) the results of the interim analysis should be available.

Note that the FDA guideline declares that all interim analyses, formal or informal, by any study participant, sponsor staff member, or data-monitoring group must be described in full even if the treatment groups are not identified. The need for statistical adjustment because of such analyses should be addressed. More details regarding interim analysis will be given later in this book.

Statistical and Clinical Inference

Statistical and clinical inferences are usually drawn based on clinical data collected from controlled randomized trials. Statistical and clinical inferences are derived from statistical tests under the assumption that the selected sample (i.e., a group of patients) is a random sample from the targeted patient population. A random sample is referred to a representative sample. However, in most clinical trials patients are not selected from intended population in a random fashion. In practice, for a clinical trial we usually select study sites (or centers) first. Patients are then recruited at each selected study site to form a sample for the intended clinical trial. Then no formal sampling theory can be applied to derive a valid statistical inference regarding the intended patient population. Consequently the clinician cannot draw a valid statistical inference to clinical practice.

It should be noted that statistical inference is only a part of induction process for the conclusions obtained from clinical trials, and it should not preclude the possibility of a meaningful clinical inference. If the inclusion and exclusion criteria are precisely stated in the study protocol before the trial is conducted, then the demographic characteristics at baseline of the patients can be used to describe the patient population from which the sample of the patients in the trial is drawn. As a result the population model described earlier can be invoked to provide a basis for a clinical inference about the patient population. This concept of clinical inference as a form of induction from sample to population is based on external validity. For example, suppose that the inclusion criteria of a single-center clinical trial allow only enrolling patients within a very narrow age range. Also suppose that there is another study with the same sample size that is a multicenter trial for investigation of the same drug. However, the second study has a much wider age range. The second study is more appropriate to make clinical inference externally simply because it is a multicenter trial with a wider age range than the first. Figure 3.4.1 provides a diagram of statistical and clinical inference with respect to the relationship among randomization for selection and assignment of patients and internal and external validity.

Rubins (1994) indicated that any single trial is unlikely to have a major impact on a physician's medical practice. Clinical inference is rarely based on the results of a single clinical trial. In order to investigate the effectiveness and safety of a therapeutic agent or a class of therapeutic agents, a series of clinical trials with the same design and concurrent control is usually conducted over patients with similar but different characteristics. Note that these clinical trials may be conducted by different investigators at different study centers in different countries. Petitti (1994) points out that the technique of the meta-analysis can provide the most conclusive evidence of clinical inference of the therapeutic agents for medical practice because of its external validity. For example, Yusuf et al. (1985) conducted a meta-analysis based on 64 randomized beta-blocker trials to conclude, once for all, the benefit of long-term use of the beta-blocker after myocardial infarction in reduction of mortality. Yusuf et al. (1985)

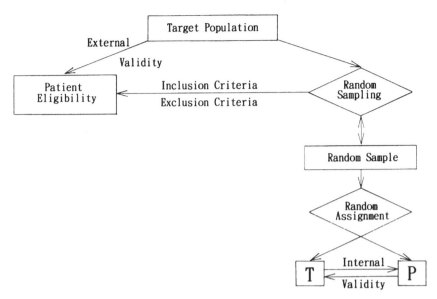

Figure 3.4.1 Diagram of statistical and clinical inference.

report that there exists no important difference in benefit among different beta-blockers. On the other hand, in the CAST trial, the results indicated that patients who received class IC antiarrhythmic agents (e.g., flecainide or encainide) after myocardial infarction had three times as high as mortality rate than those patients who received the placebo. Since the CAST trial is a single but relatively large trial, the finding was confirmed through a meta-analysis combining 10 clinical trials performed by Hines et al. (1989) in order to prove the harmful effect of the use of class IC arrhythmic agents after myocardial infarction.

In most clinical trial women with childbearing potential are excluded from the studies. This exclusion limits the generalization of the findings to the larger population. In 1993 the FDA revised its 1977 guidelines to include women of childbearing potential who were previously excluded in early drug studies of non-life-threatening diseases. This revised guideline, however, does not apply to bioequivalence trials which are required by the FDA for approval of generic drug products. As indicated by Henderson (1993), bioequivalence trials are mostly conducted in normal male volunteers aged between 18 and 50 years old whose weights are within 10% of their ideal body weight. As a result of this exclusion of female, patients, and elderly from the studies, bioequivalence trials have very limited external validity. However, under the fundamental bioequivalence assumption, clinical inference based on rather limited bioequivalence studies is still acceptable to the FDA. The fundamental bioequivalence assumption states that if two drug products are bioequivalent, then they will reach the same therapeutic effects or they are therapeutically equivalent. The legal basis of this fundamental bioequivalence assumption is from the *Drug Price Competition and Patent Term Restoration Act* passed by the United States

Congress in 1984. The FDA allows a generic drug to be used as a substitution of a brand-name drug if the generic drug product is shown to be bioequivalent to the brand-name drug. When a brand-name drug is off patent, it is expected there might be a number of generic copies for the brand-name drug approved by the FDA. Although each generic copy of the brand-name drug can be used as a substitute for the brand-name drug, the FDA does not indicate that these generic copies of the same brand-name drug can be used interchangeably. It is therefore important to investigate the overall bioequivalence and inconsistencies among all generic copies of the same brand-name. For this purpose Chow and Liu (1996) proposed the concept of meta-analysis for postapproval bioequivalence review.

3.5 OTHER ISSUES

In addition to the basic design considerations described above, some specific considerations in planning a design for a clinical trial are given below.

Single Site versus Multi-sites

In most clinical trials, multistudy sites (multicenter) are considered. This is often due to a limitation in the capacity of a single site. Besides, multistudy sites can increase patient enrollment and consequently shorten the duration of the study. Although a multicenter study has its advantages, it also suffers from some difficulties. For example, if the enrollment is too slow, the sponsor may wish to (1) terminate the inefficient study sites, (2) increase enrollments for the most aggressive sites, or (3) open new sites during the course of the trial. Each action may introduce potential biases to the study. In addition the sponsor may ship unused portions of the study drugs from the terminated sites to the newly opened sites. This can increase the chance of mixing up the randomization schedules and consequently decrease the reliability of the study. For a multicenter study the FDA requires that the treatment-by-center interaction be carefully investigated. No overall conclusion can be made across study sites if treatment-by-study-site interaction is present. In practice, it is desirable to have fewer study sites than the number of patients within each site. This is because the comparison is usually made between patients within sites. More centers may increase the chance of observing the treatment-by-study-site interaction. In addition different centers may have different standards for clinical evaluation and laboratory tests. It is therefore suggested that a central laboratory be employed whenever possible to provide appropriate and consistent measures.

Treatment Duration

In clinical trials it is desirable to collect clinical data, perform data analysis, and draw statistical inferences on the scientific questions that these clinical trials intend to address in a timely fashion. The duration of a clinical trial depends on the half-life of the study drug, the disease status, and the intended indication. For example, a 10- to 14-day treatment is usually required for antibiotic agents in order to reach the optimal therapeutic effect. For chronic diseases a longer treatment is necessary to observe significant improvements. If the intended indication is for bone loss, then the duration can be even longer. A typical clinical trial usually consists of three phases: a placebo run-in phase, an active treatment period, and a follow-up or maintenance phase. A placebo run-in phase is usually considered in order to remove effects of previous therapy or to stabilize patients' conditions prior to randomization. An active treatment period is to determine whether the study drug achieves the desired efficacy, and a follow-up or maintenance phase is to monitor the safety of the study drug.

Treatment duration is an important consideration in designing clinical trials. It has an impact on the evaluation of the study drug. An inadequate duration of treatment may not provide an unbiased and valid assessment of the true response rate of the study drug. It may also result in a high dropout rate due to the lack of efficacy. For example, the response rates at different time points may be different for antibiotic agents. It is therefore suggested that an adequate duration of treatment be considered in clinical trials to provide an unbiased and valid assessment of the study drug.

Patient Compliance

It has long been realized that patients may not follow instructions for the use of medication. As indicated by Cramer et al. (1989), overdosing, underdosing, and erratic dosing commonly occur in all patient populations regardless of the severity of illness. Patient compliance has an impact on the evaluability of patients for clinical evaluation of efficacy and safety of the study medicine. Stewart and Cluff (1972) indicate that the extent of patient default is between 20% and 80%. A typical approach to monitoring patient compliance is pill counting. Cramer et al. (1989) suggest using a medication event monitor system (MEMS) to study patients' pill-taking habits. They found that even for a once daily dose schedule, there is only 87% compliance. For the four times daily regimen, the compliance rate drops sharply to 39%. They estimated that the overall compliance rate is around 76%. In practice, to ensure that patients are eligible for clinical evaluation, a set of criteria is usually imposed on patient compliance. For example, patients are considered evaluable if they are within 80% and 120% of compliance.

Wang, Hsuan, and Chow (1996) classify the concept of compliance into to compliance and adherence. A patient is said to have a poor compliance if he or she fails to take the drug at the prescribed dose. A poor adherence is referred to as failure to take the drug at the scheduled times. Poor compliance may result in treatment failure and possible adverse reactions. It has been observed that 5% of patients have a drug-induced disease on admission to the hospital (e.g., see Seidl et al., 1966; Hurwitz, 1969). Seidl et al. (1966) also indicate that adverse drug reactions are the seventh most common cause for admission to hospital. Elaboration of other medical consequences of noncompliance can be found in Glanz et al. (1984) and Cramer et al. (1989).

Missing Value and Dropout

Another issue that is worthy of attention is possible dropouts or missing values. Dropouts can be related to the duration, the nature of the disease, and the effectiveness and toxicity of the study drug. It may be misleading to ignore the patients who dropped out prior to the maturity of the study. For example, if more dropouts occur in one treatment group, a bias may have been introduced to the trial. In practice, more patients are usually enrolled to account for possible dropouts so that the study will have sufficient evaluable patients to achieve the desired power. When there are a large number of dropouts, it is suggested that the causes of the dropouts be carefully evaluated. It can be a great concern if dropouts are related to the ineffectiveness and side effects of the study drug.

As was mentioned earlier, a clinical trial is more likely to be a multicenter trial. Matts and Lachin (1988) indicate that an adequate statistical analysis is a stratified analysis with study center as a stratum. If study center is not considered a stratum and omitted in the analysis, then such an unstratified analysis will likely produce a conservative test. In many clinical trials unplanned post hoc subgroup analyses based on some covariates (patient characteristics) are required to answer some important clinical questions, even though a stratified randomization with respect to the covariates is not performed. If the covariates used for the classification of subgroups are statistically independent of the random assignment of patients to the treatments, then the stratified analysis based on the covariates will be a valid statistical test. A typical approach is to perform an analysis of covariance (ANCOVA), which can include the following as covariates: (1) demographic factors such as age, gender, and race, (2) geographical region such as study center, and (3) some baseline characteristics such as disease severity at entry (Snedecor and Cochran, 1980). In clinical trials, it is almost impossible to collect data from all patients in order to cover all of the information regarding patient characteristics of interest. It is therefore not uncommon to have missing data in some combinations of covariates in clinical trials. One way to handle this problem of missing data is to perform the analyses only on the set of patients with the complete data. This approach is in fact a post hoc stratified analysis based on a covariate that indicates whether a patient has complete data or not. If the missing mechanism is independent of the random assignment of patients to the treatment, then the resulting analyses based only on the subset of the patients with complete data will be statistically valid. Rubin (1976) and Little and Rubin (1987) refer to the assumption of independence between the missing mechanism and treatment assignment as missing at random. However, this assumption cannot be verified or tested. Although the analyses might be valid, they are inefficient because they are based on the subset of the patients with complete data.

3.6 DISCUSSION

In practice, it should be realized that clinical trials are conducted *in* humans and are done *by* humans. Therefore some issues need to be seriously considered when planning a clinical trial. These issues include safety, compliance, and human error/bias. For example, is it safe or ethical to conduct placebo-controlled antibiotic studies? Patient compliance depends on the corporation of the patient. It depends on the duration of the study, the duration of visits, the frequency of visits, and perhaps the timing of visits. Human error/bias can be controlled through placebo control, blinding, lead-in, and education. Note that controlled studies are usually referred to as those in which the test treatment is compared with a control. For demonstration of the effectiveness and safety of a test treatment, uncontrolled studies are rarely of value in clinical research, since definitive efficacy data are unobtainable and data on adverse events may be difficult to interpret. As a result, to answer the question, in addition to the definition of patient population and the selection of clinical endpoints, controlled studies are the best way to factor out human error/bias.

Besides the basic design issues described above, clinical data quality assurance is also an important consideration when planning a clinical trial. The success of a clinical trial depends on the quality of the collected clinical data. The quality of clinical data depends on the case report forms used to capture the information. Inefficient case report forms can be disastrous to a study. Therefore effectively designed case report forms are necessary to ensure the quality of clinical data and consequently to ensure the success of the study. It is recommended that a biostatistician be involved in the design and review of the case report forms to ensure that these forms capture all of the relevant information for data analysis.

CHAPTER 4

Randomization and Blinding

4.1 INTRODUCTION

In Chapter 2 we introduced some sources of bias and variation that can occur during the conduct of clinical trials. The control of bias and variability is extremely important to ensure the integrity of clinical trials. In comparative clinical trials, randomization is usually used to control conscious or unconscious bias in the allocation of patients to treatment groups. The purpose of randomization is not only to generate comparable groups of patients who have similar characteristics but also to enable valid statistical tests for clinical evaluation of the study medicine.

The concept of randomization was first introduced in clinical research in the early 1930s for a study of sanocrysin in the treatment of patients with pulmonary tuberculosis (Amberson, 1931). However, the principle of randomization was not implemented in clinical trials until mid-1940s by the British Medical Research Council under the influence of Sir Austin Bradford Hill (1948). Since then there have been tremendous debates over the use of randomization in clinical research (e.g., see Feinstein, 1977, 1989). The primary concern is that it is not ethical for the patient not to know which treatment he or she receives, especially when one of the treatments is a placebo. However, it was not realized that before a clinical trial is conducted, no one can be 100% sure that the active treatment is indeed effective and safe for the indicated disease compared to the placebo. For many drug products it is not uncommon that the active treatment has inferior efficacy and safety than the placebo. One typical example would be the CAST study discussed in previous chapters. For another example, if randomization is not employed for comparing a surgical procedure with chemotherapy in treatment of patients with a certain cancer, then the so-called operable patients with good prognoses will more likely be assigned to surgery, while the chemotherapy will be given, as is usual, to the inoperable patients with poor prognoses. The surgical treatment would have yielded the positive results even though the surgery was not performed at all.

The use of randomization can avoid subjective assignment of treatments to patients who participate in clinical trials. Its advantage can be best illustrated by clinical studies concerning the treatment of gastric freezing for patients with peptic ulcer conducted in the 1960s (Miao, 1977; Sackett, 1989). In these studies the treatment of gastric freezing was applied to tens of thousand of patients with peptic ulcer in a nonrandom fashion. These studies showed that the gastric freezing might be a promising therapy for the disease. However, it only took one randomized trial with 160 patients, half to the real or sham freezing, to conclusively demonstrate that the treatment of gastric freezing is in fact ineffective for the treatment of peptic ulcer. Therefore Section 314.166 of the CFR requires that the method for patient treatment assignment should be described in some detail in the study protocol and report. It is recommended that for a concurrent controlled study, treatment assignment of patients be done by randomization. It should be noted that randomization in clinical trials consists of (1) random selection of a representative sample from a targeted patient population and (2) random assignment of patients in order to study the medicines.

To remove the potential bias that might occur when there are inequalities between treatment groups (e.g., demographic details or prognostic variables) allocated to different treatment groups, the use of randomization with blocking and/or stratification, if necessary, is helpful. Lachin (1988a, 1988b) provides a comprehensive summary of the various randomization models. The concept behind these randomization models allows useful randomization methods to be employed such as the complete randomization, the permuted-block randomization, and the adaptive randomization. Randomization plays an important role for the generalization of the observed clinical trials. Therefore it is recommended that a set of standard operating procedures (SOP) for the implementation of randomization be developed when conducting clinical trials. In many clinical trials bias often occurs due to preconceived ideas or perceptions acquired during the study by (1) the investigator and supporting staff who might influence reporting response to therapy or adverse events and (2) the patient who might influence compliance, cooperation, or provision of information.

In clinical trials, in addition to randomization, the technique of blinding is usually employed to avoid the risk of personal bias in comparing treatments. Basically there are several different types of blinding commonly used in clinical trials. These blindings include open label (or unblinding), single blinding, double blinding, and triple blinding. An open label study indicates that both the patient and the investigator know to which treatment group the patient is assigned, while a single blinding is referred to as that when the investigator knows but the patient does not. For a double blinding, neither the investigator nor the patient knows to which treatment group the patient is assigned. A triple blinding is an extension of the double blinding in which those monitoring outcome are unaware of treatment assignment. In practice, randomization and blinding are important to the success of clinical trials. Randomization and blinding can not only help to avoid bias but also to control variability, and consequently to achieve the desired accuracy and reliability of clinical trials.

The remainder of this chapter is organized as follows. In the next section, we introduce various randomization models. Section 4.3 covers the different randomization models. In Section 4.4 we provide a commonly employed approach for the implementation of randomization in the pharmaceutical industry. The issue regarding the generalization of controlled randomized trials is discussed in Section 4.5. The concept for the use of blinding in clinical trials is addressed in Section 4.6. A brief discussion is given in Section 4.7.

4.2 RANDOMIZATION MODELS

As was indicated in the preceding chapter, randomization ensures that patients selected from the intended patient population constitute a representative sample of the intended patient population. Therefore statistical inference can be drawn based on some probability distribution assumption of the intended patient population. The probability distribution assumption depends on the method of randomization under a randomization (population) model. As a result a study without randomization will result in the violation of the probability distribution assumption, and consequently no accurate and reliable statistical inference on the study medicine can be drawn.

Lachin (1988a) provides a comprehensive summary of the randomization basis for statistical tests under various models. His observations are discussed below.

Population Model

Cochran (1977) points out that the validity of statistical inference by which clinicians can draw conclusions for the patient population is based on the selection of a representative sample drawn from the patient population by some random procedure. This concept is called the *population model* (Lehmann, 1975; Lachin, 1988a). Suppose that for a certain disease a clinical trial is planned to investigate the efficacy and safety of a newly developed therapeutic agent compared to an inert placebo. Under the population model we can draw two samples independently with equal chance at random from the (infinitely large) patient population. One sample consists of n_T patients, and the other sample consists of n_P patients. We denote these two samples by sample T and sample P, respectively. The n_T patients in sample T will receive the newly developed agent, while the inert placebo is given to the n_P patients in sample P. If the patient population is homogeneous with respect to the inclusion and exclusion criteria specified in the protocol, we do not expect that the responses of clinical end points for a particular patient will have anything to do with those of other patients. In other words, they are statistically independent of one another. For a homogeneous population, a common (population) distribution can be used to describe the characteristics of the clinical responses. That is, they are assumed to have identical distribution. Hence the clinical responses of the n_I patients ($I = T, P$) are said to have an independent and identical distribution (i.i.d.).

Therefore optimal statistical inference can be precisely obtained. For example, with respect to hypotheses (2.6.1) regarding the detection of the difference between the new agent and placebo, the common two-sample t-test is the optimal testing procedure (Armitage and Berry, 1987).

As mentioned above, randomization in clinical trials involves random selection of the patients from the population and random assignment of patients to the treatments. Under the assumption of a homogeneous population, the clinical responses of all patients in the trial, regardless of sample T or sample P, are independent and have the same distribution. Lachin (1988a) points out that the significance level (i.e., the probability of type I error) and the power (i.e., the probability of correctly rejecting a false null hypothesis) will not be affected by random assignment of patients to the treatments as long as the patients in the trial represent a random sample from the homogeneous population. Furthermore suppose that we split a random sample into two subsamples; the statistical inferential procedures are still valid even if the one-half of the patients are assigned to the test drug and the other half to the placebo.

Invoked Population Model

In clinical trials we usually select investigators first and then select patients at each selected investigator's site. At each selected study site the investigator will usually enroll qualified patients sequentially. A qualified patient is referred to as a patient who meets the inclusion and exclusion criteria and has signed the informed consent form. As a result, neither the selection of investigators (or study centers) nor the recruitment of patient is random. However, patients who enter a trial are assigned to treatment groups at random. In practice, the collected clinical data are usually analyzed as if they were obtained under the assumption that the sample is randomly selected from a homogeneous patient population. Lachin (1988a) refers to this process as the *invoked population model* because the population model is invoked as the basis for statistical analysis as if a formal sampling procedure were actually performed. In current practice the invoked population model is commonly employed for data analysis for most clinical trials. It, however, should be noted that the invoked population model is based on the assumption that it is inherently intestable.

Note that one of the underlying assumptions for both the population model and the invoked population model is that the patient population is homogeneous. This assumption, however, is not valid in most clinical trials. In practice, we can employ the technique of stratified sampling to select samples according to some prespecified covariates to describe the differences in patient characteristics. The idea of stratification is to have homogeneous subpopulations with respect to the prespecified covariates (or patient characteristics). In many clinical trials it is almost impossible to use a few covariates to describe the differences among heterogeneous subpopulations due to the complexity of patient characteristics and disease conditions. In addition patients who are enrolled at different times may not have similar relevant demographic and baseline characteristics. In other words, the patient population is time-heterogeneous population in which the patient's characteristics are a function of the time when they enter the trial. The impact of the heterogeneity due to the recruitment time on the results of a clinical trial is well documented in the literature. For example, Byar et al. (1976) indicate that in a study conducted by the Veterans Administration Cooperative Urological Research Group in 1967, the survival rate of the patients who entered earlier in the study was worse than of those who enrolled later in the study. Therefore it is not uncommon for patient characteristics to change over time even if the population is homogeneous at one time point. The above discussion indicates that the assumption of the population model or the invoked population model may not be valid.

Randomization Model

As discussed above, for current practice, although the study site selection and patient selection are not random, the assignment of treatments to patients is usually performed based on some random mechanism. Thus treatment comparisons can be made based on the so-called randomization or permutation tests introduced in the mid-1930s (Fisher, 1935). To illustrate the concept of permutation tests, we consider the following hypothetical data set concerning endpoint changes from baselines in peak urinary flow rate (mL/s) after three months of treatment for patients with benign prostate hyperplasia who are in the test drug and placebo groups, respectively:

$$\text{Test drug}: 2.6, 0.97, 1.68;$$
$$\text{Placebo}: 1.2, -0.43.$$

An interesting question is how to determine whether there is a significant difference in endpoint change from baseline in peak urinary flow rate between the test drug and the placebo based on the above hypothetical data. Under the null hypothesis of no difference described in (2.6.1), all possible permutations according to the endpoint changes from baselines in peak urinary flow rate are equally likely (from the smallest to the largest based in the ranking). If all possible pairs of ranks for the two patients receiving placebo are all equally likely, then the sum of the ranks for the two patients in the placebo group are also equiprobable. The possible ranks and sum of the ranks for the two patients receiving placebo are given in Table 4.2.1. Since the chance is equal for all possible permutations of the ranks for the two patients in the placebo group, the probability distribution for the sum of the ranks can be obtained as given in Table 4.2.2. Since the ranks of the observed endpoint change from baseline in peak urinary flow rate for the two patients in the placebo group are 3 (for 1.2) and 1 (for -0.73), respectively, the rank sum for the placebo group is 4. As can be seen from Table 4.2.2, the p-value (i.e., the probability that the observed rank sum is due to chance or the sum of the ranks from placebo group can be

Table 4.2.1 All Possible Ranks for the Two Patients in the Placebo Group Based on Conditional Permutation

Possible Ranks	Sum of Ranks
1, 2	3
1, 3	4
1, 4	5
1, 5	6
2, 3	5
2, 4	6
2, 5	7
3, 4	7
3, 5	8
4, 5	9

at least as extreme as the observed 4) is 0.2 for a one-sided test and 0.4 for a two-sided test. Note that the possible values and the distribution of the rank sums will be the same no matter what actual observed endpoint change from baseline in peak urinary flow rate for the two patients in the placebo group are as long as the two patients are assigned to the placebo group at random. The above test is known as the Wilcoxon rank sum test (Wilcoxon, 1945). The Wilcoxon rank sum test is one of the conditional permutation tests in which permutation is performed to the confinement of random assignment of two out of five patients with prostate hyperplasia to the placebo group. As a result there are a total of 10 possible permutations of ranks for the placebo group. If we can randomly assign any five enrolled patients from 0 to 5 to receive placebo treatment, then there will be a total of 32 subsets as given in Table 4.2.3. The permutation tests over all possible subsets are called the unconditional permutation tests. The p-values for the unconditional permutation tests can be similarly computed.

The above discussion indicates that the calculation of the p-value for the Wilcoxon rank sum test does not assume any probability distribution for the

Table 4.2.2 Probability Distribution of the Sum of Ranks Based on Conditional Permutation

Sum of Ranks	Probability
3	0.1
4	0.1
5	0.2
6	0.2
7	0.2
8	0.1
9	0.1

Table 4.2.3 All Possible Unconditional Permutation for Five Subjects

Number of Subjects		Possible Permutation
Placebo	Test	
0	5	1
1	4	5
2	3	10
3	2	10
4	1	5
5	0	1
Total		$32 = 2^5$

endpoint change from baseline in peak urinary flow rate from baseline. As a matter of fact, any statistical test based on the permutation principle is assumption free. In addition, as indicated by Lachin (1988a), the family of the linear rank tests is the most general family of permutation tests (Lehmann, 1975; Randles and Wolfe, 1979). For example, the well-known Pearson chi-square statistic for comparison of two proportions is equal to $N/(N-1)$ times the chi-square statistic derived by permutation, where N is the total number of patients (Koch and Edwards, 1988). Other tests for continuous or quantitative clinical endpoints include the Wilcoxon rank sum test for two independent samples and Kruskal-Wallis test for multiple independent samples. For censored data the logrank test (Miller, 1981) and Peto-Peto-Prentice-Wilcoxon test (Kalbfleisch and Prentice, 1980) are useful. These tests are widely applied statistical procedures in clinical trials. Note that since the statistical procedures based on the concept of permutation require the enumeration of all possible permutations, it is feasible only for small samples. As the sample size increases, however, the sampling distribution of test statistics derived under permutation will approach to some known continuous distribution such as a normal distribution. In addition, as shown by Lachin (1988b), the probability distributions for the family of linear rank statistics for large samples are equivalent to those of the tests obtained under the assumption of the population model. As a result, if patients are randomly assigned to the treatments, statistical tests for evaluation of treatments should be based on permutation tests because the exact p-value can be easily calculated for small samples. For large samples the data can be analyzed by the permutation methods derived under the population model as if the patients were randomly selected from a homogeneous population.

Stratification

In most clinical trials the ultimate goal is not only to provide statistical inference on the effectiveness and safety of a test drug, compared to a control or placebo based on clinical data collected from the trials, but also to apply the results to the targeted patient population. In practice, there are many covariates such as age, gender, race, geographical locations, underlying disease severity, and others that may have an impact on the statistical inference drawn. The accuracy and reliability of the estimation of primary clinical endpoints for evaluation of the treatment effect can be affected by the heterogeneity caused by these covariates. To overcome and control such heterogeneity, a stratified randomization is found helpful. The use of stratification in clinical trials is motivated originally by the concept of blocking in agricultural experiments in the mid-1930s (Fisher, 1935). The idea is quite simple and straightforward. If a covariate is known to be the cause of heterogeneity, then the patients are stratified or blocked into several homogeneous groups (or strata) with respect to the covariate. Randomization of patients to the treatment is then performed independently within the strata. This type of randomization with strata is called *stratified randomization*. For example, the National Institute of Neurological Disorders and Stroke rt-PA stroke study group suspected that the time from the onset of stroke to the beginning of treatment of rt-PA may have a significant impact on neurologic improvement as assessed by the National Institute of Health Stroke Scale (NIHSS). As a result the study considered two strata of patients based on the time (in minutes) from the onset to the start of the treatment, namely 0 to 90 minutes and 91 to 180 minutes. For multicenter trials, stratified randomization with respect to geographical location is necessary because differences in study centers usually account for the major source of variation for many primary clinical endpoints. The idea of stratification is to keep the variability of patients within strata as small as possible and the between-strata variability as large as possible so that the inference for the treatment effect possesses the optimal precision. Another reason for the use of stratification in clinical trials is to prevent imbalance with respect to important covariates. For example, with an unstratified randomization, more males may be enrolled into the test drug group, while the placebo group may enroll more females. Hence the distribution of treatments with respect to gender is not balanced. Despite the advantages of stratified randomization, it should be noted that the stratification will eventually become more complicated and difficult to implement due to administrative complexity, increasing time and expense, and other logistic issues.

The extreme case of stratification is the technique of matching which is often employed in the case-control studies. For example, for a clinical trial comparing a test drug with a placebo, patients are to be matched in pairs with respect to some predetermined covariates such as demographic, baseline characteristics, and severity of disease. Within each pair, one patient is assigned to receive the test drug and the other patient receives the placebo. Assignment of the matched patients to treatments might not be random. Wooding (1994) points out that matching is often used as a substitution for randomization by investigators who mistrust and do not like the concept of randomization. In case-control trials, although there may be a large number of covariates, they may or may not have an impact on clinical outcomes. In practice, it is almost impossible to consider all possible covariates in a clinical trial. However, if an important covariate is missed during the process of matching, the cause of bias cannot be identified, and consequently it cannot be assessed if it truly exists. Another problem of matching in case-control trials is that the number of patients increases rapidly as the number of covariates to be considered for matching becomes large. Accordingly, the task of finding matching pairs becomes formidable (e.g., see Wooding, 1994). The primary purpose of matching is to eliminate variations of clinical endpoints caused by the differences among patients due to biological variations, as discussed in Chapter 2. As a result the method of matching attempts to consider the individual patient as a stratum and to randomize the sequence of treatment of the test drug and placebo within each stratum. Since each patient receives both treatments, the variation between patients is eliminated from the comparison between the test drug and the placebo. This type of design is called a crossover design, and it will be discussed in detail in the next chapter.

In summary, as indicated by Lachin (1988a), the chance of covariate imbalance decreases as the sample size increases. The covariate imbalance has little impact on large samples. In addition the difference in statistical power between unstratified and stratified randomization is negligible (McHugh and Matts, 1983). Furthermore a post hoc stratified analysis can always be employed to adjust bias caused by the imbalance of baseline covariates. To control covariate imbalance, Peto et al. (1976) indicate that if, during analysis, initial diagnosis (i.e., covariate) is allowed for (i.e., stratified analysis) as the different treatments are being compared, there is hardly ever need for stratification at entry in large trials. Therefore it is recommended that stratified randomization for a clinical trial be performed only with respect to those covariates that are absolutely necessary for the integrity of the study.

4.3 RANDOMIZATION METHODS

In the early 1970s, before the concept of randomization was widely accepted as an effective tool to prevent the subjective selection bias in the assignment of patients to the treatments, some systematic methods for assignment of patients to treatments under study were commonly used. These systematic methods are summarized in Table 4.3.1. As can be seen from Table 4.3.1, all of these methods are deterministic. The assignment of patients to treatments can be predicted without error. Since the investigators or patients may be aware of which treatment the patients receive, subjective bias can consciously or unconsciously occur in both the assignment of patients to treatments and the evaluation of clinical outcomes for the treatment under investigation. To prevent such bias, in this section several useful randomization methods are introduced.

Although controlled randomized trials are viewed as the state-of-the-art technology for clinical evaluation of therapeutic interventions, some investigators

Table 4.3.1 Unacceptable Methods of Assignment of Patients to Treatment

1. Assignment of patients to treatment according to the order of enrollment (every other patient is assigned to one group)
2. Assignment of patients to treatment according to patient's initial
3. Assignment of patients of treatment according to patient's birthday
4. Assignment of patients according to the dates of enrollment

still try to beat the randomization by guessing the treatments to which the patients are assigned (Karlowski et al., 1975; Brownell and Stunkard, 1982; Byington et al., 1985; Deyo et al., 1990). Hence subjective judgment for evaluation of patients' clinical outcomes always introduces potential selection bias by investigators who are aware of the treatment assignment of patients. A simple model suggested by Blackwell and Hodges (1957) and Lachin (1988a) can be adapted to assess this potential selection bias due to a wrong guess of treatment assignments by investigators. The Blackwell-Hodges diagram for selection bias is given in Table 4.3.2. This diagram is constructed under the assumption that each patient has an equal chance (50%) of being assigned to either the test drug or the placebo. Therefore, if there are a total of n patients enrolled into the study, the expected sample size for both the test drug and the placebo is equal to $n/2$. Then the total potential selection bias for evaluation of the treatment effect introduced by the investigator is represented as the (expected) difference between the observed sample means according to the treatment assignments guessed by the investigator. This expected difference can then be shown as the product of the investigator's bias in favor of the test drug times the expected bias factor which is the difference between the expected number of correct guesses and the number expected by chance. The expected bias factor is equal to one-half times the number of correct guesses minus the number of misses. Suppose that the study is double blinded and that the investigators have no other way to predict the treatment assignments but to use laboratory evaluations or some par-

Table 4.3.2 Blackwell-Hodges Diagram for Selection Bias

Investigator's Guess	Random Assignment of Equal Probability	
	Test Drug	Placebo
Test drug	a	$n/2 - b$
Placebo	$n/2 - a$	b
	$n/2$	$n/2$

Source: Blackwell and Hodges (1957).

ticular adverse events caused by the test drug. Then the probability of correctly guessing the treatment assignments is 50% for each treatment. Consequently, under this situation, the expected number of correct guesses will be the same as the number expected by chance, which is $n/2$. Hence the expected bias factor is zero. Therefore, even though the investigator might have positive bias in evaluation of patients whom he or she believe are receiving the test drug, the potential selection bias will vanish in the evaluation of the treatment effect due to the fact that the expected bias factor is zero.

Note that in addition to selection bias, an accidental bias can also occur when comparing treatments in the presence of covariate imbalances. Efron (1971) considers the effects of various randomization methods on bias for estimation of the treatment effect in a regression model assuming that important covariates are not accounted for. Gail et al. (1984) and Lachin (1988a) reported that these randomization methods will generally produce consistent estimates of the treatment effect in linear models. However, in some nonlinear models estimates of the treatment effect are biased no matter how large the sample size is (i.e., asymptotically biased) under these randomization methods. Note that for linear models, these randomization methods are equivalent in the sense that they produce estimates of treatment effect that are free of accidental bias. However, for small or finite samples, the variance of the bias varies from randomization method to randomization method. As a result the chance of accidental bias and magnitude of accidental bias vary with respect to randomization methods.

In general, randomization methods can be classified into three types according to the restriction of the randomization and the change in probability for randomization with respect to the previous treatment assignments. These types of randomization methods are the complete randomization, the permuted-block randomization, and the adaptive randomization. Randomization can be performed either by random selection or by random allocation for methods of complete and permuted-block randomization. Basically the adaptive randomization consists of treatment and covariate and response adaptive randomizations. In what follows we will describe these randomization methods and compare their relative merits and limitations whenever possible.

Complete Randomization

Simple randomization is referred to as the procedure in which no restrictions are enforced on the nature of randomization sequence except for the number of patients required for achieving the desired statistical power and the ratio of patient allocation between treatments. For a clinical trial with N patients comparing a test drug and a placebo, the method of simple randomization is called a completely binomial design (Blackwell and Hodges, 1967) or a simply complete randomization (Lachin, 1988b) if it has the following properties:

1. The chance that a patient receives either the test drug or the placebo is 50%.

2. Randomization of assignments is performed independently for each of the N patients.

The randomization codes based on the method of complete randomization can be generated either by the table of random numbers (Pocock, 1983) or by some statistical computing software such as SAS® (Statistical Analysis System, 1995). However, it should be realized that a computer cannot generate *true* random numbers but *pseudo*random numbers because only a fixed number of different long series of almost unpredictable permuted numbers are generated. Therefore Lehmann (1975) recommends that a run test be performed to verify the randomness of the generated randomization codes. In practice, however, randomization codes are preferably generated by a computer due to its speed and convenience in the maintenance of generated randomization codes. For example, the SAS® function RANBIN can be used to generate randomization codes for clinical trials with two treatment groups. For another example, suppose that a clinical trial is planned in four study centers to investigate the effectiveness and safety of a test drug as compared to an inert placebo. Ninety-six patients are intended for the study. Suppose that it is desirable to allocate patients equally in each treatment group by study center. We will consider the randomization codes given in Table 4.3.3 as generated based on complete randomization by study center. Suppose that there are three treatment groups (e.g., placebo, 100 mg, and 200 mg of the test drug), and the randomization codes can be similarly generated (see Table 4.3.4). Note that the SAS programs used for generation of the randomization codes given in Tables 4.3.3 and 4.3.4 are provided in Appendices B.1 and B.2, respectively.

In clinical trials, in the interest of balance, the assignment of an equal number of patients in treatment groups is usually considered. In practice, however, it is possible that a trial will end up with an unequal number of patients in each treatment group. Table 4.3.5 provides a distribution of the number of patients by treatment and study center for the randomization codes given in Tables 4.3.3 and 4.3.4. It can be seen that although the probability for random assignments is $1/2$ for two treatments and $1/3$ for three groups, the final sample sizes based on complete randomization are not equal for the treatment groups. In general, the treatment imbalance within each study center is more severe than that for the overall clinical trial. Lachin (1988b) provides an approximate formula for calculation of the chance of treatment imbalance for complete randomization. Based on his formula, Figure 4.3.1 plots the chance of treatment imbalance as a function of sample size for different fractions of the total sample size for the larger treatment group. It can be seen from Figure 4.3.1 that the minimum sample sizes required for a probability of less than 5% for the treatment imbalance are 386, 96, 44, and 24 when the fractions of the total sample size for the larger treatment (imbalance proportion) are 0.55, 0.60, 0.65, and 0.70, respectively. On the other hand, Figure 4.3.2 gives a graphical presentation of the fractions of the total sample size for a larger group such that would occur with the probabilities 0.005, 0.01, and 0.05 as a function of sample sizes. Figure 4.3.2 clearly shows that the fraction of the total sample size for the large group with a fixed probability of treatment imbalance is a decreasing function of sample size. Both Figures 4.3.1 and 4.3.2 demonstrate that a severe treatment imbalance based on complete randomization is unlikely when the sample size is large. For the usual statistical tests for quantitative clinical measures that can adequately be described by the normal probability model, the smallest variance for the estimate of the treatment effect can be obtained when an equal number of patients are enrolled in each treatment group. Consequently the maximum statistical power for detection of the treatment difference is achieved. In order to examine the impact on the power caused by treatment imbalance due to the complete randomization, Figure 4.3.3 provides a graph of the power for detection of a fixed treatment difference as a function of the fraction of a fixed total sample size for the larger treatment group. As can be seen from Figure 4.3.3, when the fraction of the larger group is at most 0.7, the power of the test for detection of treatment effect is hardly affected at all. In summary, although complete randomization will present a high chance of treatment imbalance, the chance of severe treatment imbalance is unlikely and moderate treatment imbalance has little impact on statistical power if the sample size of the trial exceeds 200. In practice, complete randomization is easy to implement. However, there is a high probability that it will produce unequal sample sizes among treatment groups when the total sample size is moderate (e.g., fewer than a few hundred).

Another type of simple randomization that provides equal allocation of sample size is the random allocation. Random allocation is the simplest form of restricted randomization. The method of random allocation randomly selects the $N/2$ out of a total of N patients without replacement and assigns these $N/2$ patients to receive the test drug and the other half to receive the placebo. Since there are a total of $N!/[(N/2)!]^2$ possible ways for the selection of $N/2$ patients, it is equivalent to generating a random permutation of numbers from 1 to $2N$ and assigning the first half to the test drug. Hence the SAS® procedure PLAN can be used to generate randomization codes for the method of random allocation. Table 4.3.6 provides a list of the randomization codes for a test drug and a placebo; the codes are generated based on a method of random allocation by the study center. It can be verified that for each of the four centers exactly 12 patients are assigned to each treatment group. Thus a total of 48 patients are assigned to either the test drug group or the placebo group. An SAS program for the method of random allocation for simple randomization is also provided in Appendix B.3. Although the marginal probability for assigning a patient to each of the two treatment groups is $1/2$ for the method of random allocation, the conditional probability for assignment of a patient given that the assignment of the previous patient is not equal to $1/2$ for the method of random allocation. This is because the random allocation is based on simple sampling without replacement.

Note that in a unblinded study there is no potential selection bias for complete randomization, since the expected bias factor is always zero. As indicated

Table 4.3.3 Example of Complete Randomization for Four Centers

Random codes for drug XXX, protocol XXX-014
Double-blind, randomized, placebo-control, two parallel groups

A. Hope, MD		J. Smith, MD	
Subject Number	Treatment Assignment	Subject Number	Treatment Assignment
1403001	Active drug	1401001	Placebo
1403002	Active drug	1401002	Active drug
1403003	Placebo	1401003	Active drug
1403004	Active drug	1401004	Active drug
1403005	Placebo	1401005	Placebo
1403006	Active drug	1401006	Placebo
1403007	Active drug	1401007	Active drug
1403008	Placebo	1401008	Active drug
1403009	Active drug	1401009	Placebo
1403010	Placebo	1401010	Placebo
1403011	Placebo	1401011	Placebo
1403012	Placebo	1401012	Active drug
1403013	Active drug	1401013	Placebo
1403014	Placebo	1401014	Active drug
1403015	Placebo	1401015	Placebo
1403016	Active drug	1401016	Active drug
1403017	Placebo	1401017	Active drug
1403018	Placebo	1401018	Placebo
1403019	Active drug	1401019	Active drug
1403020	Placebo	1401020	Active drug
1403021	Active drug	1401021	Placebo
1403022	Active drug	1401022	Active drug
1403023	Placebo	1401023	Active drug
1403024	Active drug	1401024	Active drug

in Lachin (1988b), the expected bias factor under the method of random allocation is an increasing function of sample size. As a result the selection bias for the method of random allocation can be very substantial as the sample size increases. Therefore it is extremely important to keep the study double blinded if the method of random allocation is employed. With respect to accidental bias caused by omitting some important covariates in estimating the treatment effect, both methods of complete randomization and random allocation are insensitive to covariate imbalance and hence are free of accidental bias when sample size is large, say over 100. For small samples the accidental bias may potentially exist for both methods. However, the accidental bias generated by the method

Table 4.3.3 Example of Complete Randomization for Four Centers (*Continued*)

Random codes for drug XXX, protocol XXX-014
Double-blind, randomized, placebo-control, two parallel groups

M. Dole, MD		C. Price, MD	
Subject Number	Treatment Assignment	Subject Number	Treatment Assignment
1402001	Placebo	1404001	Placebo
1402002	Active drug	1404002	Active drug
1402003	Placebo	1404003	Active drug
1402004	Placebo	1404004	Placebo
1402005	Active drug	1404005	Active drug
1402006	Active drug	1404006	Placebo
1402007	Active drug	1404007	Placebo
1402008	Active drug	1404008	Active drug
1402009	Active drug	1404009	Placebo
1402010	Active drug	1404010	Active drug
1402011	Active drug	1404011	Active drug
1402012	Placebo	1404012	Active drug
1402013	Placebo	1404013	Placebo
1402014	Placebo	1404014	Placebo
1402015	Active drug	1404015	Placebo
1402016	Active drug	1404016	Placebo
1402017	Placebo	1404017	Active drug
1402018	Active drug	1404018	Placebo
1402019	Active drug	1404019	Placebo
1402020	Active drug	1404020	Active drug
1402021	Placebo	1404021	Active drug
1402022	Placebo	1404022	Placebo
1402023	Active drug	1404023	Placebo
1402024	Placebo	1404024	Placebo

of random allocation is larger than that of complete randomization. As a matter of fact the accidental bias under complete randomization is the smallest among all randomization methods discussed in this section.

Permuted-Block Randomization

One of the major disadvantages of simple randomization is that treatment imbalance can occur periodically. For example, in Table 4.3.3, the randomization codes for investigator M. Dole, M.D., were generated under complete randomization with a run of seven consecutive patients (from subject 1402005 to sub-

Table 4.3.4 Example of Complete Randomization for Four Centers

Random codes for drug XXX, protocol XXX-014
Double-blind, randomized, placebo-control, three parallel groups

A. Hope, MD		J. Smith, MD	
Subject Number	Treatment Assignment	Subject Number	Treatment Assignment
1403001	Placebo	1401001	Placebo
1403002	100 mg	1401002	200 mg
1403003	100 mg	1401003	100 mg
1403004	Placebo	1401004	100 mg
1403005	200 mg	1401005	100 mg
1403006	200 mg	1401006	Placebo
1403007	100 mg	1401007	200 mg
1403008	200 mg	1401008	Placebo
1403009	100 mg	1401009	200 mg
1403010	100 mg	1401010	200 mg
1403011	100 mg	1401011	Placebo
1403012	200 mg	1401012	Placebo
1403013	Placebo	1401013	200 mg
1403014	100 mg	1401014	Placebo
1403015	Placebo	1401015	200 mg
1403016	200 mg	1401016	200 mg
1403017	Placebo	1401017	200 mg
1403018	100 mg	1401018	100 mg
1403019	200 mg	1401019	200 mg
1403020	100 mg	1401020	Placebo
1403021	Placebo	1401021	100 mg
1403022	Placebo	1401022	200 mg
1403023	Placebo	1401023	100 mg
1403024	Placebo	1401024	100 mg

ject 1402011) assigned to receive the test drug. Two observations should be noted. First, the number of patients in the treatment groups is not balanced, as was previously shown in Table 4.3.5. This may be in part explained by the fact that the randomization is performed within each center and the number of patients to be enrolled at each center is usually fewer than 50. Second, most clinical trials recruit patients sequentially. If the demographic factors or baseline characteristics change over time, then it is quite possible to have a serious covariate imbalance between treatment groups within each study center and for the entire study as well. This covariate imbalance can be potentially disastrous. For example, suppose that a clinical trial is conducted in two study centers to

Table 4.3.4 Example of Complete Randomization for Four Centers (*Continued*)

Random codes for drug XXX, protocol XXX-014
Double-blind, randomized, placebo-control, three parallel groups

M. Dole, MD		C. Price, MD	
Subject Number	Treatment Assignment	Subject Number	Treatment Assignment
1402001	Placebo	1404001	100 mg
1402002	100 mg	1404002	200 mg
1402003	100 mg	1404003	Placebo
1402004	Placebo	1404004	Placebo
1402005	100 mg	1404005	200 mg
1402006	100 mg	1404006	200 mg
1402007	200 mg	1404007	200 mg
1402008	100 mg	1404008	100 mg
1402009	100 mg	1404009	Placebo
1402010	200 mg	1404010	200 mg
1402011	Placebo	1404011	100 mg
1402012	100 mg	1404012	Placebo
1402013	200 mg	1404013	200 mg
1402014	Placebo	1404014	Placebo
1402015	200 mg	1404015	Placebo
1402016	Placebo	1404016	100 mg
1402017	Placebo	1404017	200 mg
1402018	Placebo	1404018	200 mg
1402019	100 mg	1404019	100 mg
1402020	Placebo	1404020	Placebo
1402021	100 mg	1404021	100 mg
1402022	200 mg	1404022	200 mg
1402023	200 mg	1404023	200 mg
1402024	Placebo	1404024	200 mg

evaluate the effectiveness and safety of a test drug as compared to a placebo. A complete randomization is used for the generation of randomization codes. Suppose that the randomization codes for one of the two centers contain a long run of consecutive patients who are assigned to the test drug group. Also suppose that one of the important baseline covariates is not balanced between the two treatments within the center. In this case it is extremely difficult to explain a possible difference in treatment effect between the two centers because the center effect is confounded with the effect due to this covariate. One resolution to this major disadvantage of simple randomization is periodically to enforce a balance in the number of patients assigned to each treatment. In other words,

Table 4.3.5 Sample Size by Treatment and Center for Random Codes in Tables 4.3.1 and 4.3.2.

Study Center	Active Drug	Placebo	Total	
	Table 4.3.1			
J. Smith, M.D.	14	10	24	
M. Dole, M.D.	14	10	24	
A. Hope, M.D.	12	12	24	
C. Price, M.D.	10	14	24	
Total	50	46	96	
Study Center	100 mg	200 mg	Placebo	Total
	Table 4.3.2			
J. Smith, M.D.	7	10	7	24
M. Dole, M.D.	9	6	9	24
A. Hope, M.D.	9	6	9	24
C. Price, M.D.	6	11	7	24
Total	31	33	31	96

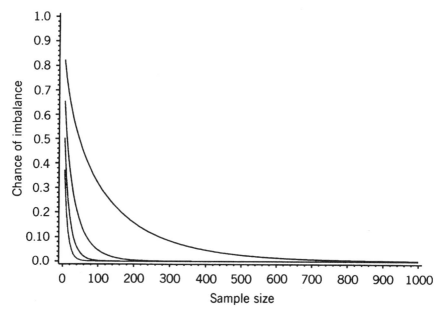

Figure 4.3.1 Chance of treatment imbalance for complete randomization as a function of sample size. Sample fraction: 0.55, 0.60, 0.65, and 0.70. (Source: Lachin, 1988b.)

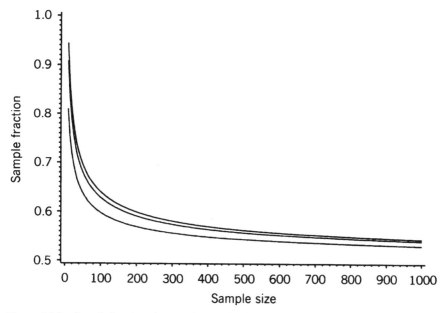

Figure 4.3.2 Sample fractions for complete randomization with chance of imbalance as a function of sample size. Chance of imbalance: 0.005, 0.01, and 0.05. (Source: Lachin, 1988b.)

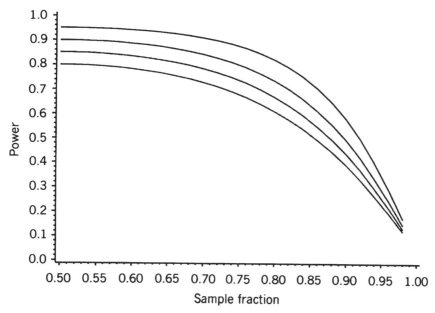

Figure 4.3.3 Power curves as a function of the sample fractions. Power is for a two-sided test at the 5% significance level. Power is 0.95, 0.90, 0.85, and 0.80 when the sample fraction is 0.5. (Source: Lachin, 1988a.)

Table 4.3.6 Example of Random Allocation for Four Centers

Random codes for drug XXX, protocol XXX-014
Double-blind, randomized, placebo-control, two parallel groups

	A. Hope, MD			J. Smith, MD	
Subject Number	Random Permutation	Treatment Assignment	Subject Number	Random Permutation	Treatment Assignment
1403001	15	Placebo	1401001	19	Placebo
1403002	12	Active drug	1401002	9	Active drug
1403003	19	Placebo	1401003	22	Placebo
1403004	1	Active drug	1401004	17	Placebo
1403005	23	Placebo	1401005	16	Placebo
1403006	11	Active drug	1401006	5	Active drug
1403007	2	Active drug	1401007	8	Active drug
1403008	20	Placebo	1401008	21	Placebo
1403009	3	Active drug	1401009	13	Placebo
1403010	22	Placebo	1401010	12	Active drug
1403011	10	Active drug	1401011	24	Placebo
1403012	16	Placebo	1401012	6	Active drug
1403013	4	Active drug	1401013	4	Active drug
1403014	6	Active drug	1401014	14	Placebo
1403015	7	Active drug	1401015	1	Active drug
1403016	13	Placebo	1401016	15	Placebo
1403017	24	Placebo	1401017	10	Active drug
1403018	9	Active drug	1401018	3	Active drug
1403019	17	Placebo	1401019	7	Active drug
1403020	21	Placebo	1401020	23	Placebo
1403021	18	Placebo	1401021	2	Active drug
1403022	8	Active drug	1401022	20	Placebo
1403023	5	Active drug	1401023	18	Placebo
1403024	14	Placebo	1401024	11	Active drug

we first divide the whole series of patients who are to enroll in the trial into several blocks with equal or unequal lengths. We then randomize the patients within each block. This method of randomization is known as the *permuted-block randomization*, and it is probably the most frequently employed method for the assignment of patients to treatments in clinical trials.

To illustrate permulated-block randomization, consider the following example. Suppose that a clinical trial is to be conducted at four centers with 24 patients in each center in order to investigate the effectiveness and safety of a test drug compared to an inert placebo. Also suppose that the permuted-block randomization with a block size of 4 patients is to be employed to prevent treatment and possible covariate imbalances. Two methods can be used to randomly

RANDOMIZATION METHODS

Table 4.3.6 Example of Random Allocation for Four Centers (*Continued*)

Random codes for drug XXX, protocol XXX-014
Double-blind, randomized, placebo-control, two parallel groups

	M. Dole, MD			C. Price, MD	
Subject Number	Random Permutation	Treatment Assignment	Subject Number	Random Permutation	Treatment Assignment
1402001	23	Placebo	1404001	4	Active drug
1402002	21	Placebo	1404002	17	Placebo
1402003	13	Placebo	1404003	15	Placebo
1402004	17	Placebo	1404004	20	Placebo
1402005	7	Active drug	1404005	5	Active drug
1402006	10	Active drug	1404006	3	Active drug
1402007	18	Placebo	1404007	14	Placebo
1402008	20	Placebo	1404008	10	Active drug
1402009	1	Active drug	1404009	11	Active drug
1402010	14	Placebo	1404010	19	Placebo
1402011	19	Placebo	1404011	21	Placebo
1402012	3	Active drug	1404012	2	Active drug
1402013	22	Placebo	1404013	22	Placebo
1402014	9	Active drug	1404014	23	Placebo
1402015	24	Placebo	1404015	16	Placebo
1402016	5	Active drug	1404016	7	Active drug
1402017	16	Placebo	1404017	9	Active drug
1402018	8	Active drug	1404018	8	Active drug
1402019	4	Active drug	1404019	1	Active drug
1402020	15	Placebo	1404020	24	Placebo
1402021	11	Active drug	1404021	18	Placebo
1402022	6	Active drug	1404022	6	Active drug
1402023	2	Active drug	1404023	12	Active drug
1402024	12	Active drug	1404024	13	Placebo

assign patients to treatments. The first method is simply to adopt the method of random allocation within each block by generating a random permutation of numbers 1–4 and assigning the first two to the test drug. This method can be easily implemented using the SAS® procedure PLAN. Table 4.3.7 provides a listing of randomization codes generated by the permuted-block randomization with random allocation within each block.

Since there are two treatments with a block size of 4, we have the following six possible permutations for random assignment of patients to treatments:

1: *TTPP*
2: *PPTT*

Table 4.3.7 Example of Permutated-Block Randomization for Four Centers and a Block Size of Four

Random codes for drug XXX, protocol XXX-014
Double-blind, randomized, placebo-control, two parallel groups

	A. Hope, MD			J. Smith, MD	
Subject Number	Random Permutation	Treatment Assignment	Subject Number	Random Permutation	Treatment Assignment
1403001	3	Placebo	1401001	1	Active drug
1403002	4	Placebo	1401002	3	Placebo
1403003	1	Active drug	1401003	2	Active drug
1403004	2	Active drug	1401004	4	Placebo
1403005	2	Active drug	1401005	4	Placebo
1403006	4	Placebo	1401006	1	Active drug
1403007	1	Active drug	1401007	2	Active drug
1403008	3	Placebo	1401008	3	Placebo
1403009	2	Active drug	1401009	1	Active drug
1403010	4	Placebo	1401010	4	Placebo
1403011	3	Placebo	1401011	2	Active drug
1403012	1	Active drug	1401012	3	Placebo
1403013	1	Active drug	1401013	4	Placebo
1403014	3	Placebo	1401014	1	Active drug
1403015	2	Active drug	1401015	2	Active drug
1403016	4	Placebo	1401016	3	Placebo
1403017	1	Active drug	1401017	1	Active drug
1403018	3	Placebo	1401018	4	Placebo
1403019	2	Active drug	1401019	2	Active drug
1403020	4	Placebo	1401020	3	Placebo
1403021	3	Placebo	1401021	2	Active drug
1403022	2	Active drug	1401022	4	Placebo
1403023	1	Active drug	1401023	3	Placebo
1403024	4	Placebo	1401024	1	Active drug

3: *TPTP*
4: *TPPT*
5: *PTPT*
6: *PTTP*

where T and P represent the test drug and the placebo, respectively. Thus there are a total of 6 blocks with 4 patients in each center. The randomization codes can then be generated by producing a random permutation of numbers from 1 to 6, where the numbers correspond to six possible permutations for random assignments of patients as described above. This method is called permuted-

Table 4.3.7 Example of Permutated-Block Randomization for Four Centers and a Block Size of Four (*Continued*)

Random codes for drug XXX, protocol XXX-014
Double-blind, randomized, placebo-control, two parallel groups

M. Dole, MD			C. Price, MD		
Subject Number	Random Permutation	Treatment Assignment	Subject Number	Random Permutation	Treatment Assignment
1402001	1	Active drug	1404001	2	Active drug
1402002	3	Placebo	1404002	3	Placebo
1402003	4	Placebo	1404003	1	Active drug
1402004	2	Active drug	1404004	4	Placebo
1402005	4	Placebo	1404005	2	Active drug
1402006	1	Active drug	1404006	4	Placebo
1402007	2	Active drug	1404007	3	Placebo
1402008	3	Placebo	1404008	1	Active drug
1402009	1	Active drug	1404009	4	Placebo
1402010	3	Placebo	1404010	3	Placebo
1402011	2	Active drug	1404011	1	Active drug
1402012	4	Placebo	1404012	2	Active drug
1402013	1	Active drug	1404013	1	Active drug
1402014	4	Placebo	1404014	4	Placebo
1402015	3	Placebo	1404015	3	Placebo
1402016	2	Active drug	1404016	2	Active drug
1402017	3	Placebo	1404017	1	Active drug
1402018	4	Placebo	1404018	3	Placebo
1402019	1	Active drug	1404019	4	Placebo
1402020	2	Active drug	1404020	2	Active drug
1402021	2	Active drug	1404021	2	Active drug
1402022	1	Active drug	1404022	4	Placebo
1402023	4	Placebo	1404023	1	Active drug
1402024	3	Placebo	1404024	3	Placebo

block randomization with random selection, and it can also be easily implemented by the SAS® procedure PLAN. Table 4.3.8 provides a listing of randomization codes generated by the permuted-block randomization with random selection. SAS programs for both methods of permuted-block randomization are also provided in Appendix B.3 and Appendix B.4, respectively. It can be verified that the randomization codes given in Tables 4.3.7 and 4.3.8 provide treatment balance not only within each study center but also for the entire study.

In addition to the assurance of treatment balance, the permuted-block randomization can account for a possible time-heterogeneous population by forcing a periodic balance. This desirable property of forced periodic balance, however,

Table 4.3.8 Example of Permutated-Block Randomization by Random Selection of Blocks for Four Centers and a Block Size of Four

Random codes for drug XXX, protocol XXX-014
Double-blind, randomized, placebo-control, two parallel groups

A. Hope, MD		M. Dole, MD	
Subject Number	Treatment Assignment	Subject Number	Treatment Assignment
1403001	Placebo	1402001	Active drug
1403002	Placebo	1402002	Placebo
1403003	Active drug	1402003	Placebo
1403004	Active drug	1402004	Active drug
1403005	Active drug	1402005	Active drug
1403006	Active drug	1402006	Active drug
1403007	Placebo	1402007	Placebo
1403008	Placebo	1402008	Placebo
1403009	Active drug	1402009	Placebo
1403010	Placebo	1402010	Active drug
1403011	Active drug	1402011	Active drug
1403012	Placebo	1402012	Placebo
1403013	Placebo	1402013	Active drug
1403014	Active drug	1402014	Placebo
1403015	Active drug	1402015	Active drug
1403016	Placebo	1402016	Placebo
1403017	Placebo	1402017	Placebo
1403018	Active drug	1402018	Placebo
1403019	Placebo	1402019	Active drug
1403020	Active drug	1402020	Active drug
1403021	Active drug	1402021	Placebo
1403022	Placebo	1402022	Active drug
1403023	Placebo	1402023	Placebo
1403024	Active drug	1402024	Active drug

becomes a disadvantage when the block size is not blinded. In the case where the block size is not blinded, the probability of correctly guessing the treatment increases at the end of each successive block. Matts and Lachin (1988) point out that if the treatment is unblinded, then such forced periodic balance provides investigators an opportunity not only to correctly guess the assignment of patients but also to alter the composition of the treatment groups. As a result, an increasing chance of correctly guessing treatment assignment and alteration of treatment composition can certainly increase the chance of introducing bias to the evaluation of the treatment effect. In addition, as block size increases, the potential

RANDOMIZATION METHODS

Table 4.3.8 Example of Permutated-Block Randomization by Random Selection of Blocks for Four Centers and a Block Size of Four (*Continued*)

Random codes for drug XXX, protocol XXX-014
Double-blind, randomized, placebo-control, two parallel groups

J. Smith, MD		C. Price, MD	
Subject Number	Treatment Assignment	Subject Number	Treatment Assignment
1401001	Active drug	1404001	Placebo
1401002	Placebo	1404002	Active drug
1401003	Active drug	1404003	Active drug
1401004	Placebo	1404004	Placebo
1401005	Placebo	1404005	Placebo
1401006	Active drug	1404006	Placebo
1401007	Placebo	1404007	Active drug
1401008	Active drug	1404008	Active drug
1401009	Placebo	1404009	Active drug
1401010	Placebo	1404010	Active drug
1401011	Active drug	1404011	Placebo
1401012	Active drug	1404012	Placebo
1401013	Active drug	1404013	Active drug
1401014	Active drug	1404014	Placebo
1401015	Placebo	1404015	Placebo
1401016	Placebo	1404016	Active drug
1401017	Active drug	1404017	Placebo
1401018	Placebo	1404018	Active drug
1401019	Placebo	1404019	Placebo
1401020	Active drug	1404020	Active drug
1401021	Placebo	1404021	Active drug
1401022	Active drug	1404022	Placebo
1401023	Active drug	1404023	Active drug
1401024	Placebo	1404024	Placebo

selection bias decreases. Although the use of random block size can reduce the selection bias, it cannot completely eliminate the bias. The only way to eliminate the selection bias is to enforce a double-blinded procedure during the entire course of study for which both investigators and patients are blinded to block size and treatment assignments. For the method of permuted-block randomization, the variance of the accidental bias, which does not depend on the number of blocks, decreases as block size increases. Although the accidental bias associated with the permuted-block randomization is negligible for large samples, it is more serious as compared to those by simple randomization for small samples.

It should be noted that permuted-block randomization is also a stratified randomization. Therefore a stratified analysis should be performed with block as a stratum to properly control the overall type I error rate and hence to provide the optimal power for detection of a possible treatment effect. In practice, the block effect is usually ignored when performing data analysis. Matts and Lachin (1988) show that the test statistic ignoring the block is equal to 1 minus the intrablock correlation coefficient times the test statistic that takes the block into account. Since the patients within the same block are usually more homogeneous than those between blocks, the intrablock correlation coefficient is often positive. As a result the tests that ignore the block will produce more conservative results. However, since most clinical trials are multicenter studies with a moderate block size, it is not clear from their discussion whether a saturated model that factors treatment, center, and block and their corresponding two-factor and three-factor interactions should be included in the analysis of variance. In addition, as given in Tables 4.3.7 and 4.3.8, there are a total of 24 strata (center-by-block combinations) with 4 patients in each stratum. It is not clear whether the Mantel-Haenszel test and linear rank statistics based on the permutation model are adequate for a complete combination of all levels of all strata with a very small number of patients in each stratum. In addition, when the block size is large, there is a high possibility that a moderate number of patients will fall in the last block where the randomization codes are not entirely used up. If the trial is also stratified according to some covariates such as study center, the number of patients in such incomplete block can become sizable. Therefore a stratified analysis can be quite complicated due to these incomplete blocks from all strata and possible presence of an intrablock correlation coefficient.

Adaptive Randomization

As discussed above, the method of complete randomization includes a constant marginal probability for the independent assignment of patients to treatments. However, to some extent it can cause an imbalance in the patient allocation to treatment. On the other hand, the methods of random allocation and permuted-block randomization are useful in forcing a balanced allocation of patients to treatments within either a fixed total sample size or a prespecified block size. These methods of restricted randomization can also maintain a constant marginal probability for the assignment of patients to treatments. In practice, in addition to enforcing a balanced allocation among treatments to some degree, it is also of interest to adjust the probability of assignment of patients to treatments during the study. This type of randomization is called *adaptive randomization* because the probability of a current patient being assigned is adjusted based on the assignment of previous patients. Unlike the other randomization methods described above, the randomization codes based on the method of adaptive randomization cannot be prepared before the study begins. This is because that the randomization process is performed at the time a patient is enrolled in the study, whereas adaptive randomization requires information on previously randomized patients. In clinical trials the method of adaptive randomization is often applied with respect to treatment, covariates, or clinical response. Therefore the adaptive randomization is also known as treatment adaptive randomization, covariate adaptive randomization, or response adaptive randomization. We will now briefly introduce these three applications of adaptive randomization.

Treatment Adaptive Randomization

The treatment adaptive randomization adjusts for the assigning probability of the current patient with respect to the number of patients who have been randomized to each treatment group. Efron (1971) first introduced the idea of biased coin randomization as a method for adjustment of assigning probability. Consider the same example discussed above. The assigning probability for the first patient is clearly $1/2$. After k patients are enrolled, k_T and k_P patients are randomized to the test drug group and the placebo group, respectively. The idea is that if more patients were randomized to the test drug group, then the next patient will be assigned to the placebo group with a probability greater than $1/2$. Similarly, if the current number of patients randomized to the test drug group is fewer than that of the placebo group, then the next patient will be assigned to the test drug group with a probability greater than $1/2$. If a treatment balance is achieved, then the next patient is assigned to either the test drug group or the placebo group with a probability of $1/2$. As an example, Pocock (1983) suggests the use of $p = 3/4$, $2/3$, $3/5$, and $5/9$, for the chance of less than 5%, for differences in the number of patients between treatment groups being 4, 6, 10, and 16, respectively.

Although the bias coin randomization attempts to achieve a treatment balance by adjusting the assigning probability with respect to the difference in the number of patients who were previously assigned, it may not be satisfactory because a constant assigning probability was used during the entire course of the study. As an alternative, Wei (1977, 1978) consider the so-called *urn randomization*, which is an extension of the biased coin randomization. For the urn randomization the probability of the assignment of the current patient is a function of the current treatment imbalance (Wei and Lachin, 1988). To illustrate the method of urn randomization, consider a urn that contains exactly A white balls and A black balls. For the assignment of a patient, draw a ball at random from the urn and replace it into the urn. If the drawn ball is a white one, then the patient is assigned to the test drug group. Otherwise, the patient is assigned to the placebo group. Therefore the assigning probability for the first patient is $1/2$. The procedure is to add B white (black) balls to the urn if the drawn ball is a black (white) one. This randomization process is repeated whenever a new patient is enrolled. From the above description, it can be seen that the assigning probability of the urn randomization is determined by A and B. Therefore a urn randomization is usually denoted by $UR(A, B)$. If at each drawing no additional ball is returned to the urn, then the urn randomization is simply a complete randomization and because with replacement, an equal assigning probability of $1/2$ is employed for the random selection. If we do not put any ball initially in the urn but use the assigning probability of $1/2$ for the first patient, then the subsequent probability of assigning a patient to the test drug after k patients have been enrolled is equal to the proportion of patients who were randomly assigned to the placebo group. This proportion is independent of the number of B balls scheduled to return to the urn.

The urn randomization can achieve a certain degree of a desired balance at the early stage of the study. This is usually accomplished by choosing appropriate numbers of A and B. As pointed out by Lachin, Matts, and Wei (1988), if the ratio of A to B is large, then the urn randomization is very similar to the complete randomization. If the investigator desires to have the treatment balance at the early stage of the study and wishes to maintain a certain degree of balance at the end of the study, then we can choose a large ratio of B to A. This nice property is especially attractive for the post hoc stratified analysis and sequential trial because the size of the post hoc–defined strata and the number of patients at the early termination are usually not known at the planning stage of the trial.

As the sample size increases, the urn randomization approaches complete randomization. As a result the expected bias factor will be very close to zero. Consequently the selection bias according to the Blackwell-Hodges model will be negligible. For finite samples the selection bias of the urn randomization is smaller than those methods of restricted randomization, though it can be very close to that of the permuted-block randomization for a sample size fewer than 10. As compared to other methods of randomization, the accidental bias caused by omitting important covariates for estimation of the treatment effect becomes negligible as the sample size increases. However, for small trials an accidental bias may still exist that cannot be ignored. As a result Lachin, Matts, and Wei (1988) recommend that the urn randomization not be employed for the trial with either the total size or the size of the smallest stratum being fewer than 10. The urn randomization with $A = 0$ and $B = 1$ has nice properties of adequate control for treatment balance. In addition it is less vulnerable to both selection and accidental bias than other methods of restricted randomization. Furthermore it is easy to implement on a computer because the assigning probability depends only on the current state of treatment allocation.

Although the urn randomization is simple, it requires a much more complicated analysis compared to other methods of randomization. This is because the urn randomization does not have an equal assigning probability for each patient. To conduct an exact permutation test based on the urn randomization, the probability of each assignment is needed in order to compute the p-value. For the large sample size Wei and Lachin (1988) derive the explicit permutation tests for the logrank and the Peto-Peto-Prentice-Wilcoxon statistics for the censored data. For the urn randomization, statistics of different strata are independent for the prospectively stratified randomization. However, they are correlated for the poststratified subgroup analyses. Wei and Lachin (1988) give an explicit expression for the conduct of a combined test over strata.

Covariate Adaptive Randomization
In certain diseases some of the prognostic factors are known to affect clinical outcomes of the treatment. Therefore it is desired to achieve a covariate balance with respect to these prognostic factors. For this purpose we may consider to employ the covariate adaptive randomization which is also known as the minimization method (e.g., see Taves, 1974; Pocock, 1983; Spilker, 1991). For an illustration of this method, consider the following hypothetical trial in which a test drug is evaluated with an inert placebo in patients with benign prostatic hyperplasia. Suppose that for patients aged over 64 years old, peak urinary flow rate less than 9 mL/s and an AUA-7 symptom score exceeding 20 will have an impact on the clinical evaluation of the test drug. The distribution of these three covariates after 106 patients and 107 patients were enrolled into the placebo and the test drug, respectively, as shown in Table 4.3.9. Suppose that the age of the next patient for randomization is 68 years old with a peak urinary flow rate of 7.4 mL/sec and an AUA-7 symptom score of 21 points. Then one can modify the frequencies of patients with respect to the categories of covariates that this patient falls into. From Table 4.3.9 the numbers of patients who satisfy the criteria (1) age older than 64 years old, (2) peak urinary flow rate less than 9 mL/s, and (3) an AUA-7 symptom score larger than 20 for the placebo group are 49, 45, and 29, respectively, while the numbers for the test drug group are 51, 44, and 30, respectively. Therefore the respective sums for the test drug group and the placebo group are 123 and 125. Since the placebo group has a smaller sum, the procedure is to assign the next patient to the placebo group. Because the minimization method described above is nonrandom, the covariate

Table 4.3.9 Frequency Distribution of Age, Peak Urinary Flow Rate, and AUA-7 Symptom Score

Covariate	Placebo	Test Drug
N	106	107
Age (years)		
<64	57	56
≥65	49	51
Peak flow rate (mL/s)		
≤9	45	44
>9	61	63
AUA-7 symptom score		
≤7	25	26
8–19	52	51
≥20	29	30

adaptive randomization can also use a probability greater than 1/2 to assign the next patient to the treatment group with a smaller sum. Pocock (1983) indicates that assigning a probability of 3/4 or 2/3 may be appropriate. The covariate adaptive randomization requires a constant update of the current status of the covariates. Hence it requires an intensive administrative effort to implement such a procedure even though the computer can alleviate the burden to some extent. As a result the covariate adaptive randomization may present a high risk of breaking blindness by either the investigators or the personnel responsible for updating the covariate imbalance status. Note that the covariate imbalance can always be adjusted by a post hoc subgroup analysis when the trial size is moderate. In practice, the covariate adaptive randomization is not recommended for trials with sample sizes greater than 100. More details on the covariate randomization can be found in Pock and Simon (1975), White and Freedman (1978), and Miller et al. (1980).

Response Adaptive Randomization
Another adaptive randomization is to adjust for the assigning probability according to the success or failure of the treatments to which previous patients were assigned. This idea was first proposed by Zelen (1969) and subsequently known as the play-the-winner (PW) rule. For the first patient enrolled in the study, an assigning probability of 1/2 is employed to either treatment. Suppose that the white ball represents the test drug (T) and the black ball represents the placebo (P). If the current patient receives treatment T and the response is a success or if the current patient receives P and the response is a failure, then put a white ball in the urn. If the current patient receives T and the response is a failure or if the current patient receives P and the response is a success, then put a black ball in the urn. When the next patient is enrolled into the trial, we randomly draw a ball without replacement from the urn. If there is no ball in the urn, then an assigning probability of 1/2 is employed to either treatment. Wei and Durham (1978) indicate that the responses of patients might not be observed before the arrival of the next patient. The urn therefore might have a very high possibility of being empty during the entire course of the trial. It turns out that the assigning probability of patients is approximately 1/2 under the play-the-winner rule which is quite similar to the method of random allocation. If the response is unavailable in a short period of time before the next patient is enrolled, Zelen (1969) suggests that one can continue to assign the same treatment if the response of the current patient is a success but switch to the other treatment if a failure is observed. This rule is called the *modified play-the-winner* (MPW) rule. Note that the MPW rule is a deterministic rather than stochastic process.

To overcome the drawback of a deterministic process, Wei and Durham (1978) suggest an alternative method known as the *randomized play-the-winner* (RPW) rule. At the beginning of the trial, an equal number (m) of white and black balls are placed in an urn, where the white balls represent the test drug and black balls represent the placebo. When a patient is enrolled into the trial, a ball is drawn at random from the urn with replacement. If the randomly selected ball is a white one, the patient is assigned to the test drug group, and otherwise, to the placebo group. If the previous patient was assigned to the test drug group and the response is a success, then additional B white balls and A black balls are put into the urn, where $B \geq A \geq 0$. If the response of the previous patient receiving the test drug is a failure, then additional A white balls and B black balls are put into the urn. Similarly, if the previous patient was assigned to the placebo group and the response is a success, then additional A white balls and B black balls are put into the urn. If the response of the previous patient receiving the placebo is a failure, then additional B white balls and A black balls are put into the urn. If the urn is empty, then a probability of $1/2$ is used. It should be noted that exactly additional A plus B balls are put into the urn whenever a response is available for the assignment of the next patient. The RPW does not require the availability of the response of the previous patients. The RPW, which is random, provides a higher probability to treat the next patient, with the better treatment based on the current result of the trial. When $A = B$, RPW is the same as the method of complete randomization. Wei and Durham (1978) show that when the ratio of B to A becomes large, the probability of assigning patients to the test drug is approximately equal to the ratio of the failure rate of the placebo to the sum of failure rates for both treatments. Therefore, if the ratio of B to A is large, then the RPW tends to assign more patients to the better treatment. In addition, Wei et al. (1990) indicate that RPW is less vulnerable to the experimental bias than other adaptive randomizations.

Since the assigning probability of the current patient to treatments adjusts for the past history of the outcomes of the previously randomized patients, the statistical analysis based on RPW is much more complicated than those based on other methods of randomization. Wei (1978) describes the permutation distribution of a test for the binary response under RPW. In addition Wei et al. (1990) study the exact conditional, exact unconditional, and approximate confidence intervals for the treatment difference in binary responses. The results indicate that the exact unconditional procedure performs much better than the conditional procedure. In addition the large sample unconditional confidence intervals derived from the likelihood statistic are not very sensitive to the adaptive randomization and perform quite satisfactory for trials with moderate sample size. The confidence intervals based on the maximum likelihood estimates behave very poorly under RPW. Therefore Wei et al. (1990) suggest that the features of response adaptive randomization be taken into account in the analysis. On the other hand, Tamura et al. (1994) perform a Bayesian analysis for a trial concerning patients with depressive disorder using RPW.

Recently several clinical trials were conducted using the play-the-winner rule. For example, a clinical trial was conducted at the University of Michigan to investigate the effectiveness and safety of extracorporeal membrane oxygenation (ECMO) in treating newborn babies with persistent pulmonary hypertension (PPH) with the conventional mechanical ventilation (CMV) as the concurrent control (Cornell et al., 1986). Past experience has shown that infants with PPH has a 80% death rate in the absence of ECMO which is an artificial heart-lung machine recycling the blood through a membrane exposed to the oxygen with a high concentration. On the other hand, ECMO is a surgical procedure with potential life-threatening complications to the infants. Since the response (either death or recovery) can be observed within a few days, the response of the previously treated infants is available before the entry of the next newborn. Consequently a RPW with $m = 1$, $A = 0$, and $B = 1$ was employed for the first infant. The result turned out to be the ECMO treatment, and the baby recovered. Then a white ball representing the ECMO treatment was put in the urn. For the next infant, although the probability of assigning infants to the ECMO treatment is 2/3 and to the CMV treatment is 1/3, the resulting treatment for the second infant is the CMV treatment with an unfortunate outcome of death. Therefore a black ball representing the CMV treatment was put back into the urn. This procedure was employed. The result turned out that the ECMO treatment was randomly assigned to the next eight consecutive infants all of whom survived. The trial was terminated at this point. Note that two more infants were also assigned to the ECMO treatment without invoking the ECMO treatment, but both of them survived too. Boston's Children Hospital Medical Center and Brigham and Women's Hospital conducted a similar adaptive trial to compare the ECMO treatment with the CMV treatment in infants with PPH (Ware, 1989). This study involved two phases. For the first phase, a permuted-block randomization with a block size of 4 was used to generate randomization codes for the trial. It was calculated at the planning stage that if 4 deaths were observed in one of the two groups, this phase would be terminated, and the study would proceed to the second phase in which all subsequent infants would be assigned to receive the other treatment. It was also predetermined that the second phase would be terminated if 4 deaths or 28 survivors were observed from the infants of both phases who were enrolled into the other treatment. For the first phase of the trial with a permuted-block randomization, 10 infants were assigned to the CMV treatment and 9 to the ECMO treatment. Four infants who were randomly assigned to the CMV treatment died, and all 9 infants receiving the ECMO treatment during the first phase survived. As a result all subsequent infants were assigned without randomization to receive the ECMO treatment. The trial was terminated at the 20th infant who enrolled into the second phase and did not survive.

For another example, Tamura et al. (1994) report that a clinical trial utilizing the RPW rule was conducted to assess the efficacy and safety of fluoxetine as compared with a placebo in patients with depressive disorder. The study was stratified according to rapid eye movement latency (REML) which is defined as the time between sleep onset and the first rapid eye movement. If REML of patients is shorter than or equal to 65 minutes, then he or she is stratified into the shortened REML; otherwise, he or she is stratified to the normal REML group. A patient is classified as a responder if the percent reduction at the final eight-week visit from baseline on the first 17 items of the Hamilton Depression scale (HAM-D-17) is at least 50%. Since a period of eight weeks is required to observe the primary endpoint and the patient accrual was rather rapid, this time delay could not allow the investigator to employ the RPW with the primary endpoint. As a result a surrogate endpoint of the percent reduction of at least 50% in HAM-D-17 in two consecutive visits after at least three weeks of therapy was used for the response adaptive randomization. Within each stratum the first six patients were assigned using the method of permuted-block randomization. Starting with the seventh patient, the RPW was initiated within each stratum with $m = 1$, $A = 0$, $B = 1$. A total of 89 patients were randomized, and yet the surrogate endpoint was only observed in 61 of the 89 randomized patients and 83 patients were included in the analysis based on the primary endpoint. Tamura et al. (1994) indicate that their experience with RPW for this trial has been generally positive despite increasing communication between the sponsor and investigators.

As indicated earlier, statistical inference depends on the statistical test used, which in turn depends on the randomization employed. It is therefore important to derive an appropriate statistical test according to the randomization employed. For example, the ECMO study conducted at the University of Michigan created a controversy over the statistical analysis used for comparing the two treatments. Recall that the 11 infants assigned to the ECMO treatment survived, and only one baby, the one assigned to the CMV treatment, died. Although the ECMO trial at the University of Michigan only involved 12 infants, the results of this study have raised many serious questions regarding complicated statistical and ethical issues. First, how does one compare two treatments with such a severe treatment imbalance (i.e., 11-ECMO versus 1-CMV)? A sample of one patient contributes very little information toward the comparison between treatments. Second, there exists no appropriate statistical test under the RPW model. Alternatively, Cornell, et al. (1986) considered the method of ranking and selection to demonstrate that the ECMO treatment is superior to the CMV treatment. However, his method does not provide p-values and confidence intervals. Wei (1988) developed a permutation test under the RPW rule. However, Begg (1990) pointed out that Wei's permutation test is inappropriate and obtained some p-values (ranging from 0.038 to 0.62) based on different analyses (also see the discussion by Wei, 1990; Pocock, 1990; Cox, 1990). For the ECMO study conducted by Harvard Medical School, there was employed a two-stage design. At the first stage, a probability of 1/2 was used to assign infants until there was statistically significant evidence that one treatment showed a superior efficacy; then all remaining infants were assigned to the superior treatment (Ware, 1989). This example demonstrates that appropriate statistical procedure must be derived for the method of randomization to be employed in clinical trials.

For the response adaptive randomization, despite its advantage in ethical terms, it is not widely accepted in clinical trials (Simon, 1991; Rosenburger and Lachin, 1993). This is probably due to the availability of appropriate statistical tests under various methods of response adaptive randomization. Rosenburger and Lachin (1993), however, provide a list of general conditions under which a response adaptive randomization can be implemented successfully given the existing methodology. These conditions are summarized below:

1. There is a single outcome or hypothesis of interest.
2. Outcomes are ascertainable in a short period of time.
3. The study has important public health consequences, but the diseases are not life-threatening.
4. The study has an adequate sample size and the composition of the sample is not likely to change over time.
5. The participants in the study have the resources to logistically implement the randomization procedure.

These general conditions may limit the application of the response adaptive randomization to clinical trials. First, a disease is a medical condition that is very complicated and usually cannot be adequately described by a single clinical outcome. Hence a clinical trial may have more than one objective based on more than one clinical outcome. Consequently the first condition seems to be very difficult to be satisfied by most clinical trials. Second, under the RPW model, despite recent developed analysis procedures for binary data (Wei, 1988), multinomial and continuous data for large samples (Rosenburger, 1993), analyses for the secondary clinical endpoints, censored data and subgroup analyses for the adjustment of covariates have not yet been developed. Finally clinical trials usually require multiple visits for an evaluation of the treatment's progress. However, statistical method for the analysis of repeated measurements has not been proposed for the RPW rule. As a result Rosenburger and Lachin (1993) conclude that the future use of the response adaptive randomization is uncertain.

4.4 IMPLEMENTATION OF RANDOMIZATION

In the pharmaceutical industry, for good clinical practice a set of standard operating procedures (SOP) for generation, implementation, and administration of randomization is usually established to ensure the integrity of clinical trials. In this section we will introduce an implementation procedure for the method of nonadaptive randomizations which is adopted by most of pharmaceutical companies for clinical research and development. In the pharmaceutical industry the department of Biostatistics and Data Management (or Biometrics) is usually responsible for the activities of statistics, programming, and clinical data management. Within the department, a drug-specific (or project-specific) team (or unit) is usually formed to oversee the development of statistics, programming, and data management during the process. This team usually consists of a biostatistician, a programmer, and a data coordinator. Note that unlike a clinical research associate or monitor, this team does not involve itself with the day-to-day activities of clinical projects. However, this team is responsible for the generation of randomization codes, case report forms design and review, clinical data management, statistical analysis, and report writing. Among these, the generation of randomization codes is the key to the success of the intended trials. Since there are many different methods for generating randomization codes as discussed above, the team is responsible for the implementation of a system that incorporates the various methods of randomization by developing computer programs. Since the system, which may contain a number of computer programs, is not designed for commercial but for internal use, it is recommended that the methods of randomization employed and the corresponding computer programs be adequately documented. Also it is desirable for a user-friendly User's Reference Manual to be developed. The User's Reference Manual should contain detailed instructions for the use of the system, references to the pseudonumber generator, methods of randomization, programs for listings of the pseudonumber generator and for the production of a listing of the randomization codes. In addition a prospective validation of the design programs should be performed according to a validation protocol before the implementation of the system. Note that the FDA requires that the results of the validation tests be documented and that the system be validated periodically.

During the development of the clinical protocol, the project clinician and biostatistician usually discuss the selection of an appropriate method of randomization and some related logistic issues for the implementation of randomization according to study objectives, primary endpoints, stratified covariates (if any), and sample size of the trial. The randomization method employed for the study should be described in detail in the study protocol without disclosure of the block size, if the permuted-block randomization is used. The study protocol and the investigator's brochure should also describe in detail a standard procedure for treatment assignment and drug dispensing. In general, patients should not receive any medication unless they have met all eligibility criteria and have signed the informed consent forms as defined in the study protocol. A formal request for randomization codes cannot be sent to the project statistician unless the study protocol has obtained an approval from an internal protocol review committee. The project statistician can then check whether the request is adequate with respect to the study protocol and design. The randomization codes will be generated according to the selected method of randomization if no concerns are raised by the project statistician. The project team is not only responsible for the generation of randomization codes but also for performing quality assurance (QA) procedures of the generated randomization codes. The QA procedures are to (1) check every generated randomization codes, (2) document the program logs for generation of the randomization codes, and (3) maintain information for the generation of the randomization codes including the seed and the first and last random numbers generated from the seed. If the randomization codes meet the requirements of the QA procedures, then a list of randomization codes is sent to the drug packaging department or some contracted laboratory for packaging the study drugs. If the trial is a triple-blind study, the project statistician and clinician or other project team members should be informed only of the generation of the randomization codes by a cover memo. If, however, the trial is a double-blind study, then the project statistician and clinician might get a copy of randomization codes upon request. The information of the randomization codes will then be locked in the database until the time at which an interim analysis or final analysis is performed. For a triple-blind study the clinical data coordinator and clinical research associate identify the patients through a sequentially assigned patient (subject) number to maintain the blindness. A patient or subject number usually contains three parts, which include the project number, the study center number, and a sequentially assigned patient number within the individual study center. Let us take, for example, a number 01401015 (i.e., 014-01-015) used to identify a patient in a clinical trial. The first three digits 014 are an identifier of the study drug XXX, the next two digits 01 represent the first study site, and the last three digits 015 indicate that the patient is the 15th patient to enroll in the study. The project team will also generate a set of dummy randomization codes for the project statistician to perform necessary programming for patient listings or case report tabulations as required by the FDA to shorten the statistical analysis after the study is completed and the database is locked.

When the drug packaging department receives the randomization codes, the study drugs are packed according to the method and instruction as stated in the protocol. The most secure method for maintaining blindness is to use identical blister packs or drug kits with identically appearing contents. Usually the drug kits have a three-part double-blind tear-off label affixed to the cover of the kit. This label has the protocol number and the preprinted patient number. Patients' initials and the time and date the drug dispensed will be recorded on each label. The time and date are important for establishing an audit trail of treatment assignment. The double-blinded tear-off portion, which will be attached to the appropriate page in the case report form, contains the actual treatment group information to which the patient is assigned. These sealed labels will not be opened unless it is required in a medical emergency when knowledge of the respective treatment may influence medical care. At the conclusion of the study, the investigators should return all used and unused study drugs to the sponsor. Usually there is a broilerplate paragraph included in the study protocol for drug accountability. Figures 4.4.1 and 4.4.2 provide flow charts of the randomization procedure discussed above.

In the pharmaceutical industry, the randomization procedure is not limited to the generation of randomization codes for treatment assignments. It can also be applied to laboratory evaluations. For example, routine hematology, blood chemistry, urinalysis, or some other special compounds such as serum hormone levels of the patients from a clinical trial are usually assayed at a centralized contracted laboratory. However, the assay of active ingredients of the study drugs are often performed by the method developed by the laboratory. Since samples from a clinical trial may be enormous due to multiple visits, assays may be required to perform at different times of a day over a period of several days due to the capacity of the laboratory. As a result a proper design for the

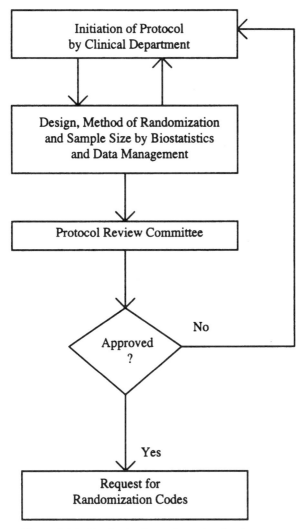

Figure 4.4.1 Randomization codes at the planning stage.

drug assay is necessary to eliminate variability due to analyst, time, and day. In addition the assay should be performed in a blinded fashion to avoid possible bias caused by the knowledge of the study drugs. Therefore randomization codes may also be generated by the same project team according to the design and method of randomization as deemed appropriate by the project statistician. The randomization codes for treatment assignments and drug assays are to be stored in the central file and cannot be released until the database is locked. The generation and implementation of randomization codes for drug assays can be logistically complicated when assays for different active ingredients or their

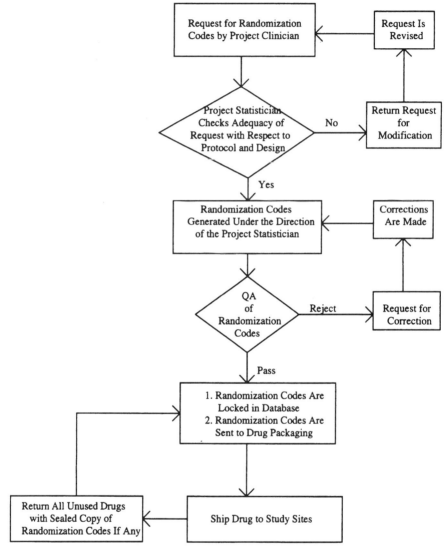

Figure 4.4.2 Generation of randomization codes.

metabolites are required to perform individually with subdivisions of the blood samples.

In some cases randomization is also used for poststudy evaluation. For example, for the evaluation of contrast agents in the enhancement of images obtained with magnetic resonance imaging (MRI), two sets of films with and without contrast agent are usually obtained. The set of films without the contrast agent is obtained before that with the contrast agent. In general, this type of trial is conducted in an open-label fashion without a concurrent control because it is almost impossible to maintain the blindness during the trial. As a result the FDA requests that the sponsors perform a blinded reader study after the trial is completed. A separate protocol for the blinded reader study is also prepared after all films are obtained from the clinical trial. The blinded reader studies are considered as adequate well-controlled studies for approval. The package insert is usually derived from the blinded reader studies rather than the actual clinical trials. For blinded reader studies, several qualified readers, who are not engaged the clinical part of the trial and are not associated with any investigators of the trial, are asked in a random but blinded fashion to evaluate the films. For this purpose, randomization codes for a blinded reader study can also be generated according to the design and randomization method specified in the protocol under the supervision of the project team.

Note that the randomization procedure described above is probably the most frequently employed procedure for conducting clinical trials in the pharmaceutical industry with sample sizes smaller than a few thousands. For most clinical trials sponsored by the NIH or other cooperative groups, the sample size can be quite large. For example, the ISIS-2 (1988) study randomized 17,187 patients with suspected acute myocardial infarction to four treatment groups and 22,071 physicians were enrolled to receive one of the four treatments in the U.S. Physician's Health Study (1989). An even larger study is the GUSTO (1993) study which enrolled a total of 41,021 patients with evolving myocardial infarction. As a result it may not be feasible to adopt the randomization procedure described above. As an alternative, a centralized randomization center may be established for random assignment of treatments either by mail or by telephone. If the time between the screening and request for random assignment is long, say a month, then the mailing system may be possible. For example, see the study conducted by the Coronary Drug Project Research Group (1973) which is described in detail in Meinert (1986). It should be noted that it is not an easy task to handle treatment assignments of more than tens of thousands of patients, especially when the time from the onset of symptom to the treatment is also considered as a crucial factor such as rt-PA for acute ischemic stroke (National Institute of Neurological Disorder and Stroke rt-PA Stroke Study Group, 1995). Hence a central administrated telephone-based assignment system should be employed. For example, ISIS-2 (1988) used a 24-hour telephone service, based in Gent and Brussels for Belgium, Berlin for Germany, Valencia for Spain, Bellinzona for Austria and Switzerland, Lyon for France, and Oxford for England and all other countries. The information of patient identifiers such as age, systolic blood pressure, hours from onset of the episode of pain that led to admission, aspirin use during the week before, and the planned treatment in hospital must be completed before a patient is randomized to receive treatments. In addition the method of minimization randomization was also adopted at Oxford for balancing the prognostic factors recorded at entry. However, on January 24, 1986, a programming error was discovered that led more patients randomized at Oxford to being allocated to the placebo infusion and placebo tablets over a period of two months (see Chapter 2 for more information on the treatments of ISIS-2). This programming error was corrected and the exact balance restored in August 1986. Similarly the GUSTO trial used a 24-hour a day, seven-day-per-week randomization center to verify patient eligibility, informed consent, and to assign treatments to more than forty thousand patients. Note that randomization can be performed through a computer networking system such as internet, ccmail, www, or any other available networking system. A computerized standard form of eligibility information and informed consent must be sent with the request to the randomization center. Then a validated computer program at the randomization center can immediately enter the data of eligibility for a patient from the mail and verify the patient's eligibility. If inclusion criteria are met and none of the exclusion criteria are observed, then a random assignment of the patient to a particular treatment can be issued in a blinded fashion and sent to the study center. Otherwise, a mail message of reasons for refusal to issue randomization codes should be sent to the study center. This randomization process is not only accomplished in seconds but also eliminates the human errors that often occur during the randomization process. It should be noted that a computer system should be validated if it is to be employed for the generation of randomization codes. In addition all personnel should have appropriate training. It is suggested that a dry run with simulated cases be done before the actual implementation of the system takes place.

4.5 GENERALIZATION OF CONTROLLED RANDOMIZED TRIALS

In most clinical trials the group of patients (or sample) who participate is just a small portion of a heterogeneous patient population with the intended disease. As indicated earlier, a well-controlled randomized clinical trial is necessary to provide an unbiased and valid assessment of the study medicine. A well-controlled randomized trial is conducted under well-controlled experimental conditions, which are usually very different from a physician's best clinical practice. Therefore it is a concern whether the clinical results observed from the well-controlled randomized clinical trial can be applied on the patient population with the disease. As a result the feasibility and generalization of well-controlled randomized trials have become an important issue in public health (Rubins, 1994). For illustration purposes, consider the following two examples.

In early 1970s a high cholesterol level was known to be a risk factor for developing coronary heart disease. To confirm this, a trial known as the *Lipids Research Clinics Coronary Primary Prevention Trials* (CPPT) was initiated by the National Heart, Lung, and Blood Institute to test the hypothesis whether lowering cholesterol can prevent the development of coronary heart disease. In the CPPT trial, a total of 4000 healthy, middle-age males were randomized to receive either the cholesterol-lowering agent cholestyramine or its matching placebo (Lipids Research Clinics Program, 1984). The primary endpoint was the incidence of coronary heart disease after a seven-year follow-up. A statistically significant reduction of 1.9% in 7-year incidence of coronary heart diseases was observed for the cholestyramine group as compared to the placebo (8.1% versus 9.8%). An expert panel recommended to extrapolate the results for the treatment of high cholesterol in populations that had never been studied and whose benefit has not yet been demonstrated (The Expert Panel, 1989; Recommendations for the Treatment of Hypercholesterolemia, 1984). Moore (1989), however, raises a serious doubt regarding the expert panel's recommendation for the treatment of patients with high cholesterol levels. Moore points out that the CPPT trial was conducted on middle-age males which cannot be applied to a general patient population with hypercholesterolemia. Another example concerning the generation of controlled randomized trials is the U.S. Physician's Health Study described earlier. The question is whether the benefit regarding fatal and nonfatal coronary heart disease, which was observed using 22,000 highly educated males aged over 40 years old, can also be observed in an average individual regardless of gender, race education, and socioeconomic background. This question is indeed a tough one to answer. We can address the question in part by performing a subgroup analysis with respect to the composition of the patients in the trial. This study led to the United States Congress passing legislation (National Institute of Health Reauthorization Bill, 1993) which requires the specification of the composition of any human studies sponsored by the NIH. More detail can be found in Wittes (1994).

One way to ensure the generalization of controlled randomized trials is to understand the process for drawing statistical and clinical inference. Basically statistical and clinical inference for the generalization of results obtained from clinical trials to other patients is a two-step process. The first step is to *internally* apply the statistical and clinical inference on the targeted population to other patients within the population. The second step is to *externally* generalize the statistical/clinical inference made on the targeted population to another patient population with different characteristics. These steps involve the concept of *population efficacy* (or safety) and *individual efficacy* (or safety) which will be illustrated below.

Note that the current conduct of clinical trials is to compare the difference in distributions of the clinical responses observed from patients under a test therapy and a standard (or reference) therapy or a placebo. This concept is referred to as population efficacy (or safety). Suppose that the distribution of a clinical response can be adequately described by a normal probability distribution. Then the population efficacy can be assessed through the comparison of the first two moments of the distributions between the test and the reference therapies. This is because a normal distribution is uniquely determined by its first two moments. The comparison of the first moment of the efficacy endpoints for the two therapies is usually referred to as *average efficacy*, while the comparison of the second moments is called the *variability of efficacy*. To provide a better understanding of average efficacy and variability of efficacy, the comparison in averages and variabilities are illustrated in Figures 4.5.1 through 4.5.3. For example, to compare the reduction in diastolic blood pressure for evalua-

162 RANDOMIZATION AND BLINDING

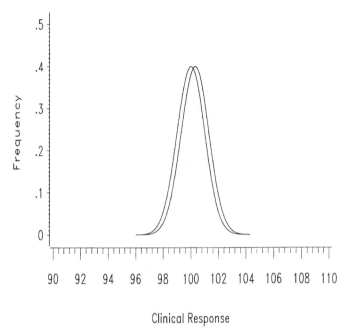

Figure 4.5.1 Population efficacy in averages and variabilities.

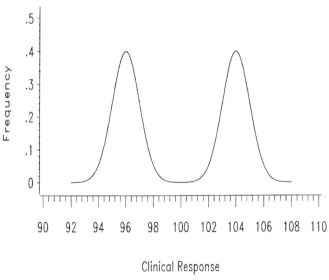

Figure 4.5.2 Population efficacy in averages and variabilities. Unequal averages and equal variabilities.

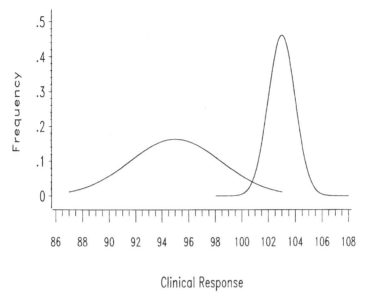

Figure 4.5.3 Population efficacy in averages and variabilities. Superior average efficacy and unequal variabilities.

tion of a new antihypertensive agent against a placebo, Figure 4.5.1 shows that the two distributions are very close in both average and variability, which indicates that there is no difference in average and variability of the reduction of diastolic blood pressure. Therefore the new agent may not be efficacious. On the other hand, Figure 4.5.2 demonstrates that the new agent is more effective in reducing blood pressure. Note that in most clinical trials with continuous primary endpoints, the objectives are often formulated as hypotheses for testing the average efficacy. As a result the population efficacy of the new therapy is often assessed through the average efficacy under the assumption of equal variability of efficacy. This assumption, which should be verified, is often ignored by both clinicians and biostatisticians. As illustrated in Figure 4.5.3, it is not uncommon that the new agent shows a better efficacy than the placebo and yet exhibits a much larger variability. Since the large variability of the new agent may cause a safety concern, it is recommended that the possible causes of the large variability be carefully examined. A large variability may be due to differences in the composition of patients such as biological variation between two populations. This will certainly have an impact on the generalization of the results to other populations. For population efficacy (or safety), we might first generalize the results to similar but slightly different populations and then, in stages, to much different populations. This concept of generalization is illustrated in Figure 4.5.4 as similarity circles. The strength of the generalization is assessed by the distance between any two points within the circle. Note that the distance is a measure of similarity between populations, which is a function of

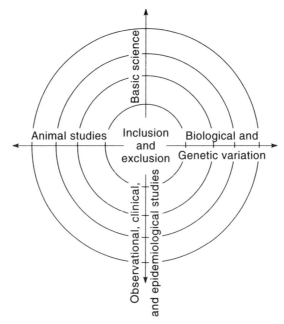

Figure 4.5.4 Generalization of clinical results as similarity circle.

factors such as basic science, animal models, biological variation, and results from other types of studies.

Note that the establishment of population efficacy does not guarantee that the results can be generalized to a patient with his or her own biological and genetic makeup, educational status, and socioeconomic status who is cared by a particular physician at a different geographical location. The reason is that the efficacy is not established within the patient. The concept for the comparison between the two distributions of the primary efficacy (or safety) endpoints obtained from the same patient under repeated administrations of the new agent and the reference is called individual efficacy (or safety). The concept of individual efficacy is not new and has been advocated by many clinical researchers. See, for example, Guyatt et al. (1986) and Sackett (1989). Guyatt et al. (1986) attempt to evaluate individual efficacy of theophylline through a N-of-1 randomized trial concerning a patient with asthma. The N-of-1 randomized trial was conducted based on the following assumptions and procedures: First, the patient and his or her attending physician determined symptoms such as shortness of breath on ordinarily daily activities, nocturnal spasms of dyspnea, and coughing as primary clinical responses for the treatment of theophylline. The patient agreed to record standardized measures of severity of these symptoms. It was also decided that a 10-day treatment would be long enough to evaluate the effectiveness of the treatments. The N-of-1 randomized trial was performed in a double-blind fashion with randomization of the order of treatments. At the end of each pair of treatment periods, the patient and physician met to examine the results (also in a blinded fashion) and decided whether to stop or to continue another pair of treatments. After administration of two pairs of treatments in a blinded and random fashion, the analysis detected a statistically significant difference between the treatments. When the randomization codes were unblinded, it was found that the patient was better on placebo than theophylline.

Repeated administrations of the test and reference therapies within the same patients made the comparison between distributions of the primary clinical endpoints within the same individual possible. If we perform this type of trial over N patients in a similar manner, then a total of N pairs of distributions of the test and reference therapies can be generated. Consequently both population and individual efficacy can be made based on these N pairs of distributions. First, within each individual, the individual average efficacy of the test therapy is assessed as the difference between averages of two distributions. In addition the individual variability of efficacy can be evaluated as the ratio of individual intrapatient variabilities between the distributions obtained under the two treatments from the same patient. As a result individual efficacy can be evaluated by comparing averages and variability of the two distributions obtained from the same patient. Since the individual average efficacy and variability of efficacy are obtained from all N patients, we can perform a statistical test to see whether the individual average efficacy and individual intrapatient variability are homogeneous across these N patients. The concept of homogeneity of individual average efficacy and individual intrapatient variability is referred to as patient-by-treatment interaction for average and variability, respectively. A patient-by-treatment interaction implies that the relative efficacy of the test therapy varies from patient to patient. Therefore, if a patient-by-treatment interaction is found, then the relative efficacy of the test agent must be assessed individually for each patient, that is, the individual efficacy. On the other hand, if the relative efficacy is not heterogeneous, then the information of individual average efficacy and individual intrasubject variability can be combined over N patients to provide a basis for population efficacy. The concepts of population and individual efficacy are motivated from population and individual bioequivalence (e.g., see Chow and Liu, 1992a, 1995b); they are important concepts for evaluation of bioequivalence between a brand-name drug product and its generic copies. However, the concept of individual efficacy (safety) has not been accepted by nor has convinced the clinical/medical community. Guyatt et al. (1986) point out that the limitations of individual efficacy include (1) it cannot be applied to a disease that can be cured in a short period of time, and (2) it cannot be assessed with the hard clinical endpoints such as death or other irreversible condition indicators.

4.6 BLINDING

Although the concept of randomization is to prevent bias from a statistically sound assessment of the study drug, it does not guarantee that there will be no bias caused by subjective judgment in reporting, evaluation, data processing, and statistical analysis due to the knowledge of the identity of the treatments. Since this subjective and judgmental bias is directly or indirectly related to treatment, it can seriously distort statistical inference on the treatment effect. In practice, it is extremely difficult to quantitatively assess such bias and its impact on the assessment of the treatment effect. In clinical trials it is therefore imperative to eliminate such bias by blocking the identity of treatments. Such an approach is referred to as *blinding*. Blinding is defined as an experimental condition in which various groups of the individuals involved with the trial are withheld from the knowledge of the treatments assigned to patients and corresponding relevant information. The blinding is also known as *masking* by some research organizations such as NIH.

For a clinical trial, if the sponsor is to monitor the study and to perform in-house data management and statistical analysis, then the clinical trial typically involves three parties: the patient, the study center or investigator, and the sponsor. The patient is the most important participant in the clinical trial. No clinical trial is possible without the patient's dedicated participation, endurance, corporation, and sacrifice. The study center, in a broad sense, is referred to as those individuals who are either directly in contact with the patient or perform various evaluations for the patient. Among these individuals is the investigator, who usually is the patient's primary care physician and members of the patient's care team such as the pathologist for histopathological evaluation, the radiologist for imaging assessment, a staff nurse who may also serve as the coordinator for the study center, the pharmacist who dispenses the study medicines, and other health care personnel at the study center including the laboratory staff and the contracted houses that perform the various laboratory evaluations for the blood or urine samples collected from the patient. Note that in clinical trials sometimes the term *investigator* may be used exchangeably with the term *study center* or *study site* in a broader sense. For a clinical trial two functional teams are usually formed by the sponsor. The first team is the clinical/medical team which consists of the project clinician (e.g., physician monitor) and clinical monitor such as the CRA. The project clinician has an overall responsibility for the success of the trial, while the responsibility of the clinical monitor is not only to monitor the conduct of the trial but also to ensure that the investigator adheres to the study protocol. The second team is the biostatistics and data management team which includes the project statistician, programmer, and data coordinator. The project statistician oversees the activities of data management, programming, and statistical aspects of the trial, while the programmer is responsible for programming support for data management, analysis, and report. The data coordinator will coordinate the activity of database setup, data entry, data verification, data query generation/resolution, database cleanup and finalization to ensure the quality of the final database.

Basically blinding in clinical trials can be classified into four types: open label, single blind, double blind, and triple blind. An open-label study is a clinical trial in which no blinding is employed. That is, both the investigator and the patient have an idea about which treatment the patient receives. Since patients may psychologically react in favor of the treatments they receive if they are aware of which treatments they receive, a serious bias will occur. For example, for the development of topical cream for the indication of some skin disorder, after revelation of the dose, two investigators were asked to give their global evaluation of a patient based on a four-point scale. Despite the fact that the procedures for global evaluation are clearly stated in the protocol, the two investigators gave a rather different evaluation for the patient simply because one of them did not believe that the drug really works at the dose for the patient and the other one is an advocate for that type of the compound for the treatment of the skin conditions. On the other hand, objective endpoints such as systolic and diastolic blood pressures or total cholesterol levels can be recorded differently if the investigators are aware of treatment assignment. Although some *hard endpoint* such as survival (mortality) or incidence of myocardial infarction are more objective than other clinical endpoints, these can still be subjective. For example, the determination of the cause of death or the diagnosis of infarction may be biased if patient's treatment is known. Therefore open-label trials are generally not recommended for comparative clinical trials. In current practice, open-label trials are not accepted as adequate well-controlled clinical trials for providing substantial evidence for approval by most regulatory agencies such as the FDA, the European Community (EC), and Ministry of Health and Welfare of Japan. However, under certain circumstances open-label trials are necessarily conducted. Spilker (1991) provides a list of situations and circumstances in which open-label trials may be conducted. As indicated in Chapter 1, in order to provide some potentially promising medications to the patients with severely debilitating or life-threatening diseases, clinical trials conducted under compassionate plea protocols or treatment IND may be open labeled. In general, open-label trials are less biased if the clinical endpoints are objective outcomes such as overall survival or the incidence of coma.

Ethical consideration is always an important factor, or perhaps the only factor that is used to determine whether a trial should be conducted in an open-label fashion. For example, phase I dose-escalating studies for determination of the maximum tolerable dose of drugs in treating terminally ill cancer patients are usually open labeled. Clinical trials for evaluation of the effectiveness and safety of a new surgical procedure are usually conducted in an open-label fashion because it clearly unethical to conduct a double-blind trial with a concurrent control group in which patients are incised under a general anesthesia to simulate the surgical procedure. Note that premarketing and postmarketing surveillance studies are usually open labeled. The purpose of premarketing surveillance studies is to collect the data of efficacy and safety with respect to the duration of exposure of a broader patient population to the test drug, while the objective of postmarketing surveillance studies is to monitor the safety and tolerability of the drug product.

By definition, a single-blind study is the one in which either the patient or investigator is blind to the assignment of the patient. In practice, a single-blind trial is referred to as a trial in which only the patient is unaware of his or her treatment assignment. As compared with open-label trials, single-blind studies offer a certain degree of control and the assurance of the validity of clinical trials. However, the investigator may bias his or her clinical evaluation by knowing which treatment the patient receives. Spilker (1991) indicates that results of single-blind trials are equivalent to those from open-label trials. Therefore, when a single-blind trial is planned, it is prudent to ask why this trial cannot be conducted in a double-blind fashion.

A double-blind trial is a trial in which neither the patients nor the investigator (study center) are aware of patient's treatment assignment. Note that the *investigator* could mean all of the health care personnel, which include the study center, contract laboratories, and other consulting experts for evaluation of effectiveness and safety of patients in a broader sense. In addition to the patients and the investigator, if all members of clinical project team of the sponsor associated with the study are also blinded, then the clinical trial is said to be triple-blinded. These members include the project clinician, the CRA, the statistician, the programmer, and the data coordinator. In addition to the patient's treatment assignment, the blindness also applies to concealment of the overall results of the trial. In practice, although the project clinician, the CRA, the statistician, the programmer, and the data coordinator usually have access to the individual patient's data, they are generally not aware of the treatment assignment for each patient. In addition the overall treatment results, if any (e.g., interim analyses), will not be made availale to the patient, the investigator, the project clinician, the CRA, the statistician, the programmer, and the data coordinator until a decision is made at an appropriate time. A triple-blind study with respect to blindness can provide the highest degree for the validity of a controlled clinical trial. Hence it provides the most conclusive unbiased evidence for the evaluation of the effectiveness and safety of the therapeutic intervention under investigation.

To ensure the success of a triple-blind study, it is recommended that the following be considered:

1. A carefully chosen study design with an appropriate randomization method.
2. A conscientiously selected concurrent control according to study objectives.
3. Adequate conduct of the trial with no apparent protocoal violations.
4. Patient compliance.
5. A sufficient power.
6. Appropriate statistical methods for data analysis.

However, the most important factor for the success of a triple-blind study is to maintain the blindness throughout the entire course of the trial. To protect the integrity of blindness, it is helpful to provide in-house training/education to all personnel related to the clinical trial, including those in the analytical laboratory or in pharmaceutical science research and development (R&D). For example, personnel in the department of analytical laboratory are responsible for the assay of blood samples for active ingredients or metabolites for patients in clinical trials, while the pharmaceutical science R&D develops the matching placebos for the clinical trial. Therefore the personnel at analytical laboratory and pharmaceutical science R&D should have a certain understanding of the concept of blindness and its implication for the integrity of clinical trials.

For a clinical trial comparing a new therapeutic agent with a concurrent control, the departments of pharmaceutical science R&D and drug supply/packaging are usually required to manufacture an identically matched control with the same dosage form. A matched placebo should be identical to the active agent in all aspects such as size, color, coating, taste, texture, shape, and odor except that it contains no active ingredient. The study drugs are then packed in an identical container such as a blister pack or a drug kit affixed with a three-part double-blind tear-off label with the study and patient number. Manufacturing of a perfectly matched control requires certain pharmaceutical techniques and packaging skills provided by both departments. Sometimes, however, a perfectly matched control may not be available due to technical difficulties for some doses. In this situation the method of administration should be modified to maintain the blindness. For example, a phase II clinical trial is to be conducted with daily dose of 100 mg, 300 mg, 600 mg, and a placebo to evaluate the dose-response relationship of a drug. To keep the blindness throughout the study, it is necessary to manufacture placebos to match the drug at different doses. Suppose that the department of pharmaceutical science R&D has difficulties in making matched placebos for tablets of 300 mg and 600 mg. However, the manufacturing of matching placebo for the smallest tablets of 100 mg is still possible. In addition, suppose that patients have difficulties in swallowing the largest tablets of 600 mg. In this case we can modify the method of administration based on 100 mg tables of the active drug and matching placebos as follows to maintain the blindness.

The first arm of 600 mg: Six 100 mg tablets of the new agent.

The second arm of 300 mg: Three 100 mg tablets of the new agent and three 100 mg matched placebo.

The third arm of 100 mg: One 100 mg tablet of the new agent and five 100 mg matched placebo.

Placebo control arm: Six 100 mg matched placebo.

Note that patients should be instructed to take all six tablets at one time (e.g., in the morning) so that the blindness will not be broken due to different time of administration. The above method is known as *multiple-placebo* or *double dummy*. This method is useful when treatments involve two different active agents or two different routes of administration. For example, a clinical trial is conducted to evaluate a once daily sustained-release formulation of an antihypertensive agent with its standard three-times-a-day (t.i.d.) immediate release formulation. The matched placebos can be made for each of two formulations. A bottle of the active sustained release formulation (e.g., bottle S) and another bottle of the placebo tablets of the immediate release formulation (e.g., bottle I) are dispensed to the patients assigned to the sustained-release formulation. The patients assigned to the group of immediate-release formulation receive a bottle of placebo tablets of the sustained-release formulation and another bottle of the active immediate-release formulation. Each patient is instructed to take a tablet from bottle S at 8:00 A.M. in the morning and a tablet from bottle I at 8:00 A.M, 2:00 P.M., and 10:00 P.M. In this case the blindness is preserved without matching tablets identically for all formulations and placebos.

Another example is ISIS-2 (1988) in which the treatments are one-hour intravenous infusions of 1.5 μ of streptokinase and one-month of 160 mg/day enteric coated aspirin. Therefore the corresponding placebo infusion and tablet were manufactured to match the active treatments as described previously in Chapter 2. However, blindness for ISIS-2 is possible because the matched placebo infusion has the same one-hour IV infusion at the same rate. On the other hand, the arm of accelerated rt-PA in the GUSTO trial (1993) had a bolus dose of 15 mg, 0.75 mg per kg of body weight, over a 30-minute period, not to exceed 50 mg; and 0.5 mg per kg, up to 35 mg, over the next 60 minutes. The active control arms used the same one-hour infusion of 1.5 μ of streptokinase as ISIS-2. Therefore, because the dose of the accelerated rt-PA had to be adjusted for body weight twice during the infusion, IV heparin had to be titrated according to the activated partial-thromboplastin time and its length of infusion was also different from other arms receiving streptokinase. As a result the GUSTO study was an open-label study. Although primary efficacy outcome is the mortality from stroke and bleeding complication as the primary safety endpoint. However, they are subjective to possible bias if the treatments are known to investigators, in particular, when classification of stroke and bleeding requires clinical judgment for some borderline cases. Due to the large size of the GUSTO study, the bias could be accumulated rapidly and become serious just from some subtle, consciously, subconsciously, or unconsciously error in clinical judgment made by an investigator. The GUSTO study, however, failed to address the bias issue due to the open label. As a result there were tremendous debates over the fact that the GUSTO was an open-label study (Rapaport, 1993; Sleight, 1993; Rider et al., 1993). In their response to the rebuttal article by the investigators of the GUSTO trial (Rider et al., 1994; Lee, 1994), Rider et al. (1994) state the essence of randomization and blindness in clinical trials: "randomization of patients is done to try to ensure that no major differences exist in baseline characteristics between treatment groups before treatments are administered, double-blinding is done to ensure that no differential effects occur after treatments are given." It is sad to see that the breach of blindness, the omitting of a rather routine and operationally and economically feasible insertion of an extra intravenous line, casts a serious shadow over the scientific validity of this originally spotless trial, and introduces an inadvertent and impossibly assessed bias.

For multiple placebos the so-called method of the *multiple-evaluator* is useful to preserve the blindness. For example, suppose that a clinical trial is conducted in three doses to assess the dose-response relationship of a contrast-enhanced agent in conjunction with magnetic resonance imaging for the diagnosis of malignant liver tumors in patients with known focal liver lesions. Since the contrast agent is administered as an IV injection by reconstruction from the vial of an active agent and the vial of a saline solution according to the body weight of each patient, the clinician who prepares and administers the contrast medium will know the dose. If the clinician is also responsible for the evaluation of the results of pre- and postcontrast MRI, bias will occur during the evaluation of films and safety data due to prior knowledge of the doses. In this case the multiple-evaluator method is helpful. At each study center, one clinician will prepare the injection according to the randomization codes in total privacy without divulging the dosing information to anyone in the study center. The clinician will then administer the contrast medium without showing the syringe to everyone. The other clinicians at the study center will evaluate the films in a totally blinded manner. The multiple-evaluator method is also useful in physical therapy. A clinical trial was conducted to evaluate the transcutaneous electrical nerve stimulation (TENS) for patients with chronic low back pain (Deyo et al., 1990). This trial employed a two-group parallel design with a real TENS and a sham TENS group. Although it is known to be very difficult to implement, in order to maintain blindness over the entire course of the study, the therapist who is responsible for instructing patients and applying TENS, asked patients not to discuss their therapy with the clinician who performed the evaluations. The clinician who performed the evaluation at baseline is different than the one at the follow-up visit. In addition the frequency of visits and duration of treatment were identical for the two groups, as were all written and verbal instructions and effort to identify ideal electrode placement.

In practice, even with the best intentions for preserving blindness throughout a study, blindness can sometimes be breached for such reasons as a distinct adverse event or the taste of the active treatment. One method to determine whether the blindness is seriously violated is to ask both patients and investigators to guess the patient's treatment assignment during the study or at the conclusion of the trial prior to unblinding. Once the guesses by patients and investigators are recorded on the case report forms and entered into the database, the degree of unblinding and its impact on introducing bias in the evaluation of treatment effect can be assessed. In what follows, some examples that may be of interest for practical use are adopted from the literature.

For the first example, a one-year double-blind placebo-controlled study was conducted by the NIH to evaluate and distinguish between the prophylactic and therapeutic effects of ascorbic acid for the common cold (Karlowski, 1975). 311 employees of NIH were randomly assigned to receive the active agent or the matched placebo based on the method of complete randomization. One hundred

Table 4.6.1 Results of Patient's Guess on Treatment for the Prophylactic Use

Patient's Guess	Actual Assignment	
	Ascorbic Acid	Placebo
Ascorbic acid	40	11
Placebo	12	39
Do not know	49	39
Total	101	89

Source: Karlowski et al. (1975).

and ninety of them completed the study. In this study, since there was no time to design, test, and manufacture a perfectly matched placebo for ascorbic acid due to the seasonal constraint, at an early stage of the study the researchers discovered that some subjects had tasted the contents of their capsules and professed to know which treatment they were taking. At the completion of the study, in order to assess the bias, a questionnaire was distributed to everyone enrolled in the study so that they could guess which treatment they had been taking. Table 4.6.1 presents the results from the 190 completed subjects for the prophylactic use. The number of correct guesses was 79 and the number of misses 23. Therefore the expected bias factor is estimated to be 28 (a half of the difference between 79 and 23). Hence considerable selection bias occurred in this study. Note that the association between the severity and the duration of symptoms and knowledge of the medication taken were also established by the researchers of this project.

The second example concerning a randomized double-blind placebo-controlled trial for evaluation of the effectiveness of an appetite suppressant in weight loss in 57 obese women (Brownell and Stunkard, 1982). At the 8th and 16th weeks of treatment, patients and physicians were asked to guess which treatment the patients had been taking. Table 4.6.2 gives the patient's guesses

Table 4.6.2 Results of Patient's Guesses on Treatment with Mean Weight Loss (kg)

Patient's Guess	Actual Assignment			
	Active Drug		Placebo	
	N	Weight Loss	N	Weight Loss
Active drug	19	9.6	3	2.6
Placebo	3	3.9	16	6.1
Do not know	2	12.2	6	5.8
Total	24	9.1	25	5.6

Source: Brownell and Stunkard (1982).

and their corresponding mean weight losses. As shown in Table 4.6.2, 71% identified their treatments, so the estimate of the expected bias factor is 14.5. In addition the weight loss for the patients with correct guesses of their treatments is statistically significantly larger than for the patients with misses (8.0 kg versus 3.3 kg with p-value < 0.05). The subgroup analysis showed a similar trend in each treatment group (9.6 kg vs. 3.9 for the active drug and 6.1 kg vs. 2.6 kg for the placebo). The treatment effect was clearly contaminated and confounded by the correct identification of the treatments by the patients. As reported by Brownell and Stunkard (1982), the IRB of this study demands detailed explanation of all possible adverse events of the active drug on the informed consent form. Therefore the patients reported that it is relatively easy to detect their actual treatment assignment. This situation is commonly seen in clinical trials for psychiatric medications. For example, clinicians are usually blinded to perform clinical evaluation for antidepressant agents. However, as pointed out by Greenberg and Fisher (1994), most antidepressant drugs can induce obvious body sensation which the placebo cannot do. Therefore the clinicians can correctly identify the patient's treatment assignment by the difference in side effects and body sensations among patients. Hence they stated "that the two conditions (active drug and placebo) are experimentally equivalent is an illusion." The use of the *active placebo* cannot completely resolve this problem because of its inactive physiological effect. In addition, since almost every psychiatric agent is virtually associated with a distinct side effect profile, blindness for the trials of psychiatric agents cannot be preserved, even if they are conducted in a triple-blind manner. Therefore Greenburg and Fisher (1994) conclude that all past studies of antidepressent effectiveness are open to question because the bias induced by the correct identification of patient's treatment is due to distinct adverse events of the active drug. In order to reduce such bias, they suggest that an active controlled trial be employed in which the test and standard marketed drugs are compared together with a placebo. They conjecture that the bias will be reduced because of different side effect profiles of the two different active drugs and the less interest in the standard treatment by the investigator.

In practice, similar issues for the maintenance of blindness are commonly seen in other therapeutic areas. For example, beta-blocker agents (e.g., propranolol) have specific pharmacologic effects such as lowering blood pressure and the heart rate and distinct adverse effects such as fatigue, nightmares, and depression. Since blood pressure and heart rate are vital signs routinely evaluated at every visit in clinical trials, if a drug such as propranolol is known to lower blood pressure and the heart rate, then preservation of blindness is a huge challenge and seems almost impossible. The Beta-Blocker Heart Attack Trial (BHAT) is a landmark, multicenter, double-blind, randomized, placebo-controlled trial designed to test the effectiveness of beta-blocker in reducing mortality during a two- to four-year period in postmyocardial infarction patients (Beta-Blocker Heart Attack Trial Research Group, 1982, 1983). At the conclusion of the trial, patients, investigators, and clinic coordinators were asked to

Table 4.6.3 Proportions (%) of Correct Guesses for Beta-Blocker Heart Attack Trial

	Propranolol	Placebo	Estimate of the Expected Bias Factor
Patient	79.9	42.8	380 ($N = 3230$)
Investigator	69.6	68.6	568 ($N = 3398$)
Clinic coordinator	67.1	70.6	669 ($N = 3552$)

guess the patient's treatment assignment. Table 4.6.3 provides the proportion of correct guesses and estimates of the expected bias factor. Apparently blindness was not totally maintained even for this landmark study with a major influence on management of care for patients who suffer myocardial infarction. Morgan (1985) suggested that to quantify possible bias, the researchers for BHAT should administer the questionnaire of guesses of patient's treatment three months into the trial rather than at the end.

4.7 DISCUSSION

As indicated earlier, randomization is integral to the success of clinical trials that address scientific and/or medical questions. However, it should be noted that in many clinical research situations, randomization may not be feasible. For example, nonrandomized observational or case-controlled studies are often conducted to study the relationship between smoking and cancer. As a result, in the report entitled *Smoking and Health* by the U.S. Surgeon General issued in 1964, seven key nonrandomized observational studies were cited as the evidence for the relationship between smoking and cancer. Note that if the randomization is not used for some medical considerations, the FDA requires that statistical justification be provided with respect to how systematic selection bias can be avoid.

It should be noted that in practice, for most clinical trials patients are enrolled into study sites in a nonrandom fashion. The selection of study sites is also a nonrandom process. Consequently the validity of statistical inference on the targeted patient population is seriously in doubt. It is therefore recommended that appropriate statistical methods be derived based on the method of the selected randomization model.

Further, although triple blinding is reserved for large cooperative, multicenter studies monitored by a committee, it has nevertheless been applied to company monitors to ensure that they remain unaware of the treatment allocation. In general practice, double blinding is the standard, since it provides the greatest probability for reducing bias.

Blocking is usually employed to ensure that the number of patients in each treatment group will be similar at certain points. For this purpose, small block sizes such as 2, 4, 6, and 8 are usually chosen. Within each block, patients

DISCUSSION

are randomly allocated to receive either the treatment or a control. In some situations, however, deliberate unequal allocation of patients between treatment groups may be desirable. For example, it may be of interest to allocate patients to the treatment and the control in a ratio of 2 to 1. This is a consideration in situations where (1) the patient population is small, (2) previous experience with the study medicine is limited, and (3) the response profile of the competitor is well known.

CHAPTER 5

Designs for Clinical Trials

5.1 INTRODUCTION

As was discussed in Chapter 3, the first step in selecting an appropriate statistical design is to determine the objective(s) of the intended clinical trials. The objective(s) of a clinical trial is usually to answer one or more scientific or medical questions related to the therapeutical intervention under study. Once the study objective(s) have been carefully defined, an appropriate (or optimal) design for the intended clinical can be chosen. Since a wrong choice of design may result in a worthless study, a good statistical design is regarded as an essential prerequisite of clinical trials. Spilker (1991) indicates that choosing the most appropriate design for a clinical trial is similar to choosing ready-made clothes. Temple (1982) indicates that the selection of the most appropriate design or the optimal design depends on the questions asked. The questions that must be asked before choosing an appropriate design include the study objective(s), the nature of the study drug, the disease status/condition under investigation, and other considerations as described in previous two chapters. Therefore the FDA suggests that a statement of the specific objectives of the study be provided. To clarify the study objective(s), the following questions are helpful.

1. What aspects are being studied?
2. Is it important to investigate other issues that may have an impact on the study drug?
3. Which control(s) might be used?

Once the objective(s) of the study is clearly stated, it is important to determine the aspects that will be studied. These aspects include the dosage form, dose, and the intended indication. For an indication under investigation, an appropriate dosage form is necessary chosen for the targeted patient population so that the drug can be delivered to the site of action efficiently for optimal therapeutic effect. In addition the selected dose may have an impact on the assessment of the effectiveness and safety of the study drug. For example, a low dose may show a better safety profile, and yet it may not produce clinically meaningful efficacy of the study drug. On the other hand, a high dose may cause a serious side effect. In some clinical trials, the dose may be required to be titrated during the study in order to reach the optimal therapeutic effect. For dose titration, the titration procedures must be clearly described in the protocol. The commonly employed titration procedures include forced titration and titration based on clinical outcome whereby the dose is titrated upward at intervals until intolerance or some specific endpoint is achieved. The issues of possible drug-to-drug interactions with food and/or other concomitant medications, the impact of patient compliance, and pharmacoeconomic outcomes such as quality of life associated with the efficacy and safety of the study drug should also be considered in choosing an appropriate statistical design. In addition, as was indicated in Chapter 3, it is also important to determine what control(s) will be used for comparative clinical trials. Different controls may serve different purposes of a clinical trial.

Selecting an appropriate statistical design is critical in clinical development during the process of drug development. In practice, when a new test agent reaches the stage for clinical development, its pharmacological/pharmacokinetic properties and the effectiveness and safety may have been studied through *in vitro* laboratory testing and *in vivo* animal studies. At this time point, however, the safety and effectiveness in humans are not known, and the test agent must be rigorously and scientifically evaluated through clinical trials within the confinement of regulations. As indicated in Chapter 1, the purposes of phase I and early phase II studies are not only to characterize the safety profile but also to determine the therapeutic range of the test agent. Since the test agent is never tested in humans, it is a challenge to acquire information regarding early safety and efficacy of the test agent. The information is extremely helpful for planning of subsequent trials. To capture as much needed information as possible, the utilization of an efficient statistical design is critical.

In recent years there has been tremendous discussion on whether the choice of study design should be based solely on medical consideration. Another interesting question raised is whether to include marketing, regulatory, and/or statistical perspectives as well. Ideally an optimal design will account for considerations from different perspectives. In practice, however, such a design may not exist. It should be noted that considerations from different perspectives always mean limitations to the choice of design. Therefore Temple (1982) points out that a study must be sufficient to its task, and design limitations should be understood before proceeding, first to see whether a better design can be found and to understand the limits on interpretation imposed by a less than optimal design, and second, so that, if necessary, the limits can be discussed with the regulatory agency and potential problems anticipated.

When planning a clinical trial, it is suggested that the relative merits and disadvantages of candidate statistical designs be compared before an appropriate design is chosen for the clinical trial. It is important to evaluate the suitability of the chosen design for addressing scientific/medical questions and/or claims. For example, if we are to choose between a crossover design and a parallel design for a clinical trial, we must first understand the nature of these two designs. For a parallel design, each patient receives one and only one treatment in random fashion, whereas for a crossover design each patient receives more than one treatment at different dosing periods. If a clinical trial is intending to investigate the residual effect that may be carried over from one treatment to another, a crossover design could be employed. Note that the Federal Register (Vol. 42, No. 5, Sec. 320.26(b) and 320.27(b), 1977) indicate that a bioequivalence trial (single dose or multiple dose) should be crossover in design, unless a parallel design or another design is more appropriate for some valid scientific reasons. On the other hand, if a clinical trial is intended to demonstrate the effectiveness and safety of a study medicine, a parallel design is more appropriate.

In the next section, we briefly introduce parallel designs including parallel-group designs and matched pairs parallel designs. Several different types of crossover designs are discussed in Section 5.3. In Section 5.4 we cover titration designs which include the up-and-down, single-stage, multiple-stage, and titration design. The concept of enriched designs is given in Section 5.5. In the last section we provide a discussion regarding the selection of an appropriate design.

5.2 PARALLEL DESIGNS

A parallel design is a complete randomized design in which each patient receives one and only one treatment in a random fashion. Basically there are two types of parallel design for comparative clinical trials, namely, group comparison (or parallel-group) designs and matched pairs parallel designs. The simplest group comparison parallel design is the two-group parallel design which compares two treatments (e.g., a treatment group vs. a control group). Each treatment group usually, but not necessarily, contains the same number of patients. An example of a three-group parallel design with a test treatment and two controls (e.g., an active control A and a placebo control B) is illustrated in Figure 5.2.1. The advantages of a parallel design are as follows:

1. It is simple.
2. It is universally accepted.
3. It is applicable to acute conditions (e.g., infection or myocardial infarction).

In addition, for ethical consideration with the control (e.g., the placebo), we can allocate patients unequally to treatment groups (in a random fashion) to allow more patients to receive the treatment (e.g., in a 2 to 1 or 3 to 1 ratio). A parallel design is probably the most commonly used design in phases II and III of

PARALLEL DESIGNS 179

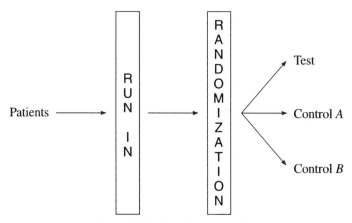

Figure 5.2.1 Parallel-group design.

clinical trials. However, it usually requires more patients than other comparative designs.

The matched pairs parallel design is a design in which each patient is *matched* with another of similar prognostic characteristics (e.g., obesity) for the disease under investigation. One patient in each pair is assigned the treatment, and the other receives the control. As compared to parallel designs, matched pairs parallel designs can reduce variability from treatment comparison. In addition a matched pairs parallel design requires a smaller patient population. Therefore it is considered a more suitable design for progressive diseases such as cancer. However, matched pairs designs suffer the disadvantages such that (1) the prognostic characteristics are not easily defined and (2) patient recruitment is usually slow. In practice, the selection bias for matched pairs designs is usually a concern in patient recruitment, which often limits its applications in clinical trials. Note that a matched pairs design is in fact an extreme case of stratification which is often considered to achieve balance in covariates or prognostic factors. When the number of covariates is large, the matched pairs design is difficult to implement. Hence the matched pairs design is not of practical interest in this case. Although at the planning stage it is almost impossible to identify all of the covariates that may have an impact on the disease, an unbiased estimate of the treatment effect can still be obtained by adjusting these covariates regardless of whether they are used for stratification or matching in order to achieve the balance in covariates.

In clinical trials, for a given clinical endpoint, basically there are two kinds of variability associated with the response. These two kinds of variability are known as the interpatient and intrapatient variabilities. Statistically the smaller these variabilities are, the more accurate and reliable the clinical results will be. For a parallel design, however, these variabilities cannot be identified because each patient recieves the same treatment during the entire course of the study. In other words, the observed variability for any comparisons between groups contains both interpatient and intrapatient variabilities that cannot be separated and estimated due to the nature of the parallel design. As a result a parallel design does not provide independent estimates of the interpatient and intrapatient variabilities. In practice, a parallel-group design is an appropriate design for comparative clinical trials if the interpatient variability is relatively small compared to the intrapatient variability. This is because a valid and efficient comparison between treatment is often assessed based on the intrapatient variability. Therefore, if the interpatient variability is relatively small compared to the intrapatient variability, the observed variability will be close to the intrapatient variability. In this case the parallel design will provide a more accurate and precise assessment of the treatment difference. Other considerations for the use of parallel designs include patient characteristics (e.g., acute or chronic and very ill or life threatening) and the nature of study medicine (e.g., potential toxicity and long elimination half-life). In some cases financial consideration may be a key factor for selecting parallel designs.

Run-in Periods

Before patients enter a clinical trial, a run-in (or lead-in) period of placebo, no active treatment, dietary control, or active maintenance therapy (e.g., diuretic and/or digoxin in heart failure studies) is usually employed prior to randomization. The inclusion of a run-in period prior to the active treatment has the following advantages:

1. It acts as a washout period to remove effects of previous therapy.
2. It can be used to obtain baseline data and to evaluate if patient fulfills study entry criteria.
3. It can be used as a training period for patients, investigators, and their staff.
4. It helps in identifying placebo responders.
5. It provides useful information regarding patient compliance.
6. It can be used to estimate and compare the magnitude of possible placebo effects between groups.

In clinical trials it is desired to have a washout period prior to an active treatment period to wear off effects of previous therapy for an unbiased and valid assessment of the study medicine. A run-in period, however, may not be suitable for patients whose conditions are acute requiring immediate treatment. It is acceptable if patients can remain without active therapy for a short period of time. In many clinical trials it is not uncommon to observe the placebo effect for many drug products. For example, for antidepressant agents, an intensive care period may significantly improve the patients' depression without any treatment. At the end of active treatment period, it is important to determine whether the observed significant effect is due to the placebo or treatment. To

PARALLEL DESIGNS 181

eliminate the possible placebo effect, it is suggested that a run-in period be included to establish patient comparability between treatment groups at baseline, and this helps to remove placebo effect from comparison at the endpoint evaluation. In clinical trials, patients' cooperation and/or their compliance to study medicine is always a concern. A run-in period can be used as a training period for patients, investigators, and their staff. For example, if the trial requires patients to complete diary cards, a run-in period provides a training period for the patients to be familiar with the diary cards. In addition it may help in identifying uncooperative patients at an early stage and provide the necessary counsel. This information is useful in improving a patient's compliance when the study moves to the active treatment period.

Note that a run-in period is usually employed based on a single-blind fashion. In other words, the participated patients are not aware of receiving a placebo. Although the inclusion of a run-in period in clinical trials has many advantages, it increases the length of a study; consequently it often requires extra study visits. This has a direct impact on the increase of cost and potentially a decrease in enthusiasm by patients and investigators.

Examples of Parallel Design in Clinical Trials

During clinical development of a drug product, parallel designs are often considered to evaluate the efficacy and safety of a monotherapy or combination therapy with other agents of the drug product. In addition a parallel design may be used to study the dose response of a drug product. For example, consider the clinical development of *Glucophage* (metformin hydrochloride). Glucophage is an oral agent for the treatment of type II noninsulin-dependent diabetes mellitus (NIDDM). Although Glucophage has been on the European market for more than 20 years, it was not available for the U.S. market until it was approved by the FDA in late December 1994. Over the past few years a number of clinical trials were conducted to further investigate the clinical pharmacology and other uses of the drug product.

To illustrate the application of parallel designs in clinical trials, consider the clinical development of Glucophage. Table 5.2.1 lists three studies of Glucophage regarding evaluation by monotherapy, combination therapy with insulin, and dose response. For the first study (Dornan et al., 1991), the objective was to test the efficacy and tolerability of Glucophage. This study was an eight-month double-blind placebo-controlled trial of Glucophage monotherapy in 60 obese patients with NIDDM. This study had a typical parallel design with a run-in period. After a dietetic review and a one-month run-in period, patients were stratified according to the level of hemoglobin A_{1C} (H_bA_{1C}) concentration and randomized to receive either Glucophage or an identical dose of placebo. The starting dose was one tablet (500 mg) daily increased at weekly intervals to three tablets daily after one month. Thereafter the dose was increased by one tablet daily at weekly intervals to a maximum of two tablets three times daily, aiming for lowering the level of fasting blood glucose less than 7 mM. Patients

Table 5.2.1 Examples of Parallel-Group Design

Study	Purpose	Sample Size	Parallel Groups	Duration (Run-in + Active)	Primary End Point
Dornan et al. (1991)	Monotherapy	60	2	1 mo + 8 mo	H_bA_{1C}, PG
Giugliano et al. (1993)	Combination therapy	50	2	4 wk + 6 mo	H_bA_{1C}, FPG
Bristol Myers Squibb (1994)	Dose response	360	6	3 wk + 11 wk	H_bA_{1C}, FPG

Note: PG = plasma glucose; FPG = fasting plasma glucose; H_bA_{1C} = hemoglobin A_{1C}.

were fasted at the beginning and end of the run-in period, and after 1, 3, 5, and 8 months of treatment they were weighted and their blood pressure was measured. In addition, blood was taken for fasting glucose, total cholesterol, triglycerides, H_bA_{1C}, and serum insulin. The results indicated that Glucophage reduced H_bA_{1C} levels from 11.7% to 10.3%, whereas the placebo treatment resulted in a rise from 11.8% to 13.3%. The mean percent reduction in H_bA_{1C} of Glucophage is 23% lower than the placebo without weight gain. In addition the final mean fasting blood glucose level was 5.1 mM (92 mg/dL) lower on Glucophage than on the placebo. The fasting glucose level fell from 13.5 (243 mg/dL) to 10.2 mM (184 mg/dL) (about 24%) on Glucophage and rose from 12.7 (229 mg/dL) to 15.3 mM (275 mg/dL) (about 17%) on the diet plus placebo. No changes or differences between groups were observed in body weight, blood pressure, C peptide, serum insulin, or triglycerides. As a result Dornan et al. (1991) concluded that Glucophage monotherapy is an effective and well-tolerated first-line treatment for obese patients with NIDDM. They also indicated that the use of Glucophage should not be restricted to very obese patients because Glucophage lowers hemoglobin A_{1C} and achieves approximately equivalent improvements in glycemic control in both mildly and moderately to severely obese patients.

Another application for the use of a parallel-group design would be the evaluation of combination therapy of the current insulin regimen with Glucophage. Giugliano et al. (1993) studied the efficacy and safety of Glucophage in the treatment of obese NIDDM patients poorly controlled by insulin after secondary failure to respond to sulfonylurea. The study is a typical parallel-group design consisting of a four-week run-in single-blind phase and a six-month double-blind treatment phase. During the placebo run-in phase, patients were given the current insulin regimen and asked to maintain their regular diet. After a six-month active treatment, Glucophage was shown to have significantly improved the glycemic and lipid control. The results indicated that after four months, the glucose level declined by 31% (4.1 mM or 73.8 mg/dL) from baseline, hemoglobin A_{1C} levels by 1.7%, and fasting insulin levels by 26%. In addition the necessary insulin dose was also reduced by more than 20% (from 90 to 71 U/d). Furthermore in the Glucophage group there were significant changes from both the baseline and placebo in levels of total cholesterol (−0.21 mM), triglycerides (−0.31 mM), and high-density lipoprotein cholesterol (+0.13 mM), and blood pressure was reduced an average of 8.8 and 4.8 mmHg versus the baseline and placebo, respectively. Therefore Giugliano et al. (1993) concluded that combination Glucophage therapy represents a safe and efficacious strategy for improving glycemic regulation and coronary artery disease risk status in patients with NIDDM which was inadequately controlled by insulin alone.

Recently, to fulfill the FDA's requirement, a study was conducted by Bristol-Myers Squibb to study the dose response of various dose levels of Glucophage compared to a placebo in patients with NIDDM. The study was a randomized double-blind placebo-controlled parallel-group study that consisted of two phases. At the end of single-blind placebo run-in period, qualified patients were randomized to one of the six double-blind treatment groups (i.e., placebo, Glucophage at 500 mg, 1000 mg, 1500 mg, 2000 mg, and 2500 mg per day) for 11 weeks. Patients assigned to the dose levels of Glucophage 500 mg/d or Glucophage 1000 mg/d began the active treatment phase at this dose level and continued on it throughout the study. Patients assigned to dose levels of Glucophage 1500 mg/d, 2000 mg/d, or 2500 mg/d underwent a forced titration during the initial three weeks of study to minimize the possibility of gastrointestinal side effects. All patients were maintained for a minimum of eight weeks on their final assigned dose level. The results suggest that there is a dose response showing Glucophage to be effective at all randomized dose levels. The dose response increased up to the 2000 mg dose but then decreased as the dose was increased from 2000 mg to 2500 mg. These results are consistent with those for FPG and H_bA_{1C} at treatment weeks 7, 11, and the end of the trial. Given the dose levels considered in this study, it is concluded that 500 mg is the minimum effective dosage ($p = 0.03$) and 2000 mg is the maximum effective dose level ($p = 0.001$) compared with the placebo.

5.3 CROSSOVER DESIGNS

A crossover design is a modified randomized block design in which each block receives more than one treatment at different dosing periods. A block can be a patient or a group of patients. Patients in each block receive different sequences of treatments. A crossover design is called a complete crossover design if each sequence contains all treatments under investigation. For a crossover design it is not necessary that the number of treatments in each sequence be greater than or equal to the number of treatments to be compared. We will refer to a crossover design as a $p \times q$ crossover design if there are p sequences of treatments administered at q different time periods. Basically a crossover design has the following advantages: (1) It allows a within-patient comparison between treatments, since each patient serves as his or her own control. (2) It removes the interpatient variability from the comparison between treatments. (3) With a proper randomization of patients to the treatment sequences, it provides the best unbiased estimates for the differences between treatments. The use of crossover designs for clinical trials has been much discussed in the literature. See, for example, Brown (1980), Huitson et al. (1982), Jones and Kenward (1989), and Chow and Liu (1992a).

For a crossover design the notions of the washout or carryover effects (or residual effects) are important for the analysis of collected clinical data. The washout period is defined as the rest period between two treatment periods for which the effect of one treatment administered at one dosing period does not carry over to the next. In a crossover design the washout period must be long enough for the treatment effect to wear off so that there is no carryover effect from one treatment period to the next. The washout period depends on the nature of the drug. A suitable washout period must be long enough to return any relevant changes that influence the clinical response to the baseline. If a drug has a long half-life or if the washout period between treatment periods is too short, the effect of the drug might persist after the end of dosing period. In this case it is necessary to distinguish the difference between the drug effect and the carryover effects. The direct drug effect is the effect that a drug product has during the period in which the drug is administered, while the carryover effect is the drug effect that persists after the end of the dosing period.

Note that crossover designs may be used in clinical trials in the following situations where (1) objective measures and interpretable data for both efficacy and safety are obtained, (2) chronic (relatively stable) diseases are under study, (3) prophylactic drugs with relatively short half-life are being investigated, (4) relatively short treatment periods are considered, (5) baseline and washout periods are feasible, and (6) an adequate number of patients for detection of the carryover effect with sufficient power that accounts for expected dropouts is feasible or extra study information is available to rule on the carryover effect. Dubey also emphasizes that appropriate analyses which can reflect the study design must be carried out when using crossover design in clinical trials.

Higher-Order Crossover Designs

The most commonly used crossover design for comparing two treatments (denoted by A and B) is a two-sequence two-period crossover design. We will refer to this design as a standard 2×2 crossover design, which is sometimes denoted by (AB, BA). For a standard 2×2 crossover design, each patient is randomly assigned to receive either sequence AB or sequence BA at two dosing periods. In other words, subjects within sequence AB (BA) receive treatment A (B) at the first dosing period and treatment B (A) at the second dosing period. The dosing periods of course are separated by a washout period of sufficient length to wear the effect due to the drug received in the first period. An example of a standard 2×2 crossover design is illustrated in Figure 5.3.1. Note that

CROSSOVER DESIGNS   185

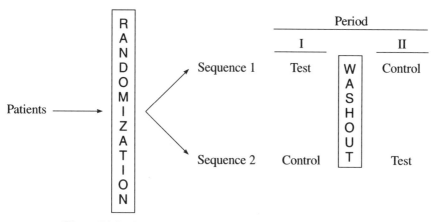

Figure 5.3.1 Standard two-sequence, two-period crossover design.

in the crossover design, the number of the treatments to be compared does not necessarily have to be equal to the number of periods. One example is a 2 × 3 crossover design for comparing two treatments as illustrated in Figure 5.3.2. In this design there are two treatments but three periods. Patients in each sequence receive one of the treatments twice at two different periods.

When the carryover effects are present, a standard 2 × 2 crossover design may not be desirable because of potential confounding effects. For example, the sequence effect, which cannot be estimated separately, is confounded (or aliased) with the carryover effects. If the carryover effects are unequal, then there exists no unbiased estimate for the direct drug effect from both periods. In addition the carryover effects cannot be precisely estimated because they can only be evaluated based on the between subject comparison. Further-

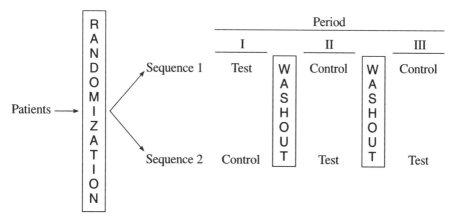

Figure 5.3.2 Two-sequence dual crossover design.

more the intrasubject variability cannot be estimated independently and directly from the observed data because each subject receives either treatment A or treatment B only once during the study. In other words, there are no replicates for each treatment within each subject. To overcome the above undesirable properties, a higher-order crossover design is usually considered (Chow and Liu, 1992a and 1992b). A higher-order crossover design is defined as a crossover design in which either the number of periods is greater than the number of treatments to be compared or the number of sequences is greater than the number of treatments to be compared. There are a number of higher-order crossover designs available in literature (Kershner and Federer, 1981; Laska, Meinser and Kushner, 1983; Laska and Meinser, 1985; Jones and Kenward, 1989). These designs, however, have their own advantages and disadvantages. An in-depth discussion can be found in Jones and Kenward (1989) and Chow and Liu (1992a, 1992b).

Table 5.3.1 lists some commonly used higher-order crossover designs. The design (AA, BB, AB, BA) is known as Balaam's design (Balaam, 1968). It is an optimal design in the class of crossover designs with two periods and two treatments. This design is formed by adding two more sequences (sequences 1 and 2) to the standard 2 × 2 crossover design (sequences 3 and 4). These two augmented sequences are AA and BB. With additional information provided by the two augmentable sequences, not only can the carryover effects be estimated using the within-subject contrasts but the intrasubject variability for both treatments can also be obtained because there are replicates for each treatment within each subject. Design (ABB, BAA) is an optimal design in the class of crossover designs with two sequences, three periods, and two treat-

Table 5.3.1 Higher-Order Crossover Designs

I. Balaam's design
   AA
   BB
   AB
   BA

II. Two-sequence dual design
   ABB
   BAA

III. Doubled (replicated) design
   AABB
   BBAA

IV. Four-sequence design
   AABB
   BBAA
   ABBA
   BAAB

ments. It can be obtained by adding an additional period to the standard 2 × 2 crossover designs. The treatments administered in the third period are the same as those in the second period. This type of designs is also known as the extended-period (extra-period) design. Note that this design is made up of a pair of dual sequences ABB and BAA, and hence it is also known as a two-sequence dual design. Two sequences whose treatments are mirror images of each other are said to be a pair of dual sequences. Jones and Kenward (1989) point out that the only crossover designs worth considering are those made up of dual sequences. Design (AABB, BBAA) is a *doubled* standard 2 × 2 crossover design (AB, BA). It is usually referred to as a replicated design. Liu (1995b), Liu and Chow (1995), and Chow (1996) indicate that a replicated design is useful in assessments of bioequivalence between drug products. Design (AABB, BBAA, ABBA, BAAB) is an optimal design in the class of the crossover designs with four sequences, four periods, and two treatments. It is also made up of two pairs of dual sequences (AABB, BBAA) and (ABBA, BAAB). Note that the frist two periods are the same as those in Balaam's design and that the last two periods are the mirror image of the first two periods. This design is much more complicated than designs (AA, BB, AB, BA) and (ABB, BAA), though it produces the maximum variance reduction for both the direct drug effect and the carryover effects among the designs considered.

Williams Designs

When there are more than two treatments to be compared, a complete crossover becomes much more complicated and may not be of practical interest based on the following considerations:

1. Potential residual effects make the assessment of efficacy and/or safety almost impossible.
2. It takes longer to complete the study.
3. Patients are likely to drop out if they are required to return frequently for tests.

Williams design can be a useful alternative. In this section, for simplicity, we will restrict our attention to those designs in which the number of periods equals the number of treatments to be compared. For comparing three treatments, there are a total of three possible pairwise comparisons between treatments: treatment 1 against treatment 2, treatment 1 against treatment 3, and treatment 2 against treatment 3. It is desirable to estimate these pairwise differences between treatments with the same degree of precision. In other words, it is desirable to have equal variances for each pairwise difference between treatments. Designs with this property are known as variance-balanced designs. It should be noted that in practice, variability associated with the selected design can vary from design to design. Thus an ideal design is one with the smallest variability such that all pairwise differences between formulations can be estimated with the same and possibly best precision. However, to achieve this goal, the design must be balanced. A design is said to be balanced if it satisfies the following conditions (Jones and Kenward, 1989):

1. Each treatment occurs only once with each subject.
2. Each treatment occurs the same number of times in each period.
3. The number of patients who receive treatment i in some period followed by treatment j in the next period is the same for all $i \neq j$.

Under the constraint of the number of periods (p) being equal to the number of formulations (t), balance can be achieved by using a complete set of *orthogonal Latin squares* (John, 1971; Jones and Kenward, 1989). When the number of treatments to be compared is large, more sequences and consequently more patients are required. This, however, may not be of practical utility. A more practical design has been proposed by Williams (1949). We will refer to this as Williams design. Williams design possesses balance property and requires fewer sequences and periods. The algorithm for constructing a Williams design with t periods and t treatments can be found in Jones and Kenward (1989) and Chow and Liu (1992a). Table 5.3.2 gives Williams designs for comparing three and four treatments. It can be seen from Table 5.3.2 that Williams design requires less sequences in order to achieve the property of variance balance as compared to the complete set of orthogonal Latin squares design. For example, for comparing four treatments, Williams design only requires 4 sequences, whereas a complete set of 4 × 4 orthogonal Latin squares requires 12 sequences.

Table 5.3.2 Williams Designs

I. Williams's design with three treatments
ACB
BAC
CBA
BCA
CAB
ABC

II. Williams's design with four treatments
ADBC
BACD
CBDA
DCAB

Balanced Incomplete Block Design

When there are a large number of treatments to be compared, a complete crossover design may not be feasible. Although a Williams design can be used, it can take a long time to complete. In practice, it is desirable to complete the study in a short period of time. In this case it is desirable to have the number of periods less than the number of treatments to be compared. Therefore it is suggested that a randomized incomplete block design be used. An incomplete block design is a randomized block design in which not all treatments are present in every block. A block is called incomplete if the number of treatments in the block is less than the number of treatments to be compared. When an incomplete block design is used, it is recommended that the treatments in each block be randomly assigned in a balanced way so that the design will possess some optimal statistical properties. This kind of design is referred to as a balanced incomplete block design. A balanced incomplete block design is an incomplete block design in which any two treatments appear together an equal number of times. Table 5.3.3 gives two examples of balanced incomplete block designs for comparing four treatments with two periods and three periods, respectively.

Note that a balanced incomplete block design possesses some good statistical properties. For example, unbiased estimates of treatment effects are available and the difference between the effects of any two treatments can be estimated with the same degree of precision.

Table 5.3.3 Balanced Incomplete Block Designs

I. Four treatments with two periods
  AB
  BC
  CD
  DA
  AC
  BD
  DB
  CA
  AD
  DC
  CB
  BA
II. Four treatments with three periods
  BCD
  CDA
  DAB
  ABC

Examples of Crossover Design in Clinical Trials

To illustrate the use of crossover designs in clinical trials, we again consider the clinical development of Glucophage. Table 5.3.4 lists three studies that have investigated the effects of Glucophage on lipids and other uses such as risk factors for cardiovascular disease.

The objective of the study conducted by Chan et al. (1993) was to compare the metabolic and hemodynamic effects of Glucophage and Glibenclamide in normotensive NIDDM patients. After a two-week run-in period on dietary treatment alone, 12 Chinese normotensive patients with uncomplicated NIDDM were randomized to receive either Glucophage or Glibenclamide for four weeks before being crossovered to the alternative treatment for an additional four weeks. Their metabolic and hemodynamic indexes including cardic output estimation by impedance cardiography were measured at the baseline and at the end of each treatment. The results indicate that at comparable degrees of glycemic control, Glucophage had the following beneficial effects compared with Glibenclamide: (1) greater weight loss (body mass index, -0.58 kg/m^2 vs. -0.12 kg/m^2), (2) greater decrease in total cholesterol (-0.7 mM vs. -0.2 mM), and (3) greater decrease in diastolic blood pressure (-12.9 mmHg vs. -6.8 mmHg). In conclusion Chan et al. (1993) indicates that the tendency to greater peripheral resistence with Glibenclamide and to lower diastolic blood pressure with Glucophage may bear on the development of hypertension in normotensive patients who are receiving long-term treatment for NIDDM.

For another application of the crossover design, Nagi and Yudkin (1993)

Table 5.3.4 Examples of Crossover Design

Study	Purpose	Sample Size	Duration (Period 1 + Washout + Period 2)	Primary Endpoint
Chan et al. (1993)	Effects on lipids	12	4 wk + 0 wk + 4 wk	Metabolic and Hemodynamic index
Nagi and Yudkin (1993)	Effects on lipids and risk factors for cardiovascular disease	27	12 wk + 2 wk + 12 wk	Insulin resistance Glycemic control Cardiovascular risk
Elkeles (1991)	Effects on lipids	35	3 mo + 6 wk + 3 mo	Serum lipids Lipoproteins Blood glucose Glycosylated hemoglobin

investigated the effects of Glucophage on glycemic control, insulin resistance, and risk factors for cardiovascular disease in NIDDM patients with different risks of cardiovascular disease. The study was conducted as a randomized double-blind placebo-controlled crossover design on 27 patients. Glucophage was administered for a total of 12 weeks, and the dose was increased stepwise from 850 mg once daily for one week to 850 mg twice daily for five weeks and to 850 mg three times daily for another six weeks. The baseline assessment took place on the day of inclusion in the study, and a similar assessment took place after 12 weeks of therapy (phase 1). After a washout period of two weeks, patients were reassessed as at entry into the trial and crossed over to the alternative treatment (phase II). The patients were reassessed finally at the end of phase 2. The results indicated that Glucophage reduced fasting plasma glucose levels by 3.08 mM, enhanced insulin sensitivity by 4.0%, and diminished triglyceride levels by 0.2 mM, total cholesterol levels by 0.52 mM, and low-density lipoprotein cholesterol levels by 0.4 mM. Nagi and Yudlin (1993) conclude that Glucophage therapy improves glycemic control by diminishing insulin resistance, enhances lipid and lipoprotein profiles, ameliorates other risk factors for cardiovascular disease independently of weight loss or improved glycemic control, and may therefore have utility in long-term reduction of coronary artery disease risk among patients with NIDDM.

Elkeles (1991) also conducted a three-month crossover trial on 35 patients with poorly controlled NIDDM to investigate the effects of Glucophage and Glibenclamide on body weight, blood glucose control, and serum lipoproteins. After six weeks of a diet that did not achieve adequate diabetic control, patients were randomized to receive either Glibenclamide 5 mg daily or Glucophage 500 mg twice a day. The dose was increased to achieve a fasting blood glucose level of 6 mmol/L or less, up to a maximum of 15 mg Glibenclamide or 3 g Glucophage daily. After three months the treatment was stopped and after six weeks of diet only again, patients were crossed over to receive the other drug. Before and after three months of treatment, blood samples were taken for serum lipids, lipoproteins, blood glucose, and glycosylated haemoglobin. Elkeles reports that Glucophage diminished hemoglobin A_{1C} levels by 2.05% and the Glibenclamide by 1.51%. Glucophage also significantly reduced both total cholesterol and low-density lipoprotein cholesterol. Elkeles points out that the improvement in H_bA_{1C} levels as well as in total cholesterol and low-density lipoprotein cholesterol reverted when Glucophage was withdrawn for six weeks. Therefore it was concluded that by reducing total cholesterol and low-density lipoprotein cholesterol over the long term, Glucophage can improve the coronary artery disease risk profile independently of its effect on glucose homeostasis.

5.4 TITRATION DESIGNS

For phase I safety and tolerance studies, Rodda et al. (1988) classify traditional designs as follows:

1. Rising single-dose design.
2. Rising single-dose crossover design.
3. Alternative-panel rising single-dose design.
4. Alternative-panel rising single-dose crossover design.
5. Parallel-panel rising multiple-dose design.
6. Alternative-panel rising multiple-dose design.

As indicated in Chapter 1, phase I studies are usually conducted in young, healthy male volunteers. The purpose of phase I studies is to obtain initial appraisement of drug safety through the evaluation of vital signs, physical health, and adverse events and frequent assessments of hematology, blood chemistry, and urine samples. The above designs are commonly employed in phase I safety and tolerance studies to efficiently provide the data that can be analyzed for generating hypotheses rather than for making definitive inference.

In medical practice, if the study medicine is intended for cancer or some life-threatening diseases, it may not be ethical to conduct phase I safety and tolerance studies on normal volunteers due to potential toxic or fatal effects. In addition results from animal studies provide little information regarding the therapeutic range for possible efficacy with tolerable safety. Therefore phase I clinical trials are necessarily conducted in patients to determine the maximum tolerable dose (MTD). In cancer clinical trials Storer (1989, 1993) defined the maximum tolerable dose (MTD) as the dose at which a specified proportion of patients (selected from a tolerance distribution) experience some objective prespecified clinical toxicity. For most phase I cancer trials, this prespecified proportion is usually taken as 1/3. A commonly employed design for determination of MTD is the so-called *up-and-down design*. During phase II of clinical development, various statistical designs can be used in describing the dose-response relationship. For example, a parallel design is the design of choice for the study of a definitive dose-response relationship if a placebo concurrent control is included. At an early stage of phase II studies, however, a titration design is sometimes considered. The purpose of a titration design is to obtain preliminary information on the dose-response relationship with the limited resources available. Note that both the up-and-down design and the titration design involve adjustment of doses either for the same patient or for subsequent patients. These designs are also referred to as dose-ranging designs.

Up-and-Down Design

As was indicated above, one of the primary objectives of phase I cancer trials is to establish MTD with an adequate precision. Therefore it is important to select an appropriate statistical design for estimation of MTD based on the following considerations:

TITRATION DESIGNS

1. The patient population is usually rather heterogeneous with some medical complications.
2. There is a limited number of patients available for trials.
3. There is a high chance of withdrawals which may or may not be related to the toxicity of the study medicine.
4. It may take a relatively long time to evaluate the toxicity exhibited in each patient.
5. The physician's evaluation of clinical toxicity may be subjective.

The principle of the up-and-down design is to start patients with a prespecified dose and adjust the dose upward (escalation) or downward (de-escalation) according to a set of predetermined rules. Under the concept of the up-and-down design, Storer (1989, 1993) classified the up-and-down designs as three single-stage designs and one two-stage design.

Single-Stage Designs

For single-stage designs, basically there are three commonly employed designs: design A, design B, and design D. For design A we start with a group of three patients who are treated with an initial dose (e.g., x_0). At the second step, if no toxicity is observed in all three patients, then the dose for the next group of three patients is escalated to the next higher dose level (e.g., x_1). Otherwise, the next group of three patients is treated at the same dose of x_0. In step 3 the dose of the next group of three patients is escalated to the next higher dose level if the prespecified clinical toxicity is observed at not more than one patient of the six patients from both steps 1 and 2. Otherwise, the trial stops. For step 4 we repeat steps 2 and 3 with two consecutive groups of three patients until the trial stops. For design A the MTD is typically defined as either the dose at which the trial stops or the next lower dose is given. Therefore the determination of MTD can basically be obtained without any statistical justification. Figure 5.4.1 provides a flowchart for Design A of this kind.

For design B we start with a single patient with an initial dose of x_0 as the first step. For the next step, if no toxicity is observed in this patient, then the next patient is treated at the same dose; otherwise, the next patient is treated with the next lower dose. At the third step, if no toxicity is observed in both consecutive patients, then the next patient is treated with the next higher dose. If a prespecified toxicity is observed in both consecutive patients, then the next patient is treated with the next lower dose. Otherwise, the trial stops. In step 4 we repeat steps 2 and 3 until the trial stops. A flowchart of this design is given in Figure 5.4.2.

For design D, similar to design A, we start with a group of three patients who are treated with an initial dose of x_0 at the first step. In the next step the dose of the next group of three patients either is escalated to the next higher level x_1 if no toxicity is observed in the three patients or stays at the same level

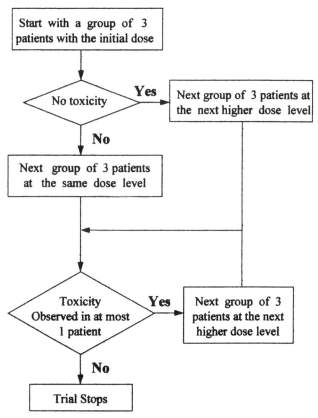

Figure 5.4.1 Flowchart for design A.

x_0 if toxicity is observed in one patient or decreases to the next lower level if toxicity is observed in more than one patient. At step 3 we continue step 2 until the trial stops. This design is summarized in Figure 5.4.3.

Two-Stage Design

Basically a two-stage design is a combination of two single-stage designs. As an illustration, we will describe a two-stage design that combines design B and design D. We will denote this design by design BD.

For design BD, at stage 1 we start with design B until the trial stops according to the stopping rule described in design B. At stage 2 we then continue the trial with design D. If a toxic response is observed in the last patient at the first stage, then the initial dose for the second stage for design D is the next lower dose level with respect to the dose of the last patient at the first stage. However, if the last patient at the first stage does not exhibit any prespecified toxicities, then the beginning dose for the second stage is the dose of the last patient at the

TITRATION DESIGNS

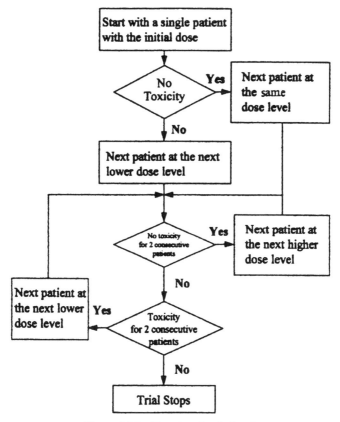

Figure 5.4.2 Flowchart for design B.

first stage. Unlike design A, the MTD under a single-stage design (e.g., design B or design D) and a two-stage design (e.g., design BD) needs to be estimated based on the data by using a formal statistical inference procedure.

The dose levels used for phase I trials are usually determined based on information from animal studies. In practice, dose levels are usually chosen to be equally spaced on the logarithmic scale. For example, Schneiderman (1967) modified the Fibonacci sequence of the diminishing multipliers of 2, 1.67, 1.5, 1.33, ... for the increment between consecutive doses. In clinical trials design A is the traditional design, and it is commonly employed for the determination of MTD. However, since design A only allows escalation rather than de-escalation, it is not a true up-and-down design. As a result there is a strong possibility that design A will not be able to detect the MTD if the starting dose is chosen to be a relatively high dose. In a simulation study Storer (1989) indicates that none of single-stage designs performs well in an arbitrary dose-response setting with fixed sample sizes. In addition design A frequently fails to provide a convergent estimate for MTD. On the other hand, the two-stage design of BD

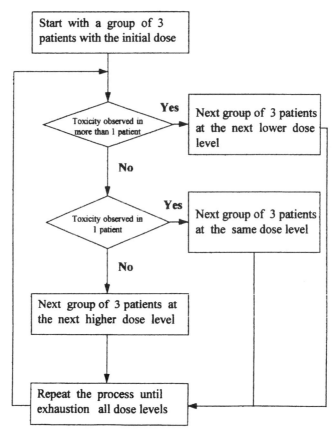

Figure 5.4.3 Flowchart for design D.

is rather robust with respect to the choice of dose levels. In addition, since the patients at the first stage reduce more bias than the same number of patients who are added to the single stage of the same design, design BD has less bias than the single-stage designs. Its bias is almost negligible even when the sample size at the second stage is as small as 18.

O'Quingley et al. (1990) argue that the traditional up-and-down designs such as designs A, B, D, and BD might not be appropriate in cancer trials. In cancer trials the treatments are usually extremely toxic at high doses and yet might provide no satisfactory efficacy at lower doses. In addition the dose toxicity curve is usually unknown. As an alternative, O'Quingley et al. (1990) suggest a continual reassessment method (CRM) based on a Bayesian approach. The CRM procedure is to update the information of the dose-response relationship through a Bayesian framework as observations on severe toxicity become available and then to use this information to concentrate the trial around the dose that might correspond to the anticipated target toxicity level. In practice, however, the number of the patients enrolled in phase I clinical trials for determination of the MTD of therapeutic cancer agents is usually too small (e.g., between 10 and 30) to be a representative (random) sample for the targeted population. As a result the continual reassessment method for estimation of the MTD under the up-and-down designs is based on a population model that is inherently untestable and can only be indirectly invoked. Further research for taking into account the randomization scheme of the design for determination of the MTD of a cancer therapeutic agent is needed. For more details regarding CRM, see Chevret (1993) and Møller (1995). For comparison between different phase I designs, see Korn, et al. (1994) and O'Quigley and Chevret (1991).

Note that in addition to point estimation, we could consider constructing confidence interval for MTD using traditional large-sample methods such as the delta method, the methods based on the Fieller's theorem, and the method of likelihood ratio. These methods, however, cannot provide an interval with a narrow width and a coverage probability close to the nominal level. Storer (1993) improves the intervals based on the delta method and Fieller's theorem by considering the distribution of their corresponding statistics on the true sampling scheme dictated by the designs. He also shows that these improved confidence interval provide correct coverage probability.

Titration Design

As indicated in Chapter 1, one of the objectives of phase II clinical trials is to sufficiently characterize the dose-response relationship, which is the most frequently asked question at the FDA Advisory Committee's meeting. The dose-response relationship defines the therapeutic range of a test drug. The therapeutic range is usually referred to as the dosage range between the minimal effective dose (MED) and the maximum tolerable dose. The MED is the lowest dose above which the efficacy of the test drug is clinically superior to and statistically significant from that of the placebo. The MED is usually demonstrated for at least one primary clinical endpoint. If the range between the MED and MTD is large, then the test drug is said to have a wide *therapeutic window*. If the MED is very close to MTD, then the test drug is considered to have a narrow therapeutic window.

In practice, although preliminary short-term safety information is available after the completion of phase I clinical trials, information on the dose-response relationship and the safety profile with respect to a moderate or long-term exposure of patients to the test drug is usually unknown at the initiation of phase II clinical trials. In order to conform to real clinical practice and to expose the patients only to the amount of dose they need, it is suggested that a titration design be used to provide a conservative and cautious approach in investigating the dose-response relationship, which can in turn be used as a preliminary estimate of the therapeutic window of the drug.

A traditional titration design is also a dose-escalation study with a set of predetermined dose levels and prespecified criteria of responders or nonresponders. A titration design starts with a placebo washout phase during which the previous medications are stopped and the placebo is administered to the patients. At the end of the placebo washout period, the baseline clinical measurements are established. Then all the patients start with the same lowest dose. A patient is considered a responder and continues to receive the same dose for the duration of the trial if he or she meets the prespecified criteria at the end of the first dosing period. If a patient fails to meet the prespecified criteria at the end of the first dosing period, the dose of the patient is then titrated up to the next higher dose provided that the patient can satisfactorily tolerate the drug. The titration process continues until all dose levels are exhausted. The determination of a responder is usually based on some objective physiological measurements. In practice, the duration of each dosing period is usually chosen to be of sufficient length so that the stabilization of the selected physiological measurements can be achieved to define the titration process.

For the analysis of the data from titration studies, several methods have been proposed in the literature. For example, Chuang (1987) considers the life-table method utilized along with a logistic linear dose-response model. As an alternative, Shih et al. (1989) and Chuang-Stein and Shih (1991) propose the method of EM algorithm. Temple (1982), however, points out that for any titration study the treatment effect is confounded with time. In addition the primary clinical measurement used to define the titration process might be highly correlated with other efficacy and/or safety endpoints. As a result the missing mechanism for these clinical endpoints are treatment related and are not at random. Consequently the methods for use of longitudinal data such as generalized estimating equations (GEE) proposed by Liang and Zeger (1986) are not appropriate (e.g., see Chuang and Stein, 1993). Therefore a limitation of titration designs is their lack of valid statistical methodologies for the analysis of continuous or ordered categorical data in adequately characterizing the dose-response relationship of a test drug.

Note that there are other variations associated with a titration design that make statistical analysis even more complicated. An example for such variation is described below. An open-label phase II clinical trial with three titration groups was conducted before availability of the results from CAST to obtain preliminary information of dose-response relationship for a new class IC antiarrhythmic agent in treatment of patients with ventricular ectopy. Each titration group had a dosing regimen of the four ascending doses displayed in Table 5.4.1. The first group consists of dose levels of a placebo and 50 mg, 75 mg, and 125 mg t.i.d. (every eight hours) of the drug, the second group includes a placebo, 50 mg, 100 mg, and 150 mg, t.i.d. of the drug, while the third group contains a placebo, 100 mg, 150 mg, and 200 mg t.i.d. of the drug. The duration of each dosing period is three days, and the patients were assigned to the three groups sequentially. The objective clinical endpoints are derived from the 24-hour Holter recording processed at a central facility. They were:

1. Hourly average of total ventricular premature contraction (VPC) counts.

2. Hourly average of single VPC counts.

TITRATION DESIGNS

Table 5.4.1 Dose Levels and Number of Patient at Each Dosing Period of the Titration Design for a Class IC Antiarrhythmic Agent

Group	Dosing Period			
	0	1	2	3
A	P(11)	50 (11)	75 (9)	125 (9)
B	P(15)	50 (15)	100 (14)	150 (12)
C	P(16)	100 (16)	150 (14)	200 (9)

Note: P = placebo; the numbers outside the parentheses are dose levels in mg, and the numbers within the parentheses are the number of patients entered each dosing interval.

3. Sum of couplets over a 24-hour period.
4. Sum of VT runs over a 24-hour period for which the pulse rate was greater than or equal to 100 beats per minutes.

A responder for an adequate suppression of ventricular ectopy is defined as (1) at least a 80% reduction in average total VPC count per hour compared to the baseline value obtained at the end of the placebo dosing period and at least a 90% reduction in the number of events of repetitive forms (couplets or nonsustained ventricular tachycardia, NSVT) or (2) at least a 90% reduction in the number of NSVT provided the number of NSVT is 10 or more during the placebo period. Patients may be withdrawn from the study according to the following efficacy and safety criteria. Efficacy criteria for worsening of ventricular ectopy is defined as either (1) $\ln(Y) \geq 3.118 + 0.646 \ln(X)$, where Y is the hourly VPC count at the end of each dosing period for the active treatment and X is the baseline VPC hourly count at the end of placebo period, or (2) an increase to 50 or more runs of VPCs if the number of runs over the 24-hour period during the placebo period is less than five or a tenfold increase in the number of runs and if it is at least five during the placebo period. The safety criteria include (1) the QRS interval being at least 180 ms, or (2) an increase of at least 40% in the corrected QC interval compared to the baseline value of the placebo period or a QC greater than 550 ms.

The dose for patients in each group was titrated upward every third day within each dosing group until an adequate suppression of ventricular ectopy according to the above criteria or patients withdrew from the study if they met either criteria for worsening of ventricular ectopy or safety criteria. It should be noted that each dosing group after the placebo period actually is a parallel group design. For example, the last dosing group is in fact compared to three parallel dosing groups: 125 mg, 150 mg, and 200 mg. On the other hand, within each group, comparison among doses is made within each patient. Consequently as an example, a comparison between 50 and 100 mg consists in a comparison between group A and C during the first dosing period and the comparison between 50 and 100 mg within the patients in group B. In addition to the complexity of the design of this study, it is also observed that (1) the study is an open-label study with a nonrandom group assignment, (2) the three dose titration groups have overlapping doses, and (3) present are possible confounding dose effects and carryover effects because of no washout period between the dosing periods. Therefore it is suggested that the conclusion and/or interpretation of the results based on inferential statistics for comparisons among doses be drawn with extremely caution.

Another issue regarding the interpretation of the information of the dose-response relationship from a titration design is the overestimation of the necessary dose. For example, the results of early titration studies may suggest that a dose of 600 mg per day or more is necessary for the cardio-selective beta-blocker atenol in the effective reduction of blood pressure. However, subsequent parallel-group, placebo-controlled studies demonstrate that a dose above 100 mg per day has no additional effect in the reduction of blood pressure. To overcome this problem, a titration design with a parallel placebo current control may be useful. For example, a phase II clinical trial was conducted to obtain initial dose-response information of an angiotensin-coverting enzyme (ACE) inhibitor captopril. Patients are randomized to either the active treatment group or placebo concurrent group. Then the titration process is performed within each group in five dosing periods with five predetermined doses 0, 25 mg, 50 mg, 100 mg, and 150 mg t.i.d. (Temple, 1982). This design is illustrated in Table 5.4.2 with the mean diastolic blood pressure (mm Hg). Within each dosing period, patients in the captopril group received the active drug at the titrated dose level and the patients in the parallel placebo concurrent control group received its matched placebo. The criterion for a clinical response was defined as a reduction of systolic blood pressure below 90 mm Hg. This design is in fact a parallel-group design with two groups. The study can be conducted in a triple-blind fashion in the sense that not the patients nor the investiga-

Table 5.4.2 Titration Design with a Parallel Placebo Concurrent Control with Diastolic Blood Pressure (mm Hg)

	Dose Level				
	0 mg	25 mg	50 mg	100 mg	150 mg
Captopril					
Observed DBD	110	100	99	96	94
Change from 0		−10	−11	−14	−16
Placebo					
Observed DBD	110	104	104	103	101
Change from 0		−6	−6	−7	−9
Difference in					
Change from 0		−4	−5	−7	−6

Source: Summarized from Temple (1982).

tor nor the sponsor know the actual treatment that is assigned to the patients, although the titration process with the corresponding doses can be made available to everyone. About 70% of patients were titrated up to 100 and 150 mg. If we only examine the results from the active treatment group as if this study had been conducted as the traditional titration design without a parallel placebo concurrent control, there is a very nice dose-response relationship in reduction of blood pressure from the baseline. In this case the results of the active group gives a wrong impression that the test drug produces a monotone increased response up to 150 mg t.i.d. However, at the same time the parallel placebo concurrent group also presents a sizable placebo effect in the reduction of blood pressure. It turns out that the treatment effect, which is the difference in reduction from baseline between captopril and placebo, reaches a plateau and remain constant after dosing period 2 during which 50 mg t.i.d. was administered. Consequently Temple (1982) suggests that 50 mg t.i.d. seems to treat most hypertensive patients well.

Note that some pharmaceutical agents might induce some undesirable but reversible safety concerns. In addition they may not be efficacious at lower doses. When conducting clinical trials with these agents, we would expect a significant number of dropouts. Therefore it is recommended that a trial with these agents begin very cautiously with a very low dose. In such a trial the criteria for titration process is based on safety rather than efficacy because the drug is unlikely to be effective at lower doses. As a result all patients who do not have the predefined safety problem will be forced to receive the next higher dose in the subsequent dosing period. This type of titration design is called the *forced dose-escalation design*. A typical example for obtaining FDA approval using the forced dose-escalation trials is the approval of Tacrine which is intended for treatment of mild to moderate dementia of the Alzheimer's type. Since Tacrine is known to induce elevation of serum alanine aminotransferase (ALT) above the upper limit of the normal range in 43% to 54% of the patients and around 28% of the patients treated with Tacrine showed an elevation of ALT exceeding three times the upper limit of the normal range, the forced dose-escalation design at six-week intervals was chosen for two adequate well-controlled studies for the approval of the drug. The design of the first adequate well-controlled randomized study is a 12-week trial that consists of two six-week double-blind phases with the placebo current control groups as shown in Table 5.4.3 (Farlow et al., 1992). Patients were first randomized to one of the six sequences. For the double-blind phase I the patients in sequences 1 and 2, 3 and 4, and 5 and 6 received placebo, 20 mg, and 40 mg per day, respectively. However, for the double-blind phase II patients in sequences 1, 3, and 5 received the same doses as those in the double-blind phase I, while the doses of the patients in sequences 2, 4, and 6 who could tolerate the doses in double-blind phase I were titrated up to 20 mg, 40 mg, and 80 mg during double-blind phase II, respectively. The second adequate well-control study is a long-term 30-week randomized double-blind trial that consists of three parallel groups for the active drug and a parallel placebo current group with a forced dose-escalation design as shown in Table

Table 5.4.3 Forced Dose-Escalation Design for 12-Week Trial of Tacrine in Alzheimer's Disease

Randomized Sequences	Double-Blind Phase I Week 1 to 6	Double-Blind Phase II Week 7 to 12
1	Placeob	Placebo
2	Placebo	Tacrine 20 mg/d
3	Tacrine 20 mg/d	Tacrine 20 mg/d
4	Tacrine 20 mg/d	Tacrine 40 mg/d
5	Tacrine 40 mg/d	Tacrine 40 mg/d
6	Tacrine 40 mg/d	Tacrine 80 mg/d

5.4.4 (Knapp, 1994). To account for anticipated increase incidence of cholingeric adverse events and dropouts at the highest dose, an unequal randomization with a ratio of 3 : 1 : 3 : 4 for groups 1, 2, 3, and 4, respectively, was used for the assignment of patients to the treatment. Patients who were randomized to group 1 received a placebo throughout the entire study. Patients who randomized to group 2 received 40 mg per day for the first six weeks. If they could tolerate the dose, the doses of those patients were escalated to 80 mg per day for the rest of the study. Patients who were randomized to group 3 received 40 mg per day during the first 6 weeks and then were titrated up to 80 mg per day between week 7 and week 12, and again titrated to 120 mg per day for the rest of the study provided that they could tolerate the dose levels. The escalation process for the patients who were randomized to group 4 was 40 mg, 80 mg, 120 mg, and 160 mg per day for the first, second, and third 6-week dosing periods, and for the rest of 12 weeks, respectively. Both trials consist of parallel groups as well as a dose titration process within each group. Therefore they can provide cross-sectional data for comparison among parallel groups as well as longitudinal data for comparison among doses based on the individual patient. However, for the 12-week trial the analysis was performed separately with the cross-sectional data collected at week 6 for the double-blind phase I and week 12 for

Table 5.4.4 Forced Dose-Escalation Design for 30-Week Trial of Tacrine in Alzheimer's Disease

| Randomized Groups | Week | | | |
	0–6	7–12	13–18	19–30
1	Placebo	Placebo	Placebo	Placebo
2	40 mg/d	80 mg/d	80 mg/d	80 mg/d
3	40 mg/d	80 mg/d	120 mg/d	120 mg/d
4	40 mg/d	80 mg/d	120 mg/d	160 mg/d

the double-blind phase II. The cross-sectional data at week 30 were analyzed for the 30-week study. Unfortunately, due to the lack of adequate statistical tools for inference of treatment effects in the presence of treatment-related withdrawal, longitudinal data were not utilized to provide useful information regarding the titration process, indeed such data can be vital for the application of Tacrine in the treatment of patients with probable Alzheimer's disease.

5.5 ENRICHMENT DESIGNS

Some therapeutic agents are likely to be effective in a specific population of patients who may have an underlying disorder that is responsive to the manipulation of dose levels of the same agent or several different agents. In practice, instead of an unselected group of patients, it is of interest to identify the patients in whom the test agent is likely to be beneficial in the early phase of the trial. This phase of manipulation of dose levels of the same therapeutic agent or test of different agents for identification of patients with drug efficacy is called the *enrichment* phase. The patients with drug efficacy identified at the enrichment phase are then randomized to receive either the efficacious dose of the test agent or the matched placebo. A design of this kind is known as an enrichment design.

An enrichment design usually consists of at least two phases. The first phase is the enrichment phase in which an open-label study with a titration design is conducted to classify patients into groups so that the agents are of some benefit to the patients. The second phase is usually randomized and double-blind, possibly with a placebo concurrent control to formally and rigorously investigate the effectiveness and safety of the test agents in these patients. The concept of enrichment design is illustrated in three recently completed clinical trials in the areas of Alzheimer's disease and arrhythmia.

The first example is a clinical trial conducted in the early stage of development of Tacrine with doses of 40 and 80 mg four times a day in treatment of the patients with probable Alzheimer's disease. As indicated by Davis et al. (1992), the reason for the enrichment design to be selected for this trial is that the clinical, biochemical, and pathological heterogeneity of the disease and clinical experience suggest that not all patients will respond to any single treatment and that those who respond might do so only within a limited dose range. This trial consisted of four phases: a six-week double-blind dose-titration enrichment phase, a two-week placebo baseline phase, a six-week randomized double-blind placebo-controlled phase, and a six-week sustained active phase, as is displayed in Figure 5.5.1. Patients who met the inclusion and exclusion criteria were enrolled into the enrichment phase of the trial which consisted of three titration sequences. Each titration sequence consisted of 3 two-week dosing periods. The dose in each titration sequence was always titrated up from 40 to 80 mg four times a day with a placebo in dosing periods 1, 2, and 3 for the titration periods 1, 2, and 3, respectively. The patients were random-

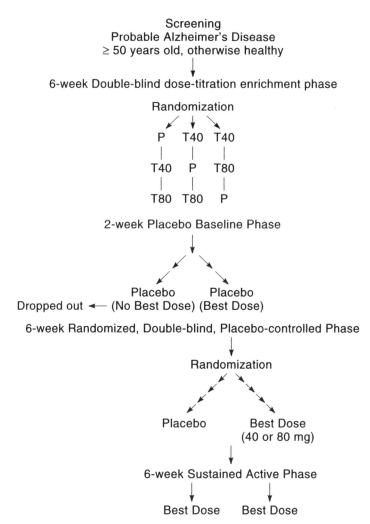

Figure 5.5.1 An enrichment design for tacrine in patients with probable Alzheimer's disease. (Source: Davis et al., 1992.)

ized into one of the three titration sequences which were conducted in double-blind fashion. The potentially therapeutic response for each patient at each dose were then assessed at the end of each two-week dosing period. The *best dose* response for a patient was defined in advance in the protocol as a reduction of at least four points from the screening value in a total score on the Alzheimer's disease assessment scale (ADAS, Folstein et al., 1975) and without intolerable side effects. Then patients with the identified *best dose* were then entered into a two-week placebo baseline period with the hope that this period would be sufficient for Tacrine to wear off from the body and for patients to return to the screening pretreatment state with comparable efficacy outcomes. At the end of the two-week placebo baseline phase, the patients with a reduction of at least four point in ADAS during the enrichment phase entered the subsequent six-week randomized double-blind parallel-group, placebo-controlled phase and were randomized with equal probability either to the active Tacrine at their best dose or to the matched placebo. The clinical endpoints measured at the end of the two-week placebo baseline phase served as the baseline for the six-week double-blind phase. Patients who completed the six-week double-blind phase then entered into the sustained active treatment phase.

This study adopted an enrichment design with three titration sequences to identify patients who are likely to respond to Tacrine at a certain dose. After a two-week washout period, the identified patients were randomized in either Tacrine at their best dose or to placebo concurrent control in a double-blind phase. It, however, should be noted that the fundamental assumption for the use of an enrichment is that the *best dose* responses are those obtained when patients are on Tacrine. In practice, it is quite possible that some patients were placebo-responders (i.e., a reduction of at least 4 points on the ADAS scale) who may not respond to Tacrine. In their paper Davis et al. (1992) did not indicate whether there were any placebo-responders in the study. In addition, although there were washout periods between the two-week dosing periods in titration sequences during the enrichment phase, the carryover effect could still exist. Thus it is impossible to estimate the treatment difference unbiasedly based on the data from all three dosing periods during the enrichment phase due to the fact that carryover effects are confounded with treatment effects. Based on the first two weeks of treatment, the enrichment design can provide an unbiased comparison between the placebo and the 40 mg. However, the other two-thirds of the information was wasted. It can be seen from the study that the carryover effect was significant, which suggests that a placebo baseline phase of two weeks was not long enough for the patients to return to the pretreatment state at the screening. In addition, since the carryover effects were confounded with the treatment effects, it is likely that the reduction of at least four points on the ADAS for the *best dose* response was in part due to the carryover effect. As a result Davis et al. (1992) admit that. "Failure to restore baseline conditions fully at the end of the washout period after dose titration makes it impossible to calculate the size of drug effect with certainty." Furthermore it is not clear how to distinguish the characteristics of the patients with the *best dose* response from those who failed to produce a *best dose*. This information is extremely important for practicing physicians who are in favor of prescribing Tacrine to patients with probable Alzheimer's disease. Consequently this study was not used as one of the two adequate well-controlled studies for approval of Tacrine.

The rationale for selecting the enrichment design in one of the trials during the development of Tacrine is that a short-term response to Tacrine is predictive of the long-term efficacy in prevention of the progression of Alzheimer's disease. The same clinical endpoints were used in both the enrichment and double-blind phases for evaluation of Tacrine's effectiveness. In practice, for other therapeutic agents, the real efficacy endpoint is mortality which requires a longer time to observe. Therefore the short-term efficacy of the agents is assessed by some other objective surrogate endpoints. It is then very important to know whether the short-term efficacy based on the surrogate endpoint is predictive of a hard endpoint such as mortality. As a result the enrichment design is usually employed for identification of the *short-term* responders at the initial stage followed by the main phase of the long-term study. Examples of this type of trial can be found in the area of arrhythmia such as in the Cardiac Arrhythmia Suppression Trial (CAST) and the Electrophysiologic Study versus Electrocardiographic Monitoring (ESVEM) trial.

CAST is a multicenter randomized placebo-controlled study sponsored by the U.S. National Heart, Lung, and Blood Institute to test the hypothesis whether the suppression of asymptomatic or mildly symptomatic ventricular arrhythmia after myocardial infarction will reduce the rate of death from arrhythmia. The active drugs included three class IC antiarrhythmic agents Encainide, Flecainide, and Morcizine with a placebo concurrent control. Since the objective of the study was to test the predictability of suppression of ventricular arrhythmia based on ventricular premature contractions (VPC) as recorded by the Holter monitor using the active drugs for mortality, an open-label enrichment design with two titration sequences involved with only active drugs was selected for this study. The patients were stratified by left ventricular ejection ($<30\%$) and time between the qualifying Holter recording and the myocardial infarction (<90 days or ≥ 90 days). Patients with an ejection fraction of at least 30% were assigned at random to receive either the sequence Encainide–Morcizine–Flecainide or the sequence Flecainide–Morcizine–Encainide. The reason for including Morcizine is its inferior efficacy in the suppression of VPC as compared to other two active agents. Each drug was tested at two dose levels. The doses of Encainde, Flecanide, and Morcizine were 35 mg and 50 mg t.i.d., 100 mg and 150 mg, b.i.d., 200 mg and 250 mg t.i.d., respectively. Since Flecainide exhibits negative inotropic properties, it was not administered to the patients with an ejection fraction less than 30%. The prespecified criteria for an adequate suppression of ventricular arrhythmia were (1) a reduction of at least 80% in VPC and (2) a reduction of at least 90% in runs of unsustained tachycardia as measured by 24-hour Holter recording 4 to 10 days after each dose was begun. The titration process for a particular patient was stopped as soon as a drug and a dose were found to yield adequate suppression. The patients whose arrhythmia were adequately suppressed were then randomized to either the *best drug* identified during the enrichment phase or to placebo for a three-year long-term follow-up. A diagram of the study's design is given in Figure 5.5.2.

The results of CAST, which showed an excessive risk of death for patients who received Encainide or Flecainide as compared to placebo, are thoroughly discussed and examined by Ruskin (1989). The enrichment phase of CAST inherited the fundamental flaw of any titration design in confounding the treatment and carryover effects. For the ethical reason of a minimal exposure of

ENRICHMENT DESIGNS

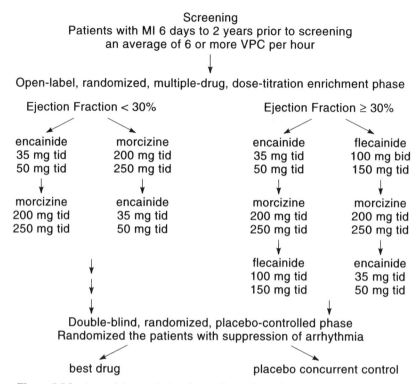

Figure 5.5.2 An enrichment design for cardiac arrhythmia suppression trial (CAST).

patients to the test agents, the titration process must be stopped as soon as a drug and a dose are found to be effective in suppression of VPC. On the other hand, the optimal drug and dose could not be found for a particular patient. The primary endpoint for CAST is the death or cardiac arrest with resuscitation due to arrhythmia. However, the analysis for the comparison of mortality rate between antiarrhythmic agents and placebo failed to take into account the occurrence of those events during the enrichment phase because of the lack of an adequate statistical tool for combining the results based on the same primary endpoint from both the enrichment phase and the randomized, placebo-controlled phase. On the other hand, other patients with adequate suppression based on VPC as a therapeutic endpoint might respond very differently from patients with arrhythmia, such as those with ventricular dysfunction or unsustained ventricular tachycardia. This can be seen from the comparison of the mortality rate assumed for sample size determination and the actual mortality rate. The mortality rate of the placebo concurrent control for the sample size determination of CAST over a period of three years was 11%, while the observed mortality of the placebo group was only about 2.2%. As a result the enrichment design in CAST produced a biased estimate of mortality rate for the placebo concurrent control and excluded death in open-label titration process during the enrichment phase in the suppression of arrhythmia for a patient population with a low risk of death.

The results of CAST indicate that the short-term efficacy measured as suppression of VPC based on noninvasive ambulatory electrocardiographic monitoring such as Holter monitor might not be a good predictor where mortality is the endpoint. Others have argued that the failure to induce ventricular tachycardia or fibrillation by some drug assessed by the invasive electrophysiologic study might be a good alternative independent predictor of recurrence of arrhythmia. Consequently the ESVEM trial sponsored by the U.S. National Institute of Heart, Lung, and Blood was the first large prospective randomized trial conducted to compare the two methods for predictability of long-term recurrence of arrhythmia by the short-term efficacy assessed by the two methods (ESVEM investigators, 1989; Mason and ESVEM investigators, 1993a, 1993b). For the assessment of predictability, a correlation was obtained using the difference in the recurrence rates of arrhythmia with the short-term efficacy by both methods. An inpatient enrichment phase was elected to identify patients in whom a test drug exhibited a short-term efficacy assessed by either of the two methods. The enrichment design is illustrated in Figure 5.5.3. Patients who met the entry criteria and a 48-hour Holter monitoring and electrophysiologic study criteria were randomized to one of two parallel groups for the two methods in order to assess the short-term drug efficacy. For the assessment of the short-term drug efficacy, the first group employed noninvasive ambulatory electrocardiographic monitoring, while the second group applied the invasive electrophysiologic study. Within each group the patients received up to six arrhythmia agents in a random order until one drug was predicted to be efficacious or until all drugs were tested. A test drug is classified as efficacious when assessed by electrocardiographic monitoring during the inpatient enrichment phase if the following efficacy criteria are met: (1) 70% reduction in mean VPC count, (2) 80% reduction in VPC pair count, (3) 90% reduction in mean ventricular tachycardia counts, and (4) absence of any runs of ventricular tachycardia longer than 15 seconds. Drug efficacy evaluated by the electrophysiologic study during the enrichment phase is defined as failure to induce a run of ventricular tachycardia longer than 15 seconds with V1V2V3 stimulation at the right ventricular apex. If a drug was proved to be efficacious for a patient during the enrichment phase, then he or she was discharged from the hospital for the long-term follow-up with the drug, and the accuracy of the prediction of efficacy was determined during the long-term follow-up. Patients in whom no drugs were proved to be effective during the enrichment phase were not randomized and were withdrawn from the study. However, the vital signs and the recurrence of arrhythmia of the withdrawn patients were monitored.

The application of an enrichment design in the ESVEM study narrowed the patient population to a very highly selective minority of sustained ventricular tachycardia associated with coronary disease (Ward and Camm, 1993). The patients had to have frequent ventricular ectopic beats by the Holter monitor and inducible ventricular tachycardia or fibrillation by electrophysiological study.

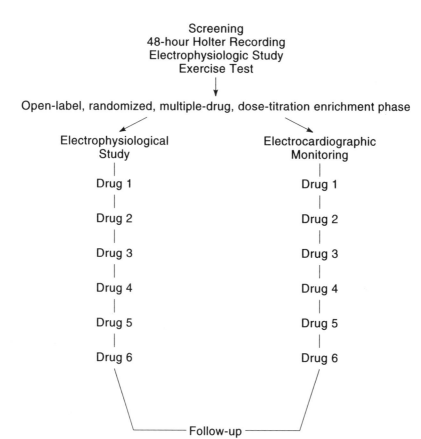

Figure 5.5.3 An enrichment design for electrophysiologic study versus electrocardiographic monitoring (ESVEM). (Source: The ESVEM Investigators, 1989).

Any meaningful clinical inference was severely limited by this constraint. In addition patients were discharged from the hospital as soon as the first drug proved to be effective. An average of 2.6 (out of 6) drug assessments were performed for each patient. Hence the patients went to the long-term follow-up on a drug that was effective but not necessarily the optimal one for the patient. Although the investigator did not know what the next drug to be tested was until the proceeding drug failed, he or she was not blinded to the drug currently being tested. Therefore bias could occur in clinical judgment or in the interpretation of the results. Mason and the ESVEM (1993b) also compared the differences between the active agent sotolol and the other drugs whose efficacy was defined as (1) the drug was tolerable during the enrichment phase, (2) the drug was predicted to be effective, (3) arrhythmia did not occur, and (4) the drug was not discontinued because of an adverse event. However, the results from this analysis (despite its intention-to-treat database) are biased because (1) the assignment of patients to drugs was not at random, (2) patients received the first drug that met the short-term efficacy criteria, and (3) there was no placebo concurrent control which was proven to be so crucial in CAST.

In summary, an enrichment design is a part of screening process that further restricts the target population to a small selective group. However, sometimes it is still not possible to distinguish this small group from other patients with the same ailment in terms of demographic and other prognostic factors. On the other hand, statistical methods for analysis based on the data from the enrichment phase and the double-blind or primary phase of the trial are not fully developed due to (1) the lack of randomization and (2) different methods of randomization for the enrichment phase. Therefore the statistical analysis and clinical interpretation for a trial using an enrichment design remain a challenge to both statisticians and clinicians.

5.6 DISCUSSION

In this chapter we discussed several basic statistical designs, the parallel design, the crossover design, the titration design, and the enrichment design, all of which are commonly employed in clinical trials at various stages of clinical development. Each design has its own merits and limitations under different circumstances. How to select an appropriate design when planning a clinical trial is an important question. The answer to this question depends on many factors, namely those summarized below:

1. Number of treatments to be compared.
2. Characteristics of the treatment.
3. Study objective(s).
4. Availability of patients.
5. Inter- and intrapatient variabilities.
6. Duration of the study.
7. Dropout rates.

For example, when choosing a design from a parallel design and a crossover design, if the intrapatient variability is the same as or larger than the interpatient variability, the inference on the difference in treatments will be the same regardless of which design is used. Actually, a crossover design in this situation would be a poor choice, since blocking results in the loss of some degrees of freedom and will actually lead to a wider confidence interval on the difference between treatments. If a clinical study compares more than three treatments, a crossover design may not be appropriate. The reasons are (1) it may be too time-consuming to complete the study, since a washout is required between treatment periods, (2) it may not be desirable to switch medications too frequently for each subject due to medical concerns, (3) too many treatment periods may increase the number of dropouts, and (4) the disease status may change from treatment period to treatment period. In this case a balanced incomplete block design is preferred. However, if we compare several test treatments with a placebo control, the within-patient comparison may not be reliable, since the patients in some sequences do not receive the placebo control. If the drug has a very long half-life, and/or it possesses a potential toxicity, or there are carryover effects, then a parallel design may be a possible choice. With this design the study avoids a possible cumulative toxicity due to the carryover effects from one treatment period to the next. In addition the study can be completed in less time compared to that of a crossover design. However, the drawback is that the comparison is made based on the interpatient variability. If the interpatient variability is large relative to the intrapatient variability, the statistical inference on the difference between treatments is not reliable. Even if the interpatient variability is relatively small, a parallel design may still require more patients in order to reach the same degree of precision achieved by a crossover design. In practice, a crossover design, which can remove the interpatient variability from the comparison between treatments, is often considered to be the design of choice if the number of treatments to be compared is small, say no more than three. If the drug has a very short half-life (i.e., there may not be carryover effects if the length of washout is long enough to eliminate the residual effects), a crossover design may be useful for the assessment of the intrapatient variability provided that the cost for adding one period is comparable to that of adding a patient. In summary, choosing an appropriate design for a clinical trial is an important issue in the development of a study protocol. The selected design may affect the data analysis and the interpretation of the results. Thus all the factors listed above should be carefully evaluated before an appropriate design is chosen.

It, however, should be noted that one of the primary assumptions for a crossover design is that the disease condition remain stable during the study. In practice, this assumption is usually not met. As a result one of the major disadvantages is that spontaneous changes in the disease condition may occur during the study. In this case, although we may establish baseline at each treatment period to eliminate the residual effect, the treatment effect may be confounded with the residual effect. Therefore a crossover design may not be feasible when there are carryover effects. Although a parallel design is not capable of identifying and removing the interpatient variability from the comparison between treatments due to its simplicity and easy implementation, it is probably the most commonly used design in clinical phase II and III studies.

CHAPTER 6

Classification of Clinical Trials

6.1 INTRODUCTION

For approval of a new drug, the FDA requires that substantial evidence of the effectiveness of the drug be provided through the conduct of adequate and well-controlled clinical trials. The characteristics of an adequate and well-controlled clinical study include adequate methods for bias reduction such as double-blinding, randomization of treatment assignments, a well-defined patient population, and adequate and valid statistical methods for data analysis.

Basically there are several different types of clinical trials, though they are not mutually exclusive. The different types of clinical trials include multicenter trials, sequential trials, active control trials, combination trials, and equivalence trials. They can be further applied in different situations depending on the objectives of the planned clinical trials. For example, a multicenter trial may be desirable because it provides replication and generalizability of clinical results to the targeted patient population. When the planned clinical trial requires a large number of patients or the duration of the study is relatively long, the investigator may wish to conduct a sequential trial. The purpose of a sequential trial is to determine whether the trial should be stopped based on results from interim analyses or administrative observations at certain time points or at points when the trial reaches a certain number of patients or events. An active control trial may be necessarily conducted to establish the efficacy of a new drug when the patients under study are very ill or have severe or life-threatening diseases. When the drug under study consists of more than one active ingredient, a combination trial is required to assess the treatment effect by taking into account potential drug-to-drug interaction. In clinical practice several drug products with different doses may have a similar therapeutic effect for a given indication. An equivalence trial is helpful in determining whether two drugs with different doses have similar efficacy and safety for treating the targeted patient population. Two drug products are said to be (therapeutically) equivalent if they have a similar therapeutic effect. Therefore two equivalent drug products can be used exchangeably as alternative drugs. In addition, for the approval of a generic drug, a bioequivalence trial is required by the FDA to show that the generic copy is bioequivalent to the innovator drug product in terms of drug absorption. The fundamental assumption of bioequivalence trials is that if two drug products are shown to be bioequivalent, they are therapeutically equivalent.

In the next section the limitations of single-site studies and the feasibility of multicenter trials are briefly discussed. Issues and concerns for the use of active control trials are given in Section 6.3. Some statistical considerations for combination trials and equivalence trials are outlined in Sections 6.4 and 6.5, respectively. A discussion is presented in Section 6.6. Note that some practical issues and concerns of interim analysis for sequential trials will be discussed in Chapter 11 and hence will not be covered in this chapter.

6.2 MULTICENTER TRIAL

When conducting a clinical trial, it may be desirable to have the study done at a single study site if (1) the study site can provide an adequate number of relatively homogeneous patients that represent the targeted patient population under study and (2) the study site has sufficient capacity, resources, and supporting staff to sponsor the study. One of the advantages for a single-site study is that it provides consistent assessment for efficacy and safety in a similar medical environment. As a result a single-study site can improve the quality and reliability of the collected clinical data and consequently the inference of the clinical results. However, a single-site study has some limitations and hence may not be feasible in many clinical trials. These limitations include the availability of patients and resources in a single site. If the intended clinical trial calls for a large number of patients, a single-site study may take a long time to complete the study, since qualified patients may not be available at the same time. Besides, even if qualified patients were available at the same time, the single-study site may not have sufficient resources to enroll these patients at the same time. In practice, qualified patients are usually enrolled sequentially at different times until the required number of patients is reached to achieve a desired power for the detection of a clinically meaningful difference.

Goldberg and Kury (1990) indicate that a single-site study may not be appropriate in situations where (1) the intended clinical trials are for relatively rare chronic diseases and (2) the clinical endpoints for the intended clinical trials are relatively rare (i.e., require a large number of patients to observe an incidence). For example, as observed by Goldberg and Kury (1990), if the intended clinical trial is to study a relatively rare chronic disease such as polycythemia vera (a disease characterized by an elevated hematocrit which has as natural consequences, stroke, hemorrhage, leukemia, and death), a single-site study is not feasible because it is unlikely that a single site is able to recruit a sufficient number of patients within a relatively short time frame to achieve the desired power for the detection of a meaningful clinical difference. Even if the single site is able to recruit the required number of patients, it will take a relatively longer time to complete the study. As Goldberg and Kury (1990) point out, if the clinical endpoint for the intended clinical trial is relatively rare such as mortality in clinical trials for acute myocardial infarction, then the single site may be required to enroll a large number of patients in order to observe a mortality. In such case, a single-site study is of little practical interest.

To overcome the disadvantages of a single-site study, the multicenter study is usually considered. A multicenter study is a single study involving several study centers (sites or investigators). In other words, a multicenter trial is a trial conducted at more than one distinct center where the data collected from these centers are intended to be analyzed as a whole. At each center an identical study protocol is used. A multicenter trial is a trial with a center or site as a natural blocking or stratified variable that provides replications of clinical results. A multicenter trial should permit an overall estimation of the treatment difference for the targeted patient population across various centers. In what follows we will discuss the impact of treatment-by-center interaction and some practical issues when planning a multicenter trial.

Treatment-by-Center Interaction

The FDA guideline suggests that individual center results should be presented for a multicenter study. In addition the FDA suggests that statistical tests for homogeneity across centers (i.e., for detecting treatment-by-center interaction) be provided. The significant level used to declare the significance of a given test for a treatment-by-center interaction should be considered in light of the sample sizes involved. Any extreme or opposite results among centers should be noted and discussed. For the presentation of the data, demographic, baseline, and postbaseline data as well as efficacy data should be presented by center, even though the combined analysis may be the primary one. Gail and Simon (1985) classify the nature of interaction as either quantitative or qualitative. A quantitative interaction between treatment and center indicates that the treatment differences are in the same direction across centers but the magnitude differs from center to center, while a qualitative interaction reveals that substantial treatment differences occur in different directions in different centers. Figure 6.2.1 depicts situations where there are no quantitative and qualitative treatment-by-center interactions. As an illustration, consider the following two examples which exhibit quantitative and qualitative treatment-by-center interactions.

As indicated by Ebbeling and Clarkson (1989), muscle soreness and elevations in serum creatine kinase (CK) can be used as a biochemical marker for skeletal muscle injury. Recently a study was conducted to evaluate the hypothesis that exposure of skeletal muscle to exercise-induced stress in combination with prior administration of a study drug (e.g., drug A) will produce a greater increase in CK than the effects of exercise alone. This study was a randomized, two-arm parallel study comparing the study drug A plus exer-

MULTICENTER TRIAL

(a) No Interaction

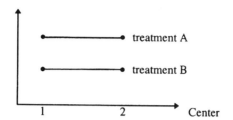

(b) Quantitative Interaction

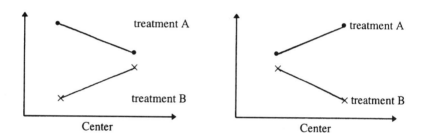

(c) Qualitative Interaction

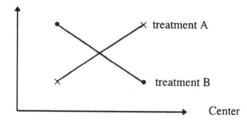

Figure 6.2.1 Treatment-by-center interaction.

cise and a placebo control plus exercise which was conducted at two distinct study centers. Table 6.2.1 lists the mean CK values at 24 hours posttreatment. Figure 6.2.2 provides a preliminary investigation of a potential treatment-by-center interaction. The plot shows that there is a potential quantitative treatment-by-center interaction. This was confirmed by an analysis of variance that includes the terms of treatment, the center, and the treatment by center (p-value = 0.08 < 0.1). An overall estimate of the treatment effect

Table 6.2.1 Mean CK Values at 24 Hours Post-Treatment

Treatment	Center 1	Center 2
Drug A	15	9
	355.87	549.78
	(69.85)	(237.27)
Placebo	16	11
	213.31	208.64
	(26.34)	(35.79)

Note: The top values and the values in the parentheses are corresponding sample sizes and standard errors.

is given by 0.43 ± 0.16 with the corresponding 95% C.I. of (0.27, 0.59) (p-value = 0.012). Note that a subgroup analysis by center reveals that there is a significant treatment effect at the second center (p-value = $0.029 < 0.05$).

As another example, consider a clinical trial for the assessment of a study drug's efficacy in treating mild to moderate hypertension. The study was conducted as a randomized placebo-controlled multicenter trial that involved 219 patients in 27 study centers. Table 6.2.2 lists the mean seated diastolic blood pressures after six weeks of treatment. A plot of mean seated diastolic pressures against study centers is given in Figure 6.2.3. In the figure the centers are

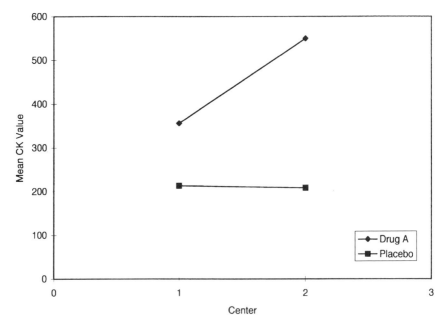

Figure 6.2.2 The mean CK values at 24 hours post-treatment.

Table 6.2.2 Mean Change From the Baseline for Seated Diastolic Blood Pressure After Six Weeks of Treatment

Site	N	Placebo	Drug	Difference
1	6	2.44	−17.44	−19.88
2	9	1.07	−13.00	−14.07
3	29	−1.74	−12.07	−10.33
4	5	−2.44	−11.73	−9.29
5	5	−1.70	−9.24	−7.54
6	5	−0.67	−7.89	−7.22
7	12	−3.89	−11.06	−7.17
8	4	−11.00	−18.00	−7.00
9	6	1.33	−4.67	−6.00
10	24	−1.18	−6.77	−5.59
11	7	−8.00	−13.50	−5.50
12	7	−6.44	−11.75	−5.31
13	8	−4.83	−10.00	−5.17
14	4	−1.00	−5.93	−4.93
15	6	−6.44	−11.33	−4.89
16	7	−7.83	−12.44	−4.61
17	8	−6.17	−10.47	−4.30
18	4	−2.07	−5.67	−3.60
19	11	−4.56	−7.93	−3.37
20	6	−11.78	−11.67	0.11
21	5	−8.67	−7.53	1.14
22	11	−14.67	−13.20	1.47
23	6	−14.22	−12.00	2.22
24	8	−13.67	−10.50	3.17
25	9	−8.00	−2.80	5.20
26	10	−8.27	0.67	8.94
27	5	−13.11	−4.00	9.11
Total	219	−5.50	−9.78	−4.36

grouped according to the magnitude of differences in mean change from the baseline. That is, the centers with a difference in mean change from baseline of −19.89 are labeled site 1 and the centers with a difference in mean change from baseline of 9.11 are labeled site 27. As can be seen, 19 centers (70.4%) show the difference in the positive direction, while 8 centers (29.6%) are in the negative direction. An analysis of variance indicates that a significant qualitative interaction between treatment and group has occurred (p-value = 0.01). If we ignore the interaction and perform an analysis of variance, an overall estimate of the treatment effect based on the difference in the mean change from the baseline is given by −4.36 ± 1.9 or (−6.26, −2.46) with a p-value less than 0.01. This positive significant result, however, is somewhat misleading because it is not reproducible in those 8 out of 27 centers. In other words, there is a relatively high chance that we may observe a totally opposite result if we are

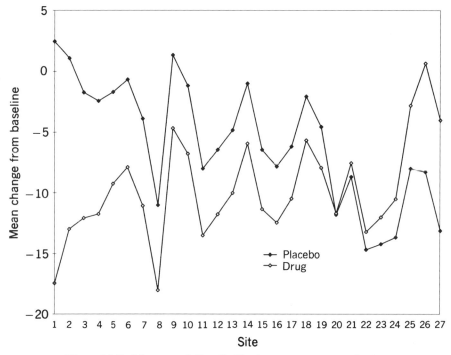

Figure 6.2.3 Mean seated diastolic blood pressures versus study site.

to randomly select a center from a pool of centers and repeat the study with the same protocol. Besides, the 8 centers involve 60 patients, who show different results, or about 27.4% of the patients under study. Therefore the reproducibility and generalizability of the results to the targeted patient population and the treatment setting is questionable.

Practical Issues

As was indicated earlier, a multicenter trial with a number of centers is often conducted to expedite the patient recruitment process. Although these centers usually follow the same study protocol to evaluate the efficacy and safety of a study drug, some design issues need to be carefully considered. These design issues include the selection of centers, the randomization of treatments, and the use of a central laboratory for laboratory evaluations. The selection of centers is important to constitute a representative sample for the targeted patient population. However, in multicenter trials the centers are usually selected based on convenience and availability. When planning a multicenter trial with a fixed sample size, it is important to determine the allocation of the centers and the number of patients in each center. For example, if the intended clinical trial calls for 100 patients, the sponsor may choose to have 5 study centers with 20 patients in each, 10 study centers with 10 patients in each, or 20 study centers with 5 patients in each. The chance for observing a significant treatment-by-center interaction for the selection of 20 centers is expected to be higher than those for the selections of 10 centers and 5 centers. If there are potential dropouts, the selection of 20 centers may result in a number of small centers (i.e., with a few patients in the center). For comparative clinical trials, the comparison between treatments is usually made between patients within centers. If there is no treatment-by-center interaction, the data can be pooled for analysis across centers. Therefore it is not desirable to have too few patients in each center. A rule-of-thumb is that the number of patients in each center should not be less than the number of centers. In this case the selection of 10 sites for a fixed sample size of 100 patients may be preferable. Some statistical justification for this rule can be found in Shao and Chow (1993). Once the centers are selected, it is also important to assign treatments to patients in a random fashion within each center. The methods of randomization described in Chapter 4 should be applied. The issue whether a central laboratory will be used for the laboratory testing of samples collected from different centers has a significant impact on the assessment of efficacy and safety of the study drug. A central laboratory provides a consistent assessment for laboratory tests. If a central laboratory is not used, the assessment of laboratory tests may differ from center to center depending on the equipment, analyst, and laboratory normal ranges used at that center. In such a case possible confounding makes it difficult to combine the laboratory values obtained from the different centers for an unbiased assessment of the safety and efficacy of the study drug.

Another practical issue of great concern in a multicenter trial is statistical analysis of the collected data from each center. As was indicated earlier, if there is no evidence of treatment-by-center interaction, the data can be pooled for analysis across centers. The analysis with combined data provides an overall estimate of the treatment effect across centers. In practice, however, if there are a large number of centers, we may observe significant treatment-by-center interaction, either quantitative or qualitative. As indicated by Gail and Simon (1985), the existence of a quantitative interaction between treatment and center does not invalidate the analysis in pooling data across centers. An overall estimate of the average treatment difference is statistically justifiable, and it provides a meaningful summary of the results across centers. On the other hand, if a qualitative interaction between treatment and center is observed, the overall or average summary statistic may be misleading and hence considered inadequate. In this case it is preferable to describe the nature of the interaction and to indicate which centers contribute toward the interaction. In practice, if there are too many centers, the trial may end up with big imbalances among centers, in that some centers may have a few patients and others a large number of patients. If there are too many small centers (with a few patients in each center), we may consider the following two approaches. The first approach is to combine these small centers to form a new *center* based on their geographical locations or some criteria prespecified in the protocol. The data can then be reanalyzed by treating the created *center* as a regular center. Another approach is to randomly assign the patients in these small centers to those larger centers and reanalyze the data. This approach is valid under the assumption that each patient in a small center has an equal chance of being treated at a large center.

As was indicated before, for approval of a new drug, two well-controlled clinical trials are conducted to provide substantial evidence of the effectiveness and safety of the new drug. Since a multicenter trial involves more than one distinct center, whether a single multicenter trial is equivalent to that of two separate trials has become an interesting question. The FDA indicates that an a priori division of a single multicenter trial into two studies is acceptable for establishing the reproducibility of drug efficacy to NDA approval. Nevius (1988) proposes a set of four conditions under which evidence from a single multicenter trial would provide sufficient statistical evidence of efficacy. These conditions are summarized below:

1. The combined analysis shows significant results.
2. There is consistency over centers in terms of direction of results.
3. There is consistency over centers in terms of producing nominally significant results in centers with sufficient power.
4. Multiple centers show evidence of efficacy after adjustment for multiple comparisons.

To address the consistency over centers, Chinchilli and Bortey (1991) propose the use of the noncentrality parameter of an F distribution as a means of testing for consistency the treatment effect across centers. As an alternative, Khatri and Patel (1992) and Tsai and Patel (1992) consider a multivariate approach assuming random center effects. Huster and Louv (1992) suggest the use of the minimax statistic as a method that can quantify the amount of evidence for reproducibility of treatment efficacy in a single multicenter trial. To describe the minimax statistic, we consider an example given by Huster and Louv (1992). Suppose that there is a four-center clinical trial. Denote the four centers by a, b, c, and d. Consider all possible divisions of these four centers into two mutually exclusive sets of centers. The minimax statistic can then be summarized as follows: First, for each of the seven divisions, find the p-value for the drug effect (adjusted for the center) for each study. Then find the maximum of these two p-values for each division. Second, find the minimum of these maximum p-values across all divisions. The rationale for choosing the maximum at the first step is that if a particular division is to show reproducibility of a drug effect, then both studies within that division should exhibit a significant drug effect. On the other hand, the rationale for choosing minimum at the second step is that the optimal division is one where the evidence against the null hypothesis from both studies in the greatest. As indicated by Huster and Louv (1992), the minimax statistic approach provides a reasonable assessment for the amount of evidence for reproducibility of treatment efficacy in a single multicenter trial. In addition it gives an objective answer to the question whether a second confirmatory study is required.

Note that the analysis of a multicenter trial is different from that of a meta-analysis. The analysis of multicenter trials combines data observed from each study center; the data are generated based on the methods prospectively specified in the same study protocol with the same method of randomization and probably at the same time. In contrast, a meta-analysis combines data retrospectively observed from a number of independent clinical trials, which may be conducted under different study protocols with different randomization schemes at different times. In either case the treatment-by-center interaction for multicenter trials or treatment-by-study interaction for meta-analyses must be carefully evaluated before pooling the data for analysis.

6.3 ACTIVE CONTROL TRIAL

For approval of a new drug, it is required by regulatory agencies that a clinical trial be conducted comparing the new drug with a control. Section 314.126 in Part 21 of CFR indicates that there are four kinds of control including placebo control, no treatment control, positive control, and historical control. A brief description of these controls is given in Section 3.3. A control is usually referred to as positive if an active treatment has been shown to be effective and safe. In practice, a positive control is also known as an active control. An adequate and well-controlled trial comparing a new drug and an active control agent is usually referred to as an active control trial.

In practice, it may not be ethical to conduct a placebo control study with very ill patients or patients with severe or life-threatening diseases in order to establish the efficacy of a new drug. In this case an active control trial is often considered as an alternative trial to evaluate the effectiveness and safety of the new drug by comparing it with an active control agent that has been shown to be effective and safe for the intended disease. In recent years the use of active control in clinical trials has become increasingly popular. Its foremost appeal over the placebo control is based on ethical considerations. In practice, it is often easier to enroll patients into a trial if they are guaranteed an active treatment.

In this section our emphasis will be placed on the primary objectives, practical issues, and considerations for conducting active control trials in clinical development.

Primary Objectives

For a test drug the primary objective of an active control trial could be (1) to establish the efficacy of the test drug, (2) to show that the test drug is equivalent to an active control agent, or (3) to demonstrate that the test drug is superior to the active control agent. For the first objective Pledger and Hall (1986) point out that active control trials offer no direct evidence of effectiveness of the test drug. The only trial that will yield direct evidence of the test drug's effectiveness is a placebo-controlled trial that compares the test drug with a placebo. If the test drug has previously proved to be efficacious compared to placebo, then the goal of the active control trial is only to compare it to the active control agent. In other words, the goal is to show that either the test drug is equivalent to the active control agent or it is superior to the active control agent. Equivalence is usually referred to as a difference less than some specified amount, while superiority is defined as a difference in the preferred direction of greater than some amount. In the situations where we are to evaluate equivalency and superiority of the test drug, the following hypotheses are often tested:

H_0: The test drug is not equivalent to the active control;
vs. H_a: The test drug is equivalent to the active control;

and

H_0: The test drug is the same as the active control;
vs. H_a: The test drug is superior to the active control.

Lamborn (1983) indicates that it is important to give a more specific definition of the scientific or medical question to be answered in active control trials. This is critical especially when different dropout rates due to intolerance are likely and/or the comparison of dose-response curves is intended. For the issue on dropout rates, Lamborn (1983) points out the example of active control trials comparing new treatments with aspirin are infeasible in any long-term trial due to the fact that aspirin has a high rate of intolerance and most new treatments are better tolerated. As a result treatment groups are likely to be comparable after a few months.

Senn (1993) examines the performance of active control trials for demonstration of equivalence between treatments in terms of the competence of the trials using a simple semi-Bayesian model. A competent trial is defined as a trial that can detect a difference between treatments where it exists. The results indicate that the posterior probability of nonequivalence given an observed difference can only be increased beyond a certain limit by reducing its incompetence. Senn (1993) indicates that active control trials are more problematic than the classical clinical trials, since they cannot be solved by simply exchanging the usual roles of null and alternative hypotheses. Makuch and Johnson (1989) suggest that active control trials should be used for showing that an experimental treatment is equivalent in efficacy to a standard therapy rather than for detecting a significant difference between treatments.

Table 6.3.1 Comparisons of Possible Outcomes for an Active Control Trial and a Placebo Control Trial

Outcome of A vs. B in an Active Control Trial	Outcomes of A vs. B vs. P in a Placebo Control Trial
AB	$AB > P$; ABP, $A > P$; PAB, $P > B$; $P > AB$; APB; ABP; PAB
BA	$BA > P$; BAP; $B > P$; PBA, $P > A$; $P > BA$; BPA; BAP; PBA
$A > B$	$A > BP$; $A > B > P$; $A > PB$; ABP, $A > B$; $AP > B$; $A > P > B$; $PA > B$; $P > A > B$
$B > A$	$B > AP$; $B > A > P$; $B > PA$; BPA, $B > A$; $BP > A$; $B > P > A$; $PB > A$; $P > B > A$

Note: (1) A = test drug; B = active control; P = placebo. (2) Left to right order indicates decreasing improvement and ">" denotes a statistically significant difference.
Source: Pledger and Hall (1986).

Issues in Active Control Trials

In practice, since active control is known to be about an effective agent, it is often assumed that showing the two treatments are equivalent in an active control study is to demonstrate that both treatments are effective. However, this assumption is not necessarily correct and cannot be verified from the data obtained from the active control trial. Pledger and Hall (1986) compare the possible outcomes of an active control trial comparing a test drug (denoted by A) and an active control agent (denoted by B) with a placebo control trial comparing A, B, and a placebo P. Table 6.3.1 summarizes the possible outcomes. It can be seen from Table 6.3.1 that the equivalence and superiority to the active control agent do not guarantee that the test drug is effective. As an example, consider the case where A is superior to B, denoted by A > B (i.e., a statistically or clinically significant difference between A and B is observed). In this case, it is possible that the actual outcome is A > B > P (i.e., both A and B are effective), A > P > B (i.e., only A is effective), or P > A > B (i.e., both A and B are ineffective). For another example, when A and B are equivalent, it may fall in one of the following possible outcomes: AB > P (i.e., both A and B are equally effective); ABP, A > P (i.e., there is no significant decreasing improvement among A, B, and P, though, A is superior to P and hence is effective); or P > AB (i.e., both A and B are ineffective).

As discussed above, showing equivalence may imply that A and B are both equally effective or equally ineffective. For some drug products such as antianxiety agents, antidepressants, antianginal agents, or appetite suppressants, the test drugs may not necessarily beat the placebo. It is then not uncommon to observe an effective agent to fail in a particular study if the study is poorly designed or conducted or if the sample size of the study is inadequate. Note that some drug products such as antibiotics or antiarrhythmics, on the other hand, almost always show a drug-placebo difference.

Considerations for Using Active Control Trials

As indicated in the FDA guideline, if the trial is an active control study intended to show equivalence between the test drug and active control, there should be assessments of (1) the response of the standard agent in the present trial compared to previous studies of similar design that included a comparison with a placebo and (2) the ability of the study to have detected differences between the treatments of a defined size, such as, by proving confidence limits for the difference between the drug and active control and/or the power to detect a difference between the treatments of specified size. Along this line, Pledger and Hall (1986) suggest that supplementary information, which is often obtained from previous trials conducted under a design similar in all important aspects such as patient population, dosage requirements, response assessment methods, and control of concomitant therapy, be provided.

Since showing the two treatments are equivalent in an active control study may imply that both are effective or that neither was, Temple (1982) recommends the following fundamental principle for active control trials.

> If we cannot be very certain that the positive control in a study would have beaten a placebo group, had one been present, the fundamental assumption of the positive control study cannot be made and that design must be considered inappropriate.

Under this principle, a poor candidate for a positive control study would be any condition in which a large spontaneous or placebo-response occurs, or in which there is great day-to-day variability, or in which effective drugs are not easily distinguished from the placebo. Temple (1982) also indicates that positive control trials may serve as primary evidence of effectiveness if the fundamental assumption of the positive control is correct, which usually cannot be verified. However, the following situations may be considered to support a positive control study. Included are situations (1) where there is a retrospective review of known placebo-controlled studies of the proposed positive control to show that the drug regularly can be shown superior to a placebo, (2) where active control trials are conducted utilizing a similar patient population and similar procedures (e.g., dose, dosage regimens, titration methods, response assessment methods, and control of concomitant therapy), and (3) where there is an estimate of the effect size of the placebo response.

The interpretation of the results from active control trials is a critical issue. If the test drug is inferior, or even indistinguishable from a standard therapy, the results are not readily interpretable. In the absence of placebo controls, one does not know if the inferior test medicine has any efficacy at all. Similarly equivalent performance may reflect simply a patient population that cannot distinguish between two treatments that differ considerably from each other, or between an active drug and a placebo. Note that in certain clinical conditions such as serious depressive states, it is difficult to evaluate the test drug in active control trials because of the delay in drug effects and the high rate

Table 6.3.2 Results of Six Trials Comparing a New Antidepressant Imipramine and a Placebo

Study	Baseline	End Point Change from Baseline			New Agent versus Imipramine	
		Placebo[a]	New Agent[a]	Imipramine[a]	p-value	Power
R301	23.9	−9.1 (36)	−10.5 (33)	−11.1 (33)	0.78	0.40
G705	26.0	−12.1 (36)	−13.0 (39)	−12.6 (30)	0.86	0.45
C311 (1)	28.1	−9.2 (13)	−8.7 (11)	−7.8 (11)	0.81	0.18
V311 (2)	29.6	−6.1 (7)	−22.3 (7)	−20.1 (8)	0.63	0.09
F313	37.6	−15.6 (8)	−15.7 (7)	−15.7 (8)	1.00	0.26
K317	26.1	−15.6 (36)	−14.9 (37)	−15.3 (32)	0.85	0.33

[a]Numbers in the parnetheses are the corresponding sample size.
Source: Temple (1983).

of spontaneous improvement, which is not readily distinguished from placebo in control trials. How much solace can one derive from a trial that shows no difference between a new putative antidepressant and a standard tricyclic? Note that for antidepressants, the European authority resists the use of placebo control trials. The U.S. FDA prefers placebo-control trials. It is then suggested that for trials including a new drug, both positive and placebo controls be used.

Examples

Temple (1983) considers six randomized double-blind trials comparing a new drug, a placebo, and a standard. For these trials the patient entry criterion was a Hamilton Depression (Ham-D) scale of at least 18. Table 6.3.2 summarizes the results of these six trials. As Table 6.3.2 indicates, the sample sizes of these six trials range from 15 to 69. Each trial showed considerable improvement in the Ham-D scale after four weeks of treatment. The two active drugs were indistinguishable (p-value $\geq$ 0.63). The six trials provided substantial evidence on the efficacy of the test drug. Note that although there is low power for detection of a 30% difference (ranging from 0.09 to 0.45), the results are highly supportive of the drug's effectiveness because the observed difference were greater than 30% from baseline which is considered a clinically meaningful difference.

A three-way comparison reveals that there are no significant differences among placebo, new agent, and the imipramine except for one study (V311(1)) showing a detectable placebo active drug difference. The results suggest that the historical assumption about imipramine may be erroneous. This example argues against the idea that positive control studies of putative antidepressants can serve as a sole basis for concluding that a drug is effective.

In another example, Temple (1983) indicates that antihypertensive drugs seem a priori a reasonable class for which positive control trials might be appropriate. These antihypertensive drugs regularly can be shown to be superior to

Table 6.3.3 Results of Five Antihypertensive Trials of Marketed Beta-Adrenergic Blocking Agents

Study	N (Drug/Placebo)	Change in Blood Pressure mm Hg (Post/Baseline)		Difference $D - P$
		Drug	Placebo	
Metoprolol				
Reeves	11/12	−17/11	0/0	−17/11
Bowen	18/17	−19/14	−11/6	−8/8
Pindolol				
Study 25	29/28	−9/7	−2/3	−7/4
Study 24	28/30	−14/11	−5/4	−9/7
Atenolol				
Curry	15/12	−17/12	−5/4	−12/8
Nadolol				
Multicenter	50/50	−9/10	−3/2	−6/8
Timolol				
Multicenter	120/144	−11/10	+1/−1	−12/8

Source: Temple (1983).

the placebo even in a fairly small trial. Table 6.3.3 shows the results of five antihypertensive drugs including metoprolol, pindolol, atenolol, nadolol, and timolol. It appears that these drugs are all distinguishable from the placebo and that all trials showed statistically significant drug-placebo difference. Note that the placebo effect can be large. For example, in the metoprolol study the placebo response of −11/6 mm Hg was about as large as the drug response in pindolol study which was 25 and not too far from the drug response in the nadolol or timolol trials. Despite the large placebo response, it may indeed prove possible to define conditions in which one can rely on evidence from positive control trials as the primary evidence of effectiveness.

6.4 COMBINATION TRIAL

As was indicated in Chapter 1, a treatment is defined as a combination therapy if it consists of more than one active ingredient. A treatment is called a combined product if it is a combination of different pharmaceutical entities such as drugs, biologics, and/or medical devices. Since the general principles of statistical designs and analyses for the assessment of the effectiveness and safety are the same for both combination therapy and combined product, in this section our discussion will be focused on the evaluation of a fixed dose for the combination of two or more active ingredients.

In recent years the search and development of combination therapy for various diseases have become a popular issue in the pharmaceutical industry (e.g., see Lasgana, 1975; Mezey, 1980). In clinical practice it is not uncommon to observe an enhanced therapeutic effect for a combination drug in which each

Table 6.4.1 **Maximum Tolerable Doses of Single Agents for Treatment of Advanced Ovarian Cancer**

Drug	Maximum Tolerable Dose (mg/m^2/week)	Organ System of Dose-Limiting Toxicity
Cisplatin	35	Renal
Carboplatin	100	Bone marrow
Cyclophosphamide	400	Bladder
		Bone marrow
Hexamethylmelamine	2200	Gastrointestinal
Dozorubicin	30	Bone marrow
Paclitaxel	250	Neutropenia

Source: Adapted from Table 1 in Simon and Korn (1991) and Table 1 in Rowinsky and Donehower (1995).

component (at a certain dose) acts through a different mechanism when applied alone. For example, the combination of a diuretic with a beta-blocker has an enhanced therapeutic effect for the treatment of patients with hypertension. In practice, however, each component of the combination drug must be administered alone at a high-dose level in order to achieve the desired therapeutic effect. In this case the component drugs may cause severe and/or sometimes fatal adverse events. Therefore it is desirable to control the synergistic effect among these components so that a combination at the lower-dose levels will reach the same or better effectiveness with less severe clinical toxicity. For example, Table 6.4.1 lists the conventional maximum tolerable doses for chemotherapeutic agents in the treatment of patients with advanced ovarian cancer. Each of these drugs induces serious organ-specific dose-limiting toxicity that jeopardizes the efficacy when they are administered alone. It is therefore imperative to search for different combinations of these agents at lower doses so that a higher equivalent cytotoxic dose can be achieved to provide a clinically meaningfully improvement in response rate and overall survival. For example, McGuire et al. (1996) report that paclitaxel and cisplatin provided not only a significantly higher response rate than a standard therapy of cisplatin plus alkylating agent cyclophosphamide (73% versus 60%, p-value = 0.01) for stage III and stage IV ovarian cancer but also a significant improvement in median survival (38 vs. 24 months, p-value < 0.001).

After the initial successful culture of *Helicobacter pylori* (or *H. pylori*) in gastric-biopsy specimens from patients with histologic gastritis in 1982 (Warren, 1982), it was recognized that more than 95% of patients with duodenal ulcers and more than 80% of patients with gastric ulcers are infected with *H. pylori* (Walsh and Peterson, 1995). Recently the Consensus Development Conference on the U.S. National Institute of Health recommended that antimicrobial therapy be used to treat patients with ulcers who are also infected with *H. pylori* (NIH Consensus Development Panel, 1994). However, it is not easy

Table 6.4.2 Combinations Used in the Eradication of *H. pylori*

Combinations	Dose	Duration	Eradication Rate (%)	Overall Adverse Events (%)
Bismuth	564 mg/4 times/day	14–15 days	89%	32%
Metronidazole	0.6–1.5 g/day	14–15 days		
Tetracycline	500 mg/4 times/day	14–15 days		
Bismuth	564 mg/4 times/day	10–14 days	84%	31%
Metronidazole	1.0–1.5 g/day	10–14 days		
Amoxicillin	1.5–2.0 g/day	10–14 days		
Metronidazole	500 mg 3 times/day	12 days	89%	13%
Amoxicillin	750 mg 3 times/day	12 days		
Ranitidine	300 mg at bedtime	6 weeks		
Clarithromycin	500 mg 3 times/day	10 days	86%	34%
Amoxicillin	750 mg 3 times/day	10 days		
Ranitidine	300 mg at bedtime	6 weeks		
Metronidazole	400 mg 3 times/day	14 days	90%	13%
Amoxicillin	750 mg 3 times/day	14 days		
Omeprazole	40 mg/day	6 weeks		
Metronidazole	500 mg 2 times/day	14 days	88%	18%
Clarithromycin	250 mg 2 times/day	14 days		
Omeprazole	20 mg 2 times/day	14 days		

Source: Table 1 in Walsh and Peterson (1995).

to identify effective agents to eradicate the infection of *H. pylori*. The standard agents such as H_2-receptor antagonists and sucralfate have no effect on *H. pylori*. In addition, bimuth such as pepto-mismol and numerous antibiotics including erythromycin, amoxicillin, and metronidazole have not been proved to provide satisfactory long-term eradication rate (Peterson, 1991). As an alternative, a different regimen for the combination of antimicrobial agents and the combination of antimicrobial and antisecretory agents has been proposed for the treatment of patients with peptic ulcer disease who are infected with *H. pylori*. Table 6.4.2 displays various combinations of antimicrobial and antisecretory agents in the eradication of *H. pylori*. As can be seen from Table 6.4.2, these different combinations consist of different microbial and antisecretory agents at different dose levels. The duration for the treatment of these combinations and the corresponding eradication rates of *H. pylori* vary from one combination to another with different overall incidence rates of adverse events. In addition to the eradication rate of *H. pylori*, other important efficacy endpoints to be evaluated include the rate of complete healing of the ulcer and the recurrence rate as documented by endoscopy.

Note that although the FDA has a specific policy for the approval of combination therapy, the assessment of the effectiveness and safety for the combination therapy is rather difficult and yet challenging because of its complexity. The development of a combination therapy involves the determination of the optimal joint region of therapeutic dosing ranges, the duration of treatment for each different drug, and the order of administration of these agents, while the assessment of the combination therapy is usually based on multiple efficacy and safety clinical endpoints. This complexity may limit the chance of a combination therapy being approved by the FDA. For example, thus far, none of the combination drugs listed in Table 6.4.2 has yet been approved by the FDA.

Fixed-Combination Prescription Drugs

For fixed-combination prescription drugs for humans, the FDA has specific regulations that are described in Section 300.50 in Part 21 of CFR. The regulation states that "Two or more drugs may be combined in a single dosage form when each component makes a contribution to the claimed effects and the dosage of each component (amount, frequency, duration) is such that the combination is safe and effective for a significant patient population." A fixed-combination prescription drug is defined as a single dosage formulated from different pharmacological agents. According to the FDA's regulation, the fixed-combination drugs are confined to a class in which different drugs can be formulated into one dosage form. However, different drugs may be administered in separate and/or different dosage forms either simultaneously or sequentially. For example, the Second International Study of Infarct Survival (ISIS-2, 1988) evaluated the effectiveness of the combined therapy of one-hour intravenous infusion of 1.5 MU of streptokinase up to 24 hours from the onset of chest pain followed by one month of oral enteric-coated aspirin at a dose of 150 mg per day for patients with suspected myocardial infarction. The dosage forms and routes of administration of the two therapeutic agents considered in ISIS2 are different and are given sequentially. It should be noted that current regulations for fixed-combination drugs do not cover the combination therapy with different dosage forms and time of administration though, in principle, the statistical design and analysis are the same for both fixed-combination prescription drugs and combination therapy.

Note that for approval of a combination therapy, the FDA also requires the evidence for which each component must make a contribution to the claimed effects of the combination be provided. As indicated in Laska and Meisner (1989) and Hung et al. (1990, 1993), the *contribution* is referred to as the contribution based on a single clinical endpoint. On the other hand, for approval of a fixed-combination prescription drug, the FDA requires evidence to be provided of the superiority of the combination over each of its components. It, however, should be noted that the combination might increase the risk of adverse events due to the unnecessary exposure of patients to some components that may not have additional clinical benefit. In addition the FDA regulation states the relationship of the combination and its constituents in terms of the claimed effects. However, it does not specify the type, form, and magnitude of the contributions of each component to the claimed effects, nor does it require the claimed effects to be contrasted with a concurrent placebo control.

Let A and B be the components of a combination drug, denoted by $A + B$, each at a fixed dose. Let μ_A, μ_B, and μ_{AB} be the average of the distributions of the clinical endpoints such as reduction in seated diastolic blood pressure, eradication rate, vascular mortality rate within a certain time interval, or median survival. Since the FDA requires that the efficacy of the combination be superior to those of each component, the hypotheses can be formulated as follows (e.g., see Laska and Meisner, 1989; Hung et al., 1990):

$$H_0: \mu_A \geq \mu_{AB} \quad \text{or} \quad \mu_B \geq \mu_{AB};$$
$$\text{vs.} \quad H_a: \mu_A < \mu_{AB} \quad \text{and} \quad \mu_B < \mu_{AB}. \quad (6.4.1)$$

Note that a larger value of μ indicates a better efficacy. Similar to the interval hypotheses described in (2.6.3), hypotheses (6.4.1) can be further decomposed into the following two one-sided hypotheses:

$$H_{01}: \mu_A \geq \mu_{AB};$$
$$\text{vs.} \quad H_{a1}: \mu_A < \mu_{AB}; \quad (6.4.2)$$

and

$$H_{02}: \mu_B \geq \mu_{AB};$$
$$\text{vs.} \quad H_{a2}: \mu_B < \mu_{AB}. \quad (6.4.3)$$

The first set of one-sided hypotheses is to verify that the combination drug $A + B$ is superior to component A. If the combination drug is indeed superior to component A, then component B does make a contribution of the claimed effect of the combination. Similarly the second set of one-sided hypotheses is to demonstrate that the combination $A + B$ is superior to component B. As a result, the null hypothesis (6.4.1) is a union of the two null hypotheses (6.4.2) and (6.4.3) which include all points in quadrants two, three, and four as well as all axes. On the other hand, the alternative hypothesis is the intersection of the two alternative hypotheses (6.4.2) and (6.4.3) which consists of all points in the first quadrant excluding the positive axes. Although hypotheses (6.4.1) can be employed to test statistically whether each component makes a contribution to the claimed effects of the combination, the type, form, and magnitude of the joint contribution by both components cannot be quantified. For example, consider the results of vascular mortality rates in days 0–35 reported by ISIS2, as shown in Table 2.4.2. The vascular mortality rates of IV placebo + active aspirin, IV streptokinase + aspirin placebo, and IV streptokinase and active aspirin are 10.7%, 10.4%, and 8.0%, respectively. The contribution of IV streptokinase to the combination is 2.7% (10.7% − 8.0%) and contribution of active aspirin to the combination is 2.4% (10.4% − 8.0%). However, without including a concurrent placebo control, it is not clear what the joint contribution of both drugs is. The mortality rate of IV placebo and aspirin placebo is 13.2%. The reduction in mortality rate of the combination therapy as compared to the concurrent placebo is 5.2% which is approximately the sum of the contribution of each component (2.7% + 2.4%). The estimated effect of combination versus placebo can be expressed as follows:

$$SA - PP = (SA - PA) + (SA - SP) + [(SP - PP) - (SA - PA)], \qquad (6.4.4)$$

where S, A, and P stand for mortality rates of IV streptokinase, oral aspirin, and the placebo. The last term on the right-hand side of (6.4.4) is the interaction between the two components, which can be measured by the difference between the difference in vascular mortality between the streptokinase and placebo for patients taking the aspirin placebo tablets and that between the streptokinase and placebo for patients taking the active aspirin tablets. In turns out that the reduction in mortality of the combination as compared to the concurrent placebo is the sum of the contributions to the reduction in mortality by each component plus the interaction between two components as demonstrated by the mortality rates from ISIS2. That is,

$$8.0\% - 13.2\% = (8.0\% - 10.7\%) + (8.0\% - 10.4\%)$$
$$+ [(10.4\% - 13.2\%) - (8.0\% - 10.7\%)]$$
$$-5.2\% = -2.7\% - 2.4\% - 0.1\%.$$

It can be seen from the above that the interaction (0.1%) between IV streptokinase and oral aspirin in the reduction of mortality is negligible. As a result the magnitude of the contributions made by each component to the reduction in vascular mortality is quite similar, and their joint contribution to vascular mortality of the combination is additive. In practice, it is recommended that a complete 2×2 factorial design, as given in panel A of Table 6.4.3, be employed to assess the efficacy and safety of the combination with the two components each at a fixed dose. One could argue that there is no need for a concurrent placebo control in the evaluation of a fixed-combination drug because the clinical benefits of monotherapy for each component have been established through prospective, randomized, double-blind, and placebo-controlled studies. This information, however, can only be served as a historical control for external validation for the assessment of a combination drug. Without a concurrent placebo control, the clinical benefit of the combination under evaluation may never be internally validated. In addition the dose levels of the combination assessed by clinical trials are usually smaller than those selected for monotherapy in order to produce synergistic effects that achieve better efficacy and safety. Therefore effectiveness of the dose levels of each component may not be adequately evaluated alone before as a single agent. As a result a concurrent placebo control must be used to investigate and quantify the effect of each component and the type, form, and magnitude of their contributions to the claimed effects of the combination under study.

Table 6.4.3 Factorial Design for Combination Therapy

Panel A: A Full 2 × 2 Factorial Design for Combination Therapy of Two Components Each at Two Dose Levels

Group	Drug A	Drug B
1	Placebo	Placebo
2	Placebo	Fixed active dose
3	Fixed active dose	Placebo
4	Fixed active dose	Fixed active dose

Panel B: A Full 2 × 2 × 2 Factorial Design for Combination Therapy of Three Components Each at Two Dose Levels

Group	Drug A	Drug B	Drug C
1	Placebo	Placebo	Placebo
2	Placebo	Placebo	Fixed active dose
3	Placebo	Fixed active dose	Placebo
4	Placebo	Fixed actice dose	Fixed active dose
5	Fixed active dose	Placebo	Placebo
6	Fixed active dose	Placebo	Fixed active dose
7	Fixed active dose	Fixed active dose	Placebo
8	Fixed active dose	Fixed active dose	Fixed active dose

Hypotheses (6.4.1), which can be tested based on the data from phase III studies, are often used to confirm the effectiveness and safety of the combination drug for each component at a particular dose level. The doses of each component in a combination drug are often determined at the stage of phase II clinical development. The combination can then be confirmed through phase III clinical trials. In practice, it usually requires sophisticated and state-of-art statistical design and analysis for identification of possible combinations of doses at which the contribution of each component to the claimed effects can be assessed.

Multilevel Factorial Design

A commonly employed design for the assessment of combination drugs or therapy is the factorial design with multilevels, which is known as a multilevel factorial design. For assessment of a combination drug with a factorial design, each component of the combination drug is referred to as a *factor*, and the doses of a component are the *levels* of the corresponding factor. Therefore a full multilevel factorial design includes a number of *treatment groups* which is made up of all possible combinations of factors and levels. The number of treatment groups equals the product of the number of levels and the number of factors. For example, in panel A of Table 6.4.3 there are two component drugs, and each component drug has two dose levels including dose 0 (or placebo) and one active fixed dose level. As a result this factorial design has two factors and

Table 6.4.4 A Full $(a + 1) \times (b + 1)$ Factorial Design for Combination Therapy of Two Components at a and b Dose Levels

Group	Drug A	Drug B
1	Placebo	Placebo
2	Placebo	Active dose 1
⋮	⋮	⋮
$b + 1$	Placebo	Active dose b
$b + 2$	Active dose 1	Placebo
$b + 3$	Active dose 1	Active dose 1
⋮	⋮	⋮
$2(b + 1)$	Active dose 1	Active dose b
$a(b + 1)$	Active dose a	Placebo
$a(b + 2)$	Active dose a	Active dose 1
⋮	⋮	⋮
$(a + 1)(b + 1)$	Active dose a	Active dose b

each has two levels, so there are a total of four treatment groups. A design of this kind is denoted by a 2×2 factorial design. As another example, panel B of Table 6.4.3 displays a $2 \times 2 \times 2$ factorial design for a combination of three component drugs where each component has two dose levels (i.e., dose 0 or a placebo and a selected active dose level). Therefore there are a total of eight treatment groups. Note that the statistical designs employed for panels A and B in Table 6.4.3 are parallel-group designs in which the number of parallel arms equals to the number of treatment groups. Therefore, when a factorial design is employed, the number of parallel groups can be as large as the number of factors and/or the number of levels of each factor increase. For example, Table 6.4.4 provides the treatment groups of clinical trials that assess a combination drug that consists of component drugs A and B. If we include dose 0 (or placebo) and a active dose levels for A and b active dose levels for B, then there are a total of $(a + 1)(b + 1)$ treatment groups. Hence the number of treatment groups would be 4, 9, 16, or 25 when the number of active dose levels for each component are 1, 2, 3, and 4, respectively. In addition a parallel-group factorial design usually requires a large sample size. As an alternative, some crossover designs such as Williams design or balanced incomplete block design might be useful. Table 6.4.5 provides the Williams design for a 2×2 factorial design with and without a concurrent placebo control group.

For a combination drug consisting of component drugs A and B, in the case where the effectiveness and safety of component A has been established of the intended indication, then the selection of the dose of component B is critical in providing an enhanced therapeutic effect. In practice, it is suggested that the dose-response relationship of component B at an established effective dose of component A be examined. In some cases, since patients who receive component A may exhibit some differential limiting clinical toxicity, it is sug-

Table 6.4.5 Williams Designs for Combination Therapy

Panel A: 2 × 2 Factorial Without Concurrent Placebo Control

Sequence	Period		
	I	II	III
1	A	B	A + B
2	B	A + B	A
3	A + B	A	B
4	A	A + B	B
5	B	A	A + B
6	A + B	B	B

Panel A: 2 × 2 Factorial With Concurrent Placebo Control

Sequence	Period			
	I	II	III	IV
1	Placebo	A + B	A	B
2	A	Placebo	B	A + B
3	B	A	A + B	Placebo
4	A + B	B	Placebo	A

gested that doses of component *B* be administered at a preselected dose after the patients are titrated to the maximum tolerable dose of component *A*. A typical example is the study of the combination therapy of cyclosporine and methotrexate for patients with severe rheumatoid arthritis (Tugwell et al., 1995). Note that the type of factorial design employed in this study is usually referred to as a partial factorial design because it does not utilize all possible combinations of dose levels from each factor as indicated in Table 6.4.6.

One important advantage of a combination therapy is that it is usually able to achieve a better clinically significantly synergistic efficacy with fewer adverse events at lower-dose levels compared to individual monotherapy of its components. In practice, although each monotherapy of the component drugs may also achieve the same efficacy at a higher dose, the corresponding safety may not be tolerable.

Global Superiority of Combination Drug

In order to identify the optimal combination of dose levels and to investigate the potential drug-to-drug interactions, it is recommended that several or all possible combinations be explored simultaneously with multilevel factorial designs. Table 6.4.7 provides a cross tabulation of a full multilevel factorial design with μ_{ij} representing the average response of the combination made of dose i of component *A* and dose j of component *B*. A separate dose-response relation-

Table 6.4.6 Designs for Two-Drug Combined Therapy With a Fixed Dose of One Component

	Panel A: The Same Fixed Dose For One Component	
Group	Drug A	Drug B
1	Active dose 1	Placebo
2	Active dose 1	Active dose 1
⋮	⋮	⋮
$b + 1$	Active dose 1	Active dose b

	Panel B: Titration of The Dose Level of One Component to The Maximum Tolerable Dose	
Group	Drug A	Drug B
1	Titrated to MTD	Placebo
2	Titrated to MTD	Active dose 1
⋮	⋮	⋮
$b + 1$	Titrated to MTD	Active dose b

ship of each component can be investigated by the first row for drug B or first column for drug A where the placebo is administered for the other component. In order to provide a useful dose-response relationship for therapeutic applications, Hung et al. (1989) suggests that the dose ranges to be investigated must include a very low dose level and a very high dose level that are not in the effective dose range. As a result, the contribution of some doses by one component when added to the other drug may not be different from the placebo, while the contribution of the other component is quite obvious. Hence with respect to the requirement described in Section 300.50 in Part 21 of CFR that each component makes a contribution to the claimed effects, Hung et al. (1990) defines the superiority of a combination drug over its component drugs in a global sense. Strict superiority of a combination drug is defined as the existence of at least one dose combination that is more effective than its components. Let d_{ij} be the minimum gain in efficacy obtained from combining dose i of drug A and dose j of drug B as compared to its component drugs at the same dose levels alone,

$$d_{ij} = \mu_{ij} - \max(\mu_{io}, \mu_{oj}), \quad i = 1, \ldots, a, j = 1, \ldots, b. \quad (6.4.5)$$

The corresponding statistical hypotheses can then be formulated as:

$$H_0: d_{ij} \leq 0 \quad \text{for every} \quad 1 \leq i \leq a \quad \text{and} \quad 1 \leq j \leq b;$$
$$\text{vs.} \quad H_a: d_{ij} > 0 \quad \text{for some} \quad 1 \leq i \leq a \quad \text{and} \quad 1 \leq j \leq b. \quad (6.4.6)$$

A combination drug is said to be superior to its components in the wide sense if

Table 6.4.7 Cross Tabulation of a Full $(a + 1) \times (b + 1)$ Factorial Design for Combination Therapy with the Mean Effects

Dose Levels for Component A	Placebo	Dose Levels for Component B				Overall
		1	2	...	b	
Placebo	μ_{00}	μ_{01}	μ_{02}	...	μ_{0b}	$\mu_{0.}$
1	μ_{10}	μ_{11}	μ_{12}	...	μ_{1b}	$\mu_{1.}$
2	μ_{20}	μ_{21}	μ_{22}	...	μ_{2b}	$\mu_{2.}$
$\vdots$	$\vdots$	$\vdots$	$\vdots$	...	$\vdots$	$\vdots$
a	μ_{a0}	μ_{a1}	μ_{a2}	...	μ_{ab}	$\mu_{a.}$
Overall	$\mu_{.0}$	$\mu_{.1}$	$\mu_{.2}$	...	$\mu_{.b}$	$\mu_{..}$

the average of all the dose combinations is superior to both the averages of the individual monotherapy doses. Let m_A represent difference between the average responses of all (i,j) combinations over the entire range of active doses and that dose i of component A when placebo is administered for component B,

$$m_A = \text{Ave}(\mu_{ij} - \mu_{io}).$$

Similarly

$$m_B = \text{Ave}(\mu_{ij} - \mu_{oj}).$$

Consequently the concept of the wide global superiority of combination drug can be stated as:

$$H_0: m_A \leq 0 \quad \text{or} \quad m_B \leq 0;$$
$$\text{vs.} \quad H_a: m_A > 0 \quad \text{and} \quad m_B > 0. \qquad (6.4.7)$$

The strict global superiority of a combination drug restricts our attention to a more effective class of combinations than either component drug alone. According to d_{ij} in the alternative hypothesis of (6.4.6), we only need to identify one μ_{ij} that is better than mean responses μ_{io} and μ_{oj} of the corresponding monotherapy of component A at dose i and of component B at dose j. However, it does not guarantees superiority of one combination over all dose levels of either monotherapy.

Consider the example given in Hung et al. (1993). Table 6.4.8 displays the mean reductions from the baseline in post-treatment supine diastolic blood pressure from a clinical trial that evaluated a combination drug with three dose levels for drug A and four dose levels for drug B, including the placebo (Hung et al., 1993). From Table 6.4.8 the mean reductions in diastolic blood pres-

Table 6.4.8 Difference From Placebo in Mean Reduction in Supine Diastolic Blood Pressure (HH mg)

Dose Levels for Component A	Dose Levels for Component B			
	Placebo	1	2	3
Placebo	0	4	4	3
1	5	9	7	8
2	5	6	6	7

Source: Hung et al. (1993).

sure of all combinations are seen to be more than those at the corresponding doses of either monotherapy. As a result the minimum gain d_{ij} are positive for all combinations, which is shown in Table 6.4.9. From Table 6.4.10 suppose that the mean reduction from the baseline in diastolic blood pressure of dose 3 for component B as monotherapy is changed from 3 to 7. Table 6.4.11 reveals that although the minimum gain of all combinations over their corresponding monotherapy is at least 0, only combinations (1, 1) and (1, 3) surpass the monotherapy of drug B at dose level 3. This result is desirable because the monotherapy of drug B at the highest dose may induce some severe adverse events, while the dose level of drug B in combination with a larger reduction is two levels lower than that of the monotherapy. As a result the combination of dose 1 of drug A with dose 1 of drug B provides a much safer margin and hence a larger benefit-to-risk ratio.

For assessment of wide global superiority, as outlined in hypotheses (6.4.7), there seem to be, on the one hand, a combination that is better than its component A and, on the other, a combination that might not be the same combination but is better than its component B. Unlike strict superiority, however, this does not guarantee that there is a combination that is better than both of its components as required by the FDA regulation. Therefore the strict global superiority meets the current regulatory requirement for approval of a combination drug

Table 6.4.9 Minimum Gain of Combinations for the Data in Table 6.4.8

Active Dose Levels for Component A	Active Dose Levels for Component B		
	1	2	3
1	4	2	3
2	1	1	2

Source: Hung et al. (1993).

Table 6.4.10 Modified Differences From Placebo in Mean Reduction in Supine Diastolic Blood Pressure (mm Hg)

Dose Levels for Component A	Dose Levels for Component B			
	Placebo	1	2	3
Placebo	0	4	4	7
1	5	9	7	8
2	5	6	6	7

product. Hung et al. (1990, 1993) propose two statistical testing procedures for hypotheses (6.4.6) under the assumption that data are normally distributed. The proposed methods reduce to the min test for hypotheses (6.4.1) proposed by Laska and Meisner (1989) when only one active dose level is included in the assessment trial of a combination drug. Note that the sampling distributions of the methods proposed by Hung et al. are quite complicated and require special tables for the significance tests. Hung (1996) extends the application of these two methods to the situations (1) where the variance of the clinical endpoint is a function of its mean and (2) where an incomplete factorial design is used.

When a combination drug consists of more than two component drugs, the concept of strict global superiority for the combination drug of two components can be easily extended. Let d_{ijk} be the minimum gain in efficacy obtained by combining dose i of drug A with dose j of drug B, and dose k of drug C over its components alone at the same dose levels,

$$d_{ijk} = \mu_{ijk} - \max(\mu_{ioo}, \mu_{ojo}, \mu_{ook}), \quad i = 1, \ldots, a, j = 1, \ldots, b, k = 1, \ldots, c, \quad (6.4.8)$$

where μ_{ijk} is the mean response of the combination of dose i of drug A, dose j of drug B, and dose k of drug C, and μ_{ioo}, μ_{ojo} and μ_{ook} represent the mean responses of its components A, B, and C alone at the same dose levels i, j, and k, respectively, where the active agents are administered with the placebos of the other two components. The strict global superiority of a three-drug combination over its components can be formulated by the following statistical hypotheses:

Table 6.4.11 Minimum Gain of Combinations for the Modified Data in Table 6.4.10

Active Dose Levels for Component A	Active Dose Levels for Component B		
	1	2	3
1	4	2	1
2	1	1	0

$$\text{H}_0: d_{ijk} \leq 0 \quad \text{for every} \quad 1 \leq i \leq a, 1 \leq j \leq b, \quad \text{and} \quad 1 \leq k \leq c;$$
$$\text{vs.} \quad \text{H}_a: d_{ijk} > 0 \quad \text{for some} \quad 1 \leq i \leq a, 1 \leq j \leq b, \quad \text{and} \quad 1 \leq k \leq c.$$
(6.4.9)

The above hypotheses can be tested to verify the existence of strict global superiority of a combination drug. However, the extension of the methods of Hung et al. (1993) for testing the above hypotheses is not straightforward. Further research is needed.

Suppose that a combination of two component drugs is developed to treat patients with benign prostatic hyperplasia (BPH). Two primary efficacy endpoints for assessment of the combination are peak urinary flow rate (mL/s) and AUA-7 symptom scores. Let c_{ij} and d_{ij} be the minimum gain of combination of dose i of drug A with dose j of drug B of peak urinary flow rate and AUA-7 symptom scores, respectively. Following the suggestion by Laska and Meisner (1990), the strict global superiority for more than one clinical endpoint is defined as that where at least one combination is superior to its component drugs for at least one clinical endpoint. Hence, if a combination is better than its component drugs for at least one clinical endpoint, then the minimum gain of the combination based on both clinical endpoints must be greater than zero. The hypotheses corresponding to the strict global superiority can be formulated as follows:

$$\text{H}_0: \min(c_{ij}, d_{ij}) \leq 0 \quad \text{for every} \quad 1 \leq i \leq a \quad \text{and} \quad 1 \leq j \leq b;$$
$$\text{vs.} \quad \text{H}_a: \min(c_{ij}, d_{ij}) > 0 \quad \text{for some} \quad 1 \leq i \leq a \quad \text{and} \quad 1 \leq j \leq b.$$
(6.4.10)

The concept of hypotheses (6.4.10) can easily be extended to an evaluation of the combination drug based on more than two endpoints. However, the definition of strict global superiority and its corresponding formulation of hypotheses and proposed statistical procedures are to verify the existence of at least one combination that is better than both of its components. Furthermore they are hypothesis testing procedures and hence cannot describe the dose-response relationship and potential drug-to-drug interaction among components. Consequently they fail to provide a way to search for combinations for which each component makes a contribution to the claimed effects should they exist.

Method of Response Surface

To overcome the drawbacks associated with the definition of the strict global superiority for combination drug, the concept of response surface methodology can provide a nice compliment to the statistical testing procedures suggested by Laska and Meisner (1989) and Hung et al. (1993). For a combination trial conducted with a factorial design, if the dose levels of each component are appropriately selected, then the technique of response surface can provide valuable information regarding (1) the therapeutic dose range of the combination drug with respect to effectiveness and safety and (2) the titration process and drug-to-drug interaction. The response surface method can empirically verify a model that adequately describes the observed data. For example, we can consider the following statistical model to describe the response:

$$Y_{ijk} = f(A_i, B_j, \theta) + e_{ijk}, \qquad (6.4.11)$$

where Y_{ijk} is the clinical response of patient k who receives dose i of drug A, denoted by A_i and dose j of dose B, denoted by B_j, θ is a vector of unknown parameters, and e_{ijk} is the random error in observing Y_{ijk}, where $k = 1, \ldots, n_{ij}$, $j = 1, \ldots, b$, $i = 1, \ldots, a$. The component $f(A_i, B_j, \theta)$ in (6.4.11) gives a mathematical description and an approximation to the true unknown response surface provided by the two drugs. When the primary clinical endpoints of interest are continuous variables, $f(A_i, B_j, \theta)$ is usually approximated by a polynomial. For a detailed description of response surface methodology, see Box and Draper (1987) and Alt, Myers, and Montgomery (1995); also see Peace (1990) for applications of response surface to a phase II development of antianginal drugs.

If the assumed mathematical model is not too complicated, then the standard statistical estimation procedures such as least squares method or maximum likelihood method can be selected to estimate the unknown parameter θ. After substitution of the parameter in $f(A_i, B_j, \theta)$ by its estimate, an estimated response surface can be obtained that provides an empirical description of the dose-response relationship either by a three-dimensional surface or by two-dimensional contours. In addition an optimal dose combination can be estimated to give an maximum clinical response if it exists and is unique. If the 95% confidence region for the optimal dose combination does not lie on both the horizontal axis representing drug A and the vertical axis representing drug B at the contours, then it is concluded that the estimated optimal combination is superior to both of its components. The technique of response surface therefore can estimate an optimal dose combination that may not be the combination of doses of both components selected for the trial given existence of such combinations. Hung (1992) suggests a procedure to identify a positive dose-response surface for combination drugs. Hence the response surface can also estimate a region in which the combinations are superior to their components. Estimation of a superior region is particularly appealing if safety is the major reason for the combination drug because a combination can be chosen from the superior region with much lower doses of both components to achieve the same superiority in efficacy but with a much better safety profile. Since μ_{ij} is estimated using the information of the entire sample, if the assumed model can adequately describe the dose-response relationship of the combination, the response surface method requires fewer patients than the procedures for testing strict global superiority proposed by Hung et al. (1993). Note that estimation of the unknown parameters, response surface, and the optimal combination dose and construction of the confidence region for the optimal combination dose are model dependent. Consequently, as indicated by Hung et al. (1990), the FDA is concerned about the application of response surface method due to the following reasons:

1. The sensitivity of the methods such as lack-of-fit tests, goodness-of-fit tests, and residual plots for verification of the adequacy of the fitted models is often questionable.
2. Even if there is no evidence for the inadequacy of the fitted model, the chance of selecting an inadequate model and the effect of such an error cannot be evaluated.
3. The response surface method is model dependent. Two different models that both adequately fit the given data may provide contradictory conclusions in any subsequent statistical analyses.

The methods for strict global superiority and response surface methodology should play crucial but complimentary roles in assessment of combination drugs. Existence of a combination superior to its components can be verified first by the model-independent statistical testing procedures proposed by Hung et al. (1993). Then the response surface technique can be applied to (1) empirically describe the dose-response relationship, (2) identify for the region of superior efficacy, and (3) estimate the optimal dose combination. Also it would be of interest to provide a scientific justification as to why the two methods yield inconsistent conclusions.

6.5 EQUIVALENCE TRIAL

As mentioned in Chapter 1, under the *Drug Price Competition and Patent Term Restoration Act* passed in 1984, the FDA is authorized to approve generic drugs based on bioequivalence trials through an ANDA. Although it is impossible to prove that two drug products are exactly the same, it is believed that if the drug absorptions of an innovator drug and its generic copies are relatively close within a narrow limit, it is reasonable to assume that they will produce similar effectiveness and safety. In other words, if generic drugs and the innovator drug are equivalent in the rate and extent to which the active ingredient or therapeutic moiety is absorbed and becomes available at the site of drug action, we may assume that they are therapeutically equivalent. This assumption is referred to as the *Fundamental Bioequivalence Assumption* (Chow and Liu, 1992a). Therefore the objective of a bioequivalence study is to assess the equivalence of a generic drug to its referenced innovator drug. This is usually done by performing an interval hypotheses testing (2.6.3) based on some pharmacokinetic parameters such as AUC (area under the blood or plasma concentration-time curve) or C_{max} (maximum concentration).

For some drug products, however, the active ingredients cannot be absorbed through the systemic and hence have negligible levels. These drugs include metered dose inhalers (MDI) which is intended for the relief of bronchospasm in patients with reversible obstructive airway disease, sucralfate for treatment of acute duodenal ulcer, topical antifungals, and vaginal antifungals. As a result bioequivalence cannot be assessed based on drug absorption. As an alternative, it is suggested that clinical trials be conducted to demonstrate therapeutical equivalence based on well-defined clinical endpoints (e.g., see Huque et al., 1989; Huque and Dubey, 1990).

For many diseases, although many standard drugs may be available, it is still of interest to develop new drugs for the same diseases with similar efficacy but a better safety profile. The new drugs would be attractive if they possess some good properties and characteristics such as that (1) they are easy to administer, (2) they are more cost-effective, (3) they require a relatively short duration of treatment, and (4) they can be used to serve as a substitution for an invasive procedure. Thus, instead of showing a superior efficacy of the new drug over the standard drugs, it is sufficient for clinical trials to provide evidence that the new drug is therapeutically equivalent to the standard treatments. For this purpose active control equivalence trials are often considered. See, for example, Dunnett and Gent (1977), Makuch and Johnson (1990), Durrleman and Simon (1990), and Blackwelder (1982).

Average and Population Equivalence

Statistical concepts for establishment of bioequivalence and therapeutical equivalence between drug products are the same. This concept is referred as to *population equivalence*, which requires equivalence in distributions of the primary clinical endpoint (for therapeutical equivalence) and pharmacokinetic parameters (for bioequivalence) between the test drug and the active control (or the reference drug). Note that in bioequivalence assessment the innovator drug is usually chosen to be the reference. Statistically, under the normality assumption, the distributions of the primary clinical endpoint for efficacy can be uniquely determined by its first two moments, namely the average and variance. Therefore, to establish equivalence between two distributions is to establish equivalences in both average and variance of the distributions. To provide a better understanding of the comparisons in average response and variability of response, equivalence in averages and variabilities are illustrated in Figures 6.5.1–6.5.3. Figure 6.5.2 presents a situation where the average of the two distributions are similar but the variability of the responses for the test product is larger than that of the reference. In this case the test product cannot provide an adequate proportion of the patients within therapeutic range, even though its average is equivalent to the reference. Therefore, to claim that two drug products are equivalent, it requires that the two drug products are equivalent in both average and variability. However, current regulation only requires that evidence of equivalence in average between drug products be provided

EQUIVALENCE TRIAL

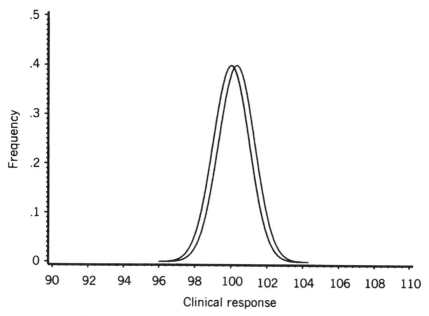

Figure 6.5.1 Equivalence in both means and variabilities.

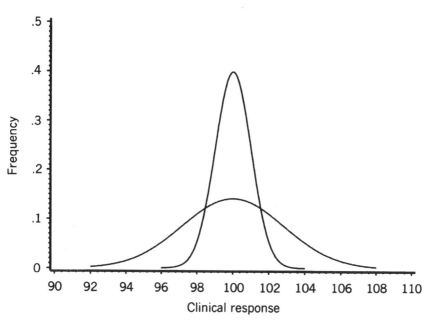

Figure 6.5.2 Equivalence in means but not in variabilities.

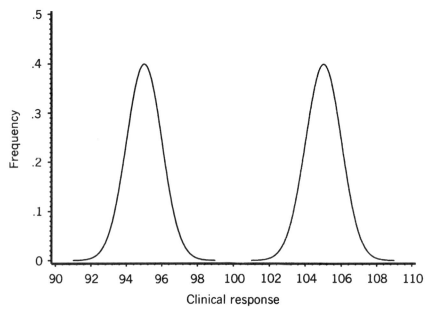

Figure 6.5.3 Equivalence in variabilities but not in the mean.

for assessment of bioequivalence or therapeutical equivalence. This concept is known as *average equivalence*. It should be noted that average equivalence does not guarantee that two equivalent drug products can be used interchangeably. To overcome this problem, the concept of *individual equivalence* which intends to account for equivalence in variability has been proposed for generic drugs (e.g., see Liu, 1994 and Chow and Liu, 1995b). This concept is not uncommon in switching drugs for assessment of therapeutical equivalence.

In Section 2.6 we have pointed out that the following point hypotheses for equality may not be appropriate for assessment of equivalence in average.

$$H_0: \mu_T = \mu_R;$$
$$\text{vs.} \quad H_a: \mu_T \neq \mu_R. \quad (6.5.1)$$

As discussed in Section 2.6, point hypotheses are often considered to evaluate whether a test drug is efficacious in clinical trials. Basically we are interested in proving that the efficacy of the test drug is different from that of the reference drug. Therefore, in practice, we would reject the null hypothesis of no difference in favor of the alternative hypothesis that there is a difference. After having concluded that there is a difference between the test drug and the reference drug, the efficacy is established by showing that there is a desired power for detection of a clinically meaningful difference. Note that the concept of power for detection of a clinical meaningful difference in point hypotheses testing has been adapted for establishment of equivalence between drug products (e.g., see Schuirmann, 1987). This method is referred to as the power approach, which is summarized below. We first select an appropriate test statistic according to the type of the primary clinical endpoints (i.e., continuous, categorical, or censored data). Then we perform the statistical test based on point hypothesis for equality. If the result of hypotheses testing for equality is not statistically significant at the preselected nominal level of significance, then we compute the power for detection of a prespecified clinically meaningful difference, such as 20% of the observed reference mean. We then conclude that the two drug products are equivalent if the empirical power computed is greater than some predetermined level, say, 80%.

Figure 6.5.4 provides the rejection region of the power approach. The term Y for the horizontal axis stands for the observed mean difference between the two drugs, and ms is the standard error of Y. Figure 6.5.4 assumes a mean difference of greater than 20% which is of clinical importance with respect to a reference mean of 100. The shape of the rejection region is an upside-down triangle that indicates a major drawback of the power approach. As the variability decreases, the rejection region of the power approach becomes smaller. When the variability is 0, the mean difference has to be also 0 in order to conclude equivalence. This implies that if the magnitude of mean differences is the same for two studies, according to the power approach, the probability of correctly concluding equivalence is smaller for the study with smaller variability than the study with

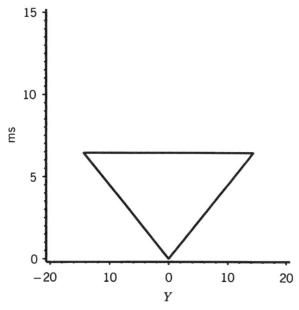

Figure 6.5.4 Rejection region of the power approach for the known symmetric equivalence limits = 20.

Table 6.5.1 Probability of Type I Error of Power Approach for Equivalence at a Nominal Level of 5%

Degrees of Freedom	Probability of Type I Error
10	0.0605
16	0.0722
20	0.0779
26	0.0847
30	0.0884
40	0.0958
50	0.1016
100	0.1188
∞	0.2000

Source: Schuirmann (1987).

larger variability. In other words, in a poorly conducted and sloppy study with more variable data it is easy to conclude equivalence than in a well-designed and carefully executed study generating estimates of efficacy with greater precision. Therefore good active control equivalence studies are penalized under the power approach. This undesirable property is a direct consequence of the incorrect use of point hypotheses for equality in the assessment of equivalence. In addition the power approach cannot control the probability of making type I error at the preselected level of 5%. Table 6.5.1 lists the probability of committing a type I error for various degrees of freedom. As the sample size increases, the probability of committing a type I error also increases. It is equal to 20% when the sample size approaches infinity. Therefore the power approach fails to protect the consumer risk (i.e., declaring equivalence when in fact the two drug products are not) at the 5% level of significance.

Statistical Methods

In early 1970s it was recognized that the usual point hypotheses for equality in (6.5.1) is not a correct set of hypotheses for assessment of equivalence (e.g., see Metzler, 1974; Westlake, 1972, 1976; Dunnett and Gent, 1977). The goal of an active control equivalence trial is to verify that two drug products are indeed equivalent. From the regulatory point of view, the manufacturers have the burden to provide substantial evidence that their products are equivalent to the reference drug. Therefore one needs to search a formulation of hypotheses in such a manner that the probability of making a type I error represents the consumer's risk, while the probability of committing a type II error reflects the producer's risk (or the probability of declaring inequivalence between the two drugs when in fact they are equivalent). It is therefore appropriate to reverse the null hypothesis of equivalence and the alternative hypothesis of inequivalence. The resulting hypotheses for equivalence are known as interval hypotheses as is given below:

$$H_0: \mu_T - \mu_R \leq L \quad \text{or} \quad \mu_T - \mu_R \geq U,$$
$$\text{vs.} \quad H_a: L < \mu_T - \mu_R < U, \qquad (6.5.2)$$

where L and U are some prespecified lower and upper equivalence limits. The above interval hypotheses can be further decomposed into two one-sided hypotheses as follows:

$$H_{01}: \mu_T - \mu_R \leq L,$$
$$\text{vs.} \quad H_{a1}: \mu_T - \mu_R > L,$$

and $\qquad (6.5.3)$

$$H_{02}: \mu_T - \mu_R \geq U,$$
$$\text{vs.} \quad H_{a2}: \mu_T - \mu_R < U.$$

A comparison of the interval hypotheses in (6.5.2) and the two one-sided hypotheses in (6.5.3) reveals that the space defined in the null hypothesis of the interval hypotheses is the union of the space of the null hypotheses in the two one-side null hypotheses, while the space defined by the alternative hpothesis in (6.5.2) is the intersection of the space of the two alternative hypotheses in (6.5.3). For each of the two one-sided hypotheses, there is a test statistic to verify whether the efficacy of the test product is not too low (high) compared to that of the reference drug. Therefore, if both null hypotheses of the two one-sided hypotheses are rejected at a prespecified nominal level of significance, then equivalence in efficacy between the test and reference product is concluded. Since this procedure involves testing two one-sided hypotheses simultaneously, it is called the two one-sided tests procedure (Schuirmann, 1987; Chow and Liu, 1992a). The rejection region for testing equivalence in average is given in Figure 6.5.5. In contrast to the rejection region of the power approach, it is an upright triangle that has some desirable properties. First of all, the equivalence between the two products will never be concluded if the standard error of the observed mean difference is above a certain value even if the observed mean difference is zero. Second, the rejection region becomes smaller as the standard error of the observed mean difference increases. Consequently poorly designed or conducted studies that produce less precise estimates are unlikely to reach the conclusion of equivalence. On the other hand, good studies that yield estimates with high precision will lead to the conclusion of equivalence even if the mean difference is large provided that it is within the equivalence limits. Since the equivalence between two drugs is concluded only if the null hypotheses of the both one-sided hypotheses are rejected at a prespecified nominal level of significance, unlike the power approach, the two one-sided tests procedure can control the overall type I error at the nominal level. However, if the variability is large or the sample size is too small, the two one-sided

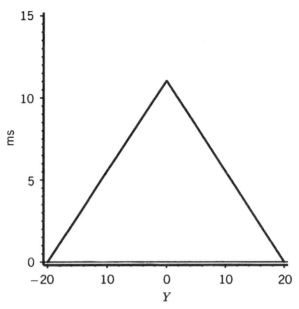

Figure 6.5.5 Rejection region of two one-sided tests procedure for known symmetric equivalence limits = 20.

tests procedure may be very conservative in concluding equivalence. Hsu et al. (1994), Brown, Hwang, and Munk (1996) propose other testing procedures for the interval hypotheses of equivalence. Although these proposed tests are more powerful than the usual two one-sided tests procedure, they are rather complicated mathematically and lack an intuitive interpretation. In addition the open nonconvex shape of the rejection region implies that any estimate of mean difference in efficacy might conclude equivalence as long as the standard error of the observed mean difference is sufficiently larger or increases to infinity. As a result their procedures might not be better than the power approach because the power approach will not conclude equivalence in the cases where the standard error of the observed mean difference is above a certain quantity due to the nature of its convex rejection region (see also Figure 6.5.4).

If the test drug is known to have less toxicity, then it suffices to verify whether the test drug is at least as efficacious as the reference drug. Therefore the hypotheses of interest will be the first set of one-sided hypotheses in (6.5.3), which is to test whether the effectiveness of the test drug is not too low compared to the reference drug (Blackwelder, 1982; Farrington and Manning, 1990). If the null hypothesis is rejected, we conclude that the test drug is at least as effective as the reference drug even though the efficacy of the test drug is below that of the reference but within the range of no clinical significance. This concept is referred to as the one-sided equivalence problem as contrasted with the two-sided equivalence problem formulated in (6.5.2). Therefore the lower equivalence limit L in (6.5.3) is usually a small negative number

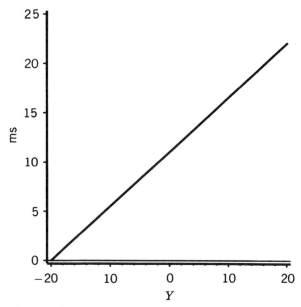

Figure 6.5.6 Rejection region of one-sided tests procedure for known lower equivalence limits = −20.

to reflect this range of no clinical significance. Figure 6.5.6 illustrates the corresponding rejection region, which indicates that if the mean difference is close to the lower equivalence limit, it requires a very small variability to conclude that the test drug is at least as efficacious as the reference drug.

As an alternative to the hypotheses testing, many researchers have suggested the use of confidence intervals in the assessment of equivalence between drug products. For example, see, Westlake (1972, 1976), Makuch and Simon (1978), Chow and Liu (1992a), Durrleman and Simon (1990), and Jennsion and Turnbull (1993). The concept and estimation procedure are quite straightforward. For the two-sided equivalence, if the probability of making a type I error is α, then the two drugs are concluded equivalent if the $100(1 - 2\alpha)\%$ confidence interval for the mean difference is completely contained within (L, U). For the one-sided equivalence problem, if the $100(1 - \alpha)\%$ lower limit for the mean difference is greater than L, then the test drug is claimed to be at least as effective as the reference drug. The confidence interval approach is more appealing than the method of hypotheses testing procedure because it can provide the magnitude and width of the mean difference between the two drugs. Note that in some situations the method of hypotheses testing and confidence interval approaches are operationally equivalent. In other words, for the two-sided equivalence problem the $100(1 - 2\alpha)\%$ confidence interval is totally within (L, U) if and only if both of the one-sided hypotheses in (6.5.3) are rejected at the α level of significance. On the other hand, for the one-sided equivalence problem the lower $100(1 - \alpha)\%$ confidence limit is greater than L if and only

Table 6.5.2 Comparison of the Results by Different Procedures for Assessing Equivalence With a Known Symmetric Limit of 0.2

Observed Difference	90% CI	H_{0E}	H_{0L}	H_{0U}	Two-sided Equivalence
0.00	(−0.10, 0.10)	Fail to reject	Reject	Reject	Yes
0.00	(−0.25, 0.25)	Fail to reject	Fail to reject	Fail to reject	No
0.05	(0.02, 0.08)	Reject	Reject	Reject	Yes
0.10	(−0.02, 0.22)	Fail to reject	Reject	Fail to reject	No

Note: All H_{0E}, H_{0L}, and H_{0U} are tested at 5% significance level and the 95% confidence interval for the observed difference of 0.05 is (0.01, 0.09).

if the first null hypothesis in (6.5.3) is rejected at the α level of significance. The decision for assessment of equivalence either by the confidence interval or the method of hypotheses testing depends on the lower and upper equivalence limits regardless if the observed difference is zero. Table 6.5.2 illustrates the consistency between the two one-sided tests procedure and confidence interval approach for the two-sided equivalence problem. As can be seen from Table 6.5.2, although the null hypothesis of equality is not rejected, the equivalence between the two drugs cannot be concluded because the 90% confidence interval may be too wide to be completely within the equivalence limits. Note that if the confidence interval is very narrow and is totally within the equivalence limits, we may conclude equivalence even though the confidence interval does not contain 0.

Equivalence Limits and Log Transformation

In practice, the selection of equivalence limits is always a challenge. If the equivalence limits are selected too narrow, it is extremely difficult to conclude equivalence even though the difference is of no clinical importance. On the other hand, if the equivalence limits are too wide, then it is easy to conclude equivalence when the two drugs can be distinguished by a clinically meaningful difference. It is important to select equivalence limits based on the response of the reference drug. For example, Huque and Dubey (1990) propose some equivalence limits for binary data such as eradication rate for antibiotics given in Table 6.5.3. It can be seen from Table 6.5.3 that the equivalence limits become narrower as the response rate of the reference drug increases. Suppose that we observed a response rate of 75% from previous clinical studies. Then the equivalence limits for the active control equivalence trial may be chosen to be ±20% according to Table 6.5.3. Based on the selected equivalence limits, the sample size can then be calculated. However, an interesting question may be raised: Should the equivalence limits be adjusted if a reference response rate of 85% is observed from the current trial? In other words, should the equivalence limits be changed to ±15%? In addition, is the sample size calculated with the limits of ±20% large enough to provide the desired power for the limits of ±15%? These

Table 6.5.3 Equivalence Limits for Binary Data

Equivalence Limits	Response Rate for the Reference Drug
±20%	50–80%
±15%	80–90%
±10%	90–95%
±5%	>95%

Source: Huque and Dubey (1990).

questions are related to the issue of selection of equivalence limits based on the reference responses obtained from the previous studies. Although the reference responses are obtained from adequate, well-controlled studies, they are not from the current trial, which may have similar but subtly different experimental environment with a seemingly similar patient population. In other words, the equivalence limits based on the reference responses from any previous studies may not be internally valid. Therefore, as an alternative, it is suggested that the equivalence limits be chosen based on the reference response obtained concurrently from the active control equivalence trial for internal validity. In this case the lower and upper equivalence limits in the interval hypotheses (6.5.2) are no longer fixed known quantities but random variables.

In most biological data the distribution of the primary clinical endpoint usually exhibits a long tail to the right. In this case a log transformation is usually recommended to remove the skewness and to achieve a symmetric and bell-shaped distribution (e.g., a normal distribution). If the data approximately follow a lognormal distribution, then, for the original scale of the data, interval hypothesis for equivalence can be expressed as the ratio of the averages of the test drug to that of the reference drug based on the original scale:

$$H_a: L < \frac{\mu_T}{\mu_R} < U, \qquad (6.5.4)$$

where $0 < L < 1 < U$. The interval hypotheses in (6.5.4) implies that the test drug is equivalent to the reference drug if its average efficacy is between $L\%$ and $U\%$ of that of the reference drug. Although L and U do not depend on the average reference response, the criterion for assessment of equivalence is based on the proportion of the estimated average response. As a result the lower and upper limits for evaluation of bioequivalence are internally valid. Logarithmic transformation of (6.5.4) reformulates the interval hypotheses from ratio to the difference as

$$H_a: \ln(L) < \ln(\mu_T) - \ln(\mu_R) < \ln(U), \qquad (6.5.5)$$

where ln denotes the natural logarithm and $\ln(L) < 0 < \ln(U)$.

For assessment of bioequivalence between a generic and a reference formulation, most regulatory agencies in the United States and the European Community request that pharmacokinetic parameters such as AUC or C_{max} be analyzed and that average bioequivalence be assessed on the log scale with $L = 80\%$ and $U = 125\%$ (i.e., $\ln(L) = -0.2231$ and $\ln(U) = 0.2231$). For clinical trials, however, the data may not follow lognormal distributions. In this case it is suggested that the equivalence limits be chosen based on the response of the reference drug for internal validity. For example, we may choose $L = s\mu_R$ and $U = b\mu_R$. Thus the two one-sided hypotheses in (6.5.3) can then be expressed as

$$H_{0L}: \mu_T - (1 - s)\mu_R \leq 0,$$
$$\text{vs.} \quad H_{aL}: \mu_T - (1 - s)\mu_R > 0,$$

and (6.5.6)

$$H_{0U}: \mu_T - (1 - b)\mu_R \geq 0,$$
$$\text{vs.} \quad H_{aU}: \mu_T - (1 - b)\mu_R < 0.$$

Consequently the two one-sided tests procedure should be based on the estimates for $\mu_T - (1 - s)\mu_R$ and $\mu_T - (1 - b)\mu_R$ and their estimated standard errors. However, the two one-sided tests procedure is currently performed with estimated equivalence limits based on the average reference response as the true unknown fixed quantities without taking into consideration of the variability of the estimated equivalence limits. Liu and Weng (1995) indicated that for the data obtained from a two-sequence, two-period crossover design, the current application of the two one-sided test procedure cannot control the probability of making a type I error. When the correlation between the two responses from the same patient approaches 1, the consumer's risk goes to 50%. In other words, even if the average efficacy of the test drug is inequivalent to the reference drug, there is still a 50% chance that the two one-sided tests procedure will conclude equivalence between the two drugs if the variability of the estimated equivalence limits are not accounted for and the correlation approaches to one. They proposed a modified two one-sided tests procedure according to the two one-sided hypotheses in (6.5.6). The results of the impact of the correlation on the consumer's risk for both procedures based on a simulation study is summarized in Table 6.5.4. The results reveal that the modified two one-sided tests procedure adequately controls the consumer's risk, while the current two one-sided tests procedure fails to do so. Similar conclusions on the probability of making a type I error can be reached if the data are obtained from parallel designs with the equivalence limits based on the average reference response concurrently from the same trial.

Table 6.5.4 Impact of Correlation on the Consumer's Risk for the Data From a Two-Sequence and Two-Period Crossover Design for Sample Size = 18 and CV = 15%

Correlation	Average Test Response	Current	Modified
0.50	80	0.0380	0.0547
	120	0.0767	0.0577
0.75	80	0.0407	0.0527
	120	0.0787	0.0557
0.90	80	0.0497	0.0510
	120	0.0840	0.0547
0.95	80	0.0727	0.0575
	120	0.0943	0.0520
0.99	80	0.1647	0.0497
	120	0.1647	0.0500
0.999	80	0.3623	0.0467
	120	0.3423	0.0440
0.9999999	80	0.5120	0.0473
	120	0.4913	0.0537

Note: The average reference response is assumed to be 100. The consumer's risk is the probability of concluding equivalence when the test and reference drugs are in fact not equivalent. $s = -b = 0.2$.
Source: Liu and Weng (1995).

Active Control Equivalence Trial

As discussed in Section 6.3, Huque and Dubey (1990) and Pledger and Hall (1990) recommend that a three-treatment trial with test drug, reference active control, and a placebo control be conducted to establish equivalence in average efficacy between the test and reference drugs. For this trial the following hypotheses are of interest:

$$H_0: \mu_T - \mu_R \leq L \quad \text{or} \quad \mu_T - \mu_R \geq U \quad \text{or} \quad \mu_T \leq \mu_P \quad \text{or} \quad \mu_R \leq \mu_P,$$
$$\text{vs.} \quad H_a: L < \mu_T - \mu_R < U \quad \text{and} \quad \mu_T > \mu_P \quad \text{and} \quad \mu_R > \mu_P, \quad (6.5.7)$$

where μ_P is the average response of the placebo.

In addition to the two one-sided hypotheses in (6.5.4) for assessment of equivalence in efficacy between the test and reference drugs, there are two more one-sided hypotheses for evaluation of the superiority in effectiveness of the test and reference active control as compared to the placebo:

$$H_{0T}: \mu_T \leq \mu_P,$$
$$\text{vs.} \quad H_{aT}: \mu_T > \mu_P,$$

and (6.5.8)

$$H_{0R}: \mu_R \leq \mu_P,$$
$$\text{vs.} \quad H_{aR}: \mu_R > \mu_P.$$

Consequently there are four one-sided hypotheses to be evaluated for an active control equivalence trial of this kind. The space of the null hypothesis in (6.5.7) is the union of the spaces of the four null hypotheses in (6.5.3) and (6.5.8), while the space of the alternative hypothesis is the interaction of the spaces of the four alternative hypotheses. Hence, if all four one-sided null hypotheses are rejected at the α level of significance, then the null hypothesis in (6.5.7) is rejected at the same α level, and it is concluded that the efficacy of both test and reference drug are equivalent and are superior to the placebo. We will refer to this procedure as the four one-sided tests procedure. This proposed procedure controls the consumer's risk at the nominal level of significance. Therefore it is undoubtedly very conservative.

6.6 DISCUSSION

In Section 6.4 we discussed the use of factorial designs for evaluation of combination drugs or combination therapy. Its application, however, should not be limited. Factorial designs are useful in the determination of optimal doses for dose-response studies. For example, for an evaluation of a possible optimal dosage regimen during phase II clinical development, it is necessary to investigate the frequency of drug administrations and dose levels at each administration simultaneously. Therefore the intended trial involves two factors: the frequency of the dosing and the magnitude of dosing. Therefore a factorial design is helpful. Ruberg (1995a, 1995b) considers a full two-factor factorial design for the assessment of the QD and BID regimens and the dose levels given in Table 6.6.1. Ruberg suggests that for the design in Table 6.6.1, a placebo group for the QD regimen might not be needed. In addition, in order to get a stronger comparison between the high dose and placebo of the BID regimen, k in Table 6.6.1 must be greater than 1.

Sometimes the treatment of a certain disease requires inpatient intravenous

Table 6.6.1 Factorial Design for Dosing Regimen and Dose Levels

	Dose Levels (Magnitude)			
Frequency	Placebo	1	2	3
QD	n	n	n	n
BID	kn	n	n	kn

Source: Ruberg (1995).

Table 6.6.2 Factorial Design for Combination of I.V. Infusion Followed by Oral Administration

Dose Level of Oral Administration	I.V. Infusion Rate			
	Placebo	1	2	3
Placebo	n	n	n	n
1	n	n	n	n
2	n	n	n	n
3	n	n	n	n

infusion of one drug as initial therapy followed by outpatient oral administration of another drug as maintenance therapy such as streptokinase and aspirin in the ISIS2 study. Then the factorial design given in Table 6.6.2 can be used in the study of a dose-response relationship. For example, the first row and first column in Table 6.6.2, where placebo of each drug is listed, gives an independent characterization of the pure dose response for different administrations of different drugs.

Although a full factorial design is useful for simultaneously evaluating the joint contribution of different components of a combination drug or therapy, the number of treatment groups can be prohibitively large. This may result in some practical issues during the conduct of the trial. For example, human error is more likely when the study involves a randomization of more than 16 parallel treatment groups for 25 study centers. In addition, due to lack of resource and financial constraints, it may be difficult to conduct a full cross-classification factorial design. A fractional factorial design is an attractive alternative for the evaluation of combination drugs (Box et al. 1978). Although fractional factorial designs are widely found in industry, they are not very common in clinical research. One of the reasons is that the general principle of a fractional factorial design may not be directly applied to meet the objectives for evaluation of combination drugs. Therefore proper modification of the fractional factorial design with valid statistical methods must be made before it can be accepted for assessment of combination therapy. Hung (1996) extended his method to the incomplete factorial design for identification of a combination superior to each component. However, it is not clear how strict superiority can be interpreted if the incomplete factorial design is disconnected as shown in Table 6.6.3 (Searle, 1970).

In combination drug trials, although the primary goal is to identify a combination whose efficacy is better than each component, the quantitative characterization of the contribution made by each component when it is administered as monotherapy is also crucial for clinical practitioners. In addition the interrelationships among the procedures proposed by Hung and others for identification of a superior combination, of dose levels between two drugs, and of each component's effect in monotherapy based on a traditional analysis of variance are quite complex. Hung et al. (1995) propose a two-stage estimation and test procedure for treatment effects of monotherapies in two-by-two factorial trials.

Table 6.6.3 An Incomplete Disconnected Factorial Design

Dose Level for Drug A	Dose Level of Drug B			
	Placebo	0	1	3
Placebo	×	×		
1	×	×		
2			×	×
3			×	×

Note: The dose × indicates the combination being studied.

For three-arm trials with combined treatment groups and two component drugs, Laska and Meisner (1989) provide tables of sample sizes for the various powers according to their min procedure which is based on the Student t-test and the Wilcoxon rank sum test for univariate clinical endpoints. They found that approximately 25% more patients per treatment group are required for evaluating the superiority of a combination drug than is usual for comparing the difference between two treatment groups. As for the combination trial with a full factorial design, the formula for calculating power is given in Hung et al. (1993), and Hung (1994) can be used to iteratively select sample sizes with a desired power. However, the relative magnitude of the sample sizes required by the procedures proposed by Hung et al. (1993) is not known, whereas the traditional F-test is used for finding the treatment difference among dose levels in monotherapy. When the objective is to examine possible interaction between two drugs, as pointed out by Byar and Piantadosi (1985), the sample size for detecting any interaction must be much larger than that for studying the main effects of monotherapy. More research is definitely needed for application to multicenter trials, group sequential methods, and interim analyses of combination drugs.

Since there can be an ethical concern regarding the inclusion of concurrent placebo control in active control equivalence trials, Huque and Dubey (1990) explore the possibility of designing a group sequential trial by performing interim analyses for the purpose of stopping the placebo arm but continuing the test and active reference control. However, in order to stop the placebo group early, based on results from interim analyses, the randomization codes of all three groups must be broken to reveal the treatment assignment of each patient. As a result bias will be unavoidably introduced into the subsequent assessment between the test and active control, and consequently the primary goal for an unbiased evaluation of the equivalence between the two drugs may be in jeopardy.

For binary data the sample size determination and equivalence limits depend on not only the assumed reference response rate but also the overall response rate (Huque and Dubey, 1990). It should be noted that the observed reference and overall response rates might differ from the assumed rates used for sample

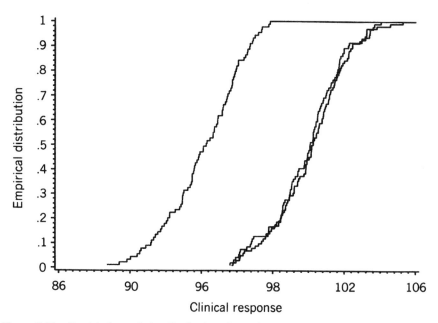

Figure 6.6.1 Empirical cumulative distribution of an active control equivalence trial with a concurrent placebo control.

size determination at the planning stage of the trial. In addition they may be different from one interim examination to another. It is, however, not clear whether the same prespecified equivalence limit should be used for all interim analyses. Durrleman and Simon (1990) and Jennison and Turnbull (1993) offer some useful thoughts on the design, conduct, monitoring and analysis of sequential active control equivalence trials. However, their methodology does not include a joint evaluation of equivalence and superiority in the efficacy of active drugs in relation to the placebo.

Note that our discussion in Section 6.5 only focused on the average equivalence rather than the population equivalence. Figure 6.6.1 illustrates the case where the cumulative distribution of the test and reference active control are equivalent, assuming that they both are (stochastically) larger than those of the placebo group. For assessment of equivalence in cumulative distribution, the hypotheses in (6.5.7) are necessarily modified as.

$$H_0: |F_T - F_R| \geq d \quad \text{or} \quad F_T \leq F_P \quad \text{or} \quad F_R \leq F_P,$$
$$\text{vs.} \quad H_a: |F_T - F_R| < d \quad \text{and} \quad F_T > F_P \quad \text{and} \quad F_R > F_P, \quad (6.6.1)$$

where F_T, F_R, and F_P are the cumulative distribution functions (CDF) of the clinical responses of test, reference active control, and concurrent placebo control, respectively. Hypotheses (6.6.1) can be further decomposed into three sets.

$$H_{0E}: |F_T - F_R| \geq d,$$
$$\text{vs.} \quad H_{aE}: |F_T - F_R| < d, \qquad (6.6.2)$$

$$H_{0T}: F_T \leq F_P,$$
$$\text{vs.} \quad H_{aT}: F_T > F_P, \qquad (6.6.3)$$

and

$$H_{0R}: F_R \leq F_P,$$
$$\text{vs.} \quad H_{aR}: F_R > F_P. \qquad (6.6.4)$$

The standard one-sided Kolmogorov-Simrnov test (Conover, 1980) can be used to verify whether the distributions of both the test and reference active control are (stochastically) larger than those of the placebo. Recently, as an alternative, Wellek (1993) proposed a method that tests the equivalence between two cumulative distributions. However, his procedure requires an estimation of a noncentrality parameter of the chi-square distribution under hypotheses (6.6.2). More research on equivalence in distributions is needed.

CHAPTER 7

Analysis of Continuous Data

7.1 INTRODUCTION

As was pointed out in Chapter 1, a well-designed protocol can ensure the success of clinical research. A protocol that is well designed focuses on all details including how the intended clinical trial is to be carried and how the data are to be collected. It also must reflect Section 312.23 of 21 CFR which requires that a statistical section, which describes how the data are to be analyzed, must be a part of the protocol. It is further important for good clinical practice to provide an accurate and reliable assessment of the efficacy and safety of the test drug under study. For this purpose appropriate and valid statistical methods must be determined for the collected data. The collected clinical data are referred to as the responses of the clinical endpoints or variables of patients under study. These clinical endpoints or variables are used to measure or evaluate the characteristics of the test drug product for treatment of patients with certain diseases under study. Basically the collected clinical data can be classified as either qualitative (categorical data) or quantitatively (numerical or measurement data); see Johnson and Bhattacharyya (1985).

When the characteristic concerns a qualitative trait that is only classified in categories and not numerically measured, the resulting data are called categorical data. For the categorical data, subjects are placed in the proper category or group, and the number of subjects in each category is enumerated. Each subject must be fit into exactly one category. In clinical trials gender and race are typical categorical variables. Note that if there is order among the categories, the resulting data are called ranked data or ordered categorical data. For example, the severity and intensity of pain are considered ranked data. For ranked data there are not necessarily equal intervals or differences between ranks. On the other hand, if the characteristic is measured on numerical scales, then the resulting data are numerical data. Note that if the numerical scales are made of distinct numbers with gaps in between such as the number of drinks daily, then the variables are referred to as discrete variables. Clinical endpoints are said to be continuous if they can take any values in an interval. Some variables such as height, weight, survival time, and diastolic and systolic blood pressures are continuous variables. The responses of a continuous variable are considered continuous data.

In general, different statistical methods are usually applied for different types of data. In this chapter, our primary emphasis will be placed on statistical analysis for continuous data. Analysis for categorical data will be discussed in the next chapter. In the next section the concept of estimation including point and interval estimates will be briefly introduced. Statistical tests such as paired t, two sample t, and F tests are given in Section 7.3. In Section 7.4 the method of variance (covariance) analysis is discussed. The use of nonparametric methods and the application of repeated measures are described in Sections 7.5 and 7.6, respectively. A brief discussion is presented in the last section.

7.2 ESTIMATION

As mentioned, clinical endpoints are used to measure or describe the characteristics of a test drug for the treatment of patients with certain diseases. Based on the observed responses of the clinical endpoints, we can draw some statistical inferences on the characteristics of the test drug. For example, reductions in seated diastolic and systolic blood pressures are often used to evaluate the effect of a test drug in the treatment of patients with mild to moderate hypertension. In this section we will focus on the estimation of some quantities such as the mean and variability of the characteristics.

Suppose that there are N subjects in a targeted patient population. For a given clinical endpoint, let $y_1, y_2, \ldots, y_N$ be the true responses of the clinical endpoint for the N subjects in the targeted patient population. Then the population mean and population variance of the clinical endpoint are given by

$$\mu = \frac{1}{N} \sum_{i=1}^{N} y_i,$$

$$\sigma = \frac{1}{N} \sum_{i=1}^{N} (y_i - \mu)^2,$$

respectively. Note that μ and σ^2 are usually unknown. In practice, we make a statistical inference on μ to assess the efficacy of the drug under study. For example, μ may represent the mean reduction in the seated diastolic blood pressure in patients with mild to moderate hypertension. As discussed in Chapter 4, in practice, since N can be very large (e.g., a few million people), it is impossible to observe all the y_i, $i = 1, \ldots, N$ values. Therefore, as an alternative, we select a random sample of size n and then draw statistical inference on μ from an analysis of information contained in the sample data provided that the

ESTIMATION

random sample is a representative of the targeted patient population. Intuitively we can use the mean and standard deviation of the n observations from the random sample to estimate μ and σ. That is, we let $\{Y_1, Y_2, \ldots, Y_n\}$ be a random sample of size n selected from the targeted patient population $\{y_1, y_2, \ldots, y_N\}$, then the realization of $\{Y_1, Y_2, \ldots, Y_n\}$ is a subset of $\{y_1, y_2, \ldots, y_N\}$, where $2 \leq n \leq N$. Then μ and σ can be estimated by the sample mean and sample standard deviation, which are given by

$$\bar{Y} = \frac{1}{n} \sum_{i=1}^{n} Y_i,$$

$$s = \left\{ \frac{1}{n-1} \sum_{i=1}^{n} (Y_i - \bar{Y})^2 \right\}^{1/2},$$

respectively. When the sample size n is sufficiently large, $\bar{Y}$ and s would be quite close to the unknown population mean μ and population standard deviation σ, respectively.

Note that if $\{Y_1, Y_2, \ldots, Y_n\}$ results from independent selections, then each has the same distribution as the population. Since random samples vary, the values of $\bar{Y}$ also vary. In other words, $\bar{Y}$ is also a random variable. The sampling distribution of $\bar{Y}$ can be obtained by the following steps:

1. List all possible samples of size n.
2. Calculate the values of $\bar{Y}$ for each sample.
3. List the distinct values of $\bar{Y}$, and calculate the corresponding probabilities by identifying all of the samples that yield the same value of $\bar{Y}$.

For the selection of a random sample of size n, there are a total of M possible selections for a population of size N, where

$$M = \binom{N}{n} = \frac{N(N-1)\cdots(N-n)}{n!}.$$

Let $\bar{Y}_k$ be the mean of the kth random sample, where $k = 1, \ldots, M$. Then the expectation of $\bar{Y}$ is the average of the sample means over all possible samples,

$$E(\bar{Y}) = \frac{1}{M} \sum_{k=1}^{M} \bar{Y}_k.$$

If

$$E(\bar{Y}) = \mu,$$

then $\overline{Y}$ is said to be an unbiased estimate of μ. Similarly the standard error of $\overline{Y}$ can be obtained as follows:

$$\sigma_{\overline{Y}} = SD(\overline{Y}) = \sqrt{\mathrm{var}(\overline{Y})} = \left\{ \frac{1}{M} \sum_{k=1}^{M} [\overline{Y}_k - E(\overline{Y})]^2 \right\}^{1/2}.$$

Since $\overline{Y}$ is the average of n random variables with common mean μ and variance σ^2, it can be verified that

$$\sigma_{\overline{Y}}^2 = \frac{1}{n} \sigma^2.$$

Note that if the population is normal with mean μ and standard deviation σ, then $\overline{Y}$ has the normal distribution with mean μ and standard deviation $\sigma/\sqrt{n}$. However, if the population is an arbitrary population with mean μ and standard deviation σ, then $\overline{Y}$ is approximately normal with mean μ and standard deviation $\sigma/\sqrt{n}$ when n is large (e.g., $n \geq 30$). As a result, since σ can be consistently estimated by s,

$$z = \frac{\overline{Y} - \mu}{s/\sqrt{n}}$$

is approximately distributed as a standard normal with mean 0 and standard deviation 1. This property is known as the *central limit theorem*. As a result, when n is large, in addition to the point estimate, we can provide an interval estimate for μ based on the central limit theorem. A large samples $100(1 - \alpha)\%$ confidence interval for μ is given by

$$\left(\overline{Y} - \frac{Z(\alpha/2)s}{\sqrt{n}}, \overline{Y} + \frac{Z(\alpha/2)s}{\sqrt{n}} \right), \qquad (7.2.1)$$

where $Z(\alpha/2)$ denotes the upper $(\alpha/2)$th quantile of the standard normal distribution (i.e., the area to the right of $Z(\alpha/2)$ is $\alpha/2$). The $100(1 - \alpha)\%$ confidence interval is a random interval that will cover the true μ with probability $(1 - \alpha)$. In other words, if random samples are repeatedly drawn from the same population and a $100(1 - \alpha)\%$ confidence interval is calculated from each sample, then about $100(1 - \alpha)\%$ of these intervals will contain the true but unknown value μ. In practice, however, since we only draw one sample (i.e., conduct the intended clinical trial once), we never know what happens in a single application. Therefore, when we say that *we are $100(1 - \alpha)\%$ confident that the confidence interval obtained from the sample will cover the true μ*, our confidence draws from the success rate of $100(1 - \alpha)\%$ in many applications.

ESTIMATION

Note that the above interval estimate is derived based on the central limit theorem with large n. In practice, n may not be large, especially when the trial involves enormous medical expenditives or the disease is rare. In this case the question *What is the sample distribution of $\bar{Y}$ when n is not large?* needs to be addressed in order to provide an interval estimate for μ. As indicated earlier, when $\bar{Y}$ is based on a random sample of size n from a normal population with mean μ and standard deviation σ, $\bar{Y}$ is distributed as a normal distribution with mean μ and standard deviation $\sigma/\sqrt{n}$. Since σ is usually unknown, an intuitive approach is to estimate σ by the sample standard deviation s. This leads to the following t statistic

$$t = \frac{\bar{Y} - \mu}{s/\sqrt{n}}.$$

The distribution of the above t statistic is known as Student's t distribution with $n - 1$ degrees of freedom. As discussed above, the Student t distribution gets closer to the standard normal as n becomes larger. For small n, although the t distributions are also symmetric about 0, they have tails that are more spread out than the standard normal distribution. Therefore, based on the t statistic, we can construct a $100(1 - \alpha)\%$ confidence interval for μ for small n as follows:

$$\left(\bar{Y} - \frac{t(\alpha/2, n-1)s}{\sqrt{n}}, \bar{Y} + \frac{t(\alpha/2, n-1)s}{\sqrt{n}} \right), \quad (7.2.2)$$

where $t(\alpha/2, n-1)$ is the upper $(\alpha/2)$th quantile of the t distribution with $n-1$ degrees of freedom.

To illustrate the estimation procedures described above, consider a clinical trial conducted in order to evaluate the safety of a injectable dosage form of a drug product (denoted by treatment A) that is compared to a placebo (denoted by treatment B) in subjects undergoing stress echocardiography. This trial was a multicenter, single-blind, and randomized study. The primary safety variables included incidence of adverse events and changes in laboratory parameters. Table 7.2.1 lists the partial data of pre- and post-treatment platelet counts of 30 patients from the three centers. Table 7.2.2 provides the summary statistics of the platelet counts. From Table 7.2.2 and using (7.2.2), we obtain the confidence intervals for the average platelet counts at the baseline (pre-treatment) and at endpoints (post-treatment) for both treatments. For example, for treatment A, since

$$\bar{Y}_A = 223066.67 \quad \text{and} \quad s_A = 15219.18,$$

the 95% confidence interval for the average pre-treatment platelet counts is given by

Table 7.2.1 Pre- and Post-Treatment Platelet Counts

Center	Subject	Pre-treatment	Post-treatment	Treatment
1	1	359,000	396,000	B
1	2	200,000	184,000	A
1	3	149,000	151,000	A
1	4	235,000	242,000	B
1	5	174,000	177,000	B
1	6	271,000	203,000	A
2	7	180,000	199,000	B
2	8	252,000	256,000	A
2	9	188,000	187,000	B
2	10	211,000	210,000	A
2	11	217,000	199,000	B
2	12	195,000	215,000	A
3	13	266,000	192,000	B
3	14	267,000	205,000	A
3	15	217,000	233,000	A
3	16	247,000	241,000	B
3	17	204,000	188,000	A
3	18	340,000	302,000	B
4	19	175,000	173,000	A
4	20	242,000	233,000	B
4	21	383,000	389,000	A
4	22	165,000	154,000	A
4	23	213,000	226,000	B
4	24	244,000	241,000	B
5	25	243,000	223,000	A
5	26	205,000	167,000	B
5	27	373,000	315,000	B
5	28	252,000	271,000	A
5	29	291,000	250,000	B
5	30	162,000	146,000	A

$$\overline{Y}_A \pm \frac{t(\alpha/2, n-1)s_A}{\sqrt{n}} = 223066.67 \pm (2.14)\left(\frac{15219.18}{\sqrt{15}}\right)$$

$$= 223066.67 \pm 8409.29$$

$$= (214657.38, 231475.96).$$

Similarly a 95% confidence interval for the average post-treatment platelet counts can be obtained. Table 7.2.3 gives 95% confidence intervals for the average pre- and post-treatment platelet counts for both treatments.

Table 7.2.2 Summary Statistics of Pre- and Post-Treatment Platelet Counts

Treatment	N	Pre-treatment	Post-treatment	Difference
A	15	223.1	213.4	−9.7
		(15.2)	(15.7)	(6.7)
B	15	251.6	237.8	−13.8
		(16.4)	(15.7)	(7.9)
A − B	15	−28.5	−24.4	4.1
		(15.8)	(15.7)	(7.3)

Note: The values are corresponding platelet counts ($\times 10^3/\mu L$) and the numbers in the parentheses are standard errors.

7.3 TEST STATISTICS

In clinical trials, as pointed out earlier, a typical approach for assessment of the efficacy of a test drug is first to demonstrate that the mean effect (i.e., μ) is significantly different from zero or a prespecified small value (e.g., μ_0) and then claim that the test drug is effective by showing that there is a desired power for detection of a clinically meaningful difference. Therefore the hypotheses of interest are given by

$$H_0: \mu = \mu_0,$$
$$\text{vs.} \quad H_a: \mu \neq \mu_0. \quad (7.3.1)$$

If the above hypotheses concerns the mean of a normal population, then we may consider the following test statistic for testing (7.3.1):

$$t = \frac{\overline{Y} - \mu_0}{s/\sqrt{n}},$$

Table 7.2.3 Confidence Intervals Based on *t*-Statistic

Treatment	95% Confidence Interval		
	Pre-treatment	Post-treatment	Difference
A	(214.7, 231.5)	(204.7, 222.1)	(−13.4, −6.0)
B	(242.5, 260.7)	(229.1, 246.5)	(−18.1, −9.5)
A − B	(−37.5, −19.5)	(−33.4, −15.4)	(−9.6, 1.3)

Note: The values in the parentheses are corresponding platelet counts ($\times 10^3/\mu L$).

which is distributed as the Student t distribution with $n-1$ degrees of freedom under H_0. Note that the alternative hypothesis given in (7.3.1) is a two-sided hypothesis. In practice, as mentioned in Chapter 2, we can consider either one of the following one-sided hypotheses:

$$H_a: \mu > \mu_0,$$

or (7.3.2)

$$H_a: \mu < \mu_0.$$

We would reject the null hypothesis of (7.3.1) at the α level of significance if

$$|t| \geq t(\alpha/2, n-1)$$

On the other hand, for the one-sided hypotheses, we can reject the null hypothesis at the α level of significance if

$$t \geq t(\alpha, n-1) \quad \text{or} \quad t \leq -t(\alpha, n-1)$$

for alternative hypotheses $H_a: \mu > \mu_0$ or $H_a: \mu < \mu_0$, respectively.

Note that the concepts of hypotheses testing and confidence interval are equivalent. For testing the hypotheses in (7.3.1), the rejection region of the level α test is

$$\left| \frac{\overline{Y} - \mu_0}{s/\sqrt{n}} \right| \geq t(\alpha/2, n-1).$$

Consequently the acceptance region is given by

$$-t(\alpha/2, n-1) < \frac{\overline{Y} - \mu_0}{s/\sqrt{n}} < t(\alpha/2, n-1),$$

which can be rewritten as

$$\overline{Y} - \frac{t(\alpha/2, n-1)s}{\sqrt{n}} < \mu_0 < \overline{Y} + \frac{t(\alpha/2, n-1)s}{\sqrt{n}}. \quad (7.3.3)$$

Note that the above expression of the acceptance region is the same as the confidence interval for μ given in (7.2.2). Expression (7.3.3) indicates that any given null hypothesis μ_0 will not be rejected at the significance level of α if μ_0 lies within the $100(1-\alpha)\%$ confidence interval of μ. Therefore, having established a $100(1 - \alpha)\%$ confidence interval for μ, all null hypotheses μ_0 lying outside this interval will be rejected at the α level of significance and all those lying inside will not be rejected. As indicated in Chapter 2, an observed p-value is always reported in accordance with the test. We would reject the null hypothesis if the p-value is less than the α level of significance. Johnson and Bhattacharyya (1985) point out that the confidence interval approach is regarded as a more comprehensive inference procedure than testing a single null hypothesis because a confidence interval statement in effect tests many null hypotheses at the same time.

Paired t Test

For some clinical trials such as noncomparative open-label studies, one of the primary objectives is to evaluate the effect before and after the treatment based on endpoint changes from baseline. In this situation a paired-t test is useful for within treatment comparison. Let Y_{Bi} and Y_{Ei} be the responses of the ith subject at baseline (or pre-treatment) and endpoint (or post-treatment), respectively. Statistically each subject is considered as a block. Therefore, although (Y_{Bi}, Y_{Ei}) are independent of one another, Y_{Bi} and Y_{Ei} within the ith subject are dependent. Consider the endpoint changes from baseline as follows:

$$D_i = Y_{Ei} - Y_{Bi},$$

where $i = 1, \ldots, n$. Since $D_i = Y_{Ei} - Y_{Bi}$, $i = 1, \ldots, n$, remove the block effect, it is reasonable to assume that they constitute a random sample from a population with mean μ_d and standard deviation σ_d, where μ_d represents the mean difference of before and after treatment. In other words,

$$\mu_d = E(D_i),$$
$$\sigma_d = \sqrt{\text{var}(D_i)}, \quad i = 1, \ldots, n. \quad (7.3.4)$$

The hypotheses of interest for a within treatment comparison are

$$\text{H}_0: \mu_d = \mu_0,$$
$$\text{vs } \text{H}_a: \mu_d \neq \mu_0. \quad (7.3.5)$$

We reject the null hypothesis at the α level of significance if

$$|t| = \left| \frac{\overline{D} - \mu_0}{s_d/\sqrt{n}} \right| \geq t(\alpha/2, n - 1), \quad (7.3.6)$$

where

$$\overline{D} = \frac{1}{n} \sum_{i=1}^{n} D_i,$$

$$s_d = \left\{ \frac{1}{n-1} \sum_{i=1}^{n} (D_i - \overline{D})^2 \right\}^{1/2},$$

and $t(\alpha/2, n-1)$ is the upper $(\alpha/2)$th quantile of a t distribution with $n-1$ degrees of freedom. Similarly we can reject the null hypothesis at the α level of significance if

$$t \geq t(\alpha, n-1) \quad \text{or} \quad t \leq -t(\alpha, n-1)$$

for alternative hypotheses $H_a: \mu_d > \mu_0$ or $H_a: \mu_d < \mu_0$, respectively.

Note that the above test for hypotheses given (7.3.5) is called a paired-t test; the paired-t test is commonly used for a within-treatment comparison between the baseline and endpoint of a clinical trial. The corresponding confidence interval for μ_d is given by

$$\left(\overline{D} - \frac{t(\alpha/2, n-1) s_d}{\sqrt{n}}, \overline{D} + \frac{t(\alpha/2, n-1) s_d}{\sqrt{n}} \right).$$

To illustrate the application of a paired-t test, consider the example of platelet counts given in the previous section. Based on difference in platelet counts between pretreatment and post-treatment, paired-t tests for treatment A and treatment B are given by

$$|t_A| = \left| \frac{\overline{D}}{s_d/\sqrt{n}} \right| = \left| \frac{-9666.67}{6702.29/\sqrt{15}} \right| = 5.59$$

and

$$|t_B| = \left| \frac{\overline{D}}{s_d/\sqrt{n}} \right| = \left| \frac{-13800.00}{7855.97/\sqrt{15}} \right| = 14.86,$$

respectively. Since t_A and t_B are both greater than $t(0.025, 14) = 2.14$, we reject the null hypothesis of no difference between pretreatment and post-treatment average platelet counts for both treatments at the α level of significance. Note that 95% confidence intervals of μ_d for treatments A and B, as given in Table 7.2.3, are $(-13.4, -6.0)$ and $(-18.1, -9.5)$ $(10^3/\mu L)$, respectively, which do not contain 0. We can conclude that there is a significant difference between the

TEST STATISTICS

pretreatment and post-treatment platelet counts for both treatments. This result is consistent with that obtained from paired-t tests.

Recall that Table 7.2.1 showed that the pre- and post-treatment platelet counts are positively correlated. Thus an analysis of changes from the baseline (i.e., post-treatment minus pretreatment) is more appropriate because it reduces the variability. As indicated in Table 7.2.2, the variability is reduced by almost a half. In addition a direct comparison suggests that the average platelet counts of treatment B are significantly greater than those of treatment A. However, Table 7.2.3 indicates that the mean decrease in platelet counts of treatment A is not statistically different from that of treatment B.

Two-Sample t Test

In comparative clinical trials the primary objective is to evaluate the efficacy and safety of a test drug as compared to a control (e.g., a placebo control or an acive control). For this purpose a parallel-group design in which patients are randomly assigned to receive either the test drug or the control is usually employed. In this design setting we basically compare two treatment groups (the test drug and the control) or two populations on the basis of independent samples (i.e., patients who are enrolled in the two treatment groups). Let Y_{ij}, $i = 1, 2$ and $j = 1, \ldots, n_i$, be random samples of sizes n_1 and n_2 from population 1 (i.e., patients under treatment of the test drug) and population 2 (i.e., patients under treatment of the control), respectively. Note that the two samples are independent. In other words, the responses under one treatment are uncorrelated to the responses under the other treatment. Suppose that the two populations are normal with means μ_1 and μ_2 and standard deviations σ_1 and σ_2, respectively. Then the efficacy of the test drug, as compared to the control, can be examined by testing the following hypotheses:

$$H_0: \mu_1 = \mu_2,$$
$$\text{vs.} \quad H_a: \mu_1 \neq \mu_2. \quad (7.3.7)$$

When sample sizes n_1 and n_2 are large, according the central limit theorem, the statistic

$$Z = \frac{(\overline{Y}_1 - \overline{Y}_2) - (\mu_1 - \mu_2)}{\sqrt{s_1^2/n_1 + s_2^2/n_2}}$$

is approximately normal with mean 0 and standard deviation 1, where $\overline{Y}_1$ and $\overline{Y}_2$ and s_1^2 and s_2^2 are the sample means and sample variances of the two populations. Therefore we can reject the null hypothesis at the α level of significance if

$$|Z| \geq Z(\alpha/2),$$

where $Z(\alpha/2)$ is the upper $(\alpha/2)$th quantile of the standard normal.

When sample sizes n_1 and n_2 are small, the above test is not valid. However, under the assumption that (1) both populations are normal and (2) $\sigma_1 = \sigma_2 = \sigma$, the test statistic

$$t = \frac{(\bar{Y}_1 - \bar{Y}_2) - (\mu_1 - \mu_2)}{s\sqrt{1/n_1 + 1/n_2}} \tag{7.3.8}$$

has Student's t distribution with $n_1 + n_2 - 2$ degrees of freedom, where

$$s = \left\{ \frac{(n_1 - 1)s_1^2 + (n_2 - 1)s_2^2}{n_1 + n_2 - 2} \right\}^{1/2},$$

which is a pooled estimate of the common standard deviation σ. Therefore we can reject the null hypothesis of (7.3.7) at the α level of significance if

$$|t| \geq t(\alpha/2, n_1 + n_2 - 2),$$

where $t(\alpha/2, n_1 + n_2 - 2)$ is the upper $(\alpha/2)$th quantile of a t distribution with $n_1 + n_2 - 2$ degrees of freedom. Similarly we can reject the null hypothesis at the α level of significance if

$$t \geq t(\alpha, n_1 + n_2 - 2) \quad \text{or} \quad t \leq -t(\alpha, n_1 + n_2 - 2)$$

for alternative hypotheses $H_a: \mu_1 > \mu_2$ or $H_a: \mu_1 < \mu_2$, respectively. Note that the above test is known as a two-sample t test. Based on this test, a $100(1 - \alpha)\%$ confidence interval for $\mu_1 - \mu_2$ is given by

$$(\bar{Y}_1 - \bar{Y}_2) \pm t(\alpha/2, n_1 + n_2 - 2)s\sqrt{\frac{1}{n_1} + \frac{1}{n_2}}.$$

In practice, σ_1 is never the same as σ_2. Whether or not to pool the data for an estimate of σ (under the assumption that $\sigma_1 = \sigma_2 = \sigma$) affects the validity of the test statistic described in (7.3.8). The relative magnitude of the two sample variances s_1^2 and s_2^2 is usually the indicator used to determine whether or not to pool. For example, if s_1^2/s_2^2 is far away from 1, then the assumption that $\sigma_1 = \sigma_2 = \sigma$ will be in doubt. In practice, as a rule of thumb, if the ratio falls within the range of

$$\frac{1}{4} \leq \frac{s_1^2}{s_2^2} \leq 4,$$

TEST STATISTICS

then we can pool the data for an estimate of the common standard deviation σ. In the case where $\sigma_1 \neq \sigma_2$, we may reject the null hypothesis if

$$|t^*| \geq t(\alpha/2, v^*),$$

where

$$t^* = \frac{(\overline{Y}_1 - \overline{Y}_2) - (\mu_1 - \mu_2)}{\sqrt{w_1 + w_2}}, \quad (7.3.9)$$

and

$$v^* = \frac{(w_1 + w_2)^2}{(w_1^2/n_1 + w_2^2/n_2)},$$

$$w_1 = s_1^2/n_1, \quad \text{and} \quad w_2 = s_2^2/n_2.$$

Based on t^*, a $100(1 - \alpha)\%$ confidence interval for $\mu_1 - \mu_2$ can be obtained as follows:

$$\overline{Y}_1 - \overline{Y}_2 \pm t(\alpha/2, v^*)\sqrt{s_1^2/n_1 + s_2^2/n_2}, \quad (7.3.10)$$

The above confidence interval is known as Satterthwaite's confidence interval.

Again, we consider the example of platelet counts to illustrate the use of a two-samle t test. The differences in platelet counts between pretreatment and post-treatment for the two treatments concern two independent samples. Therefore we can use test statistic described in (7.3.8) or (7.3.9) to test the null hypothesis of no treatment difference. As can be seen from Table 7.2.2,

$$s_A = 6702.29 \quad \text{and} \quad s_B = 7855.97.$$

Since

$$\frac{s_A^2}{s_B^2} = 0.729,$$

which is within the range (0.25, 4), we can apply the test statistic in (7.3.8) to test the null hypothesis of no treatment difference as follows:

$$|t| = \left| \frac{(\overline{Y}_A - \overline{Y}_B)}{s\sqrt{1/n_A + 1/n_B}} \right|$$

$$= \left| \frac{(-9666.67) - (-13800.00)}{\sqrt{7109121.69}} \right|$$

$$= \left| \frac{4133.33}{2666.29} \right| = 1.55,$$

which is less than $t_{0.025}(28) = 2.048$. Therefore we fail to reject the null hypothesis of no treatment difference at the $\alpha = 5\%$ level of significance. From (7.3.8) the corresponding 95% confidence interval can also be obtained, which is given by $(-9593.89, 1327.23)$. Since the 95% confidence interval contains 0, we reach the same conclusion that there is no significant difference between the two treatments. Note that since $n_A = n_B = 15$, the test statistics in (7.3.8) and (7.3.9) are essentially the same. Both tests fail to reject the null hypothesis of no treatment difference.

7.4 ANALYSIS OF VARIANCE

In previous sections we discussed the estimations and tests for comparing two treatments. Suppose that we are interested in evaluating the efficacy of several test drugs for the same indication compared to a placebo control. Although we may conduct a number of separate clinical trials whereby each compares a test drug with the placebo control, it is more efficient to conduct one clinical trial and to simultaneously compare the test drugs and the placebo. In this case the two-sample t test and its corresponding confidence interval for mean differences is not appropriate. The alternative analysis of variance (ANOVA) method can be applied to compare several population means.

One-Way Classification

Let Y_{ij} be the response of the jth subject on treatment i, where $j = 1, \ldots, n_i$ and $i = 1, \ldots, k$. Also let μ_i denote the mean of the ith treatment. Then the hypotheses for a simultaneous evaluation of the k treatments is given by

$$H_0: \mu_1 = \mu_2 = \cdots = \mu_k,$$

vs. $H_a: \mu_i \neq \mu_j$ for some $1 \leq i \neq j \leq k$. (7.4.1)

ANALYSIS OF VARIANCE

To derive a test for the above hypotheses, consider the following model:

$$Y_{ij} = \mu_i + e_{ij}$$
$$= \mu + \tau_i + e_{ij}, \qquad j = 1, \ldots, n_i, i = 1, \ldots, k, \qquad (7.4.2)$$

where μ is the overall mean, τ_i denotes the effect of the ith treatment, and e_{ij} are independent random errors in observing Y_{ij} with mean 0 and standard deviation σ. The analysis of variance concept is to partition the observations Y_{ij} into contributions from different sources. Let $\overline{Y}_{..}$ and $\overline{Y}_{i.}$ be the overall sample mean and the sample mean for the ith treatment group, respectively. Then the deviation of an individual observation from the overall sample mean can be described as

$$Y_{ij} - \overline{Y}_{..} = (\overline{Y}_{i.} - \overline{Y}_{..}) + (Y_{ij} - \overline{Y}_{i.}) \qquad (7.4.3)$$

or

$$Y_{ij} = \overline{Y}_{..} + (\overline{Y}_{i.} - \overline{Y}_{..}) + (Y_{ij} - \overline{Y}_{i.}).$$

As a result the deviation of an individual observation from the overall sample mean consists of two components: the sum of differences among the means of the treatments and the random variation in measurements within the same treatment. Expression (7.4.3) leads to the following partitions of sum of squares (SS):

$$\sum_{i=1}^{k} \sum_{j=1}^{n_i} (Y_{ij} - \overline{Y}_{..})^2 = \sum_{i=1}^{k} n_i (\overline{Y}_{i.} - \overline{Y}_{..})^2 + \sum_{i=1}^{k} \sum_{j=1}^{n_i} (Y_{ij} - \overline{Y}_{i.})^2.$$

The first term on the left-hand side is the total sum of squares, and the first and second terms on the rigtht-hand side are sum of squares due to treatment (denoted by SSA) and sum of squares due to error (denoted by SSE). Therefore we have

$$SST = SSA + SSE.$$

The mean square due to treatment and error is defined as

$$\text{MSA} = \frac{\text{SSA}}{k-1}$$

$$= \frac{1}{k-1} \sum_{i=1}^{k} n_i (\overline{Y}_{i.} - \overline{Y}_{..})^2,$$

$$\text{MSE} = \frac{\text{SSE}}{\sum_{i=1}^{k} n_i - k} = \sum_{i=1}^{k} \sum_{j=1}^{n_i} (Y_{ij} - \overline{Y}_{i.})^2.$$

Under the null hypothesis, the population means are all equal (i.e., $\mu_i = \mu$ for all i or $\tau_i = 0$ for all i), $\overline{Y}_{i.} - \overline{Y}_{..}$ is expected to be small, and consequently its mean square (MSA) is expected to be small. On the other hand, if the population means differ, then MSA is likely to be large. However, the question that is not clear is *How large a treatment mean will yield a statistically significant difference?* To determine whether the population means are statistically significantly different from one another, since the mean sum of squares due to error provides an estimate of σ^2, we consider the ratio between the treatment mean sum of squares and the mean sum of squares due to error. Thus, under the null hypothesis of (7.4.1), the ratio

$$F = \frac{\text{MSA}}{\text{MSE}}$$

$$= \frac{\text{SSA}/(k-1)}{\text{SSE}/(N-k)}$$

has an F distribution with $(k-1, N-k)$ degrees of freedom, where $N = \sum_{i=1}^{k} n_i$. Therefore we can reject the null hypothesis at the α level of significance if

$$F \geq F(\alpha, k-1, N-k),$$

where $F(\alpha, v_1, v_2)$ is the upper αth quantile of the F distribution with (v_1, v_2) degrees of freedom. Table 7.4.1 provides an analysis of variance table for comparing k treatments.

Simultaneous Confidence Intervals

Suppose that we reject the null hypothesis that there are no differences among the k treatments. Then we need to show that there is a significant difference among the k treatments. In this case it is often of interest to construct a confidence interval for $\mu_i - \mu_{i'}$ for $i \neq i'$ to determine which treatments have unequal means. Based on the mean square error from the analysis of variance, a $100(1 - \alpha)\%$ confidence interval for $\mu_i - \mu_{i'}$ where $i \neq i'$ can be constructed as follows:

ANALYSIS OF VARIANCE

Table 7.4.1 ANOVA Table for Comparing k Treatments

Source of Variation	Sum of Squares	df	Mean Squares
Treatment	$SSA = \sum_{i=1}^{k} n_i(\bar{Y}_{i.} - \bar{Y}_{..})^2$	$k-1$	$MST = \dfrac{SSA}{k-1}$
Error	$SSE = \sum_{i=1}^{k} \sum_{j=1}^{n_i} (Y_{ij} - \bar{Y}_{i.})^2$	$N-k$	$MSE = \dfrac{SSE}{N-k}$
Total	$SST = \sum_{i=1}^{k} \sum_{j=1}^{n_i} (Y_{ij} - \bar{Y}_{..})^2$	$N-1$	

Note: $N = \sum_{i=1}^{k} n_i$.

$$\bar{Y}_{i.} - \bar{Y}_{i'.} \pm t(\alpha/2, N-k) s \sqrt{\frac{1}{n_i} + \frac{1}{n_{i'}}}, \qquad (7.4.4)$$

where

$$s = \sqrt{MSE} = \sqrt{\frac{SSE}{N-k}}$$

and $t(\alpha/2, N-k)$ is the upper $(\alpha/2)$th quantile of t distribution with $N-k$ degrees of freedom. In practice, one can consider comparing means pairwisely according to (7.4.4) if the overall F test shows a significance in the means. However, for k treatments there is a total of

$$m = \binom{k}{2} = \frac{k(k-1)}{2}$$

pairwise differences $\mu_i - \mu_{i'}$, and (7.4.4) applied to all pairs yield m confidence statements in which each has a $100(1-\alpha)\%$ level of confidence. Therefore it is difficult to determine what level of confidence will be achieved for claiming that these m statements are all correct. To overcome this problem, several procedures have been developed in such a manner that the joint probability that all the statements are true is guaranteed not to exceed a predetermined level such as the α level of significance. The most commonly used method among these procedures probably is Bonferroni's simultaneous intervals, also known as multiple-t confidence intervals since the α level is adjusted for multiple comparisons among means. The $100(1-\alpha)\%$ Bonferroni's simultaneous intervals for the m pairwise differences $\mu_i - \mu_{i'}$ are given by

Table 7.4.2 Pain Relief Scores for Three Treatments

Treatment	Pain Relief (Y_{ij})	n_i	$\bar{Y}_{i.}$
Placebo	0.0, 1.0	2	0.50
Drug A	3.1, 2.7, 3.8	3	3.20
Drug B	2.3, 3.5, 2.8, 2.5	4	2.78

Note: 0.0 means no relief from pain.
Source: Wang and Chow (1995).

$$\bar{Y}_{i.} - \bar{Y}_{i'.} \pm t(\alpha/2m, N-k)s\sqrt{\frac{1}{n_i} + \frac{1}{n_{i'}}},$$

where $s = \sqrt{\text{MSE}}$ and $t(\alpha/2, N-k)$ is the upper $(\alpha/2m)$th quantile of the t distribution with $N-k$ degrees of freedom. Bonferroni's simultaneous intervals guarantee that the probability of all the m confidence statements being correct is at least $1 - \alpha$.

Consider a clinical trial on the efficacy of a newly developed drug for treating patients with migraine headaches (Wang and Chow, 1995). The clinical trial is a double-blind three-arm parallel randomized study comparing the newly developed drug (denoted by drug A), an active control agent (denoted by drug B), and a placebo control (denoted by P). The primary efficacy endpoint is the pain relief score. The pain relief scores are obtained 60 minutes after the administration of treatment based on a visual pain relief scale ranging from 0 (no relief from pain) to 10 (complete relief from pain). For the purpose of illustration, Table 7.4.2 gives the pain relief scores for each treatment. From Table 7.4.2, the sum of squares (SSE) due to error and the sum of squares (SSA) due to treatment can be obtained as

$$\text{SSE} = \sum_{i=1}^{k} \sum_{j=-1}^{n_i} (Y_{ij} - \bar{Y}_{i.})^2 = 1.95,$$

and

$$\text{SSA} = \sum_{i=1}^{k} n_i(\bar{Y}_{i.} - \bar{Y}_{..})^2 = 9.70.$$

Thus the F test is given by

ANALYSIS OF VARIANCE

Table 7.4.3 Analysis of Variance Table for Pain Relief Data

Source of Variation	df	Sum of Squares	Mean Sum of Squares	F
Treatment	2	9.70	4.85	15.15
Error	6	1.95	0.32	
Total	8	11.65		

$$F = \frac{SSA/(k-1)}{SSE/(\sum_{i=1}^{k} n_i - k)}$$

$$= \frac{9.70/2}{1.95/6} = 15.15,$$

which is greater than $F(0.05, 2, 6) = 5.14$. Hence we can reject the null hypothesis of no treatment difference, $\mu_A = \mu_B = \mu_P$ at the $\alpha = 0.05$ level of significance. The analysis of variance table is summarized in Table 7.4.3.

Since the overall F test rejects the null hypothesis of no treatment differences, we can compare the test drug A, with the active control agent (i.e., drug B) and the test drug A with the placebo P by constructing Bonferroni's simultaneous confidence intervals for $\mu_A - \mu_B$ and $\mu_A - \mu_P$. The resulting confidence intervals are given by $(-1.32, 0.48)$ and $(1.13, 3.43)$, respectively (see also Table 7.4.4). Since the simultaneous confidence interval for $\mu_A - \mu_P$ does not contain 0, it indicates that there is a significant difference between the test drug A and the placebo. On the other hand, there is no significant difference between the test drug A and the active control agent B because its simultaneous confidence interval contains 0.

Two-Way Classification

As discussed in the previous section, clinical trials are often conducted at several sites. The purpose of a multicenter trial is not only to expedite the patient enrollment but also to confirm that the result is reproducible and hence can be generalized to the targeted patient population. However, if there is treatment-

Table 7.4.4 Bonferroni's 95% Simultaneous Confidence Intervals

Contrast	95% Simultaneous Confidence Interval
$\mu_A - \mu_B$	$(-1.32, 0.48)$
$\mu_A - \mu_P$	$(\ 1.13, 3.43)$

Source: Wang and Chow (1995).

by-center interaction, the reproducibility and generalizability of the result are questionable. In practice, the following two-way classification model is considered to test the existence of the treatment-by-center interaction:

$$Y_{ijk} = \mu + \tau_i + \beta_j + (\tau\beta)_{ij} + e_{ijk}, \qquad (7.4.5)$$

where $i = 1, \ldots, a, j = 1, \ldots, b, k = 1, \ldots, n$. In the above model τ_i is the true effect of the ith treatment, β_j is the true effect of the jth center, $(\tau\beta)_{ij}$ denotes the effect of the interaction between τ_i and β_j, in and e_{ijk} is the random error is observing Y_{ijk}. For simplicity, both treatment and center effects are assumed fixed here. Since the treatment effects are defined as deviations from the overall mean, we have

$$\sum_{i=1}^{a} \tau_i = 0 \quad \text{and} \quad \sum_{j=1}^{b} \beta_j = 0.$$

Similarly the interaction effects are fixed and defined so that

$$\sum_{i=1}^{a} (\tau\beta)_{ij} = \sum_{j=1}^{b} (\tau\beta)_{ij} = 0.$$

Define $\overline{Y}_{i..}, \overline{Y}_{.j.}, \overline{Y}_{ij.}, \overline{Y}_{...}$ as the corresponding treatment, center, treatment-by-center, and overall sample mean. Then the deviation of an individual observation from the overall sample mean can be described as

$$Y_{ijk} - \overline{Y}_{...} = (\overline{Y}_{i..} - \overline{Y}_{...}) + (\overline{Y}_{.j.} - \overline{Y}_{...})$$
$$+ (\overline{Y}_{ij.} - \overline{Y}_{i..} - \overline{Y}_{.j.} + \overline{Y}_{...}) + (Y_{ijk} - \overline{Y}_{ij.}). \qquad (7.4.6)$$

As a result the deviation of an individual observation from the overall sample mean is partly due to differences among the means of the treatments, centers, treatment-by-center, and partly due to random variation in the measurements. Expression (7.4.6) leads to the following partitions of sum of squares (SS):

ANALYSIS OF VARIANCE

$$\sum_{i=1}^{a}\sum_{j=-1}^{b}\sum_{k=1}^{n}(Y_{ijk} - \overline{Y}_{...})^2$$

$$= bn\sum_{i=1}^{a}(\overline{Y}_{i..} - \overline{Y}_{...})^2 + an\sum_{j=1}^{b}(\overline{Y}_{.j.} - \overline{Y}_{...})^2$$

$$+ n\sum_{i=1}^{a}\sum_{j=1}^{b}(\overline{Y}_{ij.} - \overline{Y}_{i..} - \overline{Y}_{.j.} + \overline{Y}_{...})^2 + \sum_{i=1}^{a}\sum_{j=1}^{b}\sum_{k=1}^{n}(Y_{ijk} - \overline{Y}_{ij.})^2.$$

The first term on the left-hand side is the total sum of squares and the first two terms on the right-hand side are sums of squares due to the treatment and the center, denoted by SSA and SSC, respectively, while the last two terms are sums of squares due to treatment-by-center and random error, denoted by SS(AC) and SSE, respectively. Therefore we have

$$\text{SST} = \text{SSA} + \text{SSC} + \text{SS(AC)} + \text{SSE}.$$

Assuming that treatment and center are fixed, the expected values of the mean squares are given by

$$E(\text{MSA}) = E\left(\frac{\text{SSA}}{a-1}\right) = \sigma^2 + \frac{bn\sum_{i=1}^{a}\tau_i^2}{a-1},$$

$$E(\text{MSC}) = E\left(\frac{\text{SSC}}{b-1}\right) = \sigma^2 + \frac{an\sum_{j=1}^{b}\beta_j^2}{b-1},$$

$$E[\text{MS(AC)}] = E\left(\frac{\text{SS(AC)}}{(a-1)(b-1)}\right) = \sigma^2 + \frac{n\sum_{i=1}^{a}\sum_{j=1}^{b}(\tau\beta)_{ij}^2}{(a-1)(b-1)},$$

$$E(\text{MSE}) = E\left(\frac{\text{SSE}}{ab(n-1)}\right) = \sigma^2.$$

To test the hypotheses (H_0: $\tau_i = 0$ (i.e., no treatment effect), H_0: $\beta_j = 0$ (i.e., no center effect), and H_0: $(\tau\beta)_{ij} = 0$ (i.e., no treatment-by-center interaction), we can divide the corresponding mean square by mean square error. This ratio will follow an F distribution with appropriate numertor degrees of freedom and $ab(n-1)$ denominator degrees of freedom. Table 7.4.5 summarizes the analysis of variance table for the two-way classification fixed model.

Table 7.4.5 Analysis of Variance for the Two-Way Classification Fixed Model

Source of Variation	Sum of Squares	df	Mean Squares	F
Treatment	$SSA = bn \sum_{i=1}^{a} (\overline{Y}_{i..} - \overline{Y}_{...})^2$	$a-1$	$MSA = \dfrac{SSA}{(a-1)}$	$F_1 = \dfrac{MSA}{MSE}$
Center	$SSC = an \sum_{j=1}^{b} (\overline{Y}_{.j.} - \overline{Y}_{...})^2$	$b-1$	$MSC = \dfrac{SSC}{b-1}$	$F_2 = \dfrac{MSC}{MSE}$
Treatment *center	$SS(AC) = n \sum_{i=1}^{a} \sum_{j=1}^{b} (\overline{Y}_{ij.} - \overline{Y}_{i..} - \overline{Y}_{.j.} + \overline{Y}_{...})^2$	$(a-1)(b-1)$	$MS(AC) = \dfrac{SS(AC)}{(a-1)(b-1)}$	$F_3 = \dfrac{MS(AC)}{MSE}$
Error	$SSE = \sum_{i=1}^{a} \sum_{j=1}^{b} \sum_{k=1}^{n} (Y_{ijk} - \overline{Y}_{ij.})^2$	$ab(n-1)$	$MSE = \dfrac{SSE}{ab(n-1)}$	
Total	$SST = \sum_{i=1}^{a} \sum_{j=1}^{b} \sum_{k=1}^{n} (Y_{ijk} - \overline{Y}_{...})^2$	$abn-1$		

7.5 ANALYSIS OF COVARIANCE

In parallel-group clinical trials, patients who meet inclusion and exclusion criteria are randomly assigned to each treatment group. Under the assumption that the targeted patient population is homogeneous, we can expect that patient characteristics such as age, gender, and weight are comparable between treatment groups. If the patient population is known to be heterogeneous in terms of some demographic variables, then a stratified randomization according to these variables should be applied. At the beginning of the study, clinical data are usually collected at randomization to establish baseline values. After the administration of study drug, clinical data are often collected at each visit over the entire course of study. These clinical data are then analyzed to assess the efficacy and safety of the treatments.

As was pointed out earlier, before the analysis of endpoint values, the comparability of patient characteristics between treatments is usually examined by an analysis of variance if the variable is continuous. For the analysis of endpoint values, although the technique of analysis of variance can be directly applied, it is believed the endpoint values are usually linearly related to the baseline values. Therefore an adjusted analysis of variance should be considered to account for the baseline values. This adjusted analysis of variance is called analysis of covariance (ANCOVA).

Let X_{ij} and Y_{ij} be the baseline and endpoint values for the jth patient in the ith treatment group, where $j = 1, \ldots, n_i$ and $i = 1, \ldots, k$. To account for the baseline values, model (7.4.2) can be modified as follows:

$$Y_{ij} = \mu + \tau_i + \beta X_{ij} + e_{ij}$$

or

$$Y_{ij} = \mu + \tau_i + \beta(X_{ij} - \overline{X}_{..}) + \epsilon_{ij}. \tag{7.5.1}$$

Under the above model, the least squares estimators of μ, τ_i, and β are given by

$$\hat{\mu} = \overline{Y}_{..},$$
$$\hat{\tau}_i = \overline{Y}_{i.} - \overline{Y}_{..} - \hat{\beta}(\overline{X}_{i.} - \overline{X}_{..}),$$

and

$$\hat{\beta} = \frac{E_{XY}}{E_{XX}},$$

where $\overline{X}_{i.}$ and $\overline{Y}_{i.}$ denotes the sample means of baseline and endpoint for the ith treatment, and

$$E_{XX} = \sum_{i=1}^{k} \sum_{j=1}^{n_i} (X_{ij} - \overline{X}_{i.})^2,$$

$$E_{XY} = \sum_{i=1}^{k} \sum_{j=1}^{n_i} (X_{ij} - \overline{X}_{i.})(Y_{ij} - \overline{Y}_{i.}).$$

To derive for a test for the null hypothesis of no treatment difference as described in (7.4.1), we denote

$$S_{XX} = \sum_{i=1}^{k} \sum_{j=1}^{n_i} (X_{ij} - \overline{X}_{..})^2,$$

$$S_{XY} = \sum_{i=1}^{k} \sum_{j=1}^{n_i} (X_{ij} - \overline{X}_{..})(Y_{ij} - \overline{Y}_{..}),$$

$$S_{YY} = \sum_{i=1}^{k} \sum_{j=1}^{n_i} (Y_{ij} - \overline{Y}_{..})^2,$$

$$T_{XX} = \sum_{i=1}^{k} \sum_{j=1}^{n_i} (\overline{X}_{i.} - \overline{X}_{..})^2,$$

$$T_{XY} = \sum_{i=1}^{k} \sum_{j=1}^{n_i} (\overline{X}_{i.} - \overline{X}_{..})(\overline{Y}_{i.} - \overline{Y}_{..}),$$

$$T_{YY} = \sum_{i=1}^{k} \sum_{j=1}^{n_i} (\overline{Y}_{i.} - \overline{Y}_{..})^2.$$

Note that it can be verified that

$$E_{XX} = S_{XX} - T_{XX},$$
$$E_{XY} = S_{XY} - T_{XY},$$
$$E_{YY} = S_{YY} - T_{YY}.$$

Therefore the sum of squares due to treatment and error can be obtained as follows:

ANALYSIS OF COVARIANCE

Table 7.5.1 Analysis of Covariance

Source of Variation	Sum of Squares	df	Mean Squares
Regression	$\dfrac{(S_{XY})^2}{S_{XX}}$	1	
Treatment	SSA	$k-1$	$\text{MSA} = \dfrac{\text{SST}}{k-1}$
Error	SSE	$N-k-1$	$\text{MSE} = \dfrac{\text{SSE}}{N-k-1}$
Total	$\text{SST} = S_{YY}$	$N-1$	

Note: See the text for definitions of sums of squares and cross products.

$$\text{SSA} = S_{YY} - \frac{S_{XY}^2}{S_{XX}} - \text{SSE},$$

$$\text{SSE} = E_{YY} - \frac{E_{XY}^2}{E_{XX}}.$$

The analysis of covariance table for one-way classification is provided in Table 7.5.1. Under the null hypothesis of no treatment difference, namely $\tau_i = 0$ for all i, the test statistic

$$F = \frac{\text{SSA}/(k-1)}{\text{SSE}/(N-k-1)}$$

follows an F distribution with $(k-1, N-k-1)$ degrees of freedom, where $N = \sum_{i=1}^{k} n_i$. Therefore we can reject the null hypothesis of no treatment difference if

$$F > F(\alpha, k-1, N-k-1).$$

Note that if $\beta = 1$, then model (7.5.1) reduces to

$$Y_{ij} = \mu + \tau_i + X_{ij} + e_{ij}$$

or

$$Y_{ij} - X_{ij} = \mu + \tau_i + e_{ij}, \qquad (7.5.2)$$

where $Y_{ij} - X_{ij}$ denotes the endpoint change from the baseline for the jth patient in the ith treatment group. Therefore we can perform the usual analysis of variance on $\{Y_{ij} - X_{ij}\}$. This analysis is known as an endpoint change from the baseline analysis of covariance.

In practice, some researchers often include the endpoint change from the baseline analysis of covariance by treating the baseline as a covariate. This leads to the following model:

$$Y_{ij} - X_{ij} = \mu + \tau_i + \beta X_{ij} + e_{ij}$$

or

$$Y_{ij} - X_{ij} = \mu + \tau_i + \beta(X_{ij} - \overline{X}_{..}) + \epsilon_{ij}. \quad (7.5.3)$$

It should be noted that the above model is equivalent to model (7.5.1), since it can be rewritten as

$$Y_{ij} = \mu + \tau_i + \beta^*(X_{ij} - \overline{X}_{..}) + \epsilon_{ij}, \quad (7.5.4)$$

where $\beta^* - 1 = \beta$. In other words, under model (7.5.1), we are interested in testing the hypothesis that $H_0: \beta = 0$, which is equivalent to testing the hypothesis that $H_0: \beta^* = 1$ under model (7.5.4).

7.6 NONPARAMETRICS

In the previous section, the statistical inferences derived from the analysis of variance or the analysis of covariance are primarily based on the normality assumption of random errors. In practice, the normality assumption is not met. In this case the validity of the statistical inference drawn is questionable. As an alternative, we may consider using the method of nonparametrics to draw this statistical inference. In other words, no distribution assumptions are imposed. In this section, for illustration purposes, we will focus on the Wilcoxon signed rank test for one sample and rank sum test for shift in location due to treatment between two populations. This concept is also extended to derive the Kruskal-Wallis test for comparing k treatment groups. Distribution-free tests for other complicated design situations will not be covered in this section. However, the details of these tests can be found in Hollander and Wolfe (1973) and Lehmann (1975).

Wilcoxon Signed Rank Test

Let Y_{Bi} and Y_{Ei} be the response of the ith patient at baseline and endpoint, respectively, where $i = 1, \ldots, n$. Suppose that one of the study objectives is to show that there is a significant difference between pre- and post-treatment in patient's response. A nonparametric approach for this purpose is to consider the model

$$D_i = \theta + e_i, \qquad i = 1, \ldots, n,$$

where $D_i = Y_{Ei} - Y_{Bi}$, θ is the unknown treatment effect, and e_i are random errors which are assumed to be mutually independent and symmetric about zero. To test the null hypothesis of no treatment effect,

$$H_0: \theta = 0,$$

we may consider the following distribution-free signed rank test. First, form the absolute differences $|D_i|$, $i = 1, \ldots, n$, and let R_i denote the rank of $|D_i|$ in the joint ranking from least to greatest of $|D_1|, \ldots, |D_n|$. Define

$$\psi_i = \begin{cases} 1 & \text{if } D_i > 0, \\ 0 & \text{if } D_i < 0. \end{cases}$$

We then reject the null hypothesis at the α level of significance if

$$T^+ \leq t(\alpha_2, n) \quad \text{or} \quad T^+ \geq \frac{n(n+1)}{2} - t(\alpha_1, n),$$

where

$$T^+ = \sum_{i=1}^{n} R_i \psi_i$$

and the constant $t(\alpha, n)$ satisfies the equation

$$P\{T^+ \leq t(\alpha, n)\} = \alpha.$$

Note that the values of $t(\alpha, n)$ are given in Appendix A.6. The above distribution-free signed rank test is usually referred to as a Wilcoxon signed rank test. Under the null hypothesis, the statistics

$$T^* = \frac{T^+ - E(T^+)}{\text{var}(T^+)}$$

has an asymptotic standard normal distribution where

$$E(T^+) = \frac{n(n+1)}{4}$$

and

$$\mathrm{var}(T^+) = \frac{n(n+1)(2n+1)}{24}.$$

As a result we may reject the null hypothesis for large samples if

$$|T^*| \geq z(\alpha/2).$$

When there are ties, we can replace $\mathrm{var}(T^+)$ with the following

$$\mathrm{var}(T^+) = \frac{1}{24}\left[n(n+1)(2n+1) - \frac{1}{2}\sum_{j=1}^{g} t_j(t_j-1)(t_j+1)\right],$$

where g is the number of tied groups and t_j is the size of the tied group j.

Wilcoxon Rank Sum Test

A distribution-free approach can also be applied to test whether there is a significant difference between two treatments. Let $\{Y_{1i}, i = 1,\ldots,m\}$ and $\{Y_{2i}, i = 1,\ldots,n\}$ be two random samples from population 1 and population 2, respectively. Consider the model

$$Y_{1i} = e_i, \qquad i = 1,\ldots,m,$$

and

$$Y_{2j} = e_{m+j} + \Delta, \qquad j = 1,\ldots,n,$$

where Δ denotes the unknown shift in location due to the treatment. In other words, the null hypothesis of no treatment difference can be formulated as

$$H_0: \Delta = 0,$$
$$\text{vs.} \quad H_a: \Delta \neq 0.$$

To derive a distribution-free test for the above null hypothesis, we assume that $e_i, i = 1, \ldots, n+m$ are identically distributed and are mutually independent. The Wilcoxon rank sum test can be summarized as follows: First, we order the

$N = n + m$ observations from smallest to greatest and let R_j denote the rank of Y_{2j} in this order. Then we let

$$R = \sum_{j=1}^{n} R_j.$$

then the Wilcoxon–Mann–Whitney test statistic is given by

$$W = R - \frac{n(n+1)}{2}$$

Thus the statistic W is the sum of the ranks assigned to the Y_{2j}'s centered by $n(n+1)/2$ which follows a distribution symmetric about $nm/2$. As a result it can be verified that

$$w(1 - \alpha, m, n) = nm - w(\alpha, m, n),$$

where the constant $w(\alpha, m, n)$ satisfies

$$P\{W \leq w(\alpha, m, n)\} = \alpha.$$

Note that values of $w(\alpha, m, n)$ are given in Appendix A.5 which gives the critical values in the Mann-Whitney form of Wilcoxon rank sum statistic, that is, $W - n(n+1)/2$.

Based on the statistic W, we can reject the null hypothesis of no treatment difference at the α level of significance if

$$W \leq w(\alpha_2, m, n) \quad \text{or} \quad W \geq [nm - w(\alpha_1, m, n)],$$

where $\alpha = \alpha_1 + \alpha_2$. For a one-sided test, we reject H_a: $\Delta \leq 0$ if

$$W \leq w(\alpha, m, n),$$

and reject H_a: $\Delta > 0$ if

$$W \geq nm - w(\alpha, m, n),$$

at the α level of significance. Note that when there are ties among the N observations, W must be calculated based on average ranks.

For large samples, by the central limit theorem, the test statistic

$$W^* = \frac{W - E(W)}{[\text{var}(W)]^{1/2}}$$

has an asymptotic standard normal distribution as min(m, n) tends to infinity, where $E(W)$ and var(W) are the expected value and variance of W, which are given by

$$E(W) = \frac{nm}{2},$$

$$\text{var}(W) = \frac{mn(m+n+1)}{12}.$$

Therefore we can reject the null hypothesis of no shift in location due to treatment at the α level of significance if

$$|W^*| \geq z(\alpha/2).$$

In the case where there are ties, we can replace var(W) with

$$s^2 = \frac{mn}{12}\left[m+n+1 - \frac{\sum_{j=1}^{g} t_j(t_j^2-1)}{(m+n)(m+n-1)}\right],$$

where g is the number of tied groups and t_j is the size of tied group j (Hollander and Wolfe, 1973). Note that an untied observation is considered to be a tied group of size 1.

To illustrate the application of the Wilcoxon rank sum test, consider a clinical trial comparing the safety and efficacy of three drugs (A, B, and C) in the treatment of acute sinusitis in adults. This trial was a multicenter parallel-group randomized study. A ten-point assessment questionnaire (TAQ) is used to capture the improvement of signs and symptoms. The set of TAQ consists of ten questions. Each question is intended to capture the improvement of a specific symptom. A list of the ten symptoms is given in Table 7.6.1. For each question

Table 7.6.1 Ten Symptoms in TAQ

Item	Symptom
1	Fever
2	Nasal discharge
3	Nasal congestion
4	Cough
5	Headache
6	Facial pain
7	Facial swelling
8	Reduced activity
9	Impaired sleep
10	Impaired appetite

Table 7.6.2 TAQ Scores

Treatment	Subject Number	TAQ Score	Rank of TAQ Score
A	121	6	1.0
A	169	14	15.5
A	521	8	3.0
A	522	12	9.0
A	526	14	15.5
A	527	14	15.5
A	529	14	15.5
A	531	16	22.5
A	569	14	15.5
A	571	10	6.0
B	525	20	29.0
B	528	16	22.5
B	530	12	9.0
B	532	14	15.5
B	533	12	9.0
B	570	16	22.5
B	572	14	15.0
B	575	18	26.0
B	576	18	26.0
B	577	20	29.0
C	289	8	3.0
C	523	16	22.5
C	524	8	3.0
C	573	20	29.0
C	574	18	26.0
C	689	10	6.0
C	690	14	15.5
C	693	10	6.0
C	694	14	15.5
C	697	14	15.5

a score of 0 represents absence and a score of 2 indicates presence. Table 7.6.2 lists partial data of the total TAQ scores over the ten questions at the baseline. Based on data given in Table 7.6.2, pairwise comparisons can be made using a Wilcoxon rank sum test. The results are summarized below:

A versus B: $W_1 = \sum_{j=1}^{n} R_{1j} - 55 = 134 - 55 = 79$,
A versus C: $W_2 = \sum_{j=1}^{n} R_{2j} - 55 = 112 - 55 = 57$,
B versus C: $W_3 = \sum_{j=1}^{n} R_{3j} - 55 = 85 - 55 = 30$

where W_i denotes the sum of the ranks assigned to the Y's for the ith comparison. Since $w(0.05, 10, 10)$ is a value between 72 and 73, W_1 is the only test that is greater than $w(0.05, 10, 10)$. Therefore we reject the null hypothesis of no difference between treatments A and B. There are no significant differences between treatments A and C and between treatments B and C. Note that a normal approximation gives

A versus B: $s_1 = 76$, $W^* = -2.21$, $p = 0.027$,
A versus C: $s_2 = 98$, $W^* = -0.51$, $p = 0.610$,
B versus C: $s_3 = 125$, $W^* = 1.49$, $p = 0.135$.

Hence the normal approximation also indicates that there is a significant difference between treatments A and B.

Kruskal-Wallis Test

When there are more than two treatments to compare, a similar idea can be carried out to derive a test for the following hypothesis

$$H_0 : \mu_1 = \mu_2 = \cdots = \mu_k.$$

Let Y_{ij} be the observation for the jth patient in the ith treatment group. Consider model (7.4.2) repeated below:

$$Y_{ij} = \mu + \tau_i + e_{ij}, \quad j = 1, \ldots, n_i, i = 1, \ldots, k.$$

We first rank all $N = \sum_{i=1}^{k} n_i$ observations jointly, from least to greatest. Let r_{ij} be the rank of Y_{ij} in this joint ranking. Also for $i = 1, \ldots, k$, denote

$$R_i = \sum_{j=1}^{n_i} r_{ij}, \quad \overline{R}_i = \frac{R_i}{n_i}, \quad \text{and} \quad \overline{R} = \frac{N+1}{2}.$$

then the test statistic

$$H = \frac{12}{N(N+1)} \sum_{i=1}^{k} n_i (\overline{R}_i - \overline{R})^2$$

$$= \left(\frac{12}{N(N+1)} \sum_{i=1}^{k} \frac{\overline{R}_i^2}{n_i} \right) - 3(N+1)$$

can be used to test the null hypothesis of no treatment differences. We reject the null hypothesis at the α level of significance if

$$H \geq h(\alpha, k, (n_1, n_2, \ldots, n_k)),$$

where the constant $h(\alpha, k, (n_1, n_2, \ldots, n_k))$ satisfies

$$P\{H \geq h(\alpha, k, (n_1, n_2, \ldots, n_k))\} = \alpha.$$

The values of $h(\alpha, k, (n_1, n_2, \ldots, n_k))$ can be found in Iman, Quade, and Alexander (1975). Note that when there are ties, H should be calculated based on the average ranks.

When $\min\{n_1, \ldots, n_k\}$ tends to infinity, H is approximately distributed as a χ^2 distribution with $k-1$ degrees of freedom. Therefore for large samples we reject the null hypothesis of no treatment differences if

$$H \geq \chi^2(\alpha, k-1),$$

where $\chi^2(\alpha, k-1)$ is the upper α percentile of a χ^2 distribution with $k-1$ degrees of freedom. When there are ties, we replace H with H^* as follows:

$$H^* = \left\{1 - \left(\frac{\sum_{i=1}^{g}(t_i^3 - t_i)}{N^3 - N}\right)\right\}^{-1} H,$$

where t_i and g are defined as before.

To illustrate the application of Kruskal-Wallis test for comparing k treatments, consider the example concerning total TAQ scores described earlier. From Table 7.6.2, the Kruskal-Wallis test statistic can be easily obtained as

$$H = \left(\frac{12}{N(N+1)} \sum_{i=1}^{k} \frac{\bar{R}_i^2}{n_i}\right) - 3(N+1) = 5.215,$$

which is approximately χ^2 distributed with 2 degrees of freedom. Therefore we fail to reject the null hypothesis of no treatment differences at the 5% level of significance level since the observed p-value is given by 0.074.

7.7 REPEATED MEASURES

In clinical trials multiple assessments of a response variable are often performed at various time points (visits) from each of the patients under study. As a result the collected clinical data set may consist of repeated observations and a set of covariates (e.g., time and some patient characteristics) for each of the patients. This type of data is usually referred to as *longitudinal* data.

In many clinical trials, repeated observations after the administration of drug products are necessarily obtained to assess the efficacy and safety of the drug products under study. The objectives of repeated measures are (1) to determine whether the optimal therapeutic effect has been reached, (2) to determine whether a dose titration is necessary for good clinical practice, (3) to monitor the progress and/or health-related quality of life of patients with chronic diseases such as cancer, or (4) to study the behavior of the study drug over time (or to detect whether a potential pattern or trend in time exists).

In practice, different statistical models can be applied to address the above objectives under different model assumptions. In this section we will introduce some commonly used statistical models for the analysis of repeated measures in clinical trials. These models include the usual analysis of variance models for assessment of overall average drug effect across time points, for detection of time (visit) effect, and for determination of treatment-by-time effect and the method of generalized estimating equations (GEE) proposed by Zeger and Liang (1986) and Liang and Zeger (1986).

Assessment of Overall Average Effect Across Time

In clinical trials the ultimate goal is to determine whether there is a drug effect in terms of efficacy and safety of the drug product under study. Repeated measures occur in each patient. For simplicity, we may consider the *visit* as nested within the *patient*. Then a nested model can be used to assess the overall average drug effect across time points (visits). Let Y_{ijk} be the observation from the kth visit for the jth patient who is in the ith treatment group, where $i = 1, \ldots, a$, $j = 1, \ldots, n$, and $k = 1, \ldots, m$. The nested model used to describe Y_{ijk} is

$$Y_{ijk} = \mu + \tau_i + P_{j(i)} + e_{ijk}, \qquad (7.7.1)$$

where μ is the overall mean, τ_i and $P_{j(i)}$ denote the fixed effect of the ith treatment and the effect due to the jth patient within the ith treatment, respectively, and e_{ijk} is the random error in observing Y_{ijk}. It is assumed that e_{ijk} are independent and identically normally distributed with mean 0 and variance σ^2. Note that in the complete randomized design, $P_{j(i)}$ is often expressed as

$$P_{j(i)} = P_j + (AP)_{ij},$$

where P_j and $(AP)_{ij}$ denote the effect due to the jth patient and the effect of the interaction between the jth patient and the ith treatment.

Similarly under model (7.7.1) the deviation of an individual observation from the overall sample mean can be partitioned as follows:

$$Y_{ijk} - \overline{Y}_{...} = (\overline{Y}_{i..} - \overline{Y}_{...}) + (\overline{Y}_{ij.} - \overline{Y}_{i..}) + (Y_{ijk} - \overline{Y}_{ij.}).$$

Therefore we have

$$\text{SST} = \text{SSA} + \text{SSP}(A) + \text{SSE},$$

where SSA, SSP(A), and SSE denote sum of squares due to treatment, patient within treatment, and error, respectively, with associated degrees of freedoms given as

$$(anm - 1) = (a - 1) + a(n - 1) + an(m - 1).$$

As usual, dividing the sum of squares of error by its corresponding degrees of freedom gives population variance estimates. The analysis of variance table of model (7.7.1) is given in Table 7.7.1.

It can be seen from Table 7.7.1 that under the normality assumption, the test statistic

$$F_A = \frac{SSA/(a-1)}{SSE/[an(m-1)]}$$

is distributed as an F distribution with $(a - 1, an(m - 1))$ degrees of freedom. Therefore we reject the null hypothesis of no treatment difference; that is, H_0: $\tau_i = 0$, for all i if

$$F_A \geq F[\alpha, a - 1, an(m - 1)].$$

Similarly the effect due to patient within treatment can also be tested by

$$F_{P(A)} = \frac{SSP(A)/[a(n-1)]}{SSE/[an(m-1)]}$$

which has an F distribution with $[a(n - 1), an(m - 1)]$ degrees of freedom. Note that if there is a significant difference effect due to the patients within treatment, it indicates that the responses to treatment are not consistent from patient to patient. To further investigate the response of each patient, one might examine the patient means within each treatment based on the Newmann-Keuls range test (Keuls, 1952).

Detection of Time Effect

It is clear that model (7.7.1) does not provide any information regarding the effect due to time or visit. To examine whether there is a significant effect due to visit, we may further partition the residual sum of squares from model (7.7.1) into sum of squares due to visit and sum of squares due to error. Therefore model (7.7.1) can be modified as

$$Y_{ijk} = \mu + \tau_i + P_{j(i)} + V_k + \epsilon_{ijk}, \qquad (7.7.2)$$

where μ, τ_i, and $P_{j(i)}$ are as defined in model (7.7.1), V_k denotes the effect due to the kth visit (or assessment) and $\epsilon_{ijk} = e_{ijk} - V_k$. Similarly we can par-

Table 7.7.1 Analysis of Variance for Model (7.7.1)

Source of Variation	Sum of Squares	df	Mean Square
Treatment	$SSA = mn \sum_{i=1}^{a} (\bar{Y}_{i..} - \bar{Y}_{...})^2$	$a-1$	$MSA = \dfrac{SSA}{(a-1)}$
Patient (treatment)	$SSP(A) = m \sum_{i=1}^{a} \sum_{j=1}^{n} (\bar{Y}_{ij.} - \bar{Y}_{...})^2$	$a(n-1)$	$MSP(A) = \dfrac{SSP(A)}{a(n-1)}$
Error	$SSE = \sum_{i=1}^{a} \sum_{j=1}^{n} \sum_{k=1}^{m} (Y_{ijk} - \bar{Y}_{ij.})^2$	$an(m-1)$	$MSE = \dfrac{SSE}{an(m-1)}$
Total	$SST = \sum_{i=1}^{a} \sum_{j=1}^{n} \sum_{k=1}^{m} (Y_{ijk} - \bar{Y}_{...})^2$	$anm - 1$	

tition the deviation of an individual observation from the overall sample mean as

$$Y_{ijk} - \overline{Y}_{...} = (\overline{Y}_{i..} - \overline{Y}_{...}) + (\overline{Y}_{ij.} - \overline{Y}_{i..}) + (\overline{Y}_{..k} - \overline{Y}_{...})$$
$$+ (Y_{ijk} - \overline{Y}_{ij.} - \overline{Y}_{..k} + \overline{Y}_{...}).$$

Therefore we have

$$SST = SSA + SSP(A) + SSV + SSE,$$

where SSV denotes the sum of squares due to visit. The associated degrees of freedoms are

$$(anm - 1) = (a - 1) + a(n - 1) + (m - 1) + (an - 1)(m - 1).$$

As usual, dividing each independent sum of squares by its corresponding degrees of freedom gives the mean squares. The analysis of variance table of model (7.7.2) is given in Table 7.7.2.

As can be seen from Table 7.7.2, under the normality assumption, the test statistic

$$F_V = \frac{SSV/(m-1)}{SSE/(an-1)(m-1)}$$

is distributed as an F distribution with $[m - 1, (an - 1)(m - 1)]$ degrees of freedom. Therefore we reject the following null hypothesis:

$$H_0: v_1 = v_2 = \cdots = v_m,$$

where v_k, $k = 1, \ldots, m$, denote the mean of the kth visit.

If there is a time effect, we can use appropriate contrasts to test whether there is a linear or quadratic trend over time. Alternatively, we can consider *time* as a covariate and fit a least squares slope through each patient's data points. Then we can perform the usual analysis of variance based on these slopes by treating the slope as an outcome measure. Note that the analysis based on slopes for longitudinal lung function data is satisfactory.

Treatment-by-Time Interaction

In model (7.7.2) we consider the *visit* (or *time*) as a class variable and perform an analysis of variance to test to see whether there is a time effect. For simplicity we assume that the effect due to the patient is fixed and that there is

Table 7.7.2 Analysis of Variance for Model (7.7.2)

Source of Variation	Sum of Squares	df	Mean Squares
Treatment	$SSA = mn \sum_{i=1}^{a} (\bar{Y}_{i..} - \bar{Y}_{...})^2$	$a - 1$	$MSA = \dfrac{SSA}{(a-1)}$
Patient (treatment)	$SSP(A) = m \sum_{i=1}^{a} \sum_{j=1}^{n} (\bar{Y}_{ij.} - \bar{Y}_{i..})^2$	$a(n - 1)$	$MSP(A) = \dfrac{SSP(A)}{a(n-1)}$
Visit	$SSV = an \sum_{k=1}^{m} (\bar{Y}_{..k} - \bar{Y}_{...})^2$	$m - 1$	$MSV = \dfrac{SSV}{m-1}$
Error	$SSE = \sum_{i=1}^{a} \sum_{j=1}^{n} \sum_{k=1}^{m} (Y_{ijk} - \bar{Y}_{ij.} - \bar{Y}_{..k} + \bar{Y}_{...})^2$	$(an-1)(m-1)$	$MSE = \dfrac{SSE}{(an-1)(m-1)}$
Total	$SST = \sum_{i=1}^{a} \sum_{j=1}^{n} \sum_{k=1}^{m} (Y_{ijk} - \bar{Y}_{...})^2$	$anm - 1$	

no treatment-by-time interaction. In many clinical trials, the responses of the patients are expected to vary from one to another. To account for between-patient variability and to test for possible treatment-by-time interaction, the following model is useful:

$$Y_{ijk} = \mu + \tau_i + P_{j(i)} + V_k + (\tau V)_{ik} + \epsilon_{ijk}, \tag{7.7.3}$$

where μ, τ_i, V_k, and ϵ_{ijk} are as defined in model (7.7.2), $P_{j(i)}$ is the random effect of the jth patient within the ith treatment, and $(\tau V)_{ik}$ denotes the effect due to the interaction between the ith treatment and the kth visit (or assessment). It is assumed that $P_{j(i)}$ are i.i.d. normal with mean 0 and variance σ_S^2, which are mutually independent of ϵ_{ijk}. Similarly we can partition the deviation of an individual observation from the overall sample mean as follows:

$$\begin{aligned} Y_{ijk} - \overline{Y}_{...} &= (\overline{Y}_{i..} - \overline{Y}_{...}) + (\overline{Y}_{ij.} - \overline{Y}_{i..}) + (\overline{Y}_{..k} - \overline{Y}_{...}) \\ &+ (\overline{Y}_{i.k} - \overline{Y}_{i..} - \overline{Y}_{..k} + \overline{Y}_{...}) + (Y_{ijk} - \overline{Y}_{ij.} - \overline{Y}_{i.k} + \overline{Y}_{i..}). \end{aligned}$$

Therefore we have

$$SST + SSA + SSP(A) + SSV + SS(AV) + SSE,$$

where $SS(AV)$ denotes the sum of squares due to the interaction between treatment and visit. The associated degrees of freedoms are

$$(anm - 1) = (a - 1) + a(n - 1) + (m - 1) + (a - 1)(m - 1) + a(n - 1)(m - 1).$$

The analysis of variance table of model (7.7.3) is given in Table 7.7.3. To test the hypotheses of interest, we can divide the corresponding mean square by mean square error. This ratio will follow an F distribution with appropriate numerator degrees of freedom and denominator degrees of freedom. For example, we can reject $H_0: \tau_i = 0$ for all i (i.e., no treatment effect) if

$$F_A = \frac{MSA/(a-1)}{MSP(A)/[a(n-1)]} \geq F[\alpha, a-1, a(n-1)],$$

where $F[\alpha, a-1, a(n-1)]$ is the upper αth quantile of an F distribution with $[(a-1), a(n-1)]$ degrees of freedom. Similarly we would reject $H_0: (\tau V)_{ik} = 0$ (i.e., no treatment-by-visit interaction) if

Table 7.7.3 Analysis of Variance for Model (7.7.3)

Source of Variation	Sum of Squares	df	Mean Squares
Treatment	$SSA = nm \sum_{i=1}^{a} (\overline{Y}_{i..} - \overline{Y}_{...})^2$	$a - 1$	$MSA = \dfrac{SSA}{(a-1)}$
Patient (treatment)	$SSP(A) = m \sum_{i=1}^{a} \sum_{j=1}^{n} (\overline{Y}_{ij.} - \overline{Y}_{i..})^2$	$a(n-1)$	$MSP(A) = \dfrac{SSP(A)}{a(n-1)}$
Visit	$SSV = an \sum_{k=1}^{m} (\overline{Y}_{..k} - \overline{Y}_{...})^2$	$m - 1$	$MSV = \dfrac{SSV}{(m-1)}$
Treatment* visit	$SS(AV) = \sum_{i=1}^{a} \sum_{j=1}^{n} \sum_{k=1}^{m} (\overline{Y}_{i.k} - \overline{Y}_{i..} - \overline{Y}_{..k} + \overline{Y}_{...})^2$	$(a-1)(m-1)$	$MS(AV) = \dfrac{SS(AV)}{(a-1)(m-1)}$
Error	$SSE = \sum_{i=1}^{a} \sum_{j=1}^{n} \sum_{k=1}^{m} (Y_{ijk} - \overline{Y}_{ij.} - \overline{Y}_{i.k} + \overline{Y}_{i..})^2$	$a(n-1)(m-1)$	$MSE = \dfrac{SSE}{a(n-1)(m-1)}$
Total	$SST = \sum_{i=1}^{a} \sum_{j=1}^{n} \sum_{k=1}^{m} (Y_{ijk} - \overline{Y}_{...})^2$	$anm - 1$	

$$F_{AV} = \frac{\text{MS}(AV)/[(n-1)(m-1)]}{\text{MSE}/[a(n-1)(m-1)]}$$

$$\geq F[\alpha, (n-1)(m-1), a(n-1)(m-1)].$$

Note that an overall assessment of the average drug effect across visits is possible if the treatment-by-visit interaction is not qualitatively significant. When an interaction between treatment and visit is observed, it indicates that the differences in treatment are not consistent across visits. This inconsistency may be caused by (1) the fact that the time to reach optimal therapeutic effect varies from one drug to another, (2) the dose titration procedure (if any) is not done adequately, or (3) the disease status of the patients change over time. In this case it is suggested that the time effect or pattern be carefully evaluated in order to make valid statistical inference regarding the efficacy of the drug under study.

Method of Generalized Estimating Equations (GEE)

So far we have discussed several linear models for assessing (1) overall average drug effect across time points, (2) the detection of time effect, and (3) the effect due to treatment-by-time interaction in clinical trials with repeated measures. The primary assumptions of these linear models are (1) the outcome follows a normal distribution, (2) the outcome does not vary across time and/or across subjects, and (3) repeated observations are independent. Note that with independent observations, generalized linear models (GLMs) can be extended for time-dependent data in a variety of ways (McCullagh and Nelder, 1983). In addition the parameters in a GLM can also be assumed to vary across time as a stochastic process and/or across subjects according to a mixing distribution (Zeger and Qaqish, 1988). More recently there is Vonesh and Chinchilli's (1997) comprehensive review of the various methods for analyzing repeated measurements.

In clinical trials, however, the outcome may not follow a normal distribution and may vary across time and/or across subjects. In addition repeated outcomes for one individual are usually correlated with one another. Hence, to obtain a valid statistical inference, the analysis of longitudinal data must take into account for the correlation among repeated observations within each subject. Failure to account for the correlation may result in inefficient estimates of the parameters and/or inconsistent estimates with respect to precision across subjects. As a result the issue of correlation among repeated observations for a subject provides a challenge for analysis of longitudinal data.

To account for the correlation among a subject's repeated observations, Zeger and Liang (1992) and Diggle, Liang, and Zeger (1994) point out that the following three basic models are useful: marginal, transition, and random effects models. For a marginal model, the regression coefficients for the clinical response on covariates and the structure of the intrasubject correlation are modeled separately. The marginal expectation which is referred to as the average response over a particular stratum with the common values of covariates is a function of the explanatory variables. The regression coefficients in the marginal model hence have the same interpretation as those obtained from a cross-sectional analysis. The variance in the marginal model is the product of a variance function of the marginal mean and a scale parameter. The intrasubject correlation at two different time points is also a function of marginal means at the two time points and possible additional parameters. The approach of transition models tries to model both regression and intrasubject correlation simultaneously. Hence a common equation includes, on the same scale, both unknown parameters for the dependence of the clinical endpoints on explanatory variables and for the intrasubject correlation. The transition model then assumes that the conditioned expectation at a particular time point is a function of covariates and the proceeding responses. Consequently the prior responses are treated as explicit predictors the same way as any other explanatory variables. The concept of a random effects model is to consider subjects enrolled in the study as a representative random sample from the targeted population. As a result the regression coefficients of a subject in the trial is also a random vector that is assumed to follow a probability distribution. Ware et al. (1988) discuss the conceptual and technical differences between the marginal and transition models.

Zeger and Liang (1986) and Liang and Zeger (1986) propose a method using either the marginal or transition model in conjunction with the quasi-likelihood approach (McMullagh and Nelder, 1983; Wedderburn, 1974) in order to obtain consistent estimates of the parameters and their corresponding variances under rather weak assumptions on the correlation among a subject's repeated observations. This method is known as the generalized estimating equations (GEE). Under some assumptions for the first two moments of the joint distribution of correlated clinical responses, the regression coefficients, including the treatment effects and the correlation parameter, can be estimated consistently by solving a multivariate analogue of the quasi-score function without invoking an intractable likelihood with many nuisance parameters.

The quasi-likelihood approach is a regression methodology that requires few assumptions about the distribution of the dependent variable, and hence it can be used with a variety of outcomes. The quasi-likelihood approach is usually applied to the marginal model, which describes the relationship between the outcome and a set of explanatory variables, in order to obtain estimating equations of the parameters of interest. Based on the estimating equations, consistent estimates of the parameters and their corresponding variances can be obtained.

Let Y_{it} be the observation of the ith subject observed at time t, where $t = 1, \ldots, n_i$. Also let X_{it} be a $p \times 1$ vector of explanatory variables (or covariates) related to Y_{it}. These explanatory variables could be patient characteristics such as age and gender or time (visit). The marginal model relates the marginal expectation of Y_{it} with respect to $\mathbf{X}_{it}$ by

$$g(\mu_{it}) = \mathbf{X}'_{it}\boldsymbol{\beta},$$

where g is a known link function, $\mu_{it} = E(Y_{it})$, and $\boldsymbol{\beta}$ is a $p \times 1$ vector of unknown parameters. It is also assumed that the marginal variance of Y_{it} is a function of the marginal mean:

$$\text{var}(Y_{it}) = v(\mu_{it})\phi,$$

where v is a known function and ϕ is the overdispersion parameter that accounts for the variation of Y_{it} not explained by $v(\mu_{it})$. In addition the covariance between Y_{is} and Y_{it}, $s < t = 1, \ldots, n_i$, is assumed to be a function of the marginal means and additional parameter η:

$$\text{cov}(Y_{is}, Y_{it}) = c(\mu_{is}, \mu_{it}, \boldsymbol{\eta}),$$

where c is a known function and $\boldsymbol{\eta}$ is a $s \times 1$ vector of unknown parameters that measures the within-subject correlation. Note that the estimate of $\boldsymbol{\beta}$ is often used to describe how the averaged response (rather than one subject's response) changes with respect to the covariates. Based on the above assumptions, the consistent estimator (or quasi-likelihood estimator) is the solution of the following scorelike estimating equation system.

$$\sum_{i=1}^{K} \mathbf{D}_i' \mathbf{V}_i^{-1} \mathbf{S}_i = 0,$$

where

$$\mathbf{S}_i = \mathbf{Y}_i - \boldsymbol{\mu}_i$$

$$\mathbf{D}_i = \frac{\partial \boldsymbol{\mu}_i}{\partial \boldsymbol{\beta}}$$

with $\mathbf{Y}_i = (Y_{i1}, \ldots, Y_{in_i})'$ and $\boldsymbol{\mu}_i = (\mu_{i1}, \ldots, \mu_{in_i})'$ and

$$\mathbf{V}_i = \frac{\mathbf{A}_i^{1/2} \mathbf{R}_i(\boldsymbol{\eta}) \mathbf{A}_i^{1/2}}{\phi},$$

in which $\mathbf{A}_i$ is an $n_i \times n_i$ diagonal matrix with $g(\mu_{ij})$ as the jth diagonal element and $\mathbf{R}_i(\boldsymbol{\eta})$ is the $n_i \times n_i$ working correlation matrix for each $\mathbf{Y}_i$. Note that we do not expect $\mathbf{R}_i(\boldsymbol{\eta})$ to be correctly specified. The GEE method provides consistent estimates even when $\mathbf{R}_i(\boldsymbol{\eta})$ is incorrect.

Zeger and Liang (1992) indicate that the GEE method enjoys at least three useful properties. First of all, the estimated regression coefficients are nearly as efficient as the maximum likelihood estimates for most clinical trials. Second, despite a possible misspecification of the covariance structure for the correlated

Table 7.7.4 Measurements on 10 Girls and 10 Boys at 4 Different Ages

Gender	Subject	Age in Years			
		8	10	12	14
Girl	1	22.130	21.100	22.645	24.190
	2	22.130	22.645	25.220	26.765
	3	21.615	25.220	25.735	27.280
	4	24.705	25.735	26.250	27.795
	5	22.645	24.190	23.675	24.705
	6	21.100	22.130	22.130	23.675
	7	22.645	23.675	24.190	26.250
	8	24.190	24.190	24.705	25.220
	9	21.100	22.130	23.160	22.645
	10	17.495	20.070	20.070	20.585
Boy	1	27.280	26.250	30.370	32.430
	2	22.645	23.675	24.190	27.795
	3	24.190	23.675	25.220	28.825
	4	26.765	28.825	27.795	28.310
	5	21.100	24.705	23.675	27.280
	6	25.735	26.765	28.310	29.855
	7	23.160	23.160	25.735	27.795
	8	25.220	22.645	25.735	26.765
	9	24.190	21.615	32.430	27.280
	10	28.825	29.340	32.430	32.945

clinical responses, the estimates for regression coefficients are still consistent if the sample size is sufficiently large. Finally, the statistical inference of the regression coefficients will generally not be influenced by the covariance matrix of the clinical responses as long as the robust estimated covariance matrix of the regression coefficient estimates suggested by Liang and Zeger (1986) is used.

Example 7.7.1

For illustration purposes, consider a clinical trial that compares the effect of gender on the distance (in millimeters) from the center of the pituitary to the pteryomaxillary fissure. Table 7.7.4 lists measurements obtained from 20 children (10 girls and 10 boys) at ages of 8, 10, 12, and 14. Figure 7.7.1 plots the mean distances at ages of 8, 10, 12, and 14; and the plot suggests that there is a significant difference between boys and girls. To assess the overall average effect, age effect, and gender-by-age interaction, we consider models (7.7.1)–(7.7.3). The analyses of variance results are summarized in Table 7.7.5. From Table 7.7.5 it can be seen that there is a marginally significant gender-by-age interaction (p-value = 0.06). This quantitative interaction, however, does not involve the analysis by pooling data across ages (see also Figure 7.7.1). Model (7.7.1) provides an overall assessment of the gender effect. The result indicates that there is a significant difference between boys and girls (p-value

Table 7.7.5 Analyses of Variance for Data in Table 7.7.4

Model	Objective	Source of Variation	df	Sum of Squares	F	p-Value
7.7.1	Overall average effect	Gender	1	202.26	48.05	<0.01
		Subject (gender)	18	297.48	3.93	<0.01
7.7.2	Age effect	Gender	1	202.26	110.35	<0.01
		Subject (gender)	18	297.48	9.02	<0.01
		Age	3	148.09	26.93	<0.01
7.7.3	Gender-by-age effect	Gender	1	202.26	119.76	<0.01
		Subject (gender)	18	297.48	9.79	<0.01
		Age	3	148.09	29.23	<0.01
		Gender-by-age	3	13.27	2.62	0.06

< 0.01). In addition model (7.7.2) reveals that there is an age effect. To further compare the age effect between boys and girls, we can try to fit the linear regression for all boys and all girls and then perform an analysis of variance on the estimated slopes.

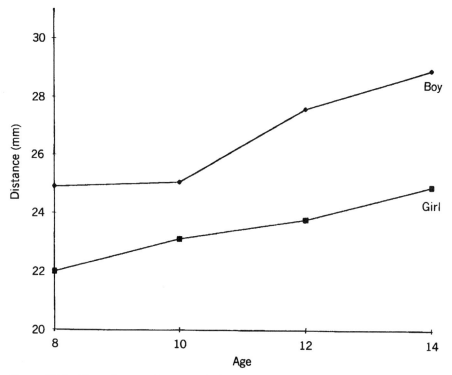

Figure 7.7.1 Mean distances from the pituitary to the pteryomaxillary fissure in boys and girls.

Table 7.7.6 Estimates, Standard Errors, and Z-Scores from the Method of GEE

Parameter	Estimate	se Naive	se Robust	Z Robust
Intercept	28.928	0.697	0.651	44.46
Gender	−4.017	0.986	0.936	−4.29
Age				
A_1	−4.017	0.551	0.459	−8.74
A_2	−3.863	0.551	0.570	−6.78
A_3	−1.339	0.551	0.775	−1.73
Gender*Age				
$(GA)_1$	1.082	0.780	0.626	1.73
$(GA)_2$	2.060	0.780	0.677	3.04
$(GA)_3$	0.206	0.780	0.808	0.26

As described in this section, the major disadvantage of the analysis of variance models (7.7.1)–(7.7.3) is that they assume that the measurements within each subject are independent. To account for the correlation among measurements within each subject, the method of GEE is also considered. To enhance efficiency of the estimate, the following working correlation structure helps describe the correlation between observations within a subject

$$\begin{bmatrix} 1.00 & 0.687 & 0.687 & 0.687 \\ 0.687 & 1.000 & 0.687 & 0.687 \\ 0.687 & 0.687 & 1.000 & 0.687 \\ 0.687 & 0.687 & 0.687 & 1.000 \end{bmatrix}.$$

The estimates, standard errors, and Z-scores obtained by the method of GEE are summarized in Table 7.7.6. In Table 7.7.6 all of the parameters are statistically significant except for A_3 (with Z-score = 1.73), $(GA)_1$ (with Z-score = 1.73), and $(GA)_3$ (with Z-score = 0.26).

7.8 DISCUSSION

As was indicated earlier, a two-sample t test is useful in comparing two treatment means and an F test can be applied to compare k treatments. There is relationship between the t test statistic and the F test statistic when $k = 2$. That is, $t^2 = F$. In other words, the F test reduces to the two-sample t test when there are only two groups evaluated in a clinical trial.

For nonparametric analysis the Wilcoxon rank sum test and the Kruskal-Wallis test are fairly robust against outliers because they are based on ranks (or relative order) rather than absolute difference of the raw data. Both Wilcoxon rank sum test and Kruskal-Wallis test can also be obtained by replacing the raw data with their corresponding ranks in the t test and sum of squares for treatment (i.e., SSA) in a one-way analysis of variance.

For the GEE method the principal distinction between the marginal model and random effects model is whether the regression coefficients describe an individual's change or an average response change with respect to the covariates. In addition the marginal model describes the covariance among repeated observations for a subject, while the random effects model explains the source of this covariance. As a result Zeger, Liang, and Albert (1988) classify these two approaches as population-average (marginal) and subject-specific (random effects) models. The population-average approach describes how the average response across subjects changes with covariates, while the subject-specific method accounts for the heterogeneity among subjects to estimate subject-specific coefficients. In addition to the GEE method for the analysis of logitudinal data in clinical trials, there are others discussed in the literature. For example, Grizzle and Allen (1969) and Hui (1984) consider fitting growth curves to repeated observations for each subject. When there are missing and/or unequally spaced examinations, Rosner and Munoz (1988) suggest that an autoregressive model be used.

The traditional approach to repeated measurements in clinical trials is to perform an analysis of variance at the last visit. This method is usually referred to as endpoint analysis or carryforward analysis. The result is compared with those obtained by the analyses of variance performed at each visit. If the results from the analysis at each visit are consistent with the endpoint analysis, then an overall conclusion on the treatment effect is drawn. However, if there is a significant difference among visits and/or the last visit, then a further analysis is performed to determine whether there is a pattern and/or trend over time.

A commonly encountered problem in clinical trials with repeated measures is missing data. In general, missing data may have two causes. On one hand, subjects may withdraw prematurely from the study at a post-treatment time point. These subjects are referred to as dropouts. As a result the data scheduled to be collected since the last visit would be completely missing. On the other hand, subjects might be lost to follow-up at a scheduled visit. However, they may return either on another unscheduled date or on the next scheduled visit. These subjects may or may not complete the study. Diggle et al. (1994) refer to those missing values due to the missed visits as intermittently missing values. Several methods have been proposed that assume the missing values to occur at random within treatment groups. See, for example, Crépeau et al. (1985), Diggle (1988), Kenward (1987), and Ridout (1991). In clinical trials it is suggested that the dropout rates and missing patterns across all visits be compared between treatments. If there is a discrepancy in dropout rates or a missing pattern, it is crucial to investigate whether a response-dependence or treatment-related problem has occurred.

CHAPTER 8

Analysis of Categorical Data

8.1 INTRODUCTION

For the assessment of the effectiveness and safety of an investigational pharmaceutical entity, ideally the scale for the primary clinical endpoint should be numerically continuous to provide an accurate and reliable assessment. In practice, however, it is impossible, or can be extremely expensive, to measure responses quantitatively. On the other hand, patients' responses to treatments can be easily documented according to the occurrence of some meaningful and well-defined event such as death, infection, or cure of a certain disease and any serious adverse events. In addition the intensity of these events can be graded according to some predefined categories. Therefore categorical data can be useful surrogate endpoints for some unobserved latent continuous variables in clinical trials. Sometimes, to provide an easy analysis and/or a better presentation of the results, continuous data are transformed to categorical data with respect to some predefined criteria. As a result many efficacy and safety endpoints in clinical trials are in the form of categorical data on either a nominal or ordinal scale. In this section we introduce the concept and methodology for analysis of categorical data through some real examples from recently published results of clinical trials. The first example concerns the U.S. Physicians' Health Study on the prevention of cardiovascular mortality and fatal and nonfatal myocardial infarction with a low dose of aspirin (325 mg on alternate days). Another primary goal of the U.S. Physicians' Study was to investigate the impact of long-term supplementation with beta carotene (50 mg on alternate day) on the incidence of malignant neoplasm and cardiovascular disease. Table 8.1.1 shows the number of subjects of malignant neoplasms for the treatment assignment of the beta carotene component (Hennekens et al., 1996). As can be seen in Table 8.1.1, the subjects were classified into two categories according to the occurrence of malignant neoplasm over an average of 12 years of treatment and follow-up. A clinical endpoint of this kind is referred to as a *binary response* on a nominal scale with two categories.

The next example is from a parallel-group, randomized, double-blind,

INTRODUCTION

Table 8.1.1 Number of Subjects With Malignant Neoplasms in the Beta Carotene Component of the U.S. Physician's Health Study

Malignant Neoplasm	Beta Carotene	Placebo
N	11,036	11,035
Yes	1273	1293
Year 1–2	120	130
Year 3–4	157	136
Year 5–9	500	567
≥10 years	496	460
No	9763	9742

Source: Hennekens et al. (1996).

placebo-controlled trial conducted by the U.S. National Institute of Neurological Disorders and Stroke (NINDS) to investigate the efficacy and safety of intravenous recombinant tissue plasminogen activator (rt-PA) for ischemic stroke. This study consisted of two parts. The first part was to examine the clinical activity of rt-PA in terms of early clinical improvement as compared to the placebo. For the second part, the primary hypothesis of interest was if there was a consistent and persuasive difference between the rt-PA and placebo groups in terms of the proportion of patients who recovered with minimal or no deficit three months after treatment. In addition to the study center, stratified randomization was performed according to time from the onset of stroke to the start of treatment (0 to 90 or 91 to 180 minutes). One of the primary clinical efficacy endpoints was early clinical improvement, which is defined as at least a four-point improvement from the baseline values in National Institutes of Health Stroke Scale (NIHSS) score or a resolution of the neurological deficit 24 hours after the onset of stroke. Hence the subjects of this study are also classified into two groups based on the defined clinical endpoint, namely those with the defined clinical improvement and those without. The defined clinical improvement is a binary response. However, unlike the occurrence of malignant neoplasm in the U.S. Physicians' Study, it is a composite clinical index derived from changes of other endpoints over the entire course of treatment. Table 8.1.2 relates the number of subjects with clinical improvement by study part and time.

In order to provide an objective, scientific, and clinically meaningful assessment of efficacy in the therapy for benign prostatic hyperplasia (BPH), the American Urological Association developed and validated a symptom index for BPH which is the sum of scores over seven individual symptoms. This symptom index is usually referred to as the AUA-7 symptom score. Table 8.1.3 lists the seven questionnaires associated with these symptom indexes. Each question in the AUA-7 symptom score consists of six categories, which are arranged in a monotone increasing order. This type of categorical data is said to be in the ordinal scale and is sometimes referred to as *ordered categorical data*. The

Table 8.1.2 Number of Subjects Showing Clinical Improvement

	rt-PA	Placebo
Part 1		
0–90 min		
N	71	68
Yes	36	31
No	35	37
91–180 min		
N	73	79
Yes	31	26
No	42	53
Part 2		
0–90 min		
N	86	77
Yes	51	30
No	35	47
90–180 min		
N	82	88
Yes	29	35
No	53	53

Source: National Institute of Neurological Disorders and rt-PA Stroke Study Group (1995).

results of the initial question 4 on frequency in the validation of discrimination between the BPH patients from the controls are displayed in Table 8.1.4. Another example of the use of the ordered categorical data is to evaluate efficacy and safety of a newly developed contrast agent in conjection with magnetic resonance imaging for diagnosis of malignant liver lesions. The primary diagnostic efficacy endpoint is the change in diagnostic confidence compared to precontrast MRI scan. This endpoint consists of a total of six categories on the ordinal scale: worsened, unchanged, minimal improvement, moderate improvement, good improvement, and excellent improvement. Table 8.1.5 provides the summary of change in diagnostic confidence by treatment group and gender from a hypothetical dataset.

The time-response profile is critical in the assessment of a new proposed therapy for the treatment of a certain illness. Evaluations of effectiveness and safety for each subject participating in the trial are performed at a number of preselected time points during entire course of the study. As a result the categorical data collected from most clinical trials are always *repeated categorical endpoints* that are not statistically independent within the same subjects. For

INTRODUCTION

Table 8.1.3 AUA Symptom Index

Questions	Not at All	Less than 1 Time in 5	Less than Half the Time	About Half the Time	More Than Half the Time	Almost Always
1. During the last month or so, how often have you a sensation of not emptying your bladder completely after you finished urinating?	0	1	2	3	4	5
2. During the last month or so, how after have you had to urinate again less than 2 hours after you finished urinating?	0	1	2	3	4	5
3. During the last month or so, how often have you stopped and started again several times when you urinated?	0	1	2	3	4	5
4. During the last month or so, how often have you found it difficult to postpone urination?	0	1	2	3	4	5
5. During the last month or so, how often have you had a weak urinary stream?	0	1	2	3	4	5
6. During the last month or so, how often have you had to push or strain to begin urination?	0	1	2	3	4	5

Table 8.1.3 (*Continued*)

Questions	None	1 Time	2 Times	3 Times	4 Times	5 or More Times
7. During the last month, how many times did you most typically get up to urinate from the time you went to bed at night until the time you got up in the morning?	0	1	2	3	4	5

AUA-7 Symptom score is the sum of scores of questions 1 to 7.
Source: Barry et al. (1992).

some known or unknown reasons, subjects might not come to the clinic for a particular scheduled visit or subjects might simply drop out and never return to the study. Therefore, repeated categorical data with missing data is the norm for most clinical trials. Koch et al., (1990) report partial results from a randomized, multicenter, placebo-controlled trial in the assessment of a new test drug for a respiratory disorder. The raw data provided by Koch et al. (1990) is an ordinal categorical endpoint of each subject's status measured at baseline and at four prescheduled visits after the initiation of treatment. This ordinal index, referred as the status score in this chapter, consists of five categories: terrible, poor, fair, good, and excellent. A summary of the transition from baseline to visit 3 in terms of frequency is provided in Table 8.1.6.

In general, different statistical methods are required for the analysis of differ-

Table 8.1.4 Partial Results of Validation for Question 4—Frequency 1

Point Score	BPH	Controls
N	73	56
0	8	16
1	18	23
2	17	7
3	14	7
4	8	2
5	8	1

Source: Barry et al. (1992).

INTRODUCTION

Table 8.1.5 Summary by Treatment for Change in Diagnostic Confidence

	Placebo		Test Drug	
	Female	Male	Female	Male
N	37	36	36	37
Worsened	4	4	1	1
Unchanged	11	9	1	2
Improvement				
Minimal	7	8	2	1
Moderate	4	6	2	3
Good	6	4	12	13
Excellent	5	5	18	17

ent categorical data types. This chapter will not only cover descriptive statistics but also interval estimation and hypotheses testing for one-sample, two-sample, and multiple-sample categorical data. These statistics are derived based on random assignment of patients to the treatment. Since, as discussed in Chapter 4, most randomization-based methods currently available are for hypothesis testing and the scope of inference is limited only to the subjects in the study, they must be adequately supplemented with a model-based method for estimation. The advantages of model-based methods include (1) the ability for investigation of the homogeneity of treatment effects across strata, (2) the flexibility for

Table 8.1.6 Summary of Frequencies of Transition of Status Score from the Baseline to Visit 3

Baseline	Terrible	Poor	Fair	Good	Excellent	Total
			Treatment: Placebo			
Terrible	0	0	0	0	0	0
Poor	4	2	3	1	1	11
Fair	4	1	9	4	2	20
Good	1	3	2	4	9	19
Excellent	2	0	0	2	3	7
Total	11	6	14	11	15	57
			Treatment: Test Drug			
Terrible	0	0	2	1	0	3
Poor	0	3	2	1	4	10
Fair	2	0	5	3	8	18
Good	0	0	0	7	5	12
Excellent	0	0	1	2	8	11
Total	2	3	10	14	25	54

adjustment of demographic and baseline characteristics, and (3) the reduction of variability (or the increase of power) by inclusion of covariates associated with the categorical endpoints. The major shortcomings of the model-based methods consists in the inability to prove the underlying assumption required for inference and a greater difficulty in the implementation and interpretation of the results. As a result it may be a good idea to combine both the randomization-based method and the model-based method for analysis of categorical data.

In the next section, statistical analyses of binary data for one sample and comparison within paired samples are introduced. Statistical inference based on binary data for independent samples are given in Section 8.3. Section 8.4 outlines statistical methods for ordered categorical data. Section 8.5 provides the methodology for combining results from the analysis of categorical data over strata. We note that most methods covered in Section 8.2 through Section 8.5 are randomization-based methods. Model-based methods such as the logistic regression model are discussed in Section 8.6. In Section 8.7 we review some methods for the analysis of repeated categorical data. Some final remarks and a brief discussion are given in Section 8.8.

8.2 STATISTICAL INFERENCE FOR ONE SAMPLE

As described in the previous section, for the NINDS trial there are two possible outcomes for each subject based on either the predefined criteria of at least a four-point improvement from the baseline in NIHSS score or a complete resolution of the neurological deficit. The two possible outcomes are either a subject improved or did not improve. For this study one of the primary objectives is to evaluate whether subjects treated with rt-PA have a greater proportion of early improvement compared to the placebo group. Before a formal statistical comparison between the rt-PA and placebo is made, a descriptive statistic for estimation of the unknown proportion of the subjects who improved in each group must be obtained.

Let Y_i denote the clinical endpoint that indicates whether a subject had an improvement in a group of n subjects treated with rt-PA, namely

$$Y_i = \begin{cases} 1, & \text{if subject had an improvement,} \\ 0, & \text{otherwise.} \end{cases} \quad (8.2.1)$$

where $i = 1, \ldots, n$. Since the n subjects in rt-PA group are different and are not related to each other, the occurrence of improvement of a subject is random and is independent of that of other subjects. In other words, Y_i are statistically independent for $i = 1, \ldots, n$. Under the assumption of independence, the population proportion of the subjects with improvement, denoted by P, can be estimated by the sample proportion of the subjects with improvement, denoted by p, which

STATISTICAL INFERENCE FOR ONE SAMPLE

is calculated as the number of subjects with improvement divided by the total number of subjects treated with rt-PA. In other words,

$$p = \frac{1}{n} \{\text{number of subjects with improvement}\} \qquad (8.2.2)$$

$$= \frac{1}{n} \sum_{i=1}^{n} Y_i$$

$$= \frac{Y}{n},$$

where

$$Y = \sum_{i=1}^{n} Y_i$$

represents the number of subjects (out of n subjects) showing improvement who were treated with rt-PA. Y is usually referred to as a binomial random variable. Since Y_i has only two possible values, 0 and 1, the sample mean $p = Y/n$ is the sample proportion of subjects with improvement. The variance of p can be estimated by

$$v(p) = \frac{1}{n} p(1-p). \qquad (8.2.3)$$

It follows that the statistic

$$Z = \frac{p - P}{\sqrt{v(p)}} \qquad (8.2.4)$$

has approximately a standard normal distribution when the sample size is large, for example, greater than 30. Based on Z, a $(1 - \alpha) \times 100\%$ confidence interval for the population proportion can be obtained as

$$p \pm z(\alpha/2)\sqrt{v(p)}, \qquad (8.2.5)$$

where $z(\alpha/2)$ is the upper $(\alpha/2)$th quantile of the standard normal distribution.

Note that the confidence interval given in (8.2.4) will not give an adequate coverage probability when the sample size is small or the number of subjects with improvement is close to 0 or to n. When n is small, Clopper and Pearson (1934) suggest using the so-called *exact confidence interval* to find the binomial proportion. This method, however, is quite conservative in the sense that the actual coverage probability is greater than the nominal level of $(1 - \alpha)100\%$. As a result this method can yield a relatively wide confidence interval. Alternatively, Blyth and Still (1983) propose a procedure for the binomial proportion. Their proposed interval provides an adequate coverage probability. In addition the proposed interval has approximately equal tail probabilities. Table 8.2.1 reproduces the lower and upper limits of the 95% confidence intervals for n from 1 to 30. When n is greater than 30, they suggest using the formula

$$\frac{1}{[n + z^2(\alpha/2)]} \left\{ (Y \pm 0.5) + z^2(\alpha/2) + z(\alpha/2) \right.$$

$$\left. \cdot \sqrt{(Y \pm 0.5) - \frac{1}{n}(Y \pm 0.5)^2 + \frac{1}{4}z^2(\alpha/2)} \right\}. \qquad (8.2.6)$$

For $Y = 0$ or 1, the lower limits are given as either 0 or $1 - (1 - \alpha)^{1/n}$ with an equivalent exception for $Y = n, n - 1$.

Suppose that from the previous reports, about 30% to 35% of the subjects with acute ischemic stroke who are not treated with rt-PA improve. Hence clinically it may be of interest to know whether the proportion of subjecs treated with rt-PA is 35%. The following hypotheses can be used to address this question:

$$H_0: P = P_0 = 0.35, \qquad (8.2.7)$$
$$\text{vs.} \quad H_a: P \neq 0.35.$$

Under the null hypothesis, $P = 0.35$, the large-sample test statistic is given by

$$Z = \frac{p - 0.35}{\sqrt{p(1-p)/n}}. \qquad (8.2.8)$$

At a preselected level of significance α, we reject the null hypothesis and conclude that the proportion of subjects with improvement in those treated with rt-PA is different from 35% if

$$|Z| > z(\alpha/2).$$

The corresponding p-value can be computed based on the standard normal distribution as follows:

$$p\text{-value} = P\{|Z| > z\}, \qquad (8.2.9)$$

where z is the observed value of Z as given in (8.2.8). Therefore the null hypothesis is rejected at the α level of significance if the p-value in (8.2.9) is smaller than α. Note that there is a relationship between hypothesis testing procedure given in (8.2.8) and the $(1 - \alpha)100\%$ confidence interval given in (8.2.5). That is, one rejects the null hypothesis if and only if the $(1 - \alpha)100\%$ confidence interval does not contain $P = 0.35$.

For small samples the exact test procedure based on p-value can be constructed. Under the null hypothesis Y follows a binomial distribution with $P = P_0 = 0.35$. We would expect an average of nP subjects treated with rt-PA to show improvement. If the observed Y is smaller than the expected number of subjects with improvement, the p-value can be calculated as

$$p\text{-value} = 2 \left\{ \sum_{y=0}^{Y} \frac{n!}{[y!(n-y)!]} (P_0)^y (1 - P_0)^{n-y} \right\}. \quad (8.2.10)$$

On the other hand, when the observed Y is larger than the expected number of subject with improvement, nP, the p-value is given by

$$p\text{-value} = 2 \left\{ 1 - \sum_{y=0}^{Y-1} \frac{n!}{[y!(n-y)!]} (P_0)^y (1 - P_0)^{n-y} \right\}. \quad (8.2.11)$$

Similarly the null hypothesis is rejected at the α significance level if the p-value (calculated based on (8.2.10) or (8.2.11)) is smaller than α.

Example 8.2.1
Consider the dataset of the NINDS trial given in Table 8.2.1; the number of subjects with early improvement for rt-PA and the placebo group are 147 and 122, respectively. Since 312 subjects were enrolled in each group, the sample proportions with early improvement for rt-PA and placebo are 47.1% (147/312) and 39.1% (122/312), respectively. As presented in Table 8.2.2, the standard errors of estimated sample proportions are 0.0283 ($\sqrt{0.47 * 0.529/312}$) and 0.0276 ($\sqrt{0.391 * 0.609/312}$), respectively. The large-sample 95% confidence intervals for populations with early improvement for rt-PA and the placebo group are (41.58%, 52.65%) and (33.68% 44.52%), respectively. The Blyth and Still method yields intervals of (41.49%, 52.82%) and (33.69%, 44.78%), which are 2.3% and 2.4% wider than those large-sample confidence intervals. The observed Z values for testing whether the proportion of improvement is 35% for rt-PA and the placebo are 4.287 ((0.471 − 0.35)/0.0283) and 1.485 ((0.391 − 0.35)/0.276), respectively. As a result, since the observed Z value for the rt-PA group is greater than $z(0.025) = 1.96$ with the associated p-value of 0.00002, we reject the null hypothesis at the 5% level of significance and conclude that the proportion of subjects with early improvement for the rt-PA group is statistically significant greater than 35%. One can reach the same conclusion

Table 8.2.1 Lower and Upper 95% Confidence Intervals for the Binomial Proportion

Sample Size	Total Responses	Lower Limit	Upper Limit	Sample Size	Total Responses	Lower Limit	Upper Limit
1	0	0.00	0.95	8	8	0.64	1.00
1	1	0.05	1.00	9	0	0.00	0.32
2	0	0.00	0.78	9	1	0.01	0.44
2	1	0.03	0.97	9	2	0.04	0.56
2	2	0.22	1.00	9	3	0.10	0.68
3	0	0.00	0.63	9	4	0.17	0.75
3	1	0.02	0.86	9	5	0.25	0.83
3	2	0.14	0.98	9	6	0.32	0.90
3	3	0.37	1.00	9	7	0.44	0.96
4	0	0.00	0.53	9	8	0.56	0.99
4	1	0.01	0.75	9	9	0.68	1.00
4	2	0.10	0.90	10	0	0.00	0.29
4	3	0.25	0.99	10	1	0.01	0.44
4	4	0.47	1.00	10	2	0.04	0.56
5	0	0.00	0.50	10	3	0.09	0.62
5	1	0.01	0.66	10	4	0.15	0.70
5	2	0.08	0.81	10	5	0.22	0.78
5	3	0.19	0.92	10	6	0.29	0.85
5	4	0.34	0.99	10	7	0.38	0.91
5	5	0.50	1.00	10	8	0.44	0.96
6	0	0.00	0.41	10	9	0.56	0.99
6	1	0.01	0.59	10	10	0.71	1.00
6	2	0.06	0.73	11	0	0.00	0.26
6	3	0.15	0.85	11	1	0.00	0.40
6	4	0.27	0.94	11	2	0.03	0.50
6	5	0.41	0.99	11	3	0.08	0.60
6	6	0.59	1.00	11	4	0.14	0.67
7	0	0.00	0.38	11	5	0.20	0.74
7	1	0.01	0.55	11	6	0.26	0.80
7	2	0.05	0.66	11	7	0.33	0.86
7	3	0.13	0.77	11	8	0.40	0.92
7	4	0.23	0.87	11	9	0.50	0.97
7	5	0.34	0.95	11	10	0.60	1.00
7	6	0.45	0.99	11	11	0.74	1.00
7	7	0.62	1.00	12	0	0.00	0.25
8	0	0.00	0.36	12	1	0.00	0.37
8	1	0.01	0.50	12	2	0.03	0.46
8	2	0.05	0.64	12	3	0.07	0.54
8	3	0.11	0.71	12	4	0.12	0.63
8	4	0.19	0.81	12	5	0.18	0.71
8	5	0.29	0.89	12	6	0.24	1.76
8	6	0.36	0.95	12	7	0.29	0.82
8	7	0.50	0.99	12	8	0.37	0.88

Table 8.2.1 (*Continued*)

Sample Size	Total Responses	Lower Limit	Upper Limit	Sample Size	Total Responses	Lower Limit	Upper Limit
12	9	0.46	0.93	15	11	0.47	0.90
12	10	0.54	0.97	15	12	0.53	0.94
12	11	0.63	1.00	15	13	0.61	0.98
12	12	0.75	1.00	15	14	0.70	1.00
13	0	0.00	0.23	15	15	0.78	1.00
13	1	0.00	0.34	16	0	0.00	0.20
13	2	0.03	0.43	16	1	0.00	0.30
13	3	0.07	0.52	16	2	0.02	0.37
13	4	0.11	0.59	16	3	0.05	0.44
13	5	0.17	0.66	16	4	0.09	0.60
13	6	0.22	0.74	16	5	0.13	0.56
13	7	0.26	0.78	16	6	0.18	0.63
13	8	0.34	0.83	16	7	0.20	0.70
13	9	0.41	0.89	16	8	0.27	0.73
13	10	0.48	0.93	16	9	0.30	0.80
13	11	0.57	0.97	16	10	0.37	0.82
13	12	0.66	1.00	16	11	0.44	0.87
13	13	0.77	1.00	16	12	0.50	0.91
14	0	0.00	0.23	16	13	0.56	0.95
14	1	0.00	0.32	16	14	0.63	0.98
14	2	0.03	0.42	16	15	0.70	1.00
14	3	0.06	0.50	16	16	0.80	1.00
14	4	0.10	0.58	17	0	0.00	0.19
14	5	0.15	0.63	17	1	0.00	0.28
14	6	0.21	0.68	17	2	0.02	0.35
14	7	0.24	0.76	17	3	0.05	0.42
14	8	0.32	0.79	17	4	0.08	0.49
14	9	0.37	0.85	17	5	0.12	0.54
14	10	0.42	0.90	17	6	0.17	0.59
14	11	0.50	0.94	17	7	0.19	0.65
14	12	0.58	0.97	17	8	0.25	0.72
14	13	0.68	1.00	17	9	0.28	0.75
14	14	0.77	1.00	17	10	0.35	0.81
15	0	0.00	0.22	17	11	0.41	0.83
15	1	0.00	0.30	17	12	0.46	0.88
15	2	0.02	0.39	17	13	0.51	0.92
15	3	0.06	0.47	17	14	0.58	0.95
15	4	0.10	0.53	17	15	0.65	0.98
15	5	0.14	0.61	17	16	0.72	1.00
15	6	0.19	0.67	17	17	0.81	1.00
15	7	0.22	0.71	18	0	0.00	0.18
15	8	0.29	0.78	18	1	0.00	0.27
15	9	0.33	0.81	18	2	0.02	0.33
15	10	0.39	0.86	18	3	0.05	0.41

Table 8.2.1 (*Continued*)

Sample Size	Total Responses	Lower Limit	Upper Limit	Sample Size	Total Responses	Lower Limit	Upper Limit
18	4	0.08	0.47	20	9	0.24	0.68
18	5	0.12	0.53	20	10	0.29	0.71
18	6	0.16	0.59	20	11	0.32	0.76
18	7	0.18	0.63	20	12	0.37	0.79
18	8	0.24	0.67	20	13	0.42	0.84
18	9	0.27	0.73	20	14	0.47	0.86
18	10	0.33	0.76	20	15	0.53	0.90
18	11	0.37	0.82	20	16	0.58	0.93
18	12	0.41	0.84	20	17	0.63	0.96
18	13	0.47	0.88	20	18	0.68	0.98
18	14	0.53	0.92	20	19	0.76	1.00
18	15	0.59	0.95	20	20	0.84	1.00
18	16	0.67	0.98	21	0	0.00	0.15
18	17	0.73	1.00	21	1	0.00	0.23
18	18	0.82	1.00	21	2	0.02	0.30
19	0	0.00	0.17	21	3	0.04	0.35
19	1	0.00	0.25	21	4	0.07	0.40
19	2	0.02	0.32	21	5	0.10	0.46
19	3	0.04	0.39	21	6	0.13	0.51
19	4	0.08	0.45	21	7	0.15	0.55
19	5	0.11	0.50	21	8	0.20	0.60
19	6	0.15	0.55	21	9	0.23	0.65
19	7	0.17	0.61	21	10	0.28	0.70
19	8	0.22	0.66	21	11	0.30	0.72
19	9	0.25	0.69	21	12	0.35	0.77
19	10	0.31	0.75	21	13	0.40	0.80
19	11	0.34	0.78	21	14	0.45	0.85
19	12	0.39	0.83	21	15	0.49	0.87
19	13	0.45	0.85	21	16	0.54	0.90
19	14	0.50	0.89	21	17	0.60	0.93
19	15	0.55	0.92	21	18	0.65	0.96
19	16	0.61	0.96	21	19	0.70	0.98
19	17	0.68	0.98	21	20	0.77	1.00
19	18	0.75	1.00	21	21	0.85	1.00
19	19	0.83	1.00	22	0	0.00	0.15
20	0	0.00	0.16	22	1	0.00	0.22
20	1	0.00	0.24	22	2	0.02	0.29
20	2	0.02	0.32	22	3	0.04	0.34
20	3	0.04	0.37	22	4	0.06	0.39
20	4	0.07	0.42	22	5	0.09	0.45
20	5	0.10	0.47	22	6	0.13	0.50
20	6	0.14	0.53	22	7	0.15	0.55
20	7	0.16	0.58	22	8	0.19	0.58
20	8	0.21	0.63	22	9	0.22	0.62

Table 8.2.1 (*Continued*)

Sample Size	Total Responses	Lower Limit	Upper Limit	Sample Size	Total Responses	Lower Limit	Upper Limit
22	10	0.26	0.66	24	7	0.13	0.50
22	11	0.29	0.71	24	8	0.17	0.54
22	12	0.34	0.74	24	9	0.20	0.59
22	13	0.38	0.78	24	10	0.23	0.63
22	14	0.42	0.81	24	11	0.26	0.66
22	15	0.45	0.85	24	12	0.31	0.69
22	16	0.50	0.87	24	13	0.34	0.74
22	17	0.55	0.91	24	14	0.37	0.77
22	18	0.61	0.94	24	15	0.41	0.80
22	19	0.66	0.96	24	16	0.46	0.83
22	20	0.71	0.98	24	17	0.50	0.87
22	21	0.78	1.00	24	18	0.54	0.89
22	22	0.85	1.00	24	19	0.59	0.91
23	0	0.00	0.14	24	20	0.63	0.94
23	1	0.00	0.21	24	21	0.69	0.97
23	2	0.02	0.27	24	22	0.74	0.98
23	3	0.04	0.32	24	23	0.80	1.00
23	4	0.06	0.39	24	24	0.87	1.00
23	5	0.09	0.43	25	0	0.00	0.13
23	6	0.12	0.48	25	1	0.00	0.19
23	7	0.14	0.52	25	2	0.01	0.25
23	8	0.18	0.57	25	3	0.03	0.30
23	9	0.21	0.61	25	4	0.06	0.36
23	10	0.25	0.64	25	5	0.08	0.40
23	11	0.27	0.68	25	6	0.11	0.44
23	12	0.32	0.73	25	7	0.13	0.48
23	13	0.36	0.75	25	8	0.16	0.52
23	14	0.39	0.79	25	9	0.19	0.56
23	15	0.43	0.82	25	10	0.22	0.60
23	16	0.48	0.86	25	11	0.25	0.64
23	17	0.52	0.88	25	12	0.30	0.68
23	18	0.57	0.91	25	13	0.32	0.70
23	19	0.61	0.94	25	14	0.36	0.75
23	20	0.68	0.96	25	15	0.40	0.78
23	21	0.73	0.98	25	16	0.44	0.81
23	22	0.79	1.00	25	17	0.48	0.84
23	23	0.86	1.00	25	18	0.52	0.87
24	0	0.00	0.13	25	19	0.56	0.89
24	1	0.00	0.20	25	20	0.60	0.92
24	2	0.02	0.26	25	21	0.64	0.94
24	3	0.03	0.31	25	22	0.70	0.97
24	4	0.06	0.37	25	23	0.75	0.99
24	5	0.09	0.41	25	24	0.81	1.00
24	6	0.11	0.46	25	25	0.87	1.00

Table 8.2.1 (*Continued*)

Sample Size	Total Responses	Lower Limit	Upper Limit	Sample Size	Total Responses	Lower Limit	Upper Limit
26	0	0.00	0.12	27	17	0.43	0.80
26	1	0.00	0.19	27	18	0.46	0.82
26	2	0.01	0.24	27	19	0.50	0.85
26	3	0.03	0.20	27	20	0.54	0.88
26	4	0.05	0.34	27	21	0.59	0.90
26	5	0.08	0.38	27	22	0.63	0.92
26	6	0.11	0.42	27	23	0.67	0.95
26	7	0.12	0.47	27	24	0.71	0.97
26	8	0.15	0.51	27	25	0.77	0.99
26	9	0.19	0.54	27	26	0.82	1.00
26	10	0.21	0.58	27	27	0.88	1.00
26	11	0.24	0.62	28	0	0.00	0.12
26	12	0.28	0.66	28	1	0.00	0.17
26	13	0.30	0.70	28	2	0.01	0.23
26	14	0.34	0.72	28	3	0.03	0.28
26	15	0.38	0.76	28	4	0.05	0.32
26	16	0.42	0.79	28	5	0.07	0.36
26	17	0.46	0.81	28	6	0.10	0.41
26	18	0.49	0.85	28	7	0.12	0.44
26	19	0.53	0.88	28	8	0.14	0.48
26	20	0.58	0.89	28	9	0.17	0.52
26	21	0.62	0.92	28	10	0.19	0.56
26	22	0.66	0.95	28	11	0.23	0.59
26	23	0.80	0.97	28	12	0.26	0.62
26	24	0.76	0.99	28	13	0.28	0.65
26	25	0.81	1.00	28	14	0.32	0.68
26	26	0.88	1.00	28	15	0.35	0.72
27	0	0.00	0.12	28	16	0.38	0.74
27	1	0.00	0.18	28	17	0.41	0.77
27	2	0.01	0.23	28	18	0.44	0.81
27	3	0.03	0.29	28	19	0.48	0.83
27	4	0.05	0.33	28	20	0.52	0.86
27	5	0.08	0.37	28	21	0.56	0.88
27	6	0.10	0.41	28	22	0.59	0.90
27	7	0.12	0.46	28	23	0.64	0.93
27	8	0.15	0.50	28	24	0.68	0.95
27	9	0.18	0.54	28	25	0.72	0.97
27	10	0.20	0.57	28	26	0.77	0.99
27	11	0.23	0.60	28	27	0.83	1.00
27	12	0.27	0.63	28	28	0.88	1.00
27	13	0.29	0.67	29	0	0.00	0.11
27	14	0.33	0.71	29	1	0.00	0.17
27	15	0.37	0.73	29	2	0.01	0.22
27	16	0.40	0.77	29	3	0.03	0.27

Table 8.2.1 (*Continued*)

Sample Size	Total Responses	Lower Limit	Upper Limit	Sample Size	Total Responses	Lower Limit	Upper Limit
29	4	0.05	0.31	30	3	0.03	0.26
29	5	0.07	0.36	30	4	0.05	0.30
29	6	0.09	0.39	30	5	0.07	0.35
29	7	0.11	0.43	30	6	0.09	0.38
29	8	0.14	0.46	30	7	0.11	0.41
29	9	0.17	0.50	30	8	0.13	0.45
29	10	0.18	0.54	30	9	0.16	0.48
29	11	0.22	0.57	30	10	0.18	0.52
29	12	0.25	0.61	30	11	0.21	0.55
29	13	0.27	0.64	30	12	0.24	0.59
29	14	0.31	0.66	30	13	0.26	0.62
29	15	0.34	0.69	30	14	0.30	0.65
29	16	0.36	0.73	30	15	0.32	0.68
29	17	0.39	0.75	30	16	0.35	0.70
29	18	0.43	0.78	30	17	0.38	0.74
29	19	0.46	0.82	30	18	0.41	0.76
29	20	0.50	0.83	30	19	0.45	0.79
29	21	0.54	0.86	30	20	0.58	0.82
29	22	0.57	0.89	30	21	0.52	0.84
29	23	0.61	0.91	30	22	0.55	0.87
29	24	0.64	0.93	30	23	0.59	0.89
29	25	0.69	0.95	30	24	0.62	0.91
29	26	0.73	0.97	30	25	0.65	0.93
29	27	0.78	0.99	30	26	0.70	0.95
29	28	0.83	1.00	30	27	0.74	0.97
29	29	0.89	1.00	30	28	0.79	0.99
30	0	0.00	0.11	30	29	0.84	1.00
30	1	0.00	0.16	30	30	0.89	1.00
30	2	0.01	0.21				

Source: Blyth and Still (1983).

by noting that the 95% confidence interval (41.58%, 52.65%) does not contain 35%. For placebo group, since the observed Z value of 1.485 is less than 1.96, we fail to reject the null hypothesis. As a result we conclude that the data of the placebo group do not provide sufficient evidence to doubt the validity of the null hypothesis that the proportion of the subjects with early improvement is equal to 35%. The same conclusion was reached based on (1) the corresponding p-value being 0.138, which is greater than 0.05, and (2) the 95% confidence interval for the placebo group of (33.69%, 44.52%) including $P = 35\%$. The exact p-value for the placebo group is given by

Table 8.2.2 Summary of the Number of Subjects with Improvement by Treatment for the NINDS Trial

	rt-PA	Placebo
N	312	312
With improvement	147	122
Proportion	47.12%	39.10%
Standard error	2.83%	2.76%
95% Confidence Interval		
Large-sample	(41.58%, 52.65%)	(33.68%, 44.52%)
Blyth-Still	(41.49%, 52.82%)	(33.69%, 44.78%)
Z-statistics	4.287	1.485
p-value		
Large-sample	0.00002	0.1375
Exact	0.00001	0.1500

$$p\text{-value} = 2\left\{1 - \sum_{y=0}^{121} \frac{312!}{[y!(312-y)!]} (0.35)^y (0.65)^{312-y}\right\}$$

$$= 0.150.$$

Similarly the exact p-value of rt-PA group is given by 0.00001. Note that in this example, both the exact and large-sample methods yield similar p-values. These results are summarized in Table 8.2.2.

In clinical trials the comparison between the pre-treatment (baseline) and the post-treatment is usually made before the assessment of efficacy and safety of the treatment. This comparison within the same subjects is usually referred to as the pre-post comparison which is also known as *change from baseline analysis*. To illustrate the pre-post comparison concept, consider regrouping the five classes of the status score of the Koch's data set given in Table 8.1.6 into two categories: unfavorable (i.e., *terrible, poor*, and *fair*) and favorable (i.e., *good* and *excellent*). The resulting endpoint is then the favorable status. Table 8.2.3 reconstructs a 2 × 2 table from Table 8.1.6 based on the favorable status. From Table 8.2.3, at baseline, it can be seen that subjects in the placebo group and the test drug group showed similar favorable status with proportions of 45.61% and 42.59%, respectively. However, at visit 3 after treatment, the proportion of subjects with favorable status in the placebo group remained the same while the proportion increased to 72.22% for the test drug group. As a result it is important to verify whether the proportion of the subjects with favorable staus within each group changes significantly after treatment.

Table 8.2.3 Summary of Frequency of Transition of Status from the Baseline at Visit 3

Baseline	Unfavorable	Favorable	Total
	Treatment: Placebo		
Unfavorable	23 (40.35%)	8 (14.04%)	31 (54.39%)
Favorable	8 (14.04%)	18 (31.58%)	26 (45.61%)
Total	31 (54.39%)	26 (45.61%)	57 (100.00%)
	Treatment: Test Drug		
Unfavorable	14 (25.93%)	17 (31.48%)	31 (57.41%)
Favorable	1 (1.85%)	22 (40.74%)	23 (42.59%)
Total	15 (27.78%)	39 (72.22%)	54 (100.00%)

Note: Unfavorable = terrible, poor, or fair; favorable = good or excellent. See Table 8.1.6.

For detection of a significant change from the baseline, the same binary endpoint is usually evaluated at the baseline and at various visits after treatment (see also Section 12.5 for McNemar's test). Hence, the resulting binary responses are correlated. In general, data of this kind can be summarized by a 2 × 2 transition table from the baseline as presented in Table 8.2.4. Basically, within each treatment, Table 8.2.4 classifies subjects into four groups: the subjects with favorable status at both the baseline and visit 3 (Y_{11}), the subjects with unfavorable status at both the baseline and visit 3 (Y_{00}), the subjects with a change from unfavorable status at baseline to favorable status at visit 3 (Y_{01}), and the subjects with a change from favorable status to unfavorable status at visit 3 (Y_{10}). Hence Y_{11} and Y_{00} represent the number of the subjects with no change in favorable status, while Y_{10} and Y_{01} are the number of subjects whose favorable status changes from the baseline. Suppose that our objective is to examine whether the proportion of the subjects with favorable status at visit 3 is the same as that at the baseline. In this case we can compare the number of subjects with favorable status at visit 3 with that at baseline, namley $Y_{.1}$ versus $Y_{1.}$. This type of comparison is known as a *marginal comparison* between visit 3 and the baseline. However the number of subjects with favorable status at visit 3 is the sum of the number of subjects with the favorable status at both the baseline and visit 3 (Y_{11}) and those with a change from the unfavorable status at baseline to the favorable status at visit 3 (Y_{01}). Similarly the number of subjects with a favorable status at the baseline is the sum of the number of subjects with favorable status at both the baseline and visit 3 (Y_{11}) and those with a change from the favorable status at the baseline to the unfavorable status at visit 3 (Y_{10}). Consequently a marginal comparison of the number of subjects with a favorable status between visit 3 and the baseline involves the difference between the number of subjects with a change from the unfavorable status at the baseline to a favorable status at visit 3 and those with a change from the

Table 8.2.4 Summary of Data for Binary Endpoint from the Baseline

Baseline	No	Yes	Total
No	$Y_{00}(P_{00})$	$Y_{01}(P_{01})$	$Y_{0.}(P_{0.})$
Yes	$Y_{10}(P_{10})$	$Y_{11}(P_{11})$	$Y_{1.}(P_{1.})$
Total	$Y_{.0}(P_{.0})$	$Y_{.1}(P_{.1})$	$Y_{..}(1)$

favorable status at the baseline to an unfavorable status at visit 3, namely $Y_{01} - Y_{10}$. The proportion can be obtained by dividing the number of subjects by the sample size.

Statistical hypotheses for comparison of proportions with favorable status between visit 3 and baseline can be expressed as

$$H_0: P_{.1} = P_{1.}, \qquad (8.2.12)$$
$$\text{vs.} \quad H_a: P_{.1} \neq P_{1.},$$

or

$$H_0: P_{01} = P_{10},$$
$$\text{vs.} \quad H_a: P_{01} \neq P_{10.}$$

Let y_{ij} and p_{ij} denote the corresponding observed number of subjects and proportions for Y_{ij} and P_{ij}, respectively. An unbiased estimator for $\theta = P_{01} - P_{10}$ is given by $p_{01} - p_{10}$. For large samples an estimate of the variance of θ can be obtained as

$$v(\theta) = \text{var}(\theta) \qquad (8.2.13)$$
$$= \frac{1}{n}[(p_{01} + p_{10}) - (p_{01} - p_{10})^2].$$

It follows that a large sample $(1 - \alpha)100\%$ confidence interval for θ is given by

$$(p_{01} - p_{10}) \pm z(\alpha/2)\sqrt{v(\theta)}. \qquad (8.2.14)$$

Under the null hypothesis of $P_{01} - P_{10} = 0$, the estimated large sample variance in (8.2.13) becomes $v(0) = (p_{01} + p_{10})/n$. As a result an asymptotic test statistic for hypotheses (8.2.12) is

$$\chi_M^2 = \frac{(p_{01} - p_{10})^2}{(p_{01} + p_{10})/n}$$

$$= \frac{(y_{01} - y_{10})^2}{(y_{01} + y_{10})}. \tag{8.2.15}$$

We reject the null hypothesis at the α level of significance if

$$\chi_M^2 > \chi^2(\alpha, 1),$$

where $\chi^2(\alpha, 1)$ is the upper αth quantile of a χ^2 random variable with one degree of freedom. However, the test statistics in (8.2.15), which is known as McNemar test (Conover, 1980), is a large-sample approximation of a continuous variable to a discrete variable. χ_M^2 may commit type I error too often with respect to the preselected nominal level when $p_{01} + p_{10} < 0.3$. An approach to overcome this issue is to employ a continuity correction as follows:

$$\chi_M^2 = \frac{(|y_{01} - y_{10}| - 1)^2}{(y_{01} + y_{10})}. \tag{8.2.16}$$

Sometimes, the test statistic with such a continuity correction is too conservative to have sufficient power for detection of a meaningful difference in marginal proportions between visit and baseline. Then one computes the exact p-value under the null hypothesis. As mentioned before, the comparison in marginal distributions between visit and baseline depends on the subjects with different status at visit and baseline, namely y_{01} and y_{10}. Given the total number of subject with discordant pairs, $y = y_{01} + y_{10}$, both probabilities of a subject with favorable status at the baseline and unfavorable status at the visit and of a subject with unfavorable status at the baseline and favorable status at the visit are equal to $1/2$ under the null hypothesis that $P_{01} = P_{10}$. It turns out that y_{01} follows a binomial distribution with a probability of success 0.5 given y number of total subjects. The exact p-value for testing hypotheses (8.2.12) can then be calculated as

$$p\text{-value} = \begin{cases} 2 \sum_{i=1}^{y_{01}} \left[\frac{y!}{i!(y-i)!} \right] (0.5)^y & \text{if } y_{01} < \frac{n}{2}, \\ 2 \left\{ 1 - \sum_{y_{01}-1}^{y} \left[\frac{y!}{i!(y-i)!} \right] (0.5)^y \right\} & \text{if } y_{01} > \frac{n}{2}, \\ 1 & \text{if } y_{01} = \frac{n}{2}. \end{cases} \tag{8.2.17}$$

Table 8.2.5 Results of Analysis in Table 8.2.3

Treatment	Method	Test-Statistic	p-Value
Placebo	Unconditionally large sample		
	No continuity correction	0	1.0000
	With continuity correction	0.0625	0.80259
	Conditional exact		1.0000
Treatment	Unconditionally large sample		
	No continuity correction	14.22	0.00016
	With continuity correction	12.50	0.00041
	Conditional exact		0.00014

The p-values by formula (8.2.17) are calculated by considering the number of subjects with discordant results to be a fixed known number. Hence this is referred to as the *conditional* exact p-value. Note that the confidence interval and test statistics given in (8.2.14) and (8.2.16) are *unconditional* large-sample procedures. In practice, the number of subjects with different statuses between the visit and the baseline is usually not known until the completion of the trial. It is in fact a random variable and thus technically not a fixed number. As an alternative, Suissa and Shuster (1991) propose an unconditional exact procedure for testing hypotheses (8.2.12). The proposed procedure requires a complicated computation of the binomial probability over the range between 0 and 1. However, they provide the critical values for different combinations of significance levels and sample sizes in addition to the required sample sizes for the different significance levels, $P_{01} - P_{10}$, and $P_{01} + P_{10}$. The exact confidence interval for $P_{01} - P_{10}$, either unconditional or conditional, has not yet been proposed.

Example 8.2.2
For the data given in Table 8.2.3, the differences in marginal proportions of favorable status between visit 3 and the baseline for the placebo and the test drug are estimated as 0 and 0.2963, respectively. The 95% large-sample confidence intervals for the difference in proportions for the placebo group and the test drug group are (−1.038, 1.038) and (0.1641, 0.4284), respectively. The results of hypothesis testing by the three procedures discussed above are summarized in Table 8.2.5. All three methods provide consistent results. For the placebo groups, there is no statistically significant difference in proportion of subjects with favorable status between visit 3 and the baseline, while the proportion of the subjects with favorable status at visit 3 for the test drug group is statistically significant higher than that at the baseline at the 5% level of significance.

8.3 INFERENCE OF INDEPENDENT SAMPLES

One of the primary goals of the NINDS trial is to compare the proportion of subjects treated with rt-PA who show early clinical improvement with that

Table 8.3.1 Data Structure of Binary Endpoint for a Parallel Two-Group Trial

Treatment	Binary Response		Total
	No	Yes	
Test drug	$Y_{10}(P_{10})$	$Y_{11}(P_{11})$	$Y_{1.}(1)$
Placebo	$Y_{20}(P_{20})$	$Y_{21}(P_{21})$	$Y_{2.}(1)$
Total	$Y_{.0}$	$Y_{.1}$	$Y_{..}$

of the subjects given a placebo. This type of comparison is usually called a between-group comparison. Comparisons among treatments in randomized, parallel-group clinical trials usually involve inference of independent samples because the clinical outcomes of a subject evaluated in terms of efficacy and safety endpoints have nothing to do with those of the subjects within the group or from another group. In this section our efforts will be directed toward statistical inference based on the binary endpoints for two independent samples.

Consider a parallel two-group clinical trial that compares a test drug in n_1 subjects with a placebo in n_2 subjects. Table 8.3.1 illustrates a simple way for summarization of a binary endpoint with two categories, *yes* and *no*. *Yes* may mean death, success, improvement, or eradication. In Table 8.3.1, Y_{i1} is the number of subjects with a yes response in the ith group with sample size $Y_{i.}$, namely $Y_i = n_i$, $i = 1, 2$; and $Y_{..}$ and $n_1 + n_2 = N$ is the total number of subjects in the trial. Since Y_{i1} are independent binomial variables within each group, the population proportion of a yes response can be estimated by the methods described in the previous section. However, descriptive measures for comparing proportions of two independent samples often involve differences in proportions, odds ratios, and relative risks.

Let y_{i1} and p_{i1} be the observed number of subjects and proportion with a yes response in the ith group. Because of independence, the difference in the population proportions of a yes response between two groups can be unbiasedly estimated by

$$d_p = p_{11} - p_{21}$$
$$= \frac{y_{11}}{n_1} - \frac{y_{21}}{n_2}, \qquad (8.3.1)$$

with an estimated large-sample variance given as

$$v(d_p) = \frac{p_{11}(1 - p_{11})}{n_1} + \frac{p_{21}(1 - p_{21})}{n_2}. \qquad (8.3.2)$$

As a result the lower and upper limits of the large-sample $(1 - \alpha)100\%$ confidence interval for the difference in population proportions can be obtained as

$$d_p \pm z(\alpha/2)\sqrt{v(d_p)}, \qquad (8.3.3)$$

where $z(\alpha/2)$ is the upper $(\alpha/2)$th quantile of the normal distribution.

Since treatments can be considered as risk factors, a measure often used to study the relationship between treatments and responses is the *odds ratio*. For example, the NINDS trial defined the odds ratio as the ratio of the odds of a subject with an early clinical improvement after exposure to treatment of rt-PA to the odds of an improvement of a subject given the placebo. The odds of a yes response in each treatment group is estimated as

$$\frac{p_{i1}}{p_{i0}} = \frac{y_{i1}}{y_{i0}}.$$

The odds ratio (OR) of a yes response for comparing the test drug against placebo is then estimated as

$$\begin{aligned} OR &= \frac{p_{11}/p_{10}}{p_{21}/p_{20}} \\ &= \frac{y_{11}/y_{10}}{y_{21}/y_{20}} \\ &= \frac{y_{11}y_{20}}{y_{10}y_{21}}. \end{aligned} \qquad (8.3.4)$$

The lower and upper limits of the large-sample $(1 - \alpha)100\%$ confidence interval for the odds ratio are given by

$$\exp\left\{\ln(OR) \pm z(\alpha/2)\sqrt{\frac{1}{y_{10}} + \frac{1}{y_{11}} + \frac{1}{y_{20}} + \frac{1}{y_{21}}}\right\}, \qquad (8.3.5)$$

where $z(\alpha/2)$ is the upper $(\alpha/2)$th quantile of the normal distribution; "exp" and "ln" denote the natural exponential and logarithm, respectively.

Another commonly employed measure for describing the relationship between treatment and its associated binary outcomes is the relative risk. The risk for obtaining a yes response for the ith group is simply the proportion of the subjects with a yes response. As a result the relative risk (RR) for the test drug to placebo can be estimated by

$$RR = \frac{p_{11}}{p_{21}}. \qquad (8.3.6)$$

An estimated large-sample variance for the logarithm of the relative risk is given as

$$v[\ln(RR)] = \frac{1-p_{11}}{y_{11}} + \frac{1-p_{21}}{y_{21}}. \quad (8.3.7)$$

It follows that the large-sample $(1 - \alpha)100\%$ confidence interval for the relative risk are given as

$$\exp\left\{\ln(RR) \pm z(\alpha/2)\sqrt{\frac{1-p_{11}}{y_{11}} + \frac{1-p_{21}}{y_{21}}}\right\}. \quad (8.3.8)$$

Note that the difference in proportion is an adequate measure for comparing two proportions when they are not close to 0, and when Y_{ij} is greater than 5, for $j = 0, 1$, and $i = 1, 2$. This measure that is symmetric about 0. On the other hand, when the proportion of a yes response is close to 0, then either the odds ratio or the relative risk is a better measure for comparing two proportions than the difference. Since the odds ratio and the relative risk range from 0 to infinity, the distributions of their estimates are skewed. If there is no difference in true proportions of a yes response between two population groups, the odds ratio or relative risk can be expected to be close to 1. When the number of subjects in each group becomes large, the estimated odds ratio and relative risk will be very close to each other. In clinical trials the difference in proportions, odds ratio, and relative risk are useful measures for comparing prospectively two independent proportions. Note that the odds ratio can also be employed for retrospective case control studies in which the difference in proportions and relative risk is appropriate.

Example 8.3.1

For the NINDS trial, estimates of the subjects experiencing early improvement for each group are given in Table 8.2.2. From these sample proportions, the difference in population proportions of the subjects with early improvement between rt-PA and the placebo is estimated by 8.01% with an estimated large-sample variance of 0.0016. This leads to a 95% confidence interval of (0.27%, 15.76%). The results are summarized in Table 8.3.2. As a result, on average, the proportion of the subjects treated with rt-PA showing an early improvement is about 8% higher than those given a placebo. Note that the 95% confidence interval does not contain zero, which indicates that significantly more subjects receiving rt-PA can show early improvement than those receiving the placebo.

For the U.S. Physicians' Health Study, the proportion of subjects with development of malignant neoplasm after an average of 12 years treatment is 11.53% for subjects receiving beta carotene and 11.72% for those receiving a placebo. The unadjusted odds ratio of malignant neoplasm for the subjects treated with

Table 8.3.2 Comparison of Subjects with Improvement in the NINDS Trial

Treatment	N	Improvement	Difference (se)	95% Confidence Interval
rt-PA	312	147 (47.12%)	0.0801 (0.0016)	(0.0027, 0.1576)
Placebo	312	122 (39.10%)		

rt-PA as compared to the placebo is 0.98 with a 95% confidence interval of (0.91, 1.07) (see Table 8.3.3). Similarly the estimate of unadjusted relative risk is 0.98, and its corresponding large-sample confidence interval is from 0.92 to 1.06. The estimated odds ratio and relative risk and their associated 95% confidence interval are consistent and numerically close. Since both 95% confidence intervals for the odds ratio and relative risk include 1, we conclude that the long-term treatment of beta carotene does not statistically significantly reduce odds or risk of development of malignant neoplasm at the 5% level of significance.

Statistical inference of binary data from a parallel two-group trial involves testing the hypotheses regarding the difference in proportions with a yes response between the test drug group and the placebo group. This can be expressed as

$$H_0: P_{11} = P_{21},$$
$$\text{vs.} \quad H_a: P_{11} \neq P_{21}. \quad (8.3.9)$$

For testing hypotheses (8.3.9), several statistical procedures are available in the literature. Among these, the method using the difference in proportions is probably the most commonly used. Under the null hypothesis the proportion of the subjects with a yes response is assumed to be equal. As a result we can combine both groups to estimate of the common proportion of the subjects with a yes response. This gives

$$p_0 = \frac{y_{11} + y_{21}}{n_1 + n_2} \quad (8.3.10)$$

with an estimated large-sample variance

Table 8.3.3 Summary of the Estimated Odds Ratio and the Relative Risk for Malignant Neoplasm due to Beta Carotene in U.S. Physicians' Health Study

Beta Carotene ($N = 11{,}036$)	Placebo ($N = 11{,}036$)	Odds Ratio (95% CI)	Relative Risk (95% CI)
1273 (11.53%)	1293 (11.72%)	0.98 (0.91, 1.07)	0.98 (0.92, 1.06)

INFERENCE OF INDEPENDENT SAMPLES 331

$$v(p_0) = \frac{p_0(1 - p_0)}{(1/n_1) + (1/n_2)}. \tag{8.3.11}$$

A test statistic for hypothesis (8.3.9) is the usual Z-statistic:

$$Z = \frac{p_{11} - p_{21}}{\sqrt{v(p_0)}}. \tag{8.3.12}$$

We then reject the null hypothesis and conclude that the proportion of the subjects receiving the test drug with a yes response is statistically, significantly different from that for subjects receiving the placebo at the α level of significance if $|Z| > z(\alpha/2)$, where $z(\alpha/2)$ is the upper $(\alpha/2)$th quantile of the standard normal distribution.

Another commonly employed method is the chi-square test which is a useful statistical method for examining the relationship between two factors. In the context of comparing the proportions between two treatments, one factor is the treatment with two levels (i.e., the active test drug and the placebo), and the other factor is the binary clinical endpoint with two categories (i.e., a *yes* response and a *no* response). For the 2 × 2 contingency table given in Table 8.3.1, the chi-square test for testing independence between the two factors is equivalent to testing the difference in proportion of the subjects with a yes response between the two groups. Under the assumption of independence, the marginal total (i.e., the number of subjects in each group $Y_1.$ and $Y_2.$) and the total number of the subjects in each category of the binary clinical endpoint (i.e., $Y_{.0}$ and $Y_{.1}$) are considered fixed. The expected frequency for the (i, j) cell is then given by

$$m_{ij} = \frac{(y_{i.})(y_{.j})}{N}, \quad i = 1, 2, j = 0, 1. \tag{8.3.13}$$

If the response has nothing to do with the treatment, we can expect the observed frequencies to be fairly close the expected frequencies. As a result the weighted sum of squares of the differences between the observed and expected frequencies, with expected frequencies as weights, can serve as a test statistic for comparing two independent proportions

$$\chi_P^2 = \sum_{i=1}^{2} \sum_{j=1}^{1} \frac{(y_{ij} - m_{ij})^2}{m_{ij}}. \tag{8.3.14}$$

The above statistic χ_P^2 is usually referred to as Pearson's chi-square test, and its computation is given in Table 8.3.4. Under the null hypothesis of no difference in proportions, χ_P^2 follows a χ^2 distribution with one degree of freedom if the

Table 8.3.4 Computation of X_P^2 Statistics for Hypothesis (8.3.6) Based on Binary Data for a Parallel Two-Group Trial

Treatment	Binary Response		Total
	No	Yes	
Test drug			
O	y_{10}	y_{11}	$y_1.$
E	m_{10}	m_{11}	
$O - E$	$y_{10} - m_{10}$	$y_{11} - m_{11}$	
$(O - E)^2/E$	$(y_{10} - m_{10})^2/m_{10}$	$(y_{11} - m_{11})^2/m_{11}$	
Placebo			
O	y_{20}	y_{21}	$y_2.$
E	m_{20}	m_{21}	
$O - E$	$y_{20} - m_{20}$	$y_{21} - m_{21}$	
$(O - E)^2/E$	$(y_{20} - m_{20})^2/m_{20}$	$(y_{21} - m_{21})^2/m_{21}$	
Total	$y._0$	$y._1$	$y_{..}$

sample size of each group is moderate (i.e., greater than 30) and the expected frequency in each cell is at least 5. We reject the null hypothesis in (8.3.9) if $\chi_P^2 > \chi^2(\alpha, 1)$, where $\chi^2(\alpha, 1)$ is the upper αth quantile of a chi-square random variable with one degree of freedom. Note that the test can be modified with a continuity correction as follows:

$$\chi_{PC}^2 = \sum_{i=1}^{2} \sum_{j=1}^{1} \frac{\{(\max[0, |y_{ij} - m_{ij}| - 0.5\}^2}{m_{ij}}. \qquad (8.3.15)$$

The above chi-square test is easy to employ. In addition it can be extended to the situation where two factors have r and c categories, respectively. The expected frequency for the (i, j)th cell can be calculated by the formula provided in (8.3.13) as the product of marginal frequencies for the ith level of one factor and the jth level of the other divided by the total sample size. The test statistics χ_P^2 now is the weighted sum of squares over all $r \times c$ cells. The test for independence can be similarly performed by comparing χ_P^2 with the upper αth quantile of a chi-square distribution with $(r-1)(c-1)$ degrees of freedom.

When sample size is small or the expected frequency in each cell is smaller than 5, approximation by the large-sample procedures such as Z-statistic or chi-square test may be inadequate. In this case we can consider the exact method for inference of comparing two proportions. Under the assumption that the marginal frequencies are fixed, there is only one cell frequency in the 2×2 contingency table given in Table 8.3.1, which is allowed to vary. If we allow the number of subjects treated with test drug having a yes response, Y_{11}, to vary, then Y_{11} follows a hypergeometric distribution with the probability for observing frequency y_{11} given as follows:

INFERENCE OF INDEPENDENT SAMPLES

$$P\{Y_{11} = y_{11}\} = \frac{(y_{1.})!(y_{2.})!(y_{.0})!(y_{.1})!}{N!(y_{10})!(y_{20})!(y_{11})!(y_{21})!}. \tag{8.3.16}$$

It follows that the two-tailed *p*-value for the Fisher's exact test is the sum of probabilities over a set of tables with $P(Y_{11})$ less than or equal to the probability calculated from the observed frequency y_{11}.

Note that the only requirement for the assumption of a hypergeometric distribution in (8.3.13) and the expected frequency of Y_{11} in (8.3.10) is randomization of subjects to receive either the test drug or the placebo. As a result the variance of Y_{11} derived from hypergeometric distribution is given as

$$v_{11} = \frac{y_{1.}y_{2.}y_{.0}y_{.1}}{N^2(N-1)}. \tag{8.3.17}$$

The randomization chi-square test statisic is then given by

$$\chi_R^2 = \frac{(y_{11} - m_{11})^2}{v_{11}}. \tag{8.3.18}$$

Koch and Edwards (1988) show that the relationship between the Pearson's and randomization chi-square test statistics can be expressed as

$$\chi_R^2 = \frac{N-1}{N} \chi_P^2. \tag{8.3.19}$$

Example 8.3.2
We continue to use the binary endpoint of early clinical improvement from the NINDS trial for illustration of the statistical testing procedures discussed above. Under the null hypothesis on the equal proportion of subjects with early clinical improvement in both groups, the common proportion is estimated as (122 + 147)/[2(312)] = 43.11% with an estimated large-sample variance of 0.001572. The Z-statistic is then given by $(0.4712 - 0.3910)/\sqrt{0.001572} = 2.0209$ which is greater than $Z(0.025) = 1.96$. The corresponding two-tailed *p*-value is 0.043, and we conclude that the proportion of the subjects treated with rt-PA showing an early clinical improvement is statistically significantly greater than that given placebo. Since the sample size of both groups is the same (312), the expected frequency of an early improvement is the same for both groups, which equals (269)(312)/624 = 134.5. Similarly the expected frequency of no improvement is 177.5 for both groups. The sum of all four expected frequencies is equal to the total sample size of 624. The test statistic χ_P^2 without the continuity correction, computed according to (8.3.11), is 4.084 which is greater than $\chi^2(0.05, 1) = 3.84$. The corresponding *p*-value is 0043. Also randomization of the chi-square gives

$$\chi_R^2 = (623/624)4.084 = 4.077$$

with a *p*-value of 0.043. However, the observed value of χ_{PC}^2 with continuity correction from (8.3.12) is 3.764 which is less than 3.84. The corresponding two-tailed *p*-value is 0.052. The Fisher's exact test also gives a two-tailed *p*-value of 0.052. Consequently, according to the chi-square test with the continuity correction or the Fisher's exact test, we fail to reject the null hypothesis of equal proportions of an early improvement for both groups. Therefore this numerical example provides an example in which the choice of continuity correction or exact test can reach different conclusion from those by large-sample approximation methods in rejection of the null hypothesis.

8.4 ORDERED CATEGORICAL DATA

In clinical trials it is not uncommon to have more discrete efficacy and safety endpoints with more than two categories. For example, although an adverse event can be classified into dichotomous groups, such as serious or nonserious adverse event, as a binary response, the intensity of an adverse event is evaluated according to the categories *mild, moderate,* or *severe*. Similar examples are seen in laboratory safety assessments. For example, we can assess the safety of a drug based on a particular laboratory parameter such as aspartate transaminase (AST) or alanine transaminase (ALT). Using the observed value of this particular parameter, subjects are usually classified into three categories: *below, within, and above* the referenced laboratory normal range. For another example, each individual symptom score in the composite AUA symptom index for benign prostatic hyperplasia (BPH) presented in Table 8.1.3 consists of six categories. These six categories evaluate individual symptoms in an increasing severity, from *not at all* as the first category to *almost always* as the last category. Although severity of a symptom is actually a continuous variable, it is, however, extremely difficult or impractical to measure the severity as a continuous variable objectively. Therefore it is an unobserved latent variable. In practice, the continuous spectrum of severity for a symptom is divided into several ordinal categories to objectively evaluate the symptom. This type of categorical data is called the *ordered categorical data*. Tables 8.1.5 and 8.1.6 provide other examples of ordered categorical data in the areas of diagnostic imaging and respiratory disorder.

The data structure of polychotomous categorical data from a parallel two-group trial is provided in Table 8.4.1, where Y_{ij} represents the number of subjects in the *j*th category for the *i*th group, $n_i = Y_{i\cdot}, j = 1, \ldots, J, i = 1, 2$. It is then of interest to evaluate whether the distribution of the subjects across categories is the same for the test drug and the placebo. For example, if the test drug is effective in treatment of subjects with benign prostatic hyperplasia, then we would expect a shift in distribution such that more subjects are in the

ORDERED CATEGORICAL DATA

Table 8.4.1 Data Structure of Polychotomous Categorical Endpoints for a Parallel Two-Group Trial

Treatment	Category				Total
	1	2	...	J	
Test drug	$Y_{11}(P_{11})$	$Y_{12}(P_{12})$	...	$Y_{1J}(P_{1J})$	$Y_{1.}(1)$
Placebo	$Y_{21}(P_{21})$	$Y_{22}(P_{22})$	...	$Y_{2J}(P_{2J})$	$Y_{2.}(1)$
Total	$Y_{.1}$	$Y_{.2}$	...	$Y_{.J}$	N

categories *not at all*, *less than 1 times in 5*, or *less than half the time*. As a consequence of randomization (nonadaptiveness) of subjects to treatments, the null hypothesis of interest can be formulated as follows:

> Ho: The distribution of subjects in response categories
> is the same as for both groups.   (8.4.1)

This hypothesis in fact tests whether there is relationship between the treatment and response categories. Under the null hypothesis and a randomization structure, Y_{ij} follow a hypergeometric distribution

$$P\{Y_{ij}\} = \frac{\prod_{i=1}^{2}(Y_{i.})! \prod_{j=1}^{J}(Y_{.j})!}{N! \prod_{i=1}^{2} \prod_{j=1}^{J}(Y_{ij})!}, \quad j = 1, \ldots, J, i = 1, 2. \quad (8.4.2)$$

Similarly to the 2 × 2 table in (8.3.13), the expected frequencies of Y_{ij} is the product of marginal frequencies in the *j*th category and the sample size of the *i*th group divided by the total sample size, namely

$$m_{ij} = \frac{y_{i.}y_{.j}}{N}, \quad i = 1, 2, j = 1, \ldots, J. \quad (8.4.3)$$

Substitution of y_{ij} and m_{ij} into (8.3.14) and summation over all 2J cells gives the Pearson's chi-square test statistic, which has a χ^2 distribution with $J - 1$ degrees of freedom if sample size of each group is moderate (>30) and the expected frequency in each cell is at least 5. To obtain the randomization chi-square statistic, we need to know the covariance between Y_{ij} and $Y_{i'j'}$, which is given by

$$\text{cov}(Y_{ij}, Y_{i'j'}) = \frac{m_{ij}(Nd_{ii'} - y_{i.})(Nd_{jj'} - y_{.j'})}{N(N-1)}, \quad (8.4.4)$$

where $d_{ii'} = 1$, if $i = i'$ and $d_{ii'} = 0$, if $i \neq i'$; and $d_{jj'} = 1$, if $j = j'$; and $d_{jj'} = 0$, if $j \neq j'$.

Let **y**, **m**, and **V** denote the vectors of the observed and expected frequencies and the covariance matrix of **y**, respectively. Also let $\mathbf{A} = [\mathbf{I}_{J-1}, \mathbf{0}_{(J-1)\times 2J}]$, where $\mathbf{I}_{J-1}$, is a $(J-1) \times (J-1)$ identity matrix and $\mathbf{0}_{(J-1)\times 2J}$ is a $(J-1) \times 2J$ matrix of 0's. Koch et al. (1982) show that the randomization chi-square can be expressed as

$$\chi_R^2 = (\mathbf{y} - \mathbf{m})'\mathbf{A}'(\mathbf{AVA}')^{-1}\mathbf{A}(\mathbf{y} - \mathbf{m})$$

$$= \frac{N-1}{N}\chi_P^2. \tag{8.4.5}$$

Both the Pearson's and randomization chi-square tests are useful for detection of the existence of general association between treatment and categorical response in either the nominal or ordinal scale. However, they cannot identify a particular relationship. Suppose that the categorical data are ordinal, one might want to see whether there is a difference in the central tendency of the distributions between two treatments. In other words, we may want to detect a location shift in response categories. Let $\mathbf{a} = (a_1, a_2, \ldots, a_J)$ be a set of scores to reflect the ordinal nature of response categories. Then the mean score for the ith group can be computed in the usual manner as we calculate the sample mean for a frequency table:

$$c_i = \sum_{j=1}^{J} \frac{a_j y_{ij}}{y_{i.}}, \quad i = 1, 2. \tag{8.4.6}$$

Because there are only two treatment groups and all marginal frequencies are assumed to be fixed, the mean score for only one treatment group is random. Hence we can only consider the mean score for the test drug. Koch and Edwards (1988) give the expected value and variance of c_1, respectively, as

$$c = \sum_{j=1}^{J} \frac{a_j y_{.j}}{N}$$

and

$$v(c_1) = \frac{(N - y_{i.})v_a}{(N-1)y_{1.}}, \tag{8.4.7}$$

where

$$v_a = \sum_{j=1}^{J} \frac{(a_j - c)^2 y_{.j}}{N}, \quad i = 1, 2.$$

From (8.4.6) and (8.4.7), the Z-statistic can be constructed to test whether there is a location shift in distribution between the two groups:

$$Z = \frac{c_1 - c}{\sqrt{v(c_1)}},$$

or equivalently

$$X = Z^2 = \frac{(c_1 - c)^2}{v(c_1)}. \tag{8.4.8}$$

If the sample size is fairly large, X approximately follows a chi-square variable with one degree of freedom. As a result we reject the null hypothesis of no location shift between two distributions at the α level of significance if $X > \chi^2(\alpha, 1)$.

The large-sample randomization test statistic for location shift can be expressed as

$$X = \frac{K(c_1 - c_2)^2}{v_a((1/y_{1.}) + (1/y_{2.}))}, \tag{8.4.9}$$

where $K = (N - 1)/N$ is the finite population correction factor.

Note that except for the finite population correction factor, X is in fact a test statistic for the detection of mean difference between the two independent samples. As a result, as in the relationship between t statistic and F statistic for continuous endpoints, the test statistics X for two independent samples given in (8.4.9) can be easily extended to test overall location shift for I independent samples, $I > 2$. The resulting test statistic is the one-way analysis of variance based on scores $\mathbf{a} = (a_1, a_2, \ldots, a_J)$ which is given as

$$X = (N - 1) \sum_{j=1}^{J} \frac{y_{i.}(c_i - c)^2}{v_a}, \tag{8.4.10}$$

where c_i and v_a are similarly defined as in (8.4.6). Note that except for the constant $(N-1)$, the numerator in (8.4.10) is the between-group sum of squares, and Nv_a represents the total sum of squares. If the sample size in each treatment group is at least moderate, X approximately follows a chi-square variable with $I - 1$ degrees of freedom.

The choice of ordinal scores affects the statistical analysis and its clinical interpretation, which usually depends on the question being asked. For example, the last question in the AUA-7 symptom index for BPH is on *how many times a subject gets up during the night to urinate*. The symptom score for this question is a discrete count such as integers 0, 1, 2, ..., and J. The scores of this kind are called integer scores. If categorical clinical endpoints represent some discrete counts or the classes are equally spaced, one might consider the use of integer scores. Alternative scores suggested by Koch and Edwards (1988) are standardized midrank scores and logrank scores. Standardized midrank scores are the expected values of order statistics of uniform distribution. Koch and Bhapkar (1982) showed that the resulting test statistic is equivalent to the Wilcoxon rank sum test (or Kruskal-Wallis test for $I > 2$) without scaling the categories. Standardized midrank scores are also referred to as modified ridit scores (van Elteren, 1960; Lehamnn, 1975). They can be assessed by means of PROC FREQ of SAS® with the statement MIDRIDIT in the option SCORES. Alternatively, one can use logrank scores or Savage scores, which are the expected values of order statistics from an exponential distribution. These scores are useful when the interest is to detect treatment difference for a distribution of the data that is L-shaped. The logrank scores can be generated by option statement of SAVAGE in PROC NPAR1WAY of SAS®.

Note that test statistic X for detecting a location shift has fewer degrees of freedom that the randomization test, χ_R^2 or Pearson's chi-squares test, χ_P^2 for the general association. In general, X is a more powerful test than either χ_R^2 or χ_P^2. The approximation of χ_R^2 or χ_P^2 by chi-square distribution depends on cell frequencies y_{ij}. Basically the test statistics for the detection of a location shift can be considered a linear rank statistics that is a linear combination of scores. Hence their large-sample approximation depends on a linear combination of y_{ij}. Therefore the sample size requirement for X for detecting a location shift is less stringent than that of χ_R^2 or χ_P^2.

Example 8.4.1

As indicated by Barry et al. (1992), the AUA symptom index should be able to discriminate the subjects from BPH from those without BPH. The data given in Table 8.1.4 summarize the results of the initial question 4 for urination frequency in two hour since the last urination in a validation study from the subjects with BPH and controls for the AUA symptom index. Pearson's chi-square is 16.874 with a p-value of 0.0047 obtained from a chi-square random variable with five degrees of freedom. The randomization chi-square statistic for general association can then be obtained by the relationship

$$\chi_R^2 = \frac{N-1}{N} \chi_P^2$$

$$= (128/129)16.874$$

$$= 16.743$$

with a corresponding p-value of 0.005. Both Pearson's and the randomization chi-square tests reject the null hypothesis at the 5% level of significance, that the response distribution with respect to urination frequency is the same as for the

Table 8.4.2 Summary of Analysis of Validation for Question 4–Frequency No. 1 in AUA Symptom Index

Category	BPH	Control	X_P^2	X_R^2	X
N	73	56			
Not at all	8 (10.96%)	16 (28.57%)	16.87	16.74	14.9
Less than 1 in 5 times	18 (24.66%)	23 (41.07%)	(0.0047)	(0.0050)	(0.0001)
Less than half the time	17 (23.29%)	7 (12.50%)			
About half the time	14 (19.18%)	7 (12.50%)			
More than half the time	8 (10.96%)	2 (3.57%)			
Almost always	8 (10.96%)	1 (1.79%)			

subjects with BPH and normal controls. However, this general association does not answer the question of whether urination frequency for the subjects with BPH is higher than normal controls. From Table 8.4.2, 41.1% of the subjects with BPH had to urinate in less than two hours since the last urination about half the time, more than half the time, or almost always, while the proportion is only 17.86% for controls. Therefore there appears to be a location shift in the distributions of categories between the two groups. Since the AUA symptom index suggests the use of integers from 0 to 5 for each of seven symptom scores, we computed the test statistic with integers scores 0, 1, 2, 3, 4, 5 for detection of location shift which turns out to be 14.90 with a p-value of 0.0001. This result confirms our suspicion that the subjects with BHP tend to urinate more frequently in two hours of the last urination than controls.

8.5 COMBINING CATEGORICAL DATA

In multicenter clinical trials the studies are conducted at different centers at which randomization schedules are independently generated. In addition subjects may be stratified based on some demographic and/or patient baseline characteristics such as gender, race, or the severity of disease. The NINDS trial is a typical clinical trial of this kind. In addition the NINDS trial is divided into two parts with additional stratification with respect to the time from the onset of the stroke to the start of treatment. These two stratification factors are (1) prerandomization during which time the randomization was performed independently and separately and (2) the time from onset to treatment. In practice, the treatment unbalance would occur by chance with respect to some important covariates that are not used as stratified factors in the randomization process. These covariates are usually called the postrandomization stratified factors. As a result we can compare the differences between the test drug and the placebo

Table 8.5.1 Data Structure of Binary Endpoint with H Strata for a Parallel Two-Group Trial

Treatment	Binary Response		Total
	No	Yes	
Test drug	$Y_{h10}(P_{h10})$	$Y_{h11}(P_{h11})$	$Y_{h1.}(1)$
Placebo	$Y_{h20}(P_{h20})$	$Y_{h21}(P_{h21})$	$Y_{h2.}(1)$
Total	$Y_{h.0}$	$Y_{h.1}$	$Y_{h..}$

Note: $h = 1, \ldots, H$.

by combining the results from a set of strata with an appropriate adjustment of the effects caused by the covariates.

Table 8.5.1 illustrates binary response data obtained from an H strata for comparing a test drug with a concurrent placebo group. If all marginal frequencies are considered fixed, similar to (8.3.13) and (8.3.17), the expected frequency and variance for the hth strata are given by, respectively,

$$m_{h11} = \frac{y_{h1.} y_{h.1}}{N_h},$$

and

$$v_h = \frac{y_{h1.} y_{h2.} y_{h.0} y_{h.1}}{N_h^2 (N_h - 1)}, \quad h = 1, \ldots, H. \quad (8.5.1)$$

Mantel and Haenszel (1959) propose the well-known Mantel-Haenszel (MH) statistic to combine the results of the difference in proportions of a yes response from H different strata. The calculation of the MH statistic is illustrated in Table 8.5.2. After the expected frequency and variance of Y_{11} are calculated for each stratum, the sum of the differences between the observed and expected frequencies and the sum of variances over all strata can be obtained. The MH statistic is the ratio of the square of the sum of the differences to the sum of variances:

$$X_{\text{MH}} = \frac{\left[\sum_{h=1}^{H} (y_{h11} - m_{h11})\right]^2}{\sum_{h=1}^{H} v_h}. \quad (8.5.2)$$

When the sum of sample size over all strata is sufficiently large, then X_{MH} approximately follows a chi-square distribution with one degree of freedom. Consequently we reject the null hypothesis of no difference between the test drug and the placebo after adjustment of covariates at the α level of significance if

COMBINING CATEGORICAL DATA   341

Table 8.5.2 Summary of Results of Binary Endpoint with H Strata

Strata	Observed Frequency	Expected Frequency	Difference	Variance
1	y_{111}	$m_{111} = y_{11.}y_{1.1}/N_1$	$y_{111} - m_{111}$	v_1
⋮				
h	y_{h11}	$m_{h11} = y_{h1.}y_{h.1}/N_h$	$y_{h11} - m_{h11}$	v_h
⋮				
H	y_{H11}	$m_{H11} = y_{H1.}y_{H.1}/N_H$	$y_{H11} - m_{H11}$	v_H
Sum	$\sum y_{h11}$	$\sum m_{h11}$	$\sum(y_{h11} - m_{h11})$	$\sum v_h$

Note: $v_h = [y_{h1.}y_{h2.}y_{h.0}y_{h.1}]/[N_h^2(N_h - 1)],\ h = 1, \ldots, H.$

$$X_{\text{MH}} > \chi^2(\alpha, 1).$$

The numerator in the MH statistic is the square of the sum of the differences between the observed and expected frequencies. Therefore the MH statistic would be more powerful if the association between treatment and response were in the same direction over all the strata. On the other hand, if the difference between the test drug and the placebo is not homogeneous across strata, the differences in (8.5.2) will cancel each other out, and resulting sum will be small. Therefore the MH statistic is not powerful in the presence of heterogeneous treatment across strata in the same direction. A test statistic for the detection of variation of treatment effects across the strata is

$$X_V = X_T - X_{\text{MH}}, \tag{8.5.3}$$

where

$$X_T = \frac{1}{v_h} \sum_{h=1}^{H} (y_{h11} - m_{h11})^2. \tag{8.5.4}$$

One can test the null hypothesis for the absence of heterogeneity in the treatment effect by comparing the observed X_V with a chi-square random variable with $H - 1$ degrees of freedom. If there is no evidence for the presence of variation of treatment differences across strata, Yusuf et al. (1985) suggests an estimate for the common odds ratio of a yes response as

$$OR_C = \exp\left\{ \frac{\sum_{h=1}^{H}(y_{h11} - m_{h11})}{\sum_{h=1}^{H} v_h} \right\}, \tag{8.5.5}$$

Table 8.5.3 Comparison Between rt-PA and a Placebo Effect in Subjects Showing Clinical Improvement for Adjustment of Part and Time from Onset to Treatment in the NINDS Trial

Part	Time	Observed Frequency	Expected Frequency	$O_h - E_h$	v_h	$(O_h - E_h)^2/v_h$
1	0–90	36	34.2230	1.7770	8.7351	0.3615
1	91–180	31	27.3750	3.6250	8.9513	1.4680
2	0–90	51	42.7362	8.2638	10.2188	6.66829
2	91–180	29	30.8706	−1.8706	10.0230	0.3491
Sum		147	135.2048	11.7952	37.9281	8.8615

Note: Observed and expected frequencies are for those subjects showing an improvement in rt-PA group.

with the corresponding lower and upper limits of a $(1 - \alpha)100\%$ confidence interval for the population common odds ratio as follows:

$$\exp\left\{\frac{\sum_{h=1}^{H}(y_{h11} - m_{h11})}{\sum_{h=1}^{H} v_h} \pm \frac{z(\alpha/2)}{\sqrt{\sum_{h=1}^{H} v_h}}\right\}. \quad (8.5.6)$$

Example 8.5.1

To illustrate the MH method for combining treatment effects, we use the data from the NINDS trial. Since the trial has two parts and the time was stratified into two intervals: 0 to 90 and 91 to 180 minutes, there are four 2 × 2 tables, one for each strata. The frequency used for calculation of the MH test is the number of subjects treated with rt-PA who showed clinical improvement. Table 8.5.3 gives the observed and expected frequencies, the difference, variance, and the square of the difference divided by variance. The sum of the differences between the observed and expected frequencies is 11.7952 and the sum of variances is 37.9281. Thus, by (8.5.2), the MH statistic is given by

$$X_{MH} = \frac{(11.7952)^2}{37.9281} = 3.6682$$

with a *p*-value of 0.055. Therefore we fail to reject the null hypothesis of no treatment effect at the 5% significance level. From Table 8.5.3, X_T is given by 8.8615. As a result

$$X_V = 8.8615 - 3.6682 = 5.1922$$

which is smaller than $\chi^2(0.05, 3) = 7.81$. Therefore there is no strong evidence to doubt the presence of significant variation of treatment effects across strata.

However, from the fifth column in Table 8.5.3, there seems to be some heterogeneity of differences among strata. The first three strata indicates that odds of clinical improvement for subject treated with rt-PA is greater than those given placebo, while the strata for the combination of part 2 and time from 91 to 180 minutes gave the result in the opposite direction. However, only the stratum with the combination of part 2 and time from 0 to 90 minutes provided a statistical significant treatment effect at the 5% level of significance. The common odds ratio is then estimated by

$$\exp\left(\frac{11.7952}{37.9281}\right) = \exp(0.3110) = 1.3648$$

with the corresponding 95% confidence interval

$$\exp\left(0.3110 \pm \frac{1.96}{\sqrt{37.9281}}\right) = (0.9928, 1.8761).$$

Note that the 95% confidence interval contains 1. This result is consistent with that from hypothesis testing that no statistically significant treatment effect was detected at the 5% level of significance.

Mantel (1963) showed that similar techniques can be extended to combine the results of discrete endpoints with a total of J ordered categories from H strata in clinical trials comparing a test drug with a placebo group. Let Y_{hij} represent the number of subjects in the jth category for the ith group from the hth stratum, $n_i = Y_{i.}, j = 1, \ldots, J, i = 1, 2, h = 1, \ldots, H$. Since subjects at different strata are different, any clinical endpoints observed from subjects at different strata are statistically independent. Therefore the probability of Y_{hij} follows a product hypergeometric distribution given by

$$P\{Y_{hij}\} = \prod_{h=1}^{H} \left\{ \frac{\prod_{i=1}^{2} (Y_{hi.})! \prod_{j=1}^{J} (Y_{h.j})!}{(N_h)! \prod_{i=1}^{2} \prod_{j=1}^{J} (Y_{hij})!} \right\}. \qquad (8.5.7)$$

Let c_{h1}, c_h, and $v(c_{h1})$ be the mean score, expected value, and variance for the hth strata as defined in (8.4.6) and (8.4.7), respectively. Then the extended Mantel-Haenszel statistic for detection of a location shift in distribution between the test drug and the placebo across strata is given by

$$X_{\text{EMH}} = \frac{\left[\sum_{h=1}^{H} (c_{h1} - c_h)\right]^2}{\sum_{h=1}^{H} v(c_{h1})}. \qquad (8.5.8)$$

One can reject the null hypothesis of no location shift after adjustment for covariates at the α level of significance if

$$X_{\text{EMH}} > \chi^2(\alpha, 1).$$

If one chooses to use standard midrank scores, then X_{EMH} is the same as the test procedure proposed by van Elteren (1960) for combining Wilcoxon rank sum statistics over a set strata, which is also referred to as the blocked Wilcoxon rank sum statistic (Lehmann, 1975). Extension to more than two treatment groups and a test for homogeneity of location shifts across strata and related issues can be found in Koch and Edwards (1988).

Similar to the Mantel-Haenszel estimator for the average odds ratio for the stratified 2×2 table given in (8.5.5), Davis and Chung (1995) suggest an MH estimator, originally proposed by Mantel (1963), to estimate the common average treatment effect across strata. Let c_{hik} be the resulting observation of the kth subject in the ith treatment from the hth strata after application of scores, $a_1, a_2, \ldots, a_J$ to each subject. Also let c_{hi} and s_{hi}^2 be the mean score and sample variance for treatment i and stratum h. Then we have

$$c_{hi} = \frac{1}{y_{hi.}} \sum_{k=1}^{y_{hi.}} c_{hik},$$

and

$$s_{hi}^2 = \frac{1}{y_{hi.} - 1} \sum_{k=1}^{y_{hi.}} (c_{hik} - c_{hi})^2. \quad (8.5.9)$$

The MH estimator of the average treatment effect proposed by Davis and Chung (1995) is given as

$$\hat{\theta}_{\text{MH}} = \frac{\sum_{h=1}^{H} w_h (c_{h1} - c_{h2})}{\sum_{h=1}^{H} w_h}, \quad (8.5.10)$$

where

$$w_h = \frac{y_{h1.} y_{h2.}}{N_h}, \quad h = 1, \ldots, H.$$

Therefore $\hat{\theta}_{\text{MH}}$ is a weighted average of the stratum-specific treatment differences with weight w_h. Davis and Chung (1995) also propose the following estimate for the variance of $\hat{\theta}_{\text{MH}}$:

$$\hat{v}(\hat{\theta}_{MH}) = \frac{1}{\left[\sum_{h=1}^{H} w_h\right]^2} \left\{ \sum_{h=1}^{H} w_h^2 \left[\frac{s_{h1}^2}{y_{h1}} + \frac{s_{h2}^2}{y_{h2}} \right] \right\}. \quad (8.5.11)$$

They indicate that $\hat{\theta}_{MH}$ is equivalent to that derived under the fixed-effects analysis of variance estimator from the main effects model. However, $\hat{\theta}_{MH}$ and $\hat{v}(\hat{\theta}_{MH})$ are obtained only under the assumption of randomization of subjects to the treatment assignment without the assumptions of normality, homoscedasticity, and additivity of effects. In addition their statistical properties are nearly as good as the optimal linear models under normality assumption.

Example 8.5.2

To demonstrate the methods for combining ordered categorical data from a set of strata, we use the data of change in diagnostic confidence in Table 8.1.5, which were obtained from a clinical trial for comparing a new contrast agent with a concurrent control. The stratum for this dataset is the gender of the subjects. We use integer score from 1 to 6 to denote categories from *worsened to excellent* in our computation. Readers can verify that the X_{EMH} is equal to 40.204 with a p-value less than 10^{-7}. Therefore at the 5% level of significance, there is a statistically significant location shift in the distribution of subjects with respect to change in diagnosis between the new contrast agent and the placebo. Table 8.5.4 provides some results for the MH estimate of the average treatment difference and its estimated large-sample variance. From Table 8.5.4,

Table 8.5.4 Estimation of Average Treatment Difference Based on the Change in Diagnostic Confidence Given in Table 8.1.6

	Stratum	
	Female	Male
Mean Score		
Test	5.13889	5.05405
Placebo	3.32432	3.33333
Difference (*D*)	1.81457	1.72072
Weight (*W*)	18.2466	18.2466
W ∗ *D*	33.1096	31.3973
Variance		
Test	1.49444	1.60811
Placebo	2.66976	2.51429
W^2 sum of variance/y_{hi}	37.8475	37.7230

$$\hat{\theta}_{MH} = \frac{33.1096 + 31.3973}{(2)(18.2466)}$$

$$= 1.7676$$

and

$$\hat{v}(\hat{\theta}_{MH}) = \frac{37.8475 + 37.7230}{(2*18.2466)^2}$$

$$= 0.05674.$$

Note that X_{EMH} can be easily obtained by using PROC FREQ of SAS® with the option statement CMH. One can also perform computation of $\hat{\theta}_{MH}$ and $\hat{v}(\hat{\theta}_{MH})$ in DATA statement of SAS®.

8.6 MODEL-BASED METHODS

In the previous sections we introduced several statistical methods for analysis of categorical data. These methods only require random assignment of subjects to treatments. As a result valid statistical inference can always be obtained from these methods for randomized clinical trials. However, since these methods are randomization-based methods that are for hypothesis testing rather than estimation, they cannot describe the relationship between the categorical response and covariates such as demographical variables, patient baseline characteristics, or stratification factors. In this section we will introduce model-based procedures as alternatives to the randomization methods for analysis of categorical data.

One of the most commonly employed model-based methods for the analysis of categorical data is logistic regression (Agresti, 1990; Hosmer and Lemeshow, 1989). The method of logistic regression is useful because that (1) it can incorporate discrete covariates such as gender as well as continuous explanatory variables such as age, (2) it provides a more powerful inference for treatment effect through the reduction of variability caused by the covariates, (3) it has flexibility that allows for an adjustment of covariates, and (5) it enables the investigation of possible interactions between covariates. Thus the method of logistic regression is usually employed to provide additional information regarding estimation, relationship between response and covariates, and homogeneity of treatment effect across different levels of covariates.

To introduce the concept of logistic regression, consider the NINDS trial. Let Y_{ghi} be the number of subjects out of n_{ghi} subjects, with time interval h to the start of treatment, who were randomly assigned to receive the ith treatment and had an early clinical improvement in study part g. Also let P_{ghi} be the corresponding probability, $g = 1, 2, h = 1, 2,$ and $i = 1, 2$. If the subjects in each stratum formed by the complete cross-classification of part, time interval,

MODEL-BASED METHODS

and treatment are randomly selected independently from the corresponding targeted population, then Y_{ghi} can be described by the following product binomial distribution:

$$P\{Y_{ghi}\} = \prod_{g=1}^{2} \prod_{h=1}^{2} \prod_{i=1}^{2} \left\{ \frac{(n_{ghi})!}{(Y_{ghi})!(n_{ghi} - Y_{ghi})!} \right\} P_{ghi}^{Y_{ghi}} (1 - P_{ghi})^{(n_{ghi} - Y_{ghi})}.$$

Define the logit of P_{ghi} as

$$\text{logit}(P_{ghi}) = \ln\left\{\frac{P_{ghi}}{1 - P_{ghi}}\right\}. \qquad (8.6.1)$$

The logit of P_{ghi} is therefore the logarithm of the odds of clinical improvement to no improvement for a subject with the time interval h who was randomized to receive treatment i in study part g. Logistic regression can then be used to investigate the relationship between the probability of improvement and a set of covariates by assuming that the logit of P_{ghi} is linearly related to the covariates,

$$\ln\left\{\frac{P_{ghi}}{1 - P_{ghi}}\right\} = \alpha + \mathbf{x}'_{ghi}\boldsymbol{\beta}, \qquad (8.6.2)$$

where α is the intercept and $\boldsymbol{\beta}$ is a vector of unknown regression coefficients corresponding to the row vector of $\mathbf{x}'_{ghi}$ representing study part g, time interval h, and treatment i. The expression in (8.6.2) is usually referred to as the logistic regression because the probability of improvement can be expressed as a function of explanatory variables by the logistic distribution as

$$P_{ghi}(\alpha + \mathbf{x}'_{ghi}\boldsymbol{\beta}) = \{1 + \exp[-(\alpha + \mathbf{x}'_{ghi}\boldsymbol{\beta})]\}^{-1}. \qquad (8.6.3)$$

Although the logistic regression assumes a linear relationship between the logit of P_{ghi} and covariates, the equation is in fact a nonlinear function. To obtain estimates of the intercept α and regression coefficients $\boldsymbol{\beta}$, the method of maximum likelihood is usually considered. The maximum likelihood estimates of α and $\boldsymbol{\beta}$ can be obtained by first substituting the probability of clinical improvement P_{ghi} in (8.6.1) by the right-hand side of (8.6.3). Then, differentiating the logarithm of the resulting equation with respect to α and $\boldsymbol{\beta}$, we can solve the resulting normal equations for α and $\boldsymbol{\beta}$. It should be noted that in general, there exist no closed forms for α and $\boldsymbol{\beta}$. Thus a variety of methods for numerical optimization are usually applied to find the maximum likelihood estimators (MLE) of α and $\boldsymbol{\beta}$. These methods usually involve the technique of iterative reweighted least squares (IRLS) such as the Newton-Raphson algorithm or the method of scoring which is available in most commercial statistical computer software packages such as SAS, BMDP, and GLIM. For further details, see Cox and Snell (1989), Agresti (1990), and Hosmer and Lemeshow (1989).

Let a and b denote the MLEs for α and β. Basically a and b are consistent estimates for α and β, the unknown intercept and regression coefficients, in the sense that a and b will be very close to α and β within a negligible distance when the sample size is sufficiently large. The joint distribution of a and b can be approximated by a multivariate normal distribution with mean vector $(\alpha, \beta)'$ and covariance matrix $\mathbf{V}$. A consistent estimate of covariance matrix of a and b can be obtained as the following inverse of the observed information matrix given by Koch and Edwards (1988):

$$v(\alpha, \beta) = \left\{ \sum_{g=1}^{2} \sum_{h=1}^{2} \sum_{i=1}^{2} n_{ghi} P_{ghi} (1 - P_{ghi})(1, \mathbf{x}_{ghi})(1, \mathbf{x}_{ghi})' \right\}^{-1}. \quad (8.6.4)$$

After the MLEs a and b are obtained, the predicted probability of clinical improvement, p_{ghi}, can be obtained by replacing the unknown parameters α and β in (8.6.3) by their MLEs a and b. Hence the number of subjects with an early clinical improvement predicted by model (8.6.2) for the stratum formed by the combination of study part g, time interval h, and treatment i is given by

$$m_{ghi} = n_{ghi} p_{ghi}. \quad (8.6.5)$$

Koch and Edwards (1988) suggest using goodness-of-fit for model (8.6.2) by way of the Pearson's test or the log-likelihood chi-square test, which are given by, respectively,

$$\chi_P^2 = \sum_{g=1}^{2} \sum_{h=1}^{2} \sum_{i=1}^{2} \left\{ \frac{(y_{ghi} - m_{ghi})^2}{m_{ghi}} + \frac{(n_{ghi} - y_{ghi})^2}{n_{ghi} - m_{ghi}} \right\}, \quad (8.6.6)$$

and

$$\chi_L^2 = \sum_{g=1}^{2} \sum_{h=1}^{2} \sum_{i=1}^{2} 2 \left\{ y_{ghi} \ln\left(\frac{y_{ghi}}{m_{ghi}}\right) + (n_{ghi} - y_{ghi}) \ln\left(\frac{n_{ghi} - y_{ghi}}{n_{ghi} - m_{ghi}}\right) \right\},$$

$$(8.6.7)$$

where $y_{ghi} \ln(y_{ghi}/m_{ghi})$ and $(n_{ghi} - y_{ghi}) \ln((n_{ghi} - y_{ghi})/(n_{ghi} - m_{ghi}))$ are defined to be 0 if $y_{ghij} = 0$ or $n_{ghi} - y_{ghi} = 0$. We conclude that a significant lack-of-fit exists for the assumed model at the α level of significance if χ_P^2 or χ_L^2 is greater

MODEL-BASED METHODS

than the upper αth quantile of a central chi-square distribution with $GHI-(s+1)$ degrees of freedom, where G, H, and I are the number of the level for each of the three covariates and s is the number of independent explanatory variables in the model. For the data of clinical improvement in Table 8.1.2, $G = H = I = 2$ because there are two study parts, two treatments (rt-PA and the placebo), and two time intervals from the onset of stroke to the start of treatment (0–90 and 91–190 minutes).

Logistic regression is a typical example that relates the random occurrence of clinical endpoints to the systematic components formed by a set of covariates through a link function (which is referred to as logit link). Other useful link functions include (1) the identity link and log link for the Poisson variable, (2) the probit link based on the normality assumption, and (3) the complementary log-link from the extreme-value distribution. Statistical inference based on other link functions can be similarly derived. For details regarding other link functions and their applications in clinical trials, see McCullagh and Nelder (1989).

Example 8.6.1

To illustrate the application of logistic regression, we again consider the data from the NINDS trial given in Table 8.1.2. In this trial there are two study parts, 1 ($g = 1$) and 2 ($g = 2$), two time intervals for the onset of stroke to the start of treatment, 0 to 90 ($h = 1$) and 91 to 180 minutes ($h = 2$), and 2 treatments for this trial, placebo ($i = 1$) and rt-PA($i = 2$). Thus there are eight logits. Based on these logits, we can fit a main effects model with terms of *part*, *time interval*, and *treatment*. Since each covariate has two levels, for simplicity, the lower level and higher level are coded as 0 and 1, respectively, to obtain the design matrix of the main effects model. Table 8.6.1 gives the design matrix for the main effects model. Table 8.6.2 summarizes the estimates of

Table 8.6.1 Design Matrix for Main Effects in the Data Set of the NINDS Trial in Table 8.1.2

	Row Time				Column	
Part	Interval	Treatment	Intercept	Part	Time Interval	Treatment
1	0–90	Placebo	1	0	0	0
1	0–90	rt-PA	1	0	0	1
1	91–180	Placebo	1	0	1	0
1	91–180	rt-PA	1	0	1	1
2	0–90	Placebo	1	1	0	0
2	0–90	rt-PA	1	1	0	1
2	91–180	Placebo	1	1	1	0
2	91–180	rt-PA	1	1	1	1

Table 8.6.2 Regression Parameters and Their Estimated Standard Errors for the Data Set of the NINDS Trial in Table 8.1.2

		Model	
Parameter	Statistic	Main Effect Only	With Interaction
Intercept	Estimate (se)	−0.2193 (0.1682)	−0.3459 (0.2282)
	p-value	0.1921	0.1295
Part	Estimate (se)	0.0302 (0.1637)	0.0488 (0.2893)
	p-value	0.8538	0.8659
Time Interval	Estimate (se)	−0.4581 (0.1633)	−0.2085 (0.2905)
	p-value	0.0050	0.4730
Treatment	Estimate (se)	0.3137 (0.1633)	0.5337 (0.2922)
	p-value	0.0548	0.0678
Part*Time interval	Estimate (se)		−0.0454 (0.3280)
	p-value		0.8900
Part*Treatment	Estimate (se)		0.0054 (0.3280)
	p-value		0.9800
Time interval* Treatment	Estimate (se)		−0.4435 (0.3273)
	p-value		0.1754
Model Chi-Square for Covariates			
Loglikelihood			
df		3	6
Chi-square		12.054	13.917
p-value		0.0072	0.0306
Score			
df		3	6
Chi-square		11.987	13.980
p-value		0.0074	0.0299

intercept and regression coefficients, their standard errors, and the corresponding two-tailed p-values. In addition to the intercept Table 8.6.2 gives the contribution of these three covariates. The chi-squares by log-likelihood method and the scoring method are 12.054 and 11.987, respectively. The corresponding p-values calculated based on a chi-square distribution with three degrees of freedom are given by 0.0072 and 0.0074, respectively. The results indicate that there is a significant contribution from the three main effects. Suppose that we are also interested in exploring the potential interaction and homogeneity of the treatment effect across study parts and in exploring the time to the start of treatment. Then we can include additional variables: part*time interaction, part*treatment interaction, and time*treatment interaction. These effects can be characterized by the product of the corresponding columns in the design matrix of the main effects model as given in Table 8.6.1. Note that the last column of Table 8.6.2 provides estimates, their standard errors, and p-values for the model with two-factor interactions. The contributions of the six explanatory variables assessed by both the log-likelihood and scoring methods are statistically significant at the 5% level of significance. However, the contribution of the three two-factor interactions based on the log-likelihood method and the scoring method are given by 1.863 and 1.993, respectively, which are not statistically significant at the 5% level of significance. For the goodness-of-fit of the main effects model, χ_P^2 and χ_L^2 test statistics gave 5.273 and 5.295 with corresponding p-values of 0.260 and 0.258, respectively. In conclusion, the main effects model is an adequate model for describing the relationship between the probability of clinical improvement and study parts, time to the start of treatment, and treatment.

The results in Table 8.6.2 indicate that the probability of an early clinical improvement for a subject with time of onset of stroke to treatment within 90 minutes is statistically significantly larger than that for a subject with time to treatment between 90 and 180 minutes. However, the treatment effect is marginally significant with a p-value of 0.0548. An analysis by strata reveals that subjects with the time to treatment within 90 minutes in the study part 2 had a significant treatment effect. In addition to the observed frequency and probability of clinical improvement, Table 8.6.3 displays the predicted frequency and probability by the fitted main effects model and their standard errors. Note that the standard error of the observed probability is larger than that of the predicted probability because of the reduction of variability induced by the covariates.

The estimated logits and their standard errors and odds of improvement with the corresponding 95% confidence intervals are given in Table 8.6.4. For the situation where the time from the onset of stroke to the start of treatment is greater than 90 minutes, the upper limit of the 95% confidence interval is less than 1. As a result the odds of clinical improvement is statistically significantly less than 1 at the 5% level of significance regardless of which treatment a subject is given for both parts of the study. When the time of onset to the start of treatment is within 90 minutes, the odds of clinical improvement is not statistically significant, since its 95% confidence interval includes 1. Note that the difference in estimated logits between rt-PA and the placebo is the same across the four (part-by-time interval) strata. This difference is exactly the same as the estimate for the regression coefficient corresponding to the treatment given by 0.3137. The common odds ratio of clinical improvement for a subject receiving rt-PA to a subject given the placebo is estimated as $\exp(0.3137) = 1.368$. The corresponding 95% confidence interval can be calculated as $\exp\{0.3137 \pm (1.96)(0.1633)\}$ which gives (0.9937, 1.8847). These results are consistent with those obtained by the Mantel-Haenszel method discussed in Example 8.5.1.

The ordinal categorical data such as the AUA-7 symptom score (Table 8.1.4), change in diagnostic confidence (Table 8.1.5), and status score (Table 8.1.6) have more than two categories in ascending order. One of the possible models for investigating the relationship between probabilities and covariates is the proportional odds model based on the cumulative probabilities (Agresti, 1990). For simplicity, consider a trial conducted for comparing a test drug with a placebo.

Table 8.6.3 Observed Frequencies and Probabilities; Predicted Frequencies and Probabilities of Clinical Improvement for the Data Set of the NINDS Trail in Table 8.1.2

Part	Time Interval	Treatment	Sample Size	Frequency		Probability (%)	
				Observed	Predicted	Observed	Predicted
1	0–90	Placebo	68	31	30.29	45.6% (6.04%)	44.5% (4.15%)
1	0–90	rt-PA	71	36	37.17	50.7% (5.93%)	52.4% (4.13%)
1	91–180	Placebo	79	26	26.91	32.9% (5.29%)	33.7% (3.70%)
1	91–180	rt-PA	73	31	29.93	42.5% (5.79%)	41.0% (4.02%)
2	0–90	Placebo	77	30	34.87	39.0% (5.56%)	45.3% (4.02%)
2	0–90	rt-PA	86	51	45.87	59.3% (5.30%)	53.1% (4.02%)
2	91–180	Placebo	88	35	30.24	39.8% (5.22%)	34.4% (3.64%)
2	91–180	rt-PA	82	29	34.22	35.4% (5.28%)	41.7% (3.90%)

MODEL-BASED METHODS

Table 8.6.4 Estimated Logits and Odds of Clinical Improvement for the Data Set of the NINDS Trial in Table 8.1.2

Part	Time Interval	Treatment	Sample Size	Logit Estimate	Standard Error	Odds of Improvement Estimate	95% CI
1	0–90	Placebo	68	−0.2194	0.1682	0.803	(0.578, 1.117)
1	0–90	rt-PA	71	0.0943	0.1654	1.099	(0.795, 1.520)
1	91–180	Placebo	79	−0.6775	0.1656	0.508	(0.367, 0.703)
1	91–180	rt-PA	73	−0.3638	0.1661	0.695	(0.502, 0.963)
2	0–90	Placebo	77	−0.1892	0.1624	0.828	(0.602, 1.138)
2	0–90	rt-PA	86	0.1245	0.1581	1.133	(0.831, 1.544)
2	91–180	Placebo	88	−0.6473	0.1613	0.523	(0.362, 0.718)
2	91–180	rt-PA	82	−0.3337	0.1606	0.716	(0.523, 0.981)

Let P_{ij} be the probability of observing the jth possible outcome (in a total of J categories of clinical responses) for subjects receiving the ith treatment, $i = 1, 2$. The cumulative logit is defined as the logarithm of the odds of the cumulative probabilities below the jth category which is given by

$$\text{logit}(P_{i1} + \ldots + P_{ij}) = \ln\left\{\frac{P_{i1} + \ldots + P_{ij}}{1 - (P_{i1} + \ldots + P_{ij})}\right\}, \tag{8.6.8}$$

where $j = 1, \ldots, J - 1$. A reasonable model is to assume a linear regression with different intercepts for each of the $J - 1$ cumulative logits and a common slope as follows:

$$\text{logit}(P_{i1} + \ldots + P_{ij}) = \alpha_j + \beta x, \tag{8.6.9}$$

where $x = 1$ if the treatment is the test drug and $x = 0$ if the treatment is the placebo and $j = 1, \ldots, J - 1$. This model assumes that the treatment effect on the odds of response below the jth category is the same for all j. In other words,

$$\text{logit}(P_{11} + \ldots + P_{1j}) - \text{logit}(P_{01} + \ldots + P_{0j})$$

$$= \ln\left\{\frac{P(Y \leq j|i=1)/P(Y > j|i=1)}{P(Y \leq j|i=0)/P(Y > j|i=0)}\right\}$$

$$= \alpha_j + \beta - \alpha_j$$

$$= \beta. \tag{8.6.10}$$

The first expression within the logarithm on the left-hand side of equation

(8.6.10) is usually referred to as the cumulative odds ratio. The model in (8.6.9) assumes that the logarithm of the cumulative odds is proportional to the difference between the values of the explanatory variables for all j. Because of this property, model (8.6.9) is also called the *proportional odds model*. Within the same treatment, we have

$$\text{logit}(P_{i1} + \ldots + P_{ij}) - \text{logit}(P_{i1} + \ldots + P_{i(j-1)}) = \alpha_j - \alpha_{j-1}, \qquad (8.6.11)$$

where $j = 1, \ldots, J - 1$. Since we assume a common slope for each cumulative logit, they are parallel to each other by an amount $\alpha_j - \alpha_{j-1}, j = 1, \ldots, J - 1$. As a result it is important to check the parallel lines assumption before the proportional odds model can be applied.

Suppose that there are s covariates. One can fit a full model to the cumulative logits without assuming common slopes. The number of parameters for the full model is $s(J - 1)$, so a proportional odds model has a total of $J - 1 + s$ parameters. Consequently the chi-square score statistic for the parallel lines assumption approximately follows a chi-square distribution with $s(J - 1) - (J - 1 + s)$ $= s(J - 2)$ degrees of freedom. Consequently one rejects the assumption of the proportional odds model if the chi-square score statistic is too large.

Example 8.6.2

The data on question 4 given in Table 8.1.4, concerning the urination frequency within two hours of the last urination in the validation study comparing subjects with BPH and normal controls, are used to illustrate the application of the proportional odds model. For this dataset, $J = 6$ and $s = 1$. The observed chi-square score statistic is 1.5378 with $4(= 6 - 2)$ degrees of freedom. Since the corresponding p-value is 0.8198, there is no evidence to suspect the validity of the parallel lines assumption. Thus the proportion odds model is applied to this dataset. Table 8.6.5 provides estimates of five intercepts, the common treatment effects, and their corresponding standard errors. The results are summarized in Table 8.6.5. From Table 8.6.5 can be seen that the treatment effect is estimated by 1.3192 with a standard error of 0.3364 which is statistically, significantly different from zero. The common cumulative ratio for observing a response below the jth category from a normal control to the BHP subjects is 3.7404 with the corresponding confidence interval of (1.9345, 7.2322).

8.7 REPEATED CATEGORICAL DATA

As indicated in the previous chapter, repeated measures are often obtained at various time points after the baseline has been established in order to evaluate the efficacy and safety of a drug under study. For drug efficacy, repeated categorical data provide valuable information regarding the time-response profile which is often used to determine (1) how long the drug needs to be admin-

Table 8.6.5 Estimates of Regression Parameters and Their Estimated Standard Errors for Question 4–Frequency No. 1 in AUA Symptom Index

Parameter	Statistics	Value
Intercept 1	Estimate	−3.5099
	se	0.5797
	p-value	<0.0001
Intercept 2	Estimate	−1.8712
	se	0.5115
	p-value	0.0003
Intercept 3	Estimate	−1.0113
	se	0.4920
	p-value	0.0398
Intercept 4	Estimate	−0.0018
	se	0.5000
	p-value	0.9971
Intercept 5	Estimate	0.8619
	se	0.5485
	p-value	0.1161
Treatment	Estimate	1.3192
	se	0.3364
	p-value	<0.0001

istered before its clinically meaningful effectiveness is achieved and (2) how long the effectiveness can be maintained by the treatment. For safety, repeated categorical measurements can also be used to characterize the time course of the occurrence of adverse events.

A commonly used approach for the evaluation of a drug is to compare the differences in changes of categorical data from the baseline between groups. For example, for data given in Table 8.2.3 which were used to illustrate the McNemar test in Section 8.2, the proportion of subjects with a favorable status in the placebo group is 45.61% at both the baseline and visit 3. As a result there is no change from the baseline at visit 3 in the proportion of favorable status for subjects receiving the placebo. On the other hand, for those subjects receiving the test drug, the proportions of favorable status are 42.59% and 72.22% at the baseline and visit 3, respectively. This indicates a 29.63% increase in proportion with favorable status for the test group. As a result it is of interest to test whether the observed 29.63% difference in the change in proportion with favorable status from the baseline between the test drug and the placebo observed at visit 3 is statistically, significantly different from zero. For this purpose we could consider the changes in marginal probabilities between visit 3 and the baseline as the response variables for comparison between treatments. This problem is referred to as the two-sample McNemar test by Feuer and Kessler (1989).

Let θ_g be the change in proportion from the baseline at a certain post-treatment visit for group g:

$$\theta_g = P_{g.1} - P_{g1.}$$
$$= P_{g01} - P_{g10}, \qquad (8.7.1)$$

where $g = T, R$. The parameter of interest is then the difference in change of proportion from the baseline between treatments. That is,

$$\delta = \theta_T - \theta_R. \qquad (8.7.2)$$

Therefore the hypotheses for the two-sample McNemar test can be formulated as follows:

$$H_0: \delta = 0,$$
$$\text{vs.} \quad H_a: \delta \neq 0. \qquad (8.7.3)$$

Similar to the one-sample McNemar test, the maximum likelihood estimator for θ_g can be obtained by replacing the unknown parameters in (8.7.1) by their corresponding observed sample proportions as

$$\hat{\theta}_g = p_{g01} - p_{g10}.$$

An estimate of the large-sample variance of $\hat{\theta}_g$ is then given by

$$\hat{v}(\hat{\theta}_g) = \frac{(p_{g01} + p_{g10}) - (p_{g01} - p_{g10})^2}{n_g}, \qquad (8.7.4)$$

where $g = T, P$. Consequently the MLE for δ and its large-sample variance are given by, respectively,

$$\hat{\delta} = \hat{\theta}_T - \hat{\theta}_R \qquad (8.7.5)$$

and

$$v(\hat{\delta}) = v(\hat{\theta}_T) + v(\hat{\theta}_R). \qquad (8.7.6)$$

Feuer and Kessler (1989) suggest using

$$X = \frac{\hat{\delta}^2}{v(\hat{\delta})} \qquad (8.7.7)$$

as the test statistic for hypotheses (8.7.3) which approximately follow a chi-square distribution with one degree of freedom. Therefore the null hypothesis of (8.7.3) is rejected at the αth level of significance if $X > \chi^2(\alpha, 1)$. The corresponding large-sample $(1 - \alpha)100\%$ confidence interval for δ is given as

$$\hat{\delta} \pm z(\alpha/2)\sqrt{v(\hat{\delta})}. \qquad (8.7.8)$$

As indicated before, many clinical trials are conducted at different centers with the same study protocol. The Mantel-Haenszel type of technique can also be applied to combine the results from different centers for the differences in changes from the baseline in marginal proportions between treatments. Denote $\hat{\delta}_h$ as the observed difference in change in proportions of favorable status from baseline between treatments for the hth center at a particular post-treatment visit. Also let $v(\hat{\delta}_h)$ be its large-sample variance, where $h = 1, \ldots, H$. Then a test statistic for testing hypotheses (8.7.3) is the ratio of the square of the sum of $\hat{\delta}_h$ to the sum of their variances:

$$X_C = \frac{\left[\sum_{h=1}^{H} \hat{\delta}_h\right]^2}{\sum_{h=1}^{H} v(\hat{\delta}_h)}. \qquad (8.7.9)$$

When the total sample size is sufficiently large, the distribution of X_C approximately follows a chi-square distribution with one degree of freedom. As a result we can reject the null hypothesis of no difference in change from the baseline in proportions between treatments at the αth level of significance if $X_C > \chi^2(\alpha, 1)$.

As described in Example 8.2.2, the binary data for the proportion with favorable status (Table 8.2.3) are obtained by re-grouping the five categories of the original data (Table 8.1.6). We can also directly compare the transition in marginal proportions between treatments without combining classes into a binary response. The two-sample McNemar test for comparing change in proportion from the baseline between groups can be extended to the categorical data with more than two classes. We first consider the one-sample case. Suppose that a categorical datum has a total of r categories. Let P_{ij} be the proportion of subjects who had a response in the ith category at the baseline and a response in the jth category at a post-treatment visit, $i, j = 1, \ldots, r$. It follows that

$$P_{i.} = \sum_{j=1}^{r} P_{ij}$$

and

$$P_{.j} = \sum_{i=1}^{r} P_{ij}$$

are the marginal proportions of the number of subjects with a response in the ith category at the baseline and that with a response in the jth category at a post-treamtent visit, respectively. Within the same group, one can test whether the marginal proportions are the same for both the baseline and the visit. The hypotheses of interest can be expressed as follows:

$$H_0: P_{i.} = P_{.i} \quad \text{for all } i,$$
$$\text{vs.} \quad H_a: P_{i.} \neq P_{.i} \quad \text{for at least one } i, i = 1, \ldots, r. \quad (8.7.10)$$

Transition in the proportions from baseline to visit can be summarized by an $(r-1) \times 1$ vector $\mathbf{d} = (d_1, d_2, \ldots, d_{r-1})'$, where $d_i = p_{i.} - p_{.i}$ is the difference in sample proportions between the baseline and a visit for the ith category, $i = 1, \ldots, r-1$. The large-sample covariance matrix for d_i can then be estimated by

$$\frac{\mathbf{W}}{n} = \frac{\{w_{ij}\}}{n},$$

where

$$w_{ij} = \begin{cases} (p_{i.} + p_{.i}) - 2p_{ii} - (p_{i.} - p_{.i})^2 & \text{if } i = j, \\ -(p_{ij} + p_{ji}) - (p_{i.} - p_{.i})(p_{j.} - p_{.j}) & \text{if } i \neq j, \end{cases} \quad (8.7.11)$$

where $1 \leq i, j \leq r - 1$. A test statistic proposed by Bhapkar (1966) for the homogeneity of marginal proportions between the baseline and the visit is given by

$$Q = n\mathbf{d}'\mathbf{W}^{-1}\mathbf{d},$$

which follows a central chi-square distribution with $r - 1$ degrees of freedom. Therefore the null hypothesis of (8.7.10) is rejected at the αth level of significance if $Q > \chi^2(\alpha, r - 1)$, where $\chi^2(\alpha, r - 1)$ is the upper αth quantile of a chi-square distribution with $r - 1$ degrees of freedom. The covariance under the null hypothesis, $P_{i.} = P_{.i}$, for all i, is given as

$$w_{ij}^0 = \begin{cases} (p_{i.} + p_{.i}) - 2p_{ii}, \\ -(p_{ij} + p_{ji}), \end{cases} \quad (8.7.13)$$

where $1 \leq i, j \leq r - 1$. Replacing $\mathbf{W}$ in the Bhapkar's Q statistic with the null covariance matrix $\mathbf{W}^0$, the resulting test statistic is the one proposed by Stuart (1955) which has the same asymptotic distribution as Bhapkar's Q.

Let $\boldsymbol{\theta}_h$ represent the vector consisting of $r - 1$ changes in marginal proportions from the baseline, where

$$\boldsymbol{\theta}_h = (P_{h1.} - P_{h.1}, \ldots, P_{h(r-1).} - P_{h.(r-1)}), \quad h = T, R. \quad (8.7.14)$$

Then $\boldsymbol{\delta} = \boldsymbol{\theta}_T - \boldsymbol{\theta}_R$ is the parameter of interest, which is the difference in changes in proportions from the baseline between treatments. The hypothesis on whether there is a difference in the change in marginal proportions from the baseline can then be formulated as

$$H_0: \boldsymbol{\delta} = 0,$$
$$\text{vs.} \quad H_a: \boldsymbol{\delta} \neq 0. \tag{8.7.15}$$

Let $\mathbf{d}_h$ be the vector of the differences obtained from the observed sample proportions for the hth group based on sample size n_h with its estimated non-null sample covariance matrix $\mathbf{W}_h$. Then the Bhapkar's Q statistic for testing hypotheses of (8.7.15) is given by

$$Q = (\mathbf{d}_T - \mathbf{d}_R)' \left[\frac{\mathbf{W}_T}{n_T} + \frac{\mathbf{W}_R}{n_R} \right]^{-1} (\mathbf{d}_T - \mathbf{d}_R). \tag{8.7.16}$$

This statistic approximately follows a chi-square distribution with $r-1$ degrees of freedom. As a result we reject the null hypothesis at the αth level of significance if $Q > \chi^2(\alpha, r-1)$.

Note that the methods for analysis of repeated categorical data discussed in this section are based on a change from the baseline in marginal proportions. Other functions of the marginal probabilities such as cumulative logits described in (8.6.8) can be used as response variables for the characterization of change from the baseline disease status.

Repeated categorical data are measurements of the same categorical endpoint on the same individuals over the course of a trial. For the same subject repeated categorical data are correlated to each other. Analysis of repeated categorical data by treating the repeated data as independent observations will result in some undesirable statistical deficiencies. First of all, the estimates of treatment effects may not be robust to the selection bias for nonrandomized trials. Second, statistical inference for an average response is not efficient, although sometimes it is valid. Finally, estimation for variability is inconsistent. Alternatively, as discussed in the previous chapter, the technique of GEE developed by Liang and Zeger (1986) and Zeger et al. (1988) can be applied for analysis of repeated categorical data using either the marginal or transition models. Miller et al. (1993) propose a procedure for the analysis of repeated categorical data with more than two classes using the GEE in connection with weighted least squares. Lipsitz et al. (1994) report a performance of the GEE in practical circumstances. Agresti (1983, 1989) provides a comprehensive review of models and methods for analysis of repeated ordered categorical response data. Zeger and Liang (1992) give an overview of methods for the analysis of longitudinal data. Recent developments on repeated categorical data can be found in Crowder and Hand (1990), Lindsey (1993), Diggle, Liang, and Zeger (1994), and Hand and Crowder (1996).

Table 8.7.1 Summary of Differences in Favorable Status for Data Set Provided in Table 8.2.3

Treatment	N	Difference in Proportion	Standard Error	Chi-square	p-Value
Test drug	54	0.2963	0.0674	19.31	<0.0001
Placebo	57	0.0000	0.0702	0.000	>0.9999
Difference	111	0.2963	0.0972	9.27	0.0023

Example 8.7.1

Again we use the binary response of favorable status to illustrate the two-sample McNemar test. Table 8.7.1 presents the results of a one-sample McNemar test using the non-null variance given in (8.7.4); the results are consistent with those given in Table 8.2.5. The MLE for the proportional difference in change of favorable status from the baseline at visit 3 is 29.63%, with an estimated standard error of 0.0972. As a result the large-sample 95% confidence interval calculated according to (8.7.8) is (10.59%, 48.68%), which does not contain zero. Hence we reject the null hypothesis at the 5% level of significance that the change in proportions at visit 3 from the baseline between treatments is the same. The conclusion is that significantly more subjects receiving the test drug than those given the placebo changed from their unfavorable status at the baseline to the favorable status at visit 3.

Next we use the status score data in the original categories for illustration of Bhapkar's procedures. Since the data are sparse in the *terrible* response category, we will combine the *terrible* and *poor* response categories. Table 8.7.2 gives the changes in marginal proportion at visit 3 from the baseline and their differences between treatments. The one-samples of Bhapkar's Q statistics are 21.82 and 8.93 for the test drug and the placebo with p-value < 0.00001 and p-value = 0.0302, respectively. The results indicate that at visit 3 there is a statistically significant change in marginal proportions from the baseline for both treatments. The two-sample of Bhapkar's Q is given by 12.05 with a p-value of 0.0072. As a result the null hypothesis of no difference in change of marginal proportions between treatments is rejected at the 5% level of significance. From Table 8.7.2 more subjects on the test drug than on the placebo changed from other statuses at the baseline to the excellent rating at visit 3.

8.8 DISCUSSION

In this chapter we provided a concise review and introduction to the concept and several useful statistical methods commonly employed for the analysis of categorical data. However, we noted that some methods appearing in the literature are not discussed. These methods include (1) methods that count data using a Poisson distribution and (2) certain model-based methods such as the

Table 8.7.2 Summary of Marginal Changes in the Data Set of Table 8.1.6

			Category			
Treatment	N	Time Point	Terrible or Poor	Fair	Good	Excellent
Test drug	54	Visit 3	9.26%	18.52%	25.92%	46.30%
		Baseline	24.07%	33.33%	22.22%	20.37%
Change			−14.81%	−14.81%	3.70%	25.93%
Bhapkar $Q = 21.82$						
with 3 df						
p-value < 0.0001						
Placebo	57	Visit	29.83%	24.56%	19.30%	26.32%
		Baseline	19.30%	35.09%	33.33%	12.28%
Change			10.63%	−10.53%	−14.03%	14.04%
Bhapkar $Q = 8.93$						
with 3 df						
p-value $= 0.0302$						
Difference in change			−25.44%	−4.28%	17.73%	11.89%
Bhapkar $Q = 12.05$						
with 3 df						
p-value $= 0.0072$						

loglinear model. For more detail on the analysis of categorical data and related topics, see Agresti (1990).

The statistical tests discussed in this chapter are mainly for testing hypothesis regarding the existence of treatment effect. These methods are often misused to establish therapeutic equivalence between a test drug and a reference drug. For testing therapeutic equivalence for the binary response, as indicated in Chapters 2 and 6, the following interval hypothese should be considered

$$H_0: P_{11} - P_{21} \geq U \quad \text{or} \quad P_{11} - P_{21} \leq L,$$
$$\text{vs.} \quad H_a: L < P_{11} - P_{21} < U.$$

We conclude that the test drug is therapeutically equivalent to the reference drug at the α level of significance if the $(1-2\alpha)100\%$ confidence interval for the difference in population proportions computed from (8.3.3) falls within the equivalence limits of (L, U). In clinical trials, however, the one-sided equivalence is more appealing because the primary objective is usually to verify whether the efficacy of the test drug is at least as good as that of the reference drug. Therefore we can conclude that the test drug is at least as effective as the reference drug and conclude that equivalence is at the α level of significance if the lower limit of the $(1 - 2\alpha)100\%$ confidence interval is greater than the lower equivalence L. For more detail on therapeutic equivalence, see Blackwelder (1982),

Huque and Dubey (1990), Durrleman, and Simon (1990), Makuch and Johnson (1990), Farrington and Manning (1990), Chow and Liu (1992a), and Jennison and Turnbull (1993).

The Bhapkar's Q for the detection of a marginal difference in the change from the baseline can be applied to categorical responses on either a nominal or an ordinal scale. However, the method does not take into account the order of categories. Agresti (1983) proposes a more powerful Mann-Whitney type statistic for testing the homogeneity of marginal proportions between the visit and the baseline when the true marginal proportions are stochastically ordered. Agresti's procedure is useful when the categorical response is ordinal and the number of ordered categories is large.

As discussed in the previous chapter, missing values or dropouts often occur in clinical trials. Little and Rubin (1987) and Little (1995) gave definitions of a hierarchy of missing value mechanisms. In general, the entire dataset in which a clinical trial is supposed to collected can be partitioned into two parts. The first part, say Y_0, concerns data observed and collected from the trial; the second part, the data, denoted by Y_m, concerns the data that were supposed to be observed but were not and hence are missing. For instance, let R be a random variable that indicates whether the data of subjects are observed or missing. The missing value mechanism is said to be missing completely at random (MCAR) if R is independent of both Y_0 and Y_m. If R is only independent of Y_m, the data are said to be missing at random (MR) or ignorable (Laird, 1988). The missing mechanism is said to be informative or unignorable if random variable R depends on Y_m. Currently, there are no well-developed methods that can satisfactorily analyze the ordered categorical data with intermittently missing values. If the mechanism for the missing values is completely random, Lachin (1988b) indicates that inference based on the observed complete data can be treated as a subgroup analysis and hence is valid but less efficient. However, the assumption of independence between R and (Y_0, Y_m) and between R and treatment is difficult to verify. Recently, however, statistical procedures were proposed by Diggle (1989) and Ridout (1991) for testing the hypothesis of completely random dropouts, and for missing patterns of the repeated categorical data by Park and Davis (1993). For repeated categorical data, the issue of premature dropout is much more complicated than that of intermittent missing data because it almost certainly suggests that the assumption of MCAR is implausible. Despite the recent research effort for the informative withdrawals by prominent statisticians such as Fitzmaurice et al. (1995) and Diggle and Kenward (1994), further work is needed on repeated categorical data.

CHAPTER 9

Censored Data and Interim Analysis

9.1 INTRODUCTION

Statistical concepts and methods for analyzing continuous and categorical endpoints in clinical trials were reviewed in the previous two chapters. Clinical endpoints for assessment of efficacy and safety of a promising therapy usually include occurrence of some predefined events such as death, the response to a new chemotherapy in treatment of some advanced cancer, the eradication of an infection caused by a certain microorganism (e.g., *Helicobacter pylori* for gastric ulcers), serious adverse events (e.g., neutropenia), or the elevation of asparate transaminase three times over the upper limit of the normal range. For these events the primary parameter of interest is usually time to the occurrence of such an event. Another parameters of interest is the median survival time. The median survival time is defined as the time at which 50% of the subjects still survive. Subjects are recruited into the trial at different calendar time point. Note that the predefined event may not be observed on the subjects who complete the scheduled duration of treatment and follow-up. On the other hand, some subjects may withdraw prematurely without observing any occurrences of the event before the end of the study. These individuals are said to be lost to follow-up. As a result we do not have any information on these subjects with respect to the event. The only information we have is that the predefined event did not occur at these subjects in their last visit (either at the end of study or at the time they dropped out from the study). The time to the occurrence of the event therefore is not known for these subjects. We refer an endpoint of this kind to as a censored endpoint. As an example, consider the most common opportunistic infection in patients with advanced stages of AIDS, which is the disseminated infection with *Mycobacterium avium* complex. Pierce et al. (1996) conducted a randomized, two-group parallel, placebo-controlled double-blind study to assess the prophylactic effect of clarithromycin 500 mg twice daily in prevention of *M. avium* complex infection and improvement of survival in patients with advanced AIDS. The entrance criteria for this study included blood cultures that were negative to *M. avium*, a Karnofsky performance score of 50 or higher, a CD4 cell count of 100 or fewer per cubic millimeter. The primary efficacy endpoints consist of the time to the detection of *M. avium*, which is defined as the time interval between randomization and the first positive blood culture, and survival, which is defined as the time interval from randomization to death from any causes. The data from the patients without infection were censored at the time of the last negative culture for the time to the detection. The survival data were censored at the time of the last contact of the alive patients. The sample size for this study is 300 patients per group to provide at least an 80% power for detection of at least a 67% reduction in the incidence of infection of *M. avium* complex for the clarithromycin group as compared with the placebo group. This trial also specified an interim analysis at the time when the first 300 patients have completed one year of therapy or when 50 patients developed *M. avium* infection, whichever, came first. The study was in fact terminated after the first interim analysis because compelling evidence of reduction in incidence of *M. avium* complex provided by clarithromycin.

Pan (1996) points out that only one out of 1000 drug candidates can make it from the stage of drug discovery to the FDA approval. The process of drug research and development is lengthy and on average takes 12 years. The cost increased from 125 million U.S. dollars in 1963 to 1975 to 450 million U.S. dollars in 1991 to 1995. Since drug research and development is not only time-consuming but also costly, it is helpful to determine whether the efficacy and safety of a new pharmaceutical entity is promising and compelling at early stage of clinical development, be that positive or negative. The information can be used to recommend early termination of the drug's development. This is especially important for trials with long-term follow-up or intended for evaluation of life-threatening or severely debilitating diseases. As a result interim analyses have become popular in clinical development. The purpose is to provide early information regarding the test drug in a more cost-effective way. If the reason for early termination is lack of efficacy or the presence of unexpected or untoward adverse experiences, then the remaining resource may be re-allocated to develop other promising drugs. Therefore interim analyses to provide statistical evidence for early termination and data monitoring have become an indispensable norm in the management of clinical trials. However, it should be noted that interim analysis and data monitoring are two different areas. They have their own functions and procedures, although they are related to each other.

Analysis of censored data and the application of interim analysis for early termination have become common practice for clinical trials. In this chapter we will first discuss different types of censoring. Some background regarding the survival function and the use of the Kaplan-Meier's method (Kaplan and Meier, 1958) for estimation of proportions based on censored data are given in Section 9.2. The logrank statistic and Gehan's generalization of Wilcoxon rank sum statistic for the censored data (Gehan, 1965a, 1965b) are outlined in Section 9.3. Section 9.4 describes the semiparametric proportional hazard model with covariates as proposed by Cox (1972). The concept of calendar and information time are illustrated in Section 9.5. Different methods for interim

ESTIMATION OF THE SURVIVAL FUNCTION

analysis are covered in Section 9.6. These methods include the group sequential methods proposed by Pocock (1977) and O'Brien and Fleming (1979), repeated confidence intervals suggested by Jennison and Turnbull (1989), the method of alpha spending function by DeMets and Lan (1994), and B-values by Lan and Wittes (1988). Some final remarks and a discussion will be provided in the last section. Whenever possible, numerical examples using published data from real trials will be used to illustrate the methods presented in this chapter.

9.2 ESTIMATION OF THE SURVIVAL FUNCTION

In conducting clinical trials, it is almost impossible and impractical to enroll all the patients into a study at the same time. Patients are usually enrolled into the trial at different time points during the study. At the end of the study or at the time of interim analysis, the event of interest may occur in some patients during the study. As a result their times to the event are actually observed. However, for the patients who completed the study or withdrew prematurely from the study without the event, their times to the events may not be observed and hence are censored. Figure 9.2.1 displays in calendar time the censoring patterns for time to a predefined event of a study, starting from the initiation to its end

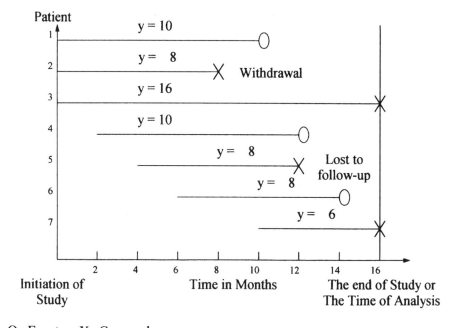

Figure 9.2.1 Censoring pattern in calender time. o: event; ×: censored.

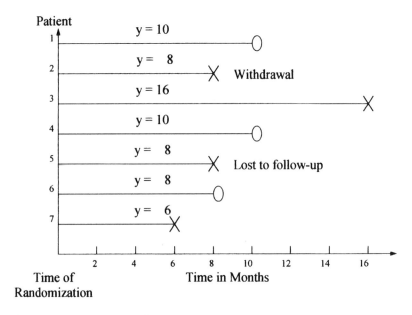

Figure 9.2.2 Censoring pattern in duration of treatment. ○: event; ×: censored.

or to the time at which an interim analysis was performed. Patient 1 enrolled at the beginning of the study and the event of interest occurred at month 10. Therefore the time to the event is 10 months. Patient 2 also entered at the beginning of the study, but for some reason this patient prematurely withdrew from the study at month 8 without the event. Hence the time to the event is unknown and censored at 8 months. Patient 3 completed the entire study of 16 months without an occurrence of the event. Therefore the time to the event of patient 3 is censored at 16 months. Patient 4 was enrolled at month 2, and the event occurred at month 12. As a result the time to the event for patient 4 is 10 months. Patient 5 entered into the trial at month 4 but lost to follow-up. Hence the time to the event for patient 5 is censored at 8 months. The time to the event is 8 months and not censored for patient 6 who entered into the trial at month 6 and experienced the event at month 14. Patient 7 was enrolled into the study at month 10. When the study ended at month 16, no occurrences of the event were observed and hence the time to the event for patient 7 is censored at 6 months. In survival analysis, it is usually of interest to obtain statistical inference on the time from initiation of treatment such as randomization to the occurrence of the predefined event. As a result it is suggested that the data be arranged in terms of the length of treatment before the occurrence of the event as illustrated in Figure 9.2.2. As can be seen from Figure 9.2.2, the times to the event for patient 1 and patient 4 are the same (10 months). Similarly the times to the event for patient 2 and patient 5 are both censored at 8 months.

Table 9.2.1 Time in Months to Progression of the Patients with Stage II or IIIA Ovarian Carcinoma by Low-Grade or Well-Differentiated Cancer

Patient Number	Time in Months	Censored	Cell Grade
1	0.92	Yes	Low
2	2.93	Yes	Low
3	5.76	Yes	Low
4	6.41	Yes	Low
5	10.16	Yes	Low
6	12.40	No	Low
7	12.93	No	Low
8	13.85	No	Low
9	14.70	No	Low
10	15.20	Yes	Low
11	23.32	No	Low
12	24.47	No	Low
13	25.33	No	Low
14	36.38	No	Low
15	39.67	No	Low
16	1.12	Yes	High
17	2.89	Yes	High
18	4.51	Yes	High
19	6.55	Yes	High
20	9.21	Yes	High
21	9.57	Yes	High
22	9.84	No	High
23	9.87	No	High
24	10.16	Yes	High
25	11.55	Yes	High
26	11.78	Yes	High
27	12.14	Yes	High
28	12.14	Yes	High
29	12.17	Yes	High
30	12.34	Yes	High
31	12.57	Yes	High
32	12.89	Yes	High
33	14.11	No	High
34	14.84	Yes	High
35	36.81	No	High

Source: Fleming et al. (1980).

Fleming et al. (1980) report the results from a study conducted in patients with limited stage II or IIIA ovarian carcinoma at the Mayo Clinic which was to determine whether the grade was related to the time over which disease progressed. Table 9.2.1 displays either the observed time in months to the progression of the disease or the time censored at the last visit. The observed time is given in the second column, while the third column provides information on whether the observed time to progression was relevant or censored. The last column indicates whether a patient had low grade or well-differentiated cancer. Table 9.2.1 provides a basic layout for the presentation of censored data which must include patient identification, time to the event, censored indicator, and other covariates such as treatment assignment and disease status.

Let Y be a continuous random variable representing the time from randomization to the occurrence of a clinically meaningful event such as the time until the detection of *M. avium* or the occurrence of death for the AIDS trial as described in the previous section. Y is usually referred to as the survival time. The cumulative distribution function (cdf) of Y, denoted by $F(y)$, is defined as the probability that a subject fails before or equal to the time y, namely

$$F(y) = P(\text{a subject fails before or equal to } y)$$
$$= P(Y \leq y), \quad 0 < y < \infty. \quad (9.2.1)$$

The cdf is a nondecreasing function of time y such that $F(0) = 0$ and $F(\infty) = 1$. The cdf of Y is usually employed to describe the mortality rate for evaluation of treatment in cardiovascular diseases, such as the Beta-blocker Heart Attack Trial (BHAT, 1982). In clinical trials, however, it is more common to use the survival function for the disease with a high mortality rate. The survival function is the probability that a subject survives longer than y, namely

$$S(y) = 1 - F(y)$$
$$= P(\text{a subject survives longer than } y)$$
$$= P(Y > y), \quad 0 < y < \infty. \quad (9.2.2)$$

Hence $S(y)$ is a nonincreasing function of time y, and $S(0) = 1$ and $S(\infty) = 0$. As a result the probability density function (pdf) of Y is the probability of failure over a very small time interval,

$$f(y) = \lim_{\Delta y \to 0} P(\text{a subject fails between } y \text{ and } y + \Delta y)/\Delta y$$
$$= \lim_{\Delta y \to 0} P(y < Y \leq y + \Delta y)/\Delta y, \quad (9.2.3)$$

Since the pdf of Y is the probability of failure over a time interval, it is greater than or equal to 0 if y is greater or equal to 0. It equals 0 if y is less than 0. In addition the area under the curve $f(y)$ is equal to 1. Comparing the pdf given in (9.2.3) and the cdf given in (9.2.1) or survival function defined in (9.2.2) reveals that $f(y)$ is the derivative of $F(y)$, which equals minus of the derivative of the survival function $S(y)$. The hazard function, denoted as $h(y)$, is the instantaneous death rate, which is the conditional probability that a subject fails over the next instant given that the subject has survived up to the beginning of the interval. Mathematically it can be expressed as

$$h(y) = \lim_{\Delta y \to 0} P(\text{a subject fails between } y \text{ and}$$

$$y + \Delta y | \text{the subject survives to } y)/\Delta y$$

$$= \lim_{\Delta y \to 0} P(y < Y \le y + \Delta y | Y > y)/\Delta y$$

$$= \frac{f(y)}{1 - F(y)}$$

$$= \frac{f(y)}{S(y)}. \qquad (9.2.4)$$

Two hazard functions $h_1(y)$ and $h_2(y)$ are said to be proportional if

$$h_1(y) = \lambda h_2(y) \qquad \text{for all } y > 0, \qquad (9.2.5)$$

where λ is a constant. From (9.2.4) and (9.2.5) the relationship between the corresponding survival functions is given by

$$S_1(y) = [S_2(y)]^\lambda \qquad \text{for all } y > 0. \qquad (9.2.6)$$

As a result, if the ratio of two hazard functions is a constant, then the one survival function can be expressed as the other survival function to the λth power.

If a predefined clinical event is observed in some subjects before the completion of the study, then their exact failure times are known. On the other hand, some subjects may withdraw prematurely without observing any occurrences of the event of interest due to some known or unknown reasons. Sometimes the event does not occur for some subjects who completed the study. As a result the time to the occurrence of the event is censored at the last known contact, and it is at least as long as the time from randomization to the time of the last contact. Let C denote the censoring time associated with the failure time Y. If C is greater than or equal to Y, then the survival time is actually observed. On the other hand, if the survival time is greater than the censoring time, then the survival time is not observed and is censored. As a result the censored data for a subject consist of a pair of responses. The first response is the observed time and the second is an indicator identifying whether the observed time is the survival time or was censored at the last contact. In other words, the data for the time to the occurrence of a predefined event obtained from n subjects of a clinical trial can be arranged as $(y_1, c_1), \ldots, (y_n, c_n)$, where y_i is the observed time for subject i and

$$c_i = \begin{cases} 1 & \text{if } y_i \text{ is the survival time,} \\ 0 & \text{if } y_i \text{ is censored.} \end{cases} \qquad (9.2.7)$$

This type of censoring mechanism is referred to as random censoring (Miller, 1981). Description for other types of censoring mechanism can be found in Miller (1981) and Lee (1992).

Since it is not easy to theoretically evaluate the true distribution of the survival time (Lee, 1992; Kalbfleisch and Prentice, 1980), in this chapter we will only cover Kaplan and Meier's nonparametric method for the survival function. For parametric methods, see Lee (1992) and Marubini and Valsecchi (1995). Let $y_{(1)} < \cdots < y_{(K)}$ be the ordered distinct failure times when the event occurs and d_k be the number of events at time $y_{(k)}$ and m_k be the number of censored observations in the interval $(y_{(k)}, y_{(k+1)})$, $k = 1, \ldots, K$. The risk set just prior to the time $y_{(k)}$ consists of the subjects who still survive and whose survival time is not censored before $y_{(k)}$. Thus, under the assumption of independent censoring, the number of subjects in the risk set just prior to the time $y_{(k)}$, denoted by n_k, is given by

$$n_k = (d_k + m_k) + \ldots + (d_K + m_K), \qquad k = 1, \ldots, K.$$

The Kaplan-Meier nonparametric estimation of the survival function at time y is given by

$$\hat{S}(y) = \prod_{y_{(k)} < y} \left(1 - \frac{d_k}{n_k} \right). \qquad (9.2.8)$$

The Kaplan-Meier estimate provides a straightforward yet intuitive interpretation of the survival function. The probability that a patient is alive (or without the event) at $y_{(k)}$ is equal to the conditional probability that the patient is alive at $y_{(k)}$, given that this patient survived through all proceeding time points when other patients failed times the probability that this patient survived all previous time points. In other words,

$$\begin{aligned}
\hat{S}(y_{(k)}) &= P(\text{surviving at } y_{(k)}) \qquad (9.2.9) \\
&= P(\text{surviving through } y_{(1)}, y_{(2)}, \ldots, y_{(k-1)}, y_{(k)}) \\
&= P(\text{surviving } y_{(k)} | \text{surviving through } y_{(1)}, y_{(2)}, \ldots, y_{(k-1)}) \\
&\quad \times P(\text{surviving through } y_{(1)}, y_{(2)}, \ldots, y_{(k-1)}) \\
&= P(\text{surviving } y_{(k)} | \text{surviving through } y_{(1)}, y_{(2)}, \ldots, y_{(k-1)}) \\
&\quad \times P(\text{surviving } y_{(k-1)} | \text{surviving through } y_{(1)}, y_{(2)}, \ldots, y_{(k-2)}) \\
&\quad \times \ldots \times P(\text{surviving } y_{(2)} | \text{surviving through } y_{(1)}) \\
&\quad \times P(\text{surviving } y_{(1)}).
\end{aligned}$$

ESTIMATION OF THE SURVIVAL FUNCTION

From (9.2.9) we see that the Kaplan-Meier estimates of survival are the same as those between two adjacent observed failure times. Then the graph of the Kaplan-Meier estimates must be a step function. According to the above discussion, the data layout for computing the Kaplan-Meier nonparametric estimates of survival consists of five columns as demonstrated in Table 9.2.2. The first column shows the ordered distinct failure times, the second column the number of events, the third column the number of censored observations, and the fourth column the number of subject at risk prior to $y_{(k)}$. The last column gives the Kaplan-Meier estimates of the survival function which are a product of the proportions of the number of surviving subjects at $y_{(k)}$ and of all proceeding distinct failure times.

The variance of the Kaplan-Meier estimate at time y can be found using Greenwood's formula:

$$v[\hat{S}(y)] = [\hat{S}(y)]^2 \sum_{y_{(k)} < y} \frac{d_k}{n_k(n_k - d_k)}. \tag{9.2.10}$$

It follows that a large-sample $(1 - \alpha)100\%$ confidence interval for $S(y)$ can be obtained as

$$\hat{S}(y) \pm z(\alpha/2)\sqrt{v[S(y)]}. \tag{9.2.11}$$

The median survival time can then be estimated as

$$q_{0.5} = \min\{y : 1 - \hat{S}(y) \geq 0.5\}. \tag{9.2.12}$$

A large-sample $(1 - \alpha)100\%$ confidence interval for the median survival time can be constructed similarly.

Example 9.2.1

The data of time to progression (in months) for 15 patients with limited stage II or stage IIIA low-grade ovarian cancer given in Table 9.2.1 are used to illustrate the computation of the Kaplan-Meier survival function in finding the probability of the patients who did not progress. In the data given in Table 9.2.3, ovarian cancer progressed in 6 out of 15 patients. At each of the first five progression time points, 0.92, 2.93, 5.76, 6.41, and 10.16 months, there was only one patient with occurrence of the event of cancer progression. The numbers of the patients in the risk set just prior to the progression time were 15, 14, 13, 12, and 11, respectively. However, between 6.41 and 10.16 months, there were 4 patients whose progression times were censored. At 10.16 months, the cancer progressed in one patient. As a result the number of patients in the risk set just prior to 10.16 months became 6. Since the progression times for the last 5 patients were censored and were longer than 10.16 months, the probability of the patients who did not progress at or after 10.16 months can be estimated by (9.2.9) as

Table 9.2.2 Data Layout for Computation of Kaplan-Meier Survival Function

Ordered Distinct Event Time	Number of Events	Number Censored in $[y_{(k)}, y_{(k+1)}]$	Number in Risk Set	$\hat{S}(y)$
$y_{(0)} = 0$	$d_0 = 0$	m_0	n_0	1
$y_{(1)}$	d_1	m_1	n_1	$1 - d_1/n_1$
$y_{(2)}$	d_2	m_2	n_2	$(1 - d_1/n_1)(1 - d_2/n_2)$
$\vdots$	$\vdots$	$\vdots$	$\vdots$	$\vdots$
$y_{(K)}$	d_K	m_K	n_K	$(1 - d_1/n_1)(1 - d_2/n_2) \cdots (1 - d_K/n_K)$

$$\hat{S}(10.16 \text{ or longer}) = \left(\frac{14}{15}\right)\left(\frac{13}{14}\right)\left(\frac{12}{13}\right)\left(\frac{11}{12}\right)\left(\frac{10}{11}\right)\left(\frac{5}{6}\right) = 0.5556.$$

The Kaplan-Meier estimates, the corresponding estimated large sample variance, and 95% confidence intervals are given in Table 9.2.4. The Kaplan-Meier estimates for patients with low-grade cancer are plotted in Figure 9.2.3 along with those for patients with well-differentiated cells. A large-sample variance of $\hat{S}(10.16$ or longer) computed by the Greenwood's formula in (9.2.10) is 0.0206. It follows that the large-sample 95% confidence interval is (0.2744, 0.8367). However, since the probability of patients free of progression at or after 10.16 months is 0.5556 which is greater than 0.5, the median time of no progression cannot be estimated for the patients with low-grade cancer. On the other hand, for the patients with well-differentiated cancer, the shortest time for the estimated probability of patients free of progression greater than 50% is 12.14 months which is the estimated median time of no progression. The corresponding large-sample 95% confidence interval is from 9.57 to 12.57.

Table 9.2.3 Computation of Kaplan-Meier Survival Function for Patients with Low-Grade Cancer in Table 9.2.1

Ordered Distinct Progression Time	Number of Events	Number of Censored in $[y_{(k)}, y_{(k+1)}]$	Number in Risk Set	$S(y)$
0	0	0	15	1
0.92	1	0	15	0.9333
2.93	1	0	14	0.8667
5.76	1	0	13	0.8000
6.41	1	0	12	0.7333
10.16	1	4	11	0.6667
15.20	1	1	6	0.5556

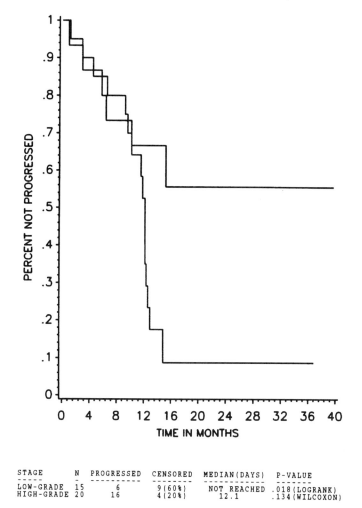

STAGE	N	PROGRESSED	CENSORED	MEDIAN(DAYS)	P-VALUE
LOW-GRADE	15	6	9(60%)	NOT REACHED	.018(LOGRANK)
HIGH-GRADE	20	16	4(20%)	12.1	.134(WILCOXON)

Figure 9.2.3 Distribution of Kaplan-Meier survival estimates on time to progression.

9.3 COMPARISON BETWEEN SURVIVAL FUNCTIONS

In clinical trials the most important inference regarding censored data is to determine whether the experimental therapy can reduce the morality rate or improve the survival rate compared to the placebo or standard treatment. Therefore statistical hypotheses are typically expressed as

$$H_0: S_1(y) = S_2(y),$$
$$\text{vs.} \quad H_a: S_1(y) \neq S_2(y). \tag{9.3.1}$$

One commonly used method for comparing two survival functions is the

Table 9.2.4 Kaplan-Meier Survival Rates for Patients with Ovarian Carcinoma by Low-Grade or Well-Differentiated Cancer

Time in Months	Censored	Estimated Proportion	Estimated Variance	Standard Error	Lower 95% Limit	Upper 95% Limit
TMT = Low Grade						
0.92	No	0.93333	0.004148	0.06440	0.80710	1.00000
2.93	No	0.86667	0.007703	0.08777	0.69464	1.00000
5.76	No	0.80000	0.010666	0.10328	0.59758	1.00000
6.41	No	0.73333	0.013037	0.11418	0.50955	0.95712
10.15	No	0.66667	0.014814	0.12171	0.42811	0.90523
12.40	Yes	0.66667				
12.93	Yes	0.66667				
13.85	Yes	0.66667				
14.70	Yes	0.66667				
15.20	No	0.55556	0.020575	0.14344	0.27441	0.83670
23.32	Yes	0.55556				
24.47	Yes	0.55556				
25.33	Yes	0.55556				
36.38	Yes	0.55556				
39.67	Yes	0.55556				

		TMT = High Grade			
1.12	No	0.95000	0.002375	0.85448	1.00000
2.89	No	0.90000	0.004500	0.76852	1.00000
4.51	No	0.85000	0.006375	0.69351	1.00000
6.55	No	0.80000	0.008000	0.62470	0.97530
9.21	No	0.75000	0.009375	0.56023	0.93977
9.57	No	0.70000	0.010500	0.49916	0.90084
9.84	Yes	0.70000			
9.87	Yes	0.70000			
10.16	No	0.64167	0.011942	0.42748	0.85585
11.55	No	0.58333	0.012962	0.36018	0.80648
11.78	No	0.52500	0.013562	0.29675	0.75325
12.14	No	0.40833	0.013497	0.18063	0.63604
12.17	No	0.35000	0.012833	0.12797	0.57203
12.34	No	0.29167	0.011747	0.07923	0.50410
12.57	No	0.23333	0.010240	0.03499	0.43167
12.89	No	0.17500	0.007972	0.00000	0.35370
14.11	Yes	0.17500			
14.84	No	0.08750	0.001993	0.00000	0.23813
36.81	Yes	0.08750			

Source: Fleming et al. (1980).

Table 9.3.1 Data Structure of Comparing Two Survival Functions at $y_{(k)}$ by the Logrank Method

	Status		Total
Treatment	Event	No Event	
Test drug	d_{1k}	$n_{1k} - d_{1k}$	n_{1k}
Placebo	d_{2k}	$n_{2k} - d_{2k}$	n_{2k}
	d_k	$n_k - d_k$	n_k

Note: $k = 1, \ldots, K$.

logrank test. The logrank test is considered to be the most powerful test for the alternative hypothesis that hazard functions are proportional, namely,

$$H_0: S_1(y) = S_2(y),$$
$$\text{vs.} \quad H_a: S_1(y) = [S_2(y)]^\lambda. \quad (9.3.2)$$

As was discussed in Section 8.5, the logrank test for comparing two independent samples is the same as the Mantel-Haenszel statistic for combining results from different strata. For logrank test statistic, a 2×2 contingency table by treatment and occurrence of events can be formed at each of the K distinct event times, $y_{(1)} < \ldots < y_{(K)}$ (see Table 9.3.1). As was mentioned in Section 8.3, under the assumption of a hypergeometric distribution, the conditional expected number of patients with occurrence of events for d_{1k} at the event time $y_{(k)}$ for the test drug is given by

$$e_{1k} = \frac{n_{1k} d_k}{n_k}, \quad (9.3.3)$$

and the conditional variance of d_{1k} is given as

$$v_{1k} = \frac{n_{1k} n_{2k} d_k (n_k - d_k)}{n_k^2 (n_k - 1)}, \quad k = 1, \ldots, K. \quad (9.3.4)$$

It turns out that the following logrank test statistic is the same as Mantel-Haenszel statistic given in (8.5.2):

$$X_{LR} = \frac{\left[\sum_{k=1}^{K} (d_{1k} - e_{1k})\right]^2}{\sum_{k=1}^{K} v_{1k}}$$

$$= \frac{(d_1 - e_1)^2}{v_1}, \quad (9.3.5)$$

Table 9.3.2 Computation of Logrank Statistic

Ordered Distinct Event Time	Observed Number of Events	Expected Number of Events	Difference	Variance
$y_{(1)}$	d_{11}	$e_{11} = n_{11}d_{11}/n_1$	$d_{11} - e_{11}$	v_{11}
$y_{(2)}$	d_{12}	$e_{12} = n_{12}d_{12}/n_2$	$d_{12} - e_{12}$	v_{12}
$\vdots$	$\vdots$	$\vdots$	$\vdots$	$\vdots$
$y_{(K)}$	d_{1K}	$e_{1K} = n_{1K}d_{1K}/n_K$	$d_{1K} - e_{1K}$	v_{1K}
	d_1	e_1	$d_1 - e_1$	v_1

Note: $v_{1k} = n_{1k}n_{2k}d_k(n_k - d_k)/[n_k^2(n_k - 1)]$.

where

$$d_1 = \sum_{k=1}^{K} d_{1k},$$

$$e_1 = \sum_{k=1}^{K} e_{1k},$$

$$v_1 = \sum_{k=1}^{K} v_{1k}.$$

Table 9.3.2 provides a summary for computation of log-rank statistic. Under the null hypothesis of equal survival functions, the logrank statistic approximately follows a central chi-square distribution with one degree of freedom when sample size is moderate. Hence the null hypothesis (9.3.1) is rejected at the αth significance level if

$$X_{LR} > \chi^2(\alpha, 1), \tag{9.3.6}$$

where $\chi^2(\alpha, 1)$ is the αth upper quantile of a central chi-square distribution with one degree of freedom. Peto et al. (1976) suggest that the relative hazard rate λ can be estimated as

$$\hat{\lambda} = \exp\left(\frac{d_1 - e_1}{v_1}\right), \tag{9.3.7}$$

where "exp" denotes the natural exponentiation. Therefore a large-sample

$(1 - \alpha)100\%$ confidence interval for λ can be obtained as

$$\exp\{[(d_1 - e_1)/v_1] \pm z(\alpha/2)/\sqrt{v_1}\}, \qquad (9.3.8)$$

where $z(\alpha/2)$ is the upper $(\alpha/2)$th quantile of a standard normal distribution.

Sometimes the randomization of patients is performed according to some predefined or natural stratified criteria such as gender or ST-segment elevation in assessment of recombinant hirubin with heparin for the treatment of acute coronary syndromes (GUSTO IIb, 1996). An extension of the logrank statistic is to combine results from strata. Suppose that there are a total of H independent strata. For each strata, d_{h1}, e_{h1}, and v_h can be calculated according to (9.3.5), $h = 1, \ldots, H$, then the Mantel-Haenszel technique can be applied to combine the results

$$X_{\text{MH}} = \frac{\left[\sum_{h=1}^{H} (d_{1h} - e_{1h})\right]^2}{\sum_{h=1}^{H} v_{1h}} \qquad (9.3.9)$$

The null hypothesis of (9.3.1) is rejected if (9.3.6) is true.

The logrank statistic is in fact a special case of a general class of nonparametric tests for comparing two survival functions which can be expressed in the form of

$$T = \sum_{k=1}^{K} w_k (d_{1k} - e_{1k}), \qquad (9.3.10)$$

where w_k is the weight assigned at the event time $y_{(k)}$. When the weight in (9.3.10) is 1, the square of T is the numerator of logrank test statistic in (9.3.5). On the other hand, if the weight is the proportion of patients in the risk set at event time $y_{(k)}$ to the total sample size, namely n_k/n, $k = 1, \ldots, K$, the resulting test statistic is Gehan's generalization of the Wilcoxon two-sample rank sum test to the censored data (Gehan, 1965a, 1965b). Since Gehan's test gives greater weights to differences that occur at the beginning of a trial, it is more powerful in the detection of differences evidently early in time (Lee, Desu, and Gehan, 1975). On the other hand, the logrank test is fully efficient (i.e., the most powerful test) when relative hazards between the two survival functions is a constant. However, it is more sensitive to differences between the two survival functions that occur late in time. Tarone and Ware (1977) suggest using the square root of the number of patients in the risk set just prior to $y_{(k)}$ as the weights for the statistic given in (9.3.10). Another generalization of Wilcoxon rank sum test suggested by Prentice (1978) is to use the following weights:

COMPARISON BETWEEN SURVIVAL FUNCTIONS

Table 9.3.3 Data Structure for comparing I Survival Functions at $y_{(k)}$ by Logrank Method

Treatment	Event	No Event	Total
Test drug 1	d_{1k}	$n_{1k} - d_{1k}$	n_{1k}
Test drug 2	d_{2k}	$n_{2k} - d_{2k}$	n_{2k}
$\vdots$	$\vdots$	$\vdots$	$\vdots$
Test drug i	d_{ik}	$n_{ik} - d_{ik}$	n_{ik}
$\vdots$	$\vdots$	$\vdots$	$\vdots$
Placebo	d_{Ik}	$n_{Ik} - d_{Ik}$	n_{Ik}
	d_k	$n_k - d_k$	n_k

(Status spans Event and No Event columns)

$$S(y) = \prod_{y_{(k)} < y}^{K} \left[1 - \frac{d_k}{n_k + 1} \right]. \qquad (9.3.11)$$

Note that the logrank statistic for comparing two survival functions can be extended to trials for assessment of more than two treatments. Suppose that there are a total of I different treatments including the placebo which is designated as the last treatment with index I. Similar to the two independent treatment groups, an $I \times 2$ contingency table by treatment and the occurrence of event can be formed at each of the K distinct event times so that $y_{(1)} < \ldots < y_{(K)}$, as illustrated in Table 9.3.3. Conditional on the marginal totals, the expected number of patients with event for d_{ik} are given by

$$e_{ik} = \frac{n_{ik} d_{ik}}{n_{ik}}, \qquad k = 1, \ldots, K, i = 1, \ldots, I. \qquad (9.3.12)$$

Therefore the sum of difference between the observed and expected number of events over all K distinct event time points for treatment i is then given by

$$\sum_{k=1}^{K} (d_{ik} - e_{ik}) = d_i - e_i, \qquad (9.3.13)$$

where

$$d_i = \sum_{k=1}^{K} d_{ik} \quad \text{and} \quad e_i = \sum_{k=1}^{K} e_{ik}, \qquad i = 1, \ldots, I.$$

The covariance between treatment i and i' for the sum of difference between the observed and expected number of events is given as

$$v_{ii'} = \sum_{k=1}^{K} \frac{(n_k n_{i'k} \delta_{ii'} - n_{ik} n_{i'k}) d_k (n_k - d_k)}{n_k^2 (n_k - 1)}, \quad 1 \leq i, i' \leq I, \quad (9.3.14)$$

where $\delta_{ii'} = 1$, if $i = i'$; $\delta_{ii'} = 0$, otherwise. Let $\mathbf{d} = (d_1, \ldots, d_{I-1})$, $\mathbf{e} = (e_1, \ldots, e_{I-1})$, and $\mathbf{V} = \{v_{ii'}, 1 \leq i, i' \leq I - 1\}$. Then the null hypothesis that $S_1 = \ldots = S_I$ is rejected if test statistic

$$X = (\mathbf{d} - \mathbf{e})' \mathbf{V}^{-1} (\mathbf{d} - \mathbf{e}) > \chi^2(\alpha, I - 1), \quad (9.3.15)$$

where $\chi^2(\alpha, I - 1)$ is the αth upper quantile of a central chi-square distribution with $I - 1$ degrees of freedom.

Example 9.3.1
Again we use the data of the time to the progression (in months) in 15 patients with ovarian cancer given in Table 9.2.1 to illustrate the computation of logrank statistic for testing hypotheses of (9.3.1) and to estimate the relative hazard rate λ as given in (9.3.7). First, we need to form all possible 2×2 contingency tables by treatment and the occurrence of event at each of distinct event times in the combined samples. For this data set, there are a total of 20 distinct time points recorded as the cancer progressed. For each 2×2 contingency table, compute the marginal totals n_{1k}, n_{2k}, d_k, $n_k - d_k$, and n_k, $k = 1, \ldots, 20$. From the marginal totals, the expected number of events and variance of d_{1k} can be computed according to (9.3.3) and (9.3.4). For example, at the first time point (0.92 month) with the occurrence of progression, $n_{11} = 15$, $n_{21} = 20$, $d_1 = 1$, $n_1 = 35$, and $n_1 - d_1 = 34$. It follows that

$$e_{11} = \frac{(1)(15)}{35} = 0.42857,$$

$$v_{11} = \frac{(15)(20)(1)(34)}{(35)^2(34)} = 0.24490.$$

The difference between d_{11} and e_{11} is equal to $1 - 0.42857 = 0.57143$. The immediate results are given in Table 9.3.4. At the bottom of Table 9.3.4 are given the sums of $d_{1k} - e_{1k}$, and v_{1k} over all distinct event time points. As a result the logrank test statistic is given by

COMPARISON BETWEEN SURVIVAL FUNCTIONS

Table 9.3.4 Computation of Logrank Test Statistic in Table 9.2.1 for Patients with Stage II or IIIA Ovarian Carcinoma

Time in Months	d_{1k}	d_{2k}	d_k	n_{1k}	n_{2k}	n_k	e_{1k}	$d_{1k} - e_{1k}$	v_{1k}
0.92	1	0	1	15	20	35	0.42857	0.57143	0.24490
1.12	0	1	1	14	20	34	0.41176	−0.41176	0.24221
2.89	0	1	1	14	19	33	0.42424	−0.42424	0.24426
2.92	1	0	1	14	18	32	0.43750	0.56260	0.24609
4.51	0	1	1	13	18	31	0.41935	−0.41935	0.24350
5.76	1	0	1	13	17	30	0.43333	0.56667	0.24556
6.41	1	0	1	12	17	29	0.41379	0.58621	0.24257
6.55	0	1	1	11	17	28	0.39286	−0.39286	0.23852
9.21	0	1	1	11	16	27	0.40741	−0.40741	0.24143
9.57	0	1	1	11	15	26	0.42308	−0.42308	0.24408
10.16	1	1	2	11	12	23	0.95652	0.04348	0.47637
11.56	0	1	1	10	11	21	0.47619	−0.47619	0.24943
11.78	0	1	1	10	10	20	0.50000	−0.50000	0.25000
12.14	0	2	2	10	9	19	1.05263	−1.05263	0.47091
12.17	0	1	1	10	7	17	0.58824	−0.58824	0.24221
12.34	0	1	1	10	6	16	0.62500	−0.62500	0.23438
12.57	0	1	1	9	5	14	0.64286	−0.64286	0.22959
12.89	0	1	1	9	4	13	0.69231	−0.69231	0.21302
14.84	0	1	1	6	2	8	0.75000	−0.75000	0.18750
15.19	1	0	1	6	1	7	0.85714	0.14286	0.12245
Sum	6	16	22					−5.33279	5.10898

$$X_{LR} = \frac{\sum_{k=1}^{20} (d_{1k} - e_{1k})^2}{\sum_{k=1}^{20} v_{1k}}$$

$$= \frac{(d_1 - e_1)^2}{v_1}$$

$$= \frac{(-5.33279)^2}{5.10898}$$

$$= 5.5664.$$

The corresponding p-value is given by 0.0183. Thus we reject the null hypothesis that the survival function for the patients with low-grade ovarian cancer is the same as that with well-differentiated ovarian cancer at the 5% level of significance.

An estimate of the relative hazard ratio can be obtained as

$$\hat{\lambda} = \exp\left[\frac{d_1 - e_1}{v_1}\right]$$

$$= \exp\left[\frac{-5.33279}{5.10898}\right]$$

$$= \exp(-1.04381)$$

$$= 0.35231.$$

The corresponding large-sample 95% confidence interval for the relative hazard λ can also be obtained as

$$\exp\{-1.04381 \pm 1.96/\sqrt{5.10898}\} = \{\exp(-1.91093), \exp(-0.17668)\}$$

$$= (0.14794, 0.83805).$$

Note that both the point estimate and the corresponding large-sample interval estimate for the relative hazard rate λ are smaller than 1. Therefore we can conclude that the number of occurrences of progression among patients with low-grade cancer is statistically lower than among those with well-differentiated cancer. In addition the time to progression for patients with low-grade cancer is also statistically longer than that with well-differentiated cancer. This can further be verified that if the weights at each distinct progression time point are chosen to be $n_k/n, k = 1, \ldots, 20$. The Gehan test statistic then is -84 with a variance of 3146. It follows that the corresponding chi-square value is 2.2428 with a p-value of 0.1342. Therefore, according to the Gehan's generalized Wilcoxon rank test, we fail to reject the null hypothesis of equal survival functions. As shown in Figure 9.2.3, the difference in survival functions between the patients with low-grade and well-differentiated cancers becomes evident only after 12 months. Therefore, as we learned earlier, the Gehan test is not as powerful as the logrank test for the detection of this difference later in time.

9.4 COX'S PROPORTIONAL HAZARD MODEL

To provide a fair and unbiased assessment of efficacy and safety of a test drug based on censored data, the Cox's proportional hazard model has become a routine statistical method since it was introduced by Cox (1972). For example, the West of Scotland Coronary Prevention Study Group (1995) reported that the reduction in risk of nonfatal myocardial infarction or death from coronary heart disease with the cholesterol-lowering agent pravastatin at 40 mg per day is 31% (95% confidence interval: 17 to 43%, p-value < 0.0001) as compared to the placebo in men with moderate hypercholesterolemia and no history of myocardial infarction. More recently the Cholesterol and Recurrent Events (CARE, 1996) trial investigators also reported that pravastatin at 40 mg per day provides a 24% reduction in risk of coronary events (95% confidence interval: 9 to 36%, p-value = 0.003) in patients with myocardial infarction but with an average cholesterol level. For both studies the reduction in risk and its corresponding 95% confidence interval were estimated through the Cox's proportional hazard model. On the other hand, in a study for evaluation of methotrexate alone, sulfasalazine and hydroxychloroquine, or a combination of all three drugs in the treatment of rheumatoid arthritis (O'Dell et al., 1995), the primary endpoint was successful completion of the two-year study. To obtain unbiased comparisons among three treatments, Cox's proportional hazard model was applied to adjust for differences in covariates at the study entry such as the erythrocyte sedimentation rate, the patient's global status, and the total joint score. Another example of application of the Cox's proportional hazard model is a recent prospective randomized trial comparing high-dose therapy and autologous bone marrow transplantation and convention chemotherapy in patients with multiple myeloma (Attal et al., 1996). In addition to a comparison between treatments, Cox's proportional hazard model was used to identify the level of beta2-microglobulin in serum as one of the prognostic factors for the overall and event-free survivals.

Instead of a direct formulation of the survival function with a constraint between 0 and 1, the dependent response variable in the Cox's proportional hazard regression model is the hazard function at time y that can be expressed as the product of a baseline hazard function and a function of covariates. Let $\mathbf{x} = (x_1, \ldots, x_p)'$ be a vector of p covariates collected from a subject. The general form of the proportional hazard regression model can be expressed as

$$h(y; \mathbf{x}) = h_0(y)\Omega(\mathbf{x}; \boldsymbol{\beta}), \quad (9.4.1)$$

where $h_0(y)$ is called the baseline hazard function and $\boldsymbol{\beta}$ is a vector of p unknown regression coefficients that relates to the hazard function at time y and p covariates.

Since the hazard and survival functions are all positive quantities, Cox (1972) suggests that function $\Omega(\mathbf{x}; \boldsymbol{\beta})$ be formulated in terms of exponentiation as follows:

$$h(y; \mathbf{x}) = h_0(y) \exp\{\boldsymbol{\beta}'\mathbf{x}\}$$

$$= h_0(y) \exp\left\{\sum_{i=1}^{p} \beta_i x_i\right\}. \quad (9.4.2)$$

Note that the regression part in the proportional hazard function, $\exp\{\boldsymbol{\beta}'\mathbf{x}\}$, does not involve time t if all covariates are independent of time. The time-independent covariates are referred to as those collected before the initiation of the study such as demographic or baseline characteristics. The covariates collected during the study may be vary at different visits, which are time-dependent variables. The baseline hazard function $h_0(y)$ is a function of time but not a function of the covariates. In addition the functional form of the baseline hazard function $h_0(y)$ is unspecified. It does explicitly express the relationship between covariates and the hazard function in a parametric form. As a result the Cox's proportional hazard regression model is a semiparametric statistical procedure.

Another important property of Cox's proportional hazard regression model is that the relative risk or hazard ratio, which is the ratio of the hazard at time y to the baseline hazard, is a function of covariates and does not vary with time if all covariates are time-independent, namely

$$\lambda(\mathbf{x}) = \frac{h(y; \mathbf{x})}{h_0(y)} = \exp\left\{\sum_{i=1}^{p} \beta_i x_i\right\}. \qquad (9.4.3)$$

The proportionality of the Cox's proportional hazard model can also be expressed in terms of the survival function as

$$S(y) = [S_0(y)]^{\lambda(\mathbf{x})}$$

$$= [S_0(y)] \exp\left\{\sum_{i=1}^{p} \beta_i x_i\right\}, \qquad (9.4.4)$$

where $S_0(y)$ is the survival function corresponding to the baseline hazard function $h_0(y)$.

For the moment suppose that the covariates of interest are the treatment indicator and gender. Let X_1 be the treatment indictor, which has a value of 1 if treatment is the test drug and has a value of 0 if treatment is the placebo. Similarly X_2 represents gender, which has a value of 1 if a subject is a female and is 0 if a subject is a male. Then model (9.4.2) can be written as

$$h(y; x_1, x_2) = h_0(y) \exp\{\beta_1 x_1 + \beta_2 x_2\}, \qquad (9.4.5)$$

where β_1 measures treatment effect after adjustment of difference in gender and β_2 reflects the prognostic effect of gender, adjusted for the treatment effect. If we are interested in comparing the test drug and the placebo in female patients, the explicit expression of the hazard for a female subject assigned to receive the test drug at time t, according to (9.4.5), is given by

$$h(y; 1, 1) = h_0(y) \exp(\beta_1 + \beta_2).$$

Therefore the relative risk with respect to the baseline hazard is

$$\lambda(1, 1) = \exp(\beta_1 + \beta_2).$$

The hazard corresponding to a female given the placebo is

$$h(y; 0, 1) = h_0(y)\exp(\beta_2)$$

with the respective relative risk

$$\lambda(0, 1) = \exp(\beta_2).$$

As a result the relative risk of the test drug to the placebo for a female subject is then given by

$$\frac{\lambda(1, 1)}{\lambda(0, 1)} = \frac{\exp(\beta_1 + \beta_2)}{\exp(\beta_2)}$$

$$= \exp(\beta_1). \quad (9.4.6)$$

We notice that first, (9.4.6) only consists of the regression coefficient with respect to treatment assignment and is not contaminated with the covariate gender. As a result it can be used to estimate the treatment effect, adjusted for difference in gender. Secondly, the relative risk of the test drug to the placebo for a female subject is a constant, which does not vary with time. This is a key assumption of Cox's proportional hazard model, which implies that

$$S_{TF}(y) = [S_{PF}(y)]^{\exp(\beta_1)},$$

and

$$S_T(y) = [S_P(y)]^{\exp(\beta_1)}, \quad (9.4.7)$$

where $S_{TF}(y)$ is the survival function at time y for a female subject assigned to the test drug and $S_{PF}(y)$ is the survival function at time y for a female subject given the placebo; $S_T(y)$ and $S_P(y)$ are similarly defined.

The second term of (9.4.7) implies that the relationship of survival functions between the test drug and the placebo for female subjects is the same as that of survival functions between the test drug and the placebo for any other covariates so long as the same constants are present in treatment groups. Let us denote $\exp(\beta_1)$ by λ. Since λ is a constant over time, the relationship of survival functions between the test drug and the placebo is the same as that specified in the alternative hypothesis of (9.2.6):

$$S_T(y) = [S_P(y)]^\lambda \qquad \text{for all } y > 0, \tag{9.4.8}$$

where $\lambda = \exp(\beta_1)$. The relationship between $S_T(y)$ and $S_P(y)$ specified in (9.3.2) and (9.4.8) is referred to as Lehmann's alternative (Lehmann, 1953). This explains why the logrank test achieves full efficiency under the assumption that the hazard ratio (i.e., relative risk) between the test drug and the placebo is a constant.

The relative risk between the test drug and the placebo in (9.4.6) is expressed as the ratio of the relative risks between hazard of each group and the baseline hazard. It, however, can be written directly as the hazard ratio of the hazard function of the test drug to that of the placebo:

$$\frac{h(y; 1, 1)}{h(y; 0, 1)} = \frac{h_0(y) \exp(\beta_1 + \beta_2)}{h_0(y) \exp(\beta_2)}$$

$$= \exp(\beta_1). \tag{9.4.9}$$

Note that the baseline hazard function of time appears in both numerator and denominator and therefore it is canceled out. As a result the relative risk does not involve the baseline hazard function, and it is not a function of time if there are no time-dependent covariates. A statistical inference of relative risk and the regression coefficients, and of the corresponding survival functions, can still be made even though we do not specify the baseline hazard function. Cox's proportional hazard regression model is quite robust in the sense that it can adequately approximate the true but unknown parametric model. In addition, unlike the logistic regression for the occurrence of the events, Cox's proportional hazard model takes the survival time and censoring pattern into account. Hence it is more efficient for the inference of the censored data than the logistic regression because Cox's proportional hazard model uses more information.

The likelihood function derived from Cox's proportional hazard model is based on the probabilities of the events of interest occurring in the subjects. It explicitly does not include the probabilities for subjects whose event times are censored. As a result the likelihood for Cox's proportional hazard model is called the partial likelihood, since it does not consist of probabilities for all subjects. Let $y_{(1)} < \ldots < y_{(K)}$ be the ordered distinct times when the event occurs and $R(y_{(k)})$ be the risk set just prior to time $y_{(k)}$. Conditional on the risk set $R(y_{(k)})$, the probability of the occurrence of the event for subject j is given as

$$\frac{\exp\left(\sum_{i=1}^{p} \beta_i x_{ij}\right)}{\sum_{j \in R(y_{(k)})} \exp\left(\sum_{i=1}^{p} \beta_i x_{ij}\right)}. \tag{9.4.10}$$

For example, in the data on time to progression of cancer for the patients with limited stage II or IIIA ovarian carcinoma, the indicator covariate X takes the value 1 for low-grade cancer and 0 for well-differentiated cancer. The hazard functions at time y are $h_0(y)$ for the patients with well-differentiated cancer and $h_0(y)\exp(\beta)$ for the patients with low-grade cancer. Therefore $\exp(\beta)$ is the relative risk of progression between lower-grade and well-differentiated cancers. The first progression occurs after 0.92 month (28 days) on a patient with low-grade cancer. The corresponding risk set just prior to 0.92 month consists of 15 patients with lower-grade cancer and 20 patients with well-differentiated cancer. As a result the probability that progression did occur at this time in a patient with lower-grade cancer is given by

$$L_1 = \frac{\exp(\beta)}{[20 + 15\exp(\beta)]}.$$

The second event of progression occurred at 1.12 months on a patient with well-differentiated cancer. The risk set just prior to 1.12 months includes 14 patients with low-grade cancer and 20 patients with well-differentiated cancer. Therefore the conditional probability of progression at 1.12 month on a patient with well-differentiated cancer is

$$L_2 = \frac{1}{[20 + 14\exp(\beta)]}.$$

We can repeat the same process until the last progression time. The partial likelihood corresponding to the Cox's proportional hazard model is the product of the conditional probabilities of the occurrence of events at each of the K distinct event times which is given by

$$L = L_1 \times L_2 \times \ldots \times L_K$$

$$= \prod_{k=1}^{K} L_k$$

$$= \prod_{k=1}^{K} \left\{ \frac{\exp\left(\sum_{i=1}^{p} \beta_i x_{ij}\right)}{\sum_{j \in R(y_{(k)})} \exp\left(\sum_{i=1}^{p} \beta_i x_{ij}\right)} \right\}. \quad (9.4.11)$$

Since the partial likelihood is constructed based on the subjects with the occurrence of the event at their respective event times, it takes the event time into account. The risk set $R(y_{(k)})$ at time $y_{(k)}$ includes the subjects whose event times are censored after $y_{(k)}$. In other words, the partial likelihood derived from Cox's proportional hazard model also contains some information of censored observations. Once the partial likelihood is derived, then the maximum likelihood estimates (MLEs) of unknown regression coefficients in the Cox's proportional hazard model can be obtained through the standard iterative method by treating the partial likelihood as the ordinary likelihood. Let b_i and $se(b_i)$ represent the MLE of β_i and its associated large-sample standard error, respectively. Then we can reject the null hypothesis that $H_0 : \beta_i = 0$ in favor of the alternative hypothesis that $H_a : \beta_i \neq 0$ if the Wald statistic

$$W_i = [b_i/se(b_i)]^2 > \chi^2(\alpha, 1), \qquad i = 1, \ldots, p, \qquad (9.4.12)$$

where $\chi^2(\alpha, 1)$ is the αth upper quantile of a central chi-square distribution with 1 degree of freedom. A large-sample $(1 - \alpha)100\%$ confidence interval for β_i can be obtained as

$$b_i \pm z(\alpha/2)se(b_i). \qquad (9.4.13)$$

The confidence interval given in (9.4.13) is based on the Wald statistics where the standard error is calculated through the information matrix evaluated at the estimated b_i. The other equivalent test for large samples is the score test proposed by Rao (1973) where the information matrix is evaluated at values of β_i specified in the null hypothesis. If the model only includes the treatment indicator and all events occur at distinct event times (i.e., no tied event times), then the score test derived under Cox's proportional hazard model is the same as that under the logrank test for a comparison between two or more survival functions, namely (9.3.5) and (9.3.15). This indicates that the log-rank test is a fully effective alternative measure of proportional hazard.

Let $\mathbf{x}_k$ be a vector of p covariates from subject k, $\mathbf{x}_{k'}$ be a vector of p covariates associated with subject k', and $\mathbf{b} = (b_1, \ldots, b_p)'$ be a vector of the p MLEs of unknown regression coefficients. Then the relative risk or hazard ratio between patient k and k' can be estimated as

$$\hat{\lambda} = \exp[(\mathbf{x}_k - \mathbf{x}_{k'})'\mathbf{b}]. \qquad (9.4.14)$$

A large-sample $(1 - \alpha)100\%$ confidence interval for λ is then given by

$$\exp\left\{(\mathbf{x}_k - \mathbf{x}_{k'})'\mathbf{b} \pm z(\alpha/2)\sqrt{(\mathbf{x}_k - \mathbf{x}_{k'})'v(\mathbf{b})(\mathbf{x}_k - \mathbf{x}_{k'})}\right\}, \qquad (9.4.15)$$

where $v(\mathbf{b})$ is the estimated large-sample covariance matrix of $\mathbf{b}$.

The validity of the statistical inference for the Cox's proportional hazard regression model depends on the assumption of proportional hazards among different values of covariates. Hence it is important to check this assumption before the application of the model. Several methods are available for detection of possible violation of proportional hazard assumption (e.g., see Kleinbaum,

1996). Two of these methods will be introduced in this chapter. The first method is the graphical approach. Recall that

$$S(y) = [S_0(y)] \exp\left(\sum \beta_i x_i\right).$$

It follows that the minus logarithm of $S(y)$ becomes

$$-\ln[S(y)] = -\left\{\ln[S_0(y)] \exp\left[\sum \beta_i x_i\right]\right\}.$$

Note that $-\ln[S(y)]$ is positive. As a result the minus logarithm of $-\ln[S(y)]$ is a linear function of $\sum \beta_i x_i$ with intercept $-\ln\{-\ln[S_0(y)]\}$, namely

$$-\ln\{-\ln[S(y)]\} = -\ln\{-\ln[S_0(y)]\} - \sum_{i=1}^{p} \beta_i x_i. \qquad (9.4.16)$$

Suppose that the only covariate in the model is the treatment indicator, which takes value 1 for the test drug and 0 for the placebo. Then, for a patient assigned to the test drug, we have

$$-\ln\{-\ln[S_T(y)]\} = -\ln\{-\ln[S_0(y)]\} - \beta.$$

On the other hand, for patients in the placebo group,

$$-\ln\{-\ln[S_P(y)]\} = -\ln\{-\ln[S_0(y)]\}.$$

It follows that the difference in $-\ln\{-\ln[S(y)]\}$ between the test drug and the placebo is a constant independent of time as demonstrated below:

$$-\ln\{-\ln[S_T(y)]\} - (\ln\{-\ln[S_0(y)]\}) = \beta. \qquad (9.4.17)$$

Under the assumption of proportional hazard between the test drug and the placebo, it can be seen from (9.4.17) that $-\ln\{-\ln[S(y)]\}$ should be parallel to each other and the distance between them is a constant. Therefore this property can be applied to check the assumption of proportional hazard. For a particular covariate we can obtain the Kaplan-Meier estimates of the survival function at each level of the covariate. Then, the minus logarithms of the minus logarithm of the Kaplan-Meier estimates can be plotted. If the curves are approximately parallel, then the proportional hazard assumption is not violated. If these curves cross each other or diverge, then the assumption is not met.

The graphical approach is the simplest and easiest way to verify the proportional hazard assumption. However, it only allows us to examine one covariate at a time. The graphical method cannot assess the proportional hazard assumption simultaneously for a group of covariates when there is more than one covariate in the model. Kleinbaum (1996) recommends that a conservative approach be taken in applying the graphical method. Unless there is a strong evidence of nonparallelism among the curves of $-\ln\{-\ln[S(y)]\}$, the proportional hazard assumption should not be rejected. However, the graphical method is still a subjective approach because it does not provide a statistical test for assessing the proportional hazard assumption for a group of covariates.

Inclusion of time-dependent covariates in Cox's proportional hazard model can provide another means of a formal statistical assessment of the proportional hazard assumption. For simplicity, we first consider one covariate for the treatment indicator X, which takes the value 1 for the test drug and 0 for the placebo. The proportional hazard assumption is then evaluated, in addition to X, by including the interaction between treatment indicator and some function of time $d(y)$ in the model. The resulting model is given by

$$h(y;x) = h_0(y)\exp\{\beta x + \theta x d(y)\}, \qquad (9.4.18)$$

where θ measures the interaction between treatment and time (i.e., the differences in treatment effects at different time points), and $d(y)$ can be any function of time. In practice, there are many choices for function $d(y)$. For simplicity, we restrict our selection to simple monotone function of time such as $d(y) = y$ or $d(y) = \ln(y)$, for $y > 0$. Similar to the ordinary regression analysis, y or $\ln(y)$ needs to be centered to avoid unnecessary difficulties that may arise during the iterative processes used in searching the MLEs. If the hazard ratios remain fairly constant within some time intervals and are markedly different from one time interval to another, then the $d(y)$ can be chosen as a step function, such as $d(y) = 0$ if $y < y_0$, and $d(y) = 1$ if $y \geq y_0$, for the two time intervals. A nonzero θ usually indicates that the hazard ratios are different over time at different values of the covariate under investigation. A positive (negative) value of θ implies that the hazard ratios increase (decrease) as time increases. As a result the hypothesis of interest here is whether the regression coefficient is different from 0, namely

$$H_0: \theta = 0,$$
$$\text{vs.} \quad H_a: \theta \neq 0. \qquad (9.4.19)$$

Let us denote $LL(\beta, \theta)$ as the log-likelihood function under the interaction model (9.4.18), and let $LL(\beta)$ be the log-likelihood function under the reduced model

$$h(y;x) = h_0(y)\exp\{\beta x\}.$$

The null hypothesis (9.4.19) is rejected if

$$-2[LL(\beta) - LL(\beta, \theta)] > \chi^2(\alpha, 1), \quad (9.4.20)$$

where $\chi^2(\alpha, 1)$ is the αth upper quantile of a central chi-square distribution with one degree of freedom. This approach can be extended to investigate the proportional hazard assumption for a set of covariates by including interaction terms between each covariate and some function of time:

$$h(y; \mathbf{x}) = h_0(y) \exp\left\{\sum_{i=1}^{p} \beta_i x_i + \sum_{i=1}^{p} \theta_i x_i d_i(y)\right\}. \quad (9.4.21)$$

The corresponding global hypothesis for the assessment of the proportional hazard simultaneously for a group of covariates is then given by

$$H_0: \theta_1 = \ldots = \theta_p = 0,$$
vs. $H_a: \theta_i \neq 0$ for at least one $i, 1 \leq i \leq p.$ (9.4.22)

Let $LL(\beta_1, \ldots, \beta_p; \theta_1, \ldots, \theta_p)$ be the log-likelihood function under the full model given in (9.4.21) and $LL(\beta_1, \ldots, \beta_p)$ be the log-likelihood function under the reduced model without the interaction terms between covariates and time given in (9.4.2). Then the null hypothesis of no time–covariate interaction is rejected if

$$-2[LL(\beta_1, \ldots, \beta_p) - LL(\beta_1, \ldots, \beta_p; \theta_1, \ldots, \theta_p)] > \chi^2(\alpha, p), \quad (9.4.23)$$

where $\chi^2(\alpha, p)$ is the αth upper quantile of a central chi-square distribution with p degrees of freedom.

When the likelihood ratio test given in (9.4.23) indicates that a total of H covariates does not satisfy the proportional hazard assumption, the stratified Cox proportional hazard model can be applied to control these H covariates. Suppose that each of these H covariates has q_h levels $h = 1, \ldots, H$. As a result the Q strata can be formed from these H covariates, where $Q = q_1 \times \ldots \times q_h$. Instead of assuming the same baseline hazard function for all covariates, the stratified Cox proportional hazard model assumes that the baseline hazard function is the same within a stratum but is different for other strata. Therefore the stratified Cox proportional hazard model is given by

$$h_q(y; \mathbf{x}) = h_{0q}(y) \exp\left\{\sum_{i=1}^{p} \beta_i x_i\right\}, \quad q = 1, \ldots, Q. \quad (9.4.24)$$

For each stratum the partial likelihood can be formulated in the same manner as (9.4.11), which is denoted L_q, $q = 1, \ldots, Q$. Since each stratum is mutually exclusive and consist of different subjects, the partial likelihood for the entire sample is the product of these Q partial likelihoods:

$$L = \prod_{q=1}^{Q} L_q. \tag{9.4.25}$$

Then the MLEs for the regression coefficients in (9.4.24) can be obtained by the standard method, and the statistical inference with respect to the covariates in the model is made in the usual way. The effects of the covariates included in the stratified Cox proportional hazard model are estimated after adjustment for other covariates including those used to form the strata. However, the effects of the covariates used for stratification cannot be estimated.

The stratified Cox's proportional hazard model in (9.4.24) assumes that the regression coefficients are the same for all strata. In other words, there is no interaction between the p covariates included in the model and those H covariates used for stratification. The stratified Cox proportional hazard model with interaction terms is given by

$$h_q(y;x) = h_{0q}(y) \exp\left\{\sum_{i=1}^{p} \beta_{iq} x_i\right\}, \quad q = 1, \ldots, Q. \tag{9.4.26}$$

There are a total of pQ regression coefficients in model (9.4.26), while model (9.4.24) includes p regression coefficients. As a result the global null hypothesis of no interaction between the covariates in the model and covariates forming strata can be performed by the usual twice of the minus difference in log-likelihood functions between (9.4.24) and (9.4.26) with $p(Q - 1)$ degrees of freedom.

Example 9.4.1
In this example we will once again use the data of progression time for patients with limited stage II or stage IIIA ovarian cancer to illustrate the application of Cox's proportional hazard model. The treatment effect is estimated as 1.1186 with a standard error of 0.4771 if treatment indicator takes the value 1 for patients with well-differentiated cancer and 0 for patients with low-grade cancer. The corresponding Wald statistics is 5.0652 with a p-value of 0.0224. Note that the score test statistic is 5.512 with a p-value of 0.0189, which is very close to the 0.0183 obtained by the logrank test in Example 9.3.1. The reason for the small discrepancy between the p-values is that there are two tied progression times at 10.16 months. A large-sample 95% confidence interval for the treatment effect is then equal to $1.1186 \pm (1.96)(0.4771) = (0.1445, 2.0928)$. As a result the relative risk (hazard ratio) of progression for a patient with

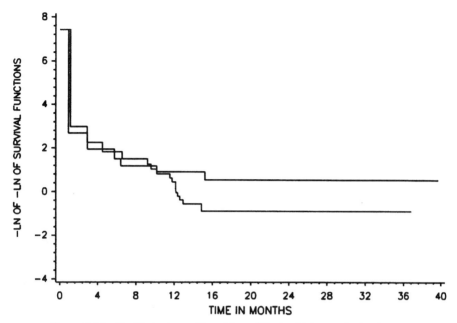

Figure 9.4.1 Doubled minus of logarithm of Kaplan-Meier survival estimates.

well-differentiated cancer compared to others with low-grade cancer is estimated as exp(1.1186) = 3.0606 with the corresponding 95% confidence interval (exp(0.1445), exp(2.0928)) = (1.1554, 9.1076). Therefore the risk of progression in patients with well-differentiated cancer is about three times as high as that with low-grade cancer. However, from the 95% confidence interval, the variability of the estimated relative risk is quite large.

Figure 9.4.1 shows the $-\ln\{-\ln[S(y)]\}$ curves for patients with low-grade and with well-differentiated cancer. Since the two curves intersect each other many times before 12 months, the proportional hazard assumption may be seriously in doubt. It is therefore suggested that a formal statistical test be performed, such as given in (9.4.20), to verify the assumption. Note that the median progression time is about 12 months. As can be seen from the figure, the two curves of minus logarithm of minus logarithm of survival function become roughly parallel to each other after 12 months. To reflect this phenomenon, we define the function as a function of time:

$$d(y) = \begin{cases} 0 & \text{for low-grade cancer,} \\ y - 12 & \text{well-differentiated cancer.} \end{cases}$$

The interaction θ between treatment and time in model (9.4.18) is estimated as 0.2438 with a 95% confidence interval from 0.0251 to 0.4625. Twice the minus of the log-likelihood is 123.528 for model (9.4.18) and is 128.610 for the reduced model. The difference is 5.082 with a p-value of 0.0248 which is significant at the 5% level. These results indicate that the risk for progression of ovarian cancer in a well-differentiated case of cancer is statistically greater than that of low-grade cancer, and this risk increases over time compared to the patients with low-grade cancer.

Example 9.4.2

Kalbfleisch and Street (1990) report the results from a clinical trial on cyclosporine and methotrexate (CSP + MTX) therapy versus methotrexate (MTX) alone followed by an infusion of marrow from an HLA-identical family member in patients with severe aplastic anemia. One of the primary endpoints is the time (in days) to sever (stage 2) acute graft versus host disease (AGVHD), death, or last contact. The data are reproduced in Table 9.4.1. The treatment indicator is coded as 0 for CSP + MTX and 1 for MTX. In addition to treatment, other covariates include age in years at the time of transplantation and laminar airflow isolation (LAF). Figure 9.4.2 gives the Kaplan-Meier survival estimates for the two treatments. The horizontal axis is truncated at 70 days because all events occurred before day 50 and event times of 41 patients (64%) were censored after day 49, more than 5 patients were censored with event times greater than 1,000 days. From Figure 9.4.2 can be seen that the addition of CSP to MTX prolonged the time to severe AGVHD, or death over MTX alone. In addition Kalbfleisch and Street (1990) report that age is not only an important prognostic factor for the time to severe AGVHD or death but also interaction between treatment and age exists in favor of older patients (greater than or equal to 26 years older). On the other hand, LAF is not a significant prognostic factor, so it will be omitted from further discussion.

It seems that according to Figure 9.4.2, the hazard ratio of MTX to CSP + MTX increases over time. In order to investigate our conjecture of increasing hazard further and its relationship with age, as well as to follow the suggestion by Kalbfleisch and Street (1990), in addition to the original age we categorize age into three groups by the following two indicator variables AGE1 and AGE2:

$$AGE1 = \begin{cases} 1 & \text{if age is between 16 and 25 inclusively,} \\ 0 & \text{otherwise,} \end{cases}$$

and

$$AGE2 = \begin{cases} 1 & \text{if age is greater than 25,} \\ 0 & \text{otherwise.} \end{cases}$$

The proportional hazard assumption is examined using the method described above by inclusion of time-dependent covariates in the model; these are the interaction between time-independent covariates and some function of time. For

Table 9.4.1 Time in Days to Severe (Stage 2) Acute Graft Versus Host Disease or Death for Patients with Severe Aplastic Anemia

Patient Number	Treatment	Age in Years	LAF	Time in Days	Censored
1	CSP + MTX	40	0	3	Censored
2	CSP + MTX	21	1	8	AGHVD
3	CSP + MTX	18	1	10	AGHVD
4	CSP + MTX	42	0	12	Censored
5	CSP + MTX	23	0	16	AGHVD
6	CSP + MTX	21	0	17	AGHVD
7	CSP + MTX	13	1	22	AGHVD
8	CSP + MTX	29	0	64	Censored
9	CSP + MTX	15	1	65	Censored
10	CSP + MTX	34	1	77	Censored
11	CSP + MTX	14	1	82	Censored
12	CSP + MTX	10	1	98	Censored
13	CSP + MTX	27	0	155	Censored
14	CSP + MTX	9	1	189	Censored
15	CSP + MTX	19	1	199	Censored
16	CSP + MTX	14	1	247	Censored
17	CSP + MTX	23	0	324	Censored
18	CSP + MTX	13	1	356	Censored
19	CSP + MTX	34	1	378	Censored
20	CSP + MTX	27	1	408	Censored
21	CSP + MTX	5	1	411	Censored
22	CSP + MTX	23	1	420	Censored
23	CSP + MTX	37	1	449	Censored
24	CSP + MTX	35	1	490	Censored
25	CSP + MTX	32	1	528	Censored
26	CSP + MTX	32	1	547	Censored
27	CSP + MTX	38	1	691	Censored
28	CSP + MTX	18	1	767	Censored
29	CSP + MTX	20	0	1111	Censored
30	CSP + MTX	12	0	1173	Censored
31	CSP + MTX	12	0	1213	Censored
32	CSP + MTX	29	0	1357	Censored
1	MTX	35	1	9	AGHVD
2	MTX	27	1	11	AGHVD
3	MTX	22	0	12	AGHVD
4	MTX	21	1	20	AGHVD
5	MTX	30	1	20	AGHVD
6	MTX	7	0	22	AGHVD
7	MTX	36	1	25	AGHVD
8	MTX	38	1	25	AGHVD
9	MTX	20	0	25	Censored
10	MTX	25	0	28	AGHVD

Table 9.4.1 (*Continued*)

Patient Number	Treatment	Age in Years	LAF	Time in Days	Censored
11	MTX	28	0	28	AGHVD
12	MTX	17	1	31	AGHVD
13	MTX	21	1	35	AGHVD
14	MTX	25	1	35	AGHVD
15	MTX	35	1	46	AGHVD
16	MTX	19	0	49	AGHVD
17	MTX	21	1	104	Censored
18	MTX	19	1	106	Censored
19	MTX	15	1	156	Censored
20	MTX	26	1	218	Censored
21	MTX	11	0	230	Censored
22	MTX	14	1	231	Censored
23	MTX	15	1	316	Censored
24	MTX	27	1	393	Censored
25	MTX	2	0	395	Censored
26	MTX	3	0	428	Censored
27	MTX	14	1	469	Censored
28	MTX	18	1	602	Censored
29	MTX	23	0	681	Censored
30	MTX	9	1	690	Censored
31	MTX	11	1	1112	Censored
32	MTX	11	0	1180	Censored

Note: CSP + MTX = cyclosporine and methotrexate, MTX = methotrexate, LAF = laminar airflow isolation.
Source: Kalbfleisch and Street (1990).

this example we choose the function of time to be the identity function centered at the median of the following time of 173 days:

$$d(y) = y - 173.$$

Table 9.4.2 provides the twice minus of log-likelihood for the various models based on treatment, AGE1, AGE2, and their interactions with time using $d(y)$ as defined above. Comparison of the twice minus of log-likelihood from models with treatment only and with treatment and treatment-by-time interaction indicates an increase of 6.232 (152.587 − 146.355) with a p-value of 0.0119 in chi-square with one degree of freedom. Therefore the hazard ratio between two treatments is not constant over time. In addition the regression coefficient corresponding to the interaction is estimated as 0.1582, which is positive. As a result the relative risk of progression for MTX to CSP + MTX increases over time. Both AGE1 and AGE2 are also significant prognostic factors because they increase the likelihood by an amount of 9.433 from the model of treatment alone (152.587 − 143.154). Their joint contribution is statistically sig-

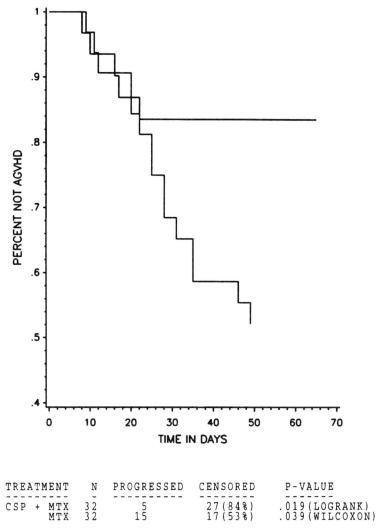

Figure 9.4.2 Distribution of Kaplan-Meier estimates on time to AGHVD.

nificant at the 5% level of significance (p-value < 0.01). It is also of interest to check whether the interaction between treatment and time can be ignored after the inclusion of treatment, AGE1, and AGE2 in the model. The difference in the twice minus of the log-likelihood between the two models is 6.766 (143.154 − 136.388) with a p-value of 0.0093. To examine whether interactions between AGE1 and AGE2 with time exist, we test the joint contribution of two time-dependent covariates. The twice minus of the log-likelihood function is 135.744 which slightly increases by an amount of 0.644 with a p-value of 0.7247. As a result the model with treatment, AGE1, AGE2, and treatment-

Table 9.4.2 Doubled Minus of the Log-likelihood for Various Models in Table 9.4.1 with Age Categorized into Three Groups

Model	df	$-2LR$	Chi-square
Without covariate		158.247	
TMT	1	152.587	5.661
TMT, TMT*$g(y)$	2	146.355	11.894
TMT, AGE1, AGE2	3	143.154	15.095
TMT, AGE1, AGE2, TMT*$g(y)$	4	136.388	21.857
TMT, AGE1, AGE2, AGE1*$g(y)$	4	143.108	15.140
TMT, AGE1, AGE2, AGE2*$g(y)$	4	143.148	15.099
TMT, AGE1, AGE2, TMT*$g(y)$, AGE1*$g(y)$	5	136.303	21.946
TMT, AGE1, AGE2, TMT*$g(y)$, AGE2*$g(y)$	5	135.905	22.234
TMT, AGE1, AGE2, AGE1*$g(y)$, AGE2*$g(y)$	5	142.938	15.309
TMT, AGE1, AGE2, TMT*$g(y)$, AGE1*$g(y)$, AGE2*$g(y)$	6	135.744	22.503

by-time interaction seems to be an adequate model. Table 9.4.3 provides the estimates, their corresponding large-sample standard errors for the model with treatment, AGE1, and AGE2 along with the model with inclusion of additional treatment-by-time interaction. Since the estimated coefficient for treatment-by-time interaction is 0.1608, the increasing hazard ratio of MTX to CSP + MTX is statistically significant over time even after the adjustment for age. While the estimates for AGE1 and AGE2 for both models are quite similar between the two models, the estimate for a treatment effect is markedly different in magnitude. For the model without treatment-by-time interaction, the estimated treatment effect is 1.1651 with a large-sample standard error of 0.5372. Hence the hazard ratio is estimated as 3.2062 with a large-sample 95% confidence interval from 1.1187 to 9.1890. On the other hand, for the model with inclusion of treatment-by-time interaction, the estimated treatment effect is inflated to 25.7644 with a large-sample standard error 12.7056. Hence the hazard ratio is estimated as 1.546×10^{11} with a large-sample 95% confidence interval from

Table 9.4.3 Estimates of Models with Treatment, AGE1, AGE2, and Treatment-by-$g(y)$ Interaction for Data in Table 9.4.1

Statistics	TMT	AGE1	AGE2	TMT*$g(y)$
Estimate	1.1651	1.9071	1.6776	
se	0.5372	0.7707	0.8097	
p-value	0.0301	0.0133	0.0383	
Estimate	25.7644	1.9331	1.7798	0.1608
se	12.7056	0.7710	0.8123	0.0813
p-value	0.0426	0.0122	0.0284	0.0480

Table 9.4.4 Doubled Minus of Log-likelihood for Various Models in Table 9.4.1 Without Age Categorized into Three Groups

Model	df	$-2LR$	Chi-square
Without covariate		158.247	
TMT	1	152.587	5.661
TMT, TMT*$g(y)$	2	146.355	11.894
TMT, AGE	2	147.418	10.929
TMT, AGE, TMT*$g(y)$	3	140.448	17.800
TMT, AGE, TMT*AGE	3	142.006	16.241
TMT, AGE, TMT*AGE, TMT*$g(y)$	4	134.236	24.012

2.3665 to 1.111×10^{22}. It seems that the addition of the product term between treatment and time causes instability in the estimation of the treatment effect.

For investigation of possible interaction between treatment and age, Table 9.4.4 presents the twice minus of the log-likelihood from six different models: successive significant contributions of age after treatment (chi-square = 5.661, p-value = 0.0119), treatment-by-time interaction after treatment and age (chi-square 6.97, p-value = 0.0083), and treatment-by-age interaction after treatment, age, and treatment-by-time interaction (chi-square = 134.236, p-value = 7.77, p-value = 0.0053). Because contributions of all four covariates, as assessed by log-likelihood, are all statistically significant at the 5% level, the resulting model is a four-parameter model. Table 9.4.5 displays the estimates of the regression coefficients and their large-sample standard errors for the four-parameter model and for the model without inclusion of treatment-by-time interaction,

Table 9.4.5 Comparison of Estimates of Regression Coefficients With and Without Age Centered at 21 Years Old in Table 9.4.1

Model	Statistics	TMT	AGE	TMT*AGE	TMT*$g(y)$
Age not centered	Estimate	−1.6470	−0.1641	0.1293	
	se	0.1293	0.1008	0.0569	
	p-value	0.2088	0.1035	0.0232	
Age centered	Estimate	1.0678	−0.1641	0.1293	
	se	0.5445	0.1008	0.0569	
	p-value	0.0499	0.1035	0.0232	
Age not centered	Estimate	24.1590	−0.1758	0.1398	0.1708
	se	12.6835	0.1022	0.0580	0.0812
	p-value	0.0568	0.0854	0.0159	0.0353
Age centered	Estimate	27.0956	−0.1758	0.1398	0.1708
	se	12.6770	0.1022	0.0580	0.0812
	p-value	0.0324	0.0854	0.0159	0.0353

Table 9.4.6 Results of the Estimates of Age Effects Stratified for Treatment in Table 9.4.1

Statistics	AGE1	AGE2
Estimate	1.9309	1.7902
se	0.7712	0.8121
p-value	0.0123	0.0275

both obtained when age is both centered and not centered at the median age of 21 years old. Although the log-likelihood is the same whether age is centered or not, estimates of regression coefficients may be drastically different. For the model without treatment-by-time interaction, the estimated treatment effect is −1.6470 when age is not centered. On the other hand, when age is centered, the estimated treatment effect is 1.0678, which is very close to the 1.1651 obtained from the model with treatment, AGE1, and AGE2. This example illustrates that the collinearity between age and treatment-by-age interaction with age not centered gives not only an erroneous estimate of the treatment effect but also a wrong sign. Other estimates, however, are not affected regardless of whether or not the age is centered. Furthermore the magnitude and sign of estimates for age and treatment-by-age interaction are very consistent for both three-parameter and four-parameter models. A negative estimate of age effect indicates that the risk of severe AGVHD or death is significantly smaller in older patients than younger patients. The hazard ratio of MTX to CSP + MTX increases over time as demonstrated by a significant positive estimated treatment-by-time interaction. Again the estimated treatment effect is inflated when treatment-by-time interaction is in the model. A significant treatment-by-time interaction indicates possible different baseline hazard functions between the two treatments. A stratified Cox proportional hazard model was fitted to the data using treatment as a stratification factor and AGE1 and AGE2 as covariates in the model. The results are given in Table 9.4.6. Their estimates and corresponding large-sample standard errors are very similar to those provided in Table 9.4.3 for both models without and with treatment-by-time interaction.

9.5 CALENDAR TIME AND INFORMATION TIME

At the planning stage the sample size is usually calculated based on information regarding the expected difference between treatments, its corresponding variability, and the desired statistical power for a predetermined risk of type I error. After the study is initiated, patients are enrolled to receive the treatment for a fixed length of time until he or she reaches either the end of the study or the time of analysis with survival as one of the primary endpoints. The Postmenopausal Estrogen/Progestin Interventions (PEPI), for example, is a trial cofounded by the U.S. National Institutes of Health to evaluate the effects of unopposed estrogen and combined estrogen-progestin therapy on four major

Table 9.5.1 Treatments Evaluated in PEPI

Treatment	Estrogen	Progestin
1	Premarin: 0.625 mg daily	Placebo
2	Premarin: 0.625 mg daily	Provera: 10 mg days 1–12 Placebo: 13–28
3	Premarin: 0.625 mg daily	Provera: 25 mg daily
4	Premarin: 0.625 mg daily	Micronized progestrone: 200 mg days 1–12
5	Placebo	Placebo

Source: Espeland (1995).

cardiovascular disease risk factors in postmenopausal women (Espeland et al., 1995). These four risk factors include plasma HDL-cholesterol, systolic blood pressure, two-hour postoral glucose serum insulin, and plasma fibrinogen. Table 9.5.1 gives the five treatments to be assessed in the PEPI. In this trial 168 women per treatment for a total sample size of 840 women was determined to provide a minimum of 92% statistical power for detecting a difference of at least 5 mg/dl in mean change in HDL. The sample size was selected to control an overall type I error rate of 5% with Bonferroni adjustment for all pairwise comparisons among treatments. The sample size was also chosen to account for a 10% of lost to follow-up and 27% of dropouts. This study was conducted in seven clinical centers in the United States. Each center was expected to recruit and randomize 120 women. Based on the past experience, it was anticipated that enrollment and randomization of all women could be completed in a 12-month recruiting period, beginning in December 1989. Although by the end of November 1990 recruiting activities were stopped, a decision was made to allow the women who entered the screening process to be enrolled. As a result the last woman was randomized in February 1991. It turned out that a total of 875 women were actually randomized and enrolled into the study. After randomization, each subject returned to one of the seven clinical centers after 3, 6, and 12 months for the first year and at 6-month intervals thereafter for a total of three years.

The length of actual trial of the PEPI was four years and three months. When there is a gap between the time of recruiting of a subject and the time of randomization after screening for eligibility, the number of subjects actually randomized is usually larger than the planned sample size, though for this study there was only about 4% more than the planned 840 subjects. In addition the length of the recruiting period for the PEPI was within the general limits of the expected time frames.

Since despite a well-prepared protocol and enrollment plan, lots of unexpected problems and logistic issues can occur during the recruiting period, the duration of the recruiting period of clinical trials is generally much longer than the anticipated length. The target of the planned sample size cannot be reached if appropriate actions are not taken. Therefore, after randomization of the first patient, the trial information starts to accumulate as time progresses. A full

100% of information should be accumulated by the time the last patient completes the scheduled follow-up.

As illustrated by the following examples, this is not always the case. In particular, for clinical trials with survival as one of the primary endpoints, the planned sample size is not even observed. For example, the Beta-blocker Heart Attack Trial (BHAT, 1982) was a randomized, double-blind, placebo-controlled trial sponsored by the U.S. National Heart, Lung, and Blood Institute to evaluate the effect of long-term use of propranolol on possible reduction of mortality of patients with a myocardial infarction 5 to 21 days within randomization. A total of 3837 patients were recruited between June 1978 and October 1980. The protocol was planned to have an average three-year follow-up period that was scheduled to end in June 1982. Since the primary endpoint was the mortality rate, the sample size was actually the expected deaths by June 1982, which was postulated to be 628 in the protocol. However, this number was never observed because the BHAT was terminated early in October 1981 after 318 deaths. Clearly unexpected harmful events can force investigators to terminate the trial earlier. Another example was the Cardiac Arrhythmia Suppression Trial (CAST, 1989) which was stopped early because of a significantly higher number of deaths in patients treated with flecainide or encainide compared to the placebo group. The CAST was designed to enroll a total of 4400 patients to achieve at least 80% power for detection of a 30% reduction in death from arrhythmia provided by the active drugs, under the assumption of a 11% mortality over three years for placebo. The preliminary report indicated that the CAST was not designed to prove that an antiarrhythmic drug can cause harm. However, the drugs did, and the trial was terminated early in April 1989. Both the original 425 arrhythmia events specified in the protocol and 300 lately revised in March 1989 for early termination were never observed because the CAST was stopped after 1455 patients received blind therapy and only fewer than 10% of the total expected events were observed.

Let N be the total sample size stated in the protocol for the provision of a desired power using a particular alternative hypothesis at a specified significance level. It is expected to recruit N subjects in the time interval $(0, T_c)$, where T_c is the maximum length of duration in calendar time for completion of the study. From the above discussion, T_c and N_c (the final sample size of any clinical trials) are random variables. The relationship between the duration and sample size of a clinical trial can be explained by the concept of calendar time and information time (Lan and DeMets, 1989; Lan and Zucker, 1993). For simplicity this concept is illustrated through statistical inference for a population mean by the following hypotheses:

$$H_0: \mu = \mu_0,$$
$$\text{vs.} \quad H_a: \mu \neq \mu_0.$$

As was mentioned earlier, the variance of any estimates of μ obtained from the data is a measure of the precision of the estimate. The smaller variance is, the higher the precision in the estimation and the more information regarding parameter μ obtained. As a result the statistical information about μ provided by its estimator $\hat{\mu}$ computed from the data is defined as the inverse of the variance of $\hat{\mu}$. Let $Y_1, \ldots, Y_N$ be i.i.d. clinical responses with population mean μ and variance σ^2. Then the sample mean $\bar{Y}$ is the most commonly employed unbiased estimator for μ. Its variance is given by $\text{var}(\bar{Y}) = \sigma^2/N$. In other words, the information about the unknown population mean μ provided by the sample mean is defined as

$$I = \frac{N}{\sigma^2}, \qquad (9.5.1)$$

which is the sample size times the inverse of the population variance. If $\sigma^2 = 1$, then the information about μ provided by the sample $(Y_1, \ldots, Y_N)$ is simply the sample size N.

The experimental units in most of clinical trials are subjects. As a result the information regarding the effectiveness and safety of the drug under study is also conceptually measured in terms of the number of patients who complete the study. The more subjects that complete the trial, the more clinical information is provided. The statistical information defined above therefore can also be interpreted as the clinical information. Thus the planned sample size specified in the protocol can be viewed as the minimum information required in order to achieve the desired power for detecting a minimum treatment effect at a predetermined risk of type I error.

As illustrated in the previous examples, almost all the clinical trials are longitudinal in nature in which responses of subjects are evaluated at prescheduled visits during the course of the study until either subjects complete the trial or they withdraw prematurely from the study for various reasons. The information defined above is based on the number of the patients who complete the study. In other words, the contribution to the information is 1 if the patient completes the study and is 0 otherwise. However, the information provided by the patients who discontinue the trial before the completion of the study is also valuable and useful for the assessment of efficacy, safety, and quality of life of the drug under study. The intention-to-treat analysis requires one to include all measurements at all time points by every patient because the information provided by the discontinued patients is not 0 but between 0 and 1.

Lan and Zucker (1993) give a definition of information for a particular subject for the inference of population mean slope under a linear random effects model as

$$I_i = \frac{1}{\sigma_\theta^2 \{1 + [R/SS(t_i)]\}}, \qquad (9.5.2)$$

where

$$R = \frac{\sigma_e^2}{\sigma_\theta^2},$$

σ_e^2 is the variance for the within-subject error, σ_θ^2 is the variance for random subject-specific slope,

$$SS(t_i) = \sum_{j=1}^{J_i} (t_{ij} - \bar{t}_i)^2$$

and t_{ij} represent the time point j for subject i where the clinical endpoints were measured, $j = 1, \ldots, J_i$; $i = 1, \ldots, N$. The total information of the entire clinical trial I is then the sum of information from each individual as given by

$$I = \sum_{i=1}^{N} I_i.$$

$SS(t_i)$ in (9.5.2) is the correct sum of squares for the time points of subject i with the length of follow-up equal to J_i. In addition, $t_{i1} < \ldots < t_{iJ_i}$. It follows that as J_i increases, more responses are measured and $SS(t_i)$ gets larger. Assuming that $\sigma_\theta^2 = 1$ if $SS(t_i) = 0$, then the individual information $I_i = 0$. However, as $SS(t_i)$ tends to infinity, the individual information becomes 1. The contribution by each subject to the total information depends on the number of measurements for the primary endpoints made by the individual. The number of measurements for the primary endpoints is, in general, proportional to the length of the duration that an individual stays in the trial. The longer a patient stays in the study, the more information is provided by the patient. The contribution by an individual patient is 1 if he or she completes the study.

As was indicated before, the entire duration of a clinical trial roughly consists of a fixed recruiting period plus either a fixed or an open-ended follow-up period. Subjects are enrolled into the study during different time points, which may be at a different entry rate during the recruiting period. This type of accrual of patients is referred to as the staggering entry. In addition patients may withdraw from the study prematurely during the follow-up period for some known or unknown reasons which may or may not be related to the treatments under the study. As a result, at a given calendar time point t, the information is different for each subject in the study. However, the total information can still be obtained at t by enumeration of the observed number of patients by the length of their durations in the follow-up period.

Suppose that a clinical trial plans to enroll a total of N patients with a maximum duration T_c in the calendar time scale. Let $n(t)$ be the number of subjects who complete the study at calendar time t. For the purpose of illustration, we use first definition of individual information which is 1 if a subject completes the trial. Therefore the total information at calendar time t, $I(t)$ is just $n(t)$. $I(t)$ represents the amount of information accumulated by calendar time t. Similarly the total information at the maximum duration T_c, $I(T_c)$, is $N(T_c)$ which is the maximum information expected to obtain at the maximum duration of the trial. It follows that the information time at calendar time t is defined as the proportion of the information available as of calendar t to the total information provided at the maximum duration T_c (Lan and Zucker, 1993). As a result the relationship between calendar time and information time $s(t)$ is given by

$$s(t) = \frac{n(t)}{N(T_c)}, \qquad 0 \leq t \leq T_c. \tag{9.5.4}$$

From (9.5.4) the information time is between 0 and 1. Increment of the number of subjects who complete the study is discrete. In other words, the information time $s(t)$ defined in (9.5.4) is also discrete despite the fact that calendar time is continuous. Let t_k be the calendar time at which k subjects complete the study, $k = 0, 1, \ldots, N$. Then the corresponding information time is given as

$$\begin{aligned} s_k &= s(t_k) \\ &= \frac{n(t_k)}{N(T_c)}, \qquad k = 0, \ldots, N. \end{aligned} \tag{9.5.5}$$

We can transform the discrete information times defined in (9.5.5) into a continuous time by the following relationship:

$$s = \begin{cases} 0 & \text{if } s < s_1 = \frac{1}{N}, \\ s_k & \text{if } s \in [s_k, s_{k+1}). \end{cases} \tag{9.5.6}$$

In other words, the explicit expression for the relationship between calendar time and information time is given as

$$s(t) = \begin{cases} 0 & \text{if } t < t_1, \\ s_k & \text{if } t \in [t_k, t_{k+1}). \end{cases} \tag{9.5.7}$$

Formulation (9.5.7) states that the information time from calendar time t_k, when the kth subject completes the study, to just prior to t_{k+1}, when the $(k+1)$th subject just about completes the study, is equal to $s_k = n(t_k)/N(T_c)$, the fraction of information available at calendar time t_k relative to the total information $N(T_c)$ at the maximum duration T_c.

The definition of information time and relationship between the information time and calendar time assumes that both the maximum duration T_c and maximum information $N(T_c)$ are known and that the total information at the maximum duration T_c is equal to the maximum information. However, as illustrated by the PEPI, CAST, and BHAT studies, both maximum information and maximum duration are random variables. If a clinical trial is allowed to continue until all N subjects, the predetermined sample size, have completed the scheduled follow-up period, then it is referred to as the maximum information trial. On the other hand, if a clinical trial is terminated at the maximum duration T_c, then it is referred to as the maximum duration trial. Since a maximum information trial continues until the last of all the predetermined number of patients completes the study, the total duration of the trial must be a random variable. Similarly for a maximum duration trial, the total information is a random variable.

Example 9.5.1

The results of BHAT (BHAT Investigators, 1982; Lan and DeMets, 1989) are used to illustrate the concept of calendar time and information time. The BHAT used the logrank test statistic given in (9.3.5) to compare the mortality of 1912 patients assigned to propranolol with that of 1921 patients given the placebo who were enrolled between June 1978 and October 1980. Unlike most test statistics for continuous or categorical responses, the information provided by the logrank test statistic is a function of the number of expected deaths. When the BHAT was designed, it was expected to have 628 deaths by June 1982, the scheduled end of the follow-up period. However, the maximum information of 628 deaths was never observed because the BHAT was terminated early in October 1981 due to convincing evidence of reduction in mortality provided by propranolol. The maximum information was revised to 400 when the data were available later in September 1982. The Data Monitoring Committee (DMC) met six times in May 1979, October 1979, March 1980, October 1980, April 1981, and October 1981. The corresponding number of deaths were 56, 77, 126, 177, 247, and 318. Since the BHAT enrolled the first patient in June 1978 and was originally scheduled to end the follow-up in June 1982, $T_c = 48$ months. $I(48)$ was originally planned to be 628 and later was revised to be 400. By the time when the BHAT was terminated by October 1981, the total duration of the trial in calendar time was 40 months which is about 83% of the maximum duration of 48 months specified in the protocol. There were a total of 318 deaths in October 1981. Hence $t_{318} = 40$ months. If the maximum information $I(48) = 628$ deaths, the corresponding information time at calendar time $t_{318} = 40$ months is $s(40) = 0.51$. On the other hand, if $I(48) = 400$, then $s(40) = 0.80$ which is quite close to 0.80, the cumulative proportion of 40 months with respect to the scheduled maximum duration of 48 months. Table 9.5.2 gives a summary of calendar times and information times for each of the six DMC meetings.

Table 9.5.2 Calendar Time and Information Time for Beta-Blocker Heart Attack Trial

Time of DMCM	Calendar Time	Cumulative Proportion in Calendar Time	Cumulative Deaths	$D = 628$ Information Time	$D = 400$ Information Time
6/1978	0 month	0	0	0	0
5/1979	11 months	0.23	56	0.09	0.14
10/1979	16 months	0.33	77	0.12	0.19
3/1980	21 months	0.44	126	0.20	0.32
10/1980	28 months	0.58	177	0.28	0.44
4/1981	34 months	0.71	247	0.39	0.62
10/1981	40 months	0.83	318	0.51	0.80
6/1982	48 months	1.00	628	1.00	1.00

Source: Lan and DeMets (1989).
Note: DMCM: Data-monitoring committee meeting.

9.6 GROUP SEQUENTIAL METHODS

With almost no exception, clinical trials are longitudinal in nature, and it is not impossible to enroll and randomize all required patients on the same day either. The data of clinical trials are accumulated sequentially over time. It is therefore ethical and in the best interest of patients, as well as scientific and economic, to allow monitoring of information for management of the study and for decision making for possible early termination based on convincing evidence of either benefit or harm of the drug under investigation. The rationale for interim analyses of accumulating data in clinical trials was established by the Greenberg Report (Heart Special Project Committee, 1968) three decades ago. Since then, development of statistical methodology and decision processes for implementation of data monitoring and interim analyses for early termination has attracted a lot of attention. This section is to provide an overview of some of the most commonly employed statistical methods for interim analyses and data-monitoring process. Literature in this area is huge, and this overview is by no means comprehensive. Books on this field include Armitage (1975) and Whitehead (1992) for theoretical background and methodology development and Peace (1992) with emphasis on the biopharmaceutical applications. Armitage et al. (1969), Haybittle (1971), Peto et al. (1976), Pocock (1977), and O'Brien and Fleming (1979) provide some well-known group sequential methods. Lan and DeMets (1983) introduce the alpha spending function. In addition DeMets and Lan (1994) give a review of interim analyses through the alpha spending function. PMA Biostatistics and Medical Ad hoc Committee on interim analysis also published a position paper of the U.S. Pharmaceutical Manufacturing Association interim analysis in the pharmaceutical industry (PMA, 1993). Issues 5 and 6 of volume 12 of *Statistics in Medicine* (1993) published the proceedings of a workshop on *Practical Issues in Data Monitoring of Clinical Trials*, which was sponsored by the four institutes of the U.S. National Institutes of Health held at Bethesda, Maryland, on January 27–28, 1992. Finally, with respect to cancer trials, a workshop on *Early Stopping Rules in Cancer Clinical Trials* was held at Robinson College, Cambridge, England, on April 13–15, 1993. The proceedings for this workshop was published in Issue 13–14 of volume 13 of *Statistics in Medicine* (1994).

We consider a randomized, triple-blind, two parallel-group trial that compares an antihypertensive agent with a matching placebo. One of the primary endpoints is change from the baseline in diastolic blood pressure (mm Hg). The hypotheses can be expressed as follows:

$$H_0: \mu_T = \mu_P,$$
$$\text{vs.} \quad H_a: \mu_T \neq \mu_P, \quad (9.6.1)$$

where μ_T and μ_P represent the population mean change from the baseline in diastolic blood pressure, respectively, for the test drug and the placebo. The intuitive test statistic for the null hypothesis of (9.6.1) is the Z-statistic which is given by

$$Z = \frac{\overline{Y}_T - \overline{Y}_P}{se(Y_T) + se(Y_P)}, \quad (9.6.2)$$

where $\overline{Y}_T$ and $\overline{Y}_P$ are the sample means in the change from the baseline in diastolic pressure computed from the patient receiving test drug and placebo, respectively, and s.e.$(\overline{Y}_T)$ and s.e.$(\overline{Y}_P)$ are the standard errors of Y_T and Y_P. When the sample size is at least moderate, the standard normal distribution provides an adequate approximation to the distribution of Z. Hence the null hypothesis of (9.6.1) is rejected at the 5% level of significance if the absolute value of the observed Z-value is greater than 1.96, which is the 2.5% upper quantile of a standard normal distribution.

The concept of the group sequential procedures is fairly simple. The number of planned interim analyses can be determined in advance and specified in the protocol. Let N be the total planned sample size with equal allocation to the two treatments. Suppose that K interim analyses are planned with equal increments of accumulating data. Then we can divide the duration of the clinical trial into K intervals. Within each stage, the data of $n = N/K$ patients are accumulated. At the end of each interval, an interim analysis will be performed using Z-statistic, denoted by Z_i, in (9.6.2) with the data accumulated up to that point. Two decisions will be made based on the result of each interim analysis. First, the trial is continued if

$$|Z_i| \leq z_i, \quad i = 1, \ldots, K - 1, \quad (9.6.3)$$

where z_i are critical values known as the group sequential boundaries. However, we declare that we fail to reject the null hypothesis of (9.6.1) if

GROUP SEQUENTIAL METHODS

Table 9.6.1 Data Structure of Group Sequential Methods

Number of Interim Analysis (K)	Randomization		Information Time	Z-Statistic
	Test Drug	Placebo		
1	$n/2$	$n/2$	$1/K(n)$	Z_1
2	$2n/2$	$2n/2$	$2/K(2n)$	Z_2
3	$3n/2$	$3n/2$	$3/K(3n)$	Z_3
$\vdots$	$\vdots$	$\vdots$	$\vdots$	$\vdots$
K	$Kn/2$	$Kn/2$	$1(Kn)$	Z_K

$$|Z_i| \leq z_i \quad \text{for all } i = 1, \ldots, K. \quad (9.6.4)$$

Nevertheless, the null hypothesis is rejected, and we can terminate the trial if at any of the K interim analyses

$$|Z_i| > z_i, \quad i = 1, \ldots, K. \quad (9.6.5)$$

For example, at the end of the first interval, an interim analysis is carried out with the data of n patients. If (9.6.3) is true, we continue the trial to the second planned interim analysis. Otherwise, we reject the null hypothesis, and we can stop the trial. After the trial continues to the end of the third interval, the accumulated data of $3n$ patients will be used for the third interim analysis. Data of all patients will be utilized for the final interim analysis after equation (9.6.3) is satisfied at all previous $K - 1$ interim analyses. The trial will be terminated at the final analysis regardless if (9.6.3) is true or not. If (9.6.3) is true at the final analysis, then we can declare that the data from the trial do not provide sufficient evidence to doubt the validity of the null hypothesis. Otherwise, the null hypothesis is rejected, and we can conclude that there is statistically significant difference in change from the baseline of diastolic blood pressure between the test drug and the placebo. The data structure for the group sequential procedures is illustrated in Table 9.6.1. As can be seen from Table 9.6.1, the interim analysis is scheduled after equal information on n patients is accumulated. Therefore the time scale for the interim analyses is the information time. For this example each interim analysis is planned based on information time at $1/K, 2/K, 3/K, \ldots, (K - 1)/K$, and 1, which is called the K-stage group sequential procedure.

In contrast to the fixed sample where only one final analysis is performed, K analyses are carried out for the K-stage group sequential procedure. Suppose that the nominal significance level for each of the K interim analyses is still 5%. Then because of repeated testing based on the accumulated data, the overall significance level is inflated. In other words, when there is no difference in the

Table 9.6.2 Repeated Significance Tests on Accumulative Data

Number of Repeated Tests at 5% Level	Overall Significance Level
1	0.05
2	0.08
3	0.11
4	0.13
5	0.14
10	0.19
20	0.25
50	0.32
100	0.37
1000	0.53
∞	1.00

Source: Armitage et al. (1969).

change from the baseline of diastolic blood pressure between the test drug and the placebo, the probability of declaring at least one significance result increases due to K interim analyses. For the five planned interim analyses as shown in Table 9.6.2 (Armitage, et al., 1969), the overall type I error rate inflates to 14% instead of 5%. In other words, even though there is no difference between the test drug and the placebo, the odds of at least one false positive finding increase from 1 out of 20 to 1 in 7.

Various methods have been proposed to maintain the overall significance level at the prespecified nominal level. One of the early methods was proposed by Hybittle (1971) and Peto et al. (1976). They proposed using 3.0 as the group sequential boundaries for all interim analyses except for the final analysis for which they suggested 1.96. In other words,

$$z_i = \begin{cases} 3.0 & \text{if } i = 1, \ldots, K - 1, \\ 1.96 & \text{if } i = K. \end{cases}$$

Therefore their method can be summarized as follows:

Step 1. At each of the K interim analyses, compute Z_i, $i = 1, \ldots, K - 1$.

Step 2. If the absolute value of Z_i crosses 3.0, then reject the null hypothesis and recommend a possible early termination of the trial; otherwise, continue the trial to the next planned interim analysis and repeat steps 1 and 2.

Step 3. For the final analysis, use 1.96 for the boundary. The trial stops here regardless if the null hypothesis is rejected.

GROUP SEQUENTIAL METHODS

Table 9.6.3 Group Sequential Boundaries

Number of Interim Analyses (K)	Group Sequential Boundaries		
	Pocock	O'Brien-Fleming	
	Value	Multiplier	Value
1	1.960	$\sqrt{1/i}$	1.960
2	2.178	$\sqrt{2/i}$	1.978
3	2.289	$\sqrt{3/i}$	2.004
4	2.361	$\sqrt{4/i}$	2.024
5	2.413	$\sqrt{5/i}$	2.040
6	2.453	$\sqrt{6/i}$	2.053
7	2.485	$\sqrt{7/i}$	2.063
8	2.512	$\sqrt{8/i}$	2.072
9	2.535	$\sqrt{9/i}$	2.080
10	2.555	$\sqrt{10/i}$	2.086

Note: Two-sided: 0.05; One-sided: 0.025. $i = 1, \ldots, K$.
Source: Modified from Jennison and Turnball (1989, 1991).

The Haybittle and Peto's method is very simple. However, it is a procedure with ad hoc boundaries that are independent of any planned interim analyses and stages. Pocock (1977) proposes different group sequential boundaries that depend on the number of planned interim analyses. However, his boundaries are constant at each stage of an interim analysis. For example, when the number of planned interim analyses is 5, Pocock (1977) suggests using 2.413 as the group sequential boundary for each of the five interims. However, if $K = 4$, then Pocock's boundary decreases to 2.361. The implementation of Pocock's group sequential procedure is the same as that of the Haybittle and Peto steps 1 to 3 above. The Pocock's group sequential boundaries are given in Table 9.6.3. Since limited information is included in the early stages of the interim analyses, O'Brien and Fleming (1979) suggest setting conservative boundaries for the interim analyses scheduled to be carried out in an early phase of the trial. Their group sequential boundaries not only depend on the number of the planned interim analysis but also are a function of their stages. As a result the O'Brien-Fleming boundaries can be calculated as:

$$z_{iK} = \frac{c_K \sqrt{K}}{\sqrt{i}}, \qquad 1 \leq i \leq K, \tag{9.6.6}$$

where c_K is the critical value for a total of K planned interim analyses also provided in Table 9.6.3. Again let us suppose that five planned interim analyses are scheduled. Then $c_5 = 2.04$ and boundaries for each stage of these five interim analyses are given as

$$z_{i5} = \frac{2.04\sqrt{5}}{\sqrt{i}}, \qquad 1 \le i \le 5.$$

For example, the O'Brien-Fleming boundary for the first interim analysis is equal to $(2.04)(\sqrt{5}) = 4.561$. The O'Brien-Fleming boundaries for the other four interim analyses can be similarly computed as 3.225, 2.633, 2.280, and 2.040, respectively. A graphical comparison of the group sequential boundaries among these three methods is given in Figure 9.6.1. It is evident from Figure

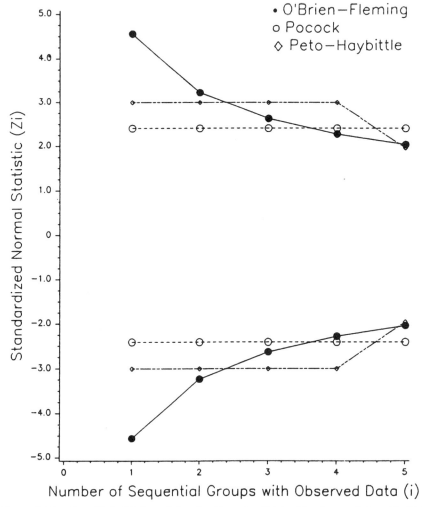

Figure 9.6.1 Two-sided 5% boundaries for Pocock, O'Brien-Fleming, and Peto-Haybittle methods. (Source: Lan and DeMets, 1994.)

9.6.1 that the O'Brien-Fleming boundaries are so conservative, that the early trial results must be extreme for any prudent and justified decision-making in recommendation of a possible early termination when very limited information is available. On the other hand, in the late phase of the trial when the accumulated information approaches the required maximum information, their boundaries also become quite close to the critical value when no interim analysis has been planned. As a result the O'Brien-Fleming method does not require a significant increase in the sample size for what has been already planned. Therefore the O'Brien-Fleming group sequential boundaries have become one of the most employed procedures for the planned interim analyses of clinical trials. Jennison and Turnbull (1989) suggest performing interim analyses by application of repeated confidence intervals. The repeated confidence intervals take the usual form of a confidence interval but with critical values replaced by the group sequential boundaries given in Table 9.6.3. The ith repeated confidence interval for $\mu_T - \mu_P$, the mean difference in change from the baseline of diastolic blood pressure between the test drug and the placebo, is given by

$$(\bar{Y}_T - \bar{Y}_P) \pm z_{iK}[se(\bar{Y}_T) + se(\bar{Y}_P)], \qquad i \leq i \leq K, \qquad (9.6.7)$$

where z_{iK} are the group sequential boundaries. The implementation of the repeated confidence interval approach is similar to the group sequential procedure described above. For a two-sided alternative, an early termination of the ith interim analysis is recommended if the ith repeated confidence interval does not contain 0. Otherwise, the trial continues. The repeated confidence interval approach is appealing because it combines aspects of sequential estimation and testing for a full exploration of the data at each interim analysis. The repeated confidence interval approach can also be applied to the equivalence problem (Jennison and Turnbull, 1993). Gould (1995b) extends the group sequential procedure to the bioequivalence problem.

The implementation of any group sequential procedures requires someone to specify the number of planned interim analyses in advance. It also dictates equal increments of information (number of patients) for each stage. As illustrated in Table 9.6.1, the time scale for scheduling interim analyses is the information time. However, the data-monitoring committee meetings for the review of the data and interim results are scheduled on calendar time. Unless the rate of accumulating information is uniform, it is quite difficult to convert information time to the calendar time for prescheduling the DMC meetings. DeMets and Lan (1994) describe in details this problem from their experiences with BHAT. Moreover it is not uncommon that an occurrence of some unexpected beneficial or harmful effect demands, for ethical and scientific reasons, changes in the frequency of the planned interim analyses and prescheduled DMC meetings. Under these circumstances the group sequential procedures become rather inflexible to limit the decisions and actions that the DMC must take for the best interest of the patients. Lan and DeMets (1983) in their landmark paper introduce the concept of the alpha spending function to overcome the drawbacks of traditional group sequential procedures.

In the standard group sequential procedures, the boundaries are determined by critical values so that the sum of probabilities of exceeding these values at the prescheduled discrete information time points is equal to the predetermined total probability of type I error. The idea of the alpha spending function proposed by Lan and DeMets (1983) is to spend (i.e., distribute) the total probability of false positive risk as a continuous function of the information time. As a result, if the total information scheduled to accumulate over the maximum duration T is known, the boundaries can also be computed as a continuous function of the information time. This continuous function of the information time is referred to as the alpha spending function, denoted by $\alpha(s)$. The alpha spending function is an increasing function of information time. It is 0 when information time is 0, and it is equal to the overall significance level when information time is 1. In other words, $\alpha(0) = 0$ and $\alpha(1) = \alpha$. Let s_1 and s_2 be two information times, $0 < s_1 < s_2 \leq 1$. Also denote $\alpha(s_1)$ and $\alpha(s_2)$ as their corresponding value of alpha spending function at s_1 and s_2. Then $0 < \alpha(s_1) < \alpha(s_2) \leq \alpha$. $\alpha(s_1)$ is the probability of type I error one wishes to spend at information time s_1. With respect to $\alpha(s_1)$, the boundary $z(s_1)$ can then be computed by the following probability statement:

$$P\{Z(s_1) > z(s_1)\} = \alpha(s_1).$$

Suppose that the trial fails to terminate at s_1 and continues to accumulate information up to s_2 when we perform the second interim analysis. The cumulative false positive probability is given by

$$\begin{aligned}\alpha(s_2) &= P\{Z(s_1) > z(s_1) \text{ or } Z(s_2) > z(s_2)\} \\ &= P\{Z(s_1) > z(s_1)\} + P\{Z(s_1) \leq z(s_1) \text{ and } Z(s_2) > z(s_2)\} \\ &= \alpha(s_1) + P\{Z(s_1) \leq z(s_1) \text{ and } Z(s_2) > z(s_2)\}.\end{aligned} \quad (9.6.7)$$

Then $\alpha(s_2) - \alpha(s_1)$ is the proportion of the probability of type I error one is willing to allocate to the additional accumulation of information $s_2 - s_1$. The boundary $z(s_2)$ can be obtained through numerical integration by the following formulation:

$$\alpha(s_2) - \alpha(s_1) = P\{Z(s_1) \leq z(s_1) \text{ and } Z(s_2) > z(s_2)\}. \quad (9.6.8)$$

This relationship is given in Figure 9.6.2, which provides a graphical representation of the alpha spending function. Table 9.6.4 provides different forms of alpha spending functions whose boundaries for 5 interim looks are given in Table 9.6.5.

As an example, consider the O'Brien-Fleming group sequential procedure with a total of five planned interim analyses with equal increment of information for one-sided alternative at the 2.5% overall significance level. The five boundaries are given as $z(0.2) = 4.56$, $z(0.4) = 3.23$, $z(0.6) = 2.63$, $z(0.8) = 2.28$, and $z(1.0) = 2.04$. The cumulative probabilities of type I error is given

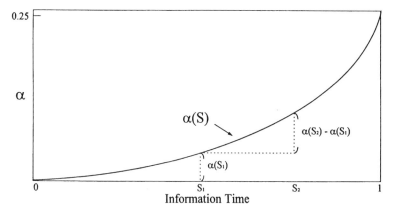

Figure 9.6.2 The alpha spending function $\alpha(s)$.

in Table 9.6.6. As a result the cumulative false positive error rate at the five information time points are given as

$P\{Z(0.2) > 4.56\} = 2.6 \times 10^{-6}$
$\quad P\{Z(0.2) > 4.56 \text{ or } Z(0.4) > 3.23\}$
$\quad\quad = P\{Z(0.2) > 4.56\} + P\{Z(0.2) \leq 4.56 \text{ and } Z(0.4) > 3.23\}$
$\quad\quad = 2.6 \times 10^{-6} + 0.000574 = 0.0006,$
$\quad P\{Z(0.2) > 4.56 \text{ or } Z(0.4) > 3.23 \text{ or } Z(0.6) > 2.63\}$
$\quad\quad = P\{Z(0.2) > 4.56 \text{ or } Z(0.4) > 3.23\}$
$\quad\quad\quad + P\{Z(0.2) \leq 4.56 \text{ and } Z(0.4) \leq 3.23 \text{ and } Z(0.6) > 2.63\}$
$\quad\quad = 0.0006 + 0.0039 = 0.0045,$
$\quad P\{Z(0.2) > 4.56 \text{ or } Z(0.4) > 3.23 \text{ or } Z(0.6) > 2.63 \text{ or } Z(0.8) > 2.28\}$
$\quad\quad = P\{Z(0.2) > 4.56 \text{ or } Z(0.4) > 3.23 \text{ or } Z(0.6) > 2.63\}$
$\quad\quad\quad + P\{Z(0.2) \leq 4.56 \text{ and } Z(0.4) \leq 3.23 \text{ and } Z(0.6) \leq 2.63$
$\quad\quad\quad\quad \text{ and } Z(0.8) > 2.28\}$
$\quad\quad = 0.0045 + 0.0080 = 0.0125,$
$\quad P\{Z(0.2) > 4.56 \text{ or } Z(0.4) > 3.23 \text{ or } Z(0.6) > 2.63 \text{ or } Z(0.8) > 2.28$
$\quad\quad\quad \text{ or } Z(0.1) > 2.04\}$
$\quad\quad = P\{Z(0.2) > 4.56 \text{ or } Z(0.4) > 3.23 \text{ or } Z(0.6) > 2.63$
$\quad\quad\quad \text{ or } Z(0.8) > 2.28\}$
$\quad\quad\quad + P\{Z(0.2) \leq 4.56 \text{ and } Z(0.4) \leq 3.23 \text{ and } Z(0.6) \leq 2.63$
$\quad\quad\quad\quad \text{ and } Z(0.8) \leq 2.28 \text{ and } Z(1.0) > 2.04\}$
$\quad\quad = 0.0125 + 0.0125 = 0.0250$

Table 9.6.4 Different Forms of Alpha Spending Functions

	Approximation
$\alpha_1(s) = 2\{1 - \Phi[z(\alpha/2)/\sqrt{s}]\}$	O'Brien-Fleming
$\alpha_2(s) = \alpha \ln[1 + (e - 1)s]$	Pocock
$\alpha_3(s) = \alpha s^\theta,\ \theta > 0$	Lan-DeMets-Kim
$\alpha_4(s) = \alpha[(1 - e^{-\zeta s})/(1 - e^{-\zeta})],\ \zeta \neq 0,$	Hwang-Shih

When 40% of information of the trial is accumulated, with respect to the O'Brien-Fleming boundaries $Z(0.2) = 4.56$ and $z(0.4) = 3.23$, the allowable probability of type I error is only 0.0006. However, the spendable false positive rate is 0.0039 when the third interim analysis is performed at $s = 0.6$. According to Table 9.6.6 in the O'Brien-Fleming group sequential procedure spends very little probability of type I error for the interim analyses performed early in the trial. On the other hand, successive increments of the false positive rate in the Pocock group sequential method is a decreasing function of information as illustrated in Table 9.6.6. Consequently most of the significance level is spent at an early stage of the trial. Figure 9.6.3 graphically compares the alpha spending functions of the Pocock and O'Brien-Fleming group sequential procedures.

In a maximum duration trial, as noted above, the total information is a random variable, and it is not observed until the trial ends at the maximum duration T_c. It is not uncommon to revise the maximum information specified in the protocol as demonstrated by some of well-known trials sponsored by the U.S. National Institutes of Health such as BHAT (1982) or CAST (1989). As a result information time cannot be used for computing the boundaries. Lan and DeMets (1983) suggest using a monotone function $g(t)$ to convert the calendar time into information time. However, sometimes $g(t)$ is not known and must be estimated. One such approximation is t/T, the ratio of the current calendar time to the maximum duration. Suppose that the trial has progressed to the calendar time t_k with information accumulated to s_k by previous interim analyses performed at some discrete information time s_i, $i = 1, \ldots, k$. Although the total

Table 9.6.5 Examples of Boundaries by Alpha Spending Function

Interim Analysis(s)	O'Brien-Fleming	$\alpha_1(s)$	Pocock	$\alpha_2(s)$	$\alpha_3(s)[\theta = 1]$
1(0.2)	4.56	4.90	2.41	2.44	2.58
2(0.4)	3.23	3.35	2.41	2.43	2.49
3(0.6)	2.63	2.68	2.41	2.41	2.41
4(0.8)	2.28	2.29	2.41	2.40	2.34
5(1.0)	2.04	2.03	2.41	2.39	2.28

Note: Number of interim Analyses = 5. Two-sided: 0.05; One-sided: 0.025

Table 9.6.6 Cumulative Probability of Type I Error

Interim Analysis(s)	Group Sequential Boundaries					
	Pocock			O'Brien-Fleming		
	Value	$\alpha(s)$	Increment	Value	$\alpha(s)$	Increment
1(0.2)	2.41	0.0079	0.0079	4.56	0.0000	2.6×10^{-6}
2(0.4)	2.41	0.0138	0.0059	3.23	0.0006	0.000574
3(0.6)	2.41	0.0183	0.0045	2.63	0.0045	0.0039
4(0.8)	2.41	0.0219	0.0036	2.28	0.0125	0.0080
5(1.0)	2.41	0.0250	0.0031	2.04	0.0250	0.0125

information is not known as t_k, the covariance between $Z(s_i)$ and $Z(s_k)$ can be estimated by $I(t_i)$ to $I(t_k)$, the ratio of information available at calendar time t_i to t_k. Therefore the boundary for the interim analysis at s_k can be computed using the method described above.

The implementation of the alpha spending function requires an advance

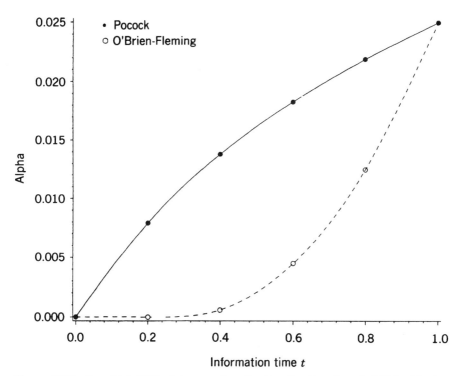

Figure 9.6.3 One-sided 2.5% alpha spending function for Pocock and O'Brien-Fleming boundaries. (Source: Lan and DeMets, 1994.)

selection and specification of the spending function in the protocol. One cannot change and choose another spending function in the middle of a trial. Geller (1994) suggests that the spending function be convex and have the property that the same value of a test statistic is more compelling as the sample sizes increase. Since it is flexible and has no requirement for total information and equal increment of information, there is a danger of abuse of the alpha spending function in increasing the frequency of interim analyses as the results approach the boundary. However, DeMets and Lan (1994) reported that altering the frequency of interim analyses has very little impact on the overall significance level if an O'Brien-Fleming-type or a Pocock-type continuous spending function is adopted.

Pawitan and Hallstrom (1990) study the alpha spending function for CAST with the use of a permutation test. A permutation test is conceptually simple, and it provides an exact test for small sample sizes. In addition it is valid for complicated stratified analysis in which the exact sampling distribution is, in general, unknown and large-sample approximation may not be adequate. Consider the one-sided alternative. For the kth interim analyses, under the assumption of no treatment effect, the null joint permutation distribution of the test statistics $(Z_1, \ldots, Z_K)$ can be obtained by random permutation of treatment assignments on the actual data. Let $(Z_{1b}^*, \ldots, Z_{Kb}^*)$, $b = 1, \ldots, B$, be the statistics computed from B treatment assignments and B be the total number of possible permutations. Given $\alpha(s_1)$, $\alpha(s_2) - \alpha(s_1)$, $\ldots$, $\alpha(s_k) - \alpha(s_{K-1})$, the probabilities of type I error allowed at successive interim analyses, the one-sided boundaries $z_1, \ldots, z_K$ can be determined by the formulas

$$\frac{\text{Number of } (Z_1^* > z_1)}{B} = \alpha(s_1),$$

$$\frac{\text{Number of } (Z_1^* > z_1 \text{ or } Z_2^* > z_2)}{B} = \alpha(s_2) - \alpha(s_1),$$

$$\vdots$$

$$\frac{\text{Number of } (Z_1^* > z_1 \text{ or } Z_2^* > z_2, \ldots, \text{ or } Z_K^* > z_K)}{B} = \alpha(s_K) - \alpha(s_{K-1}). \quad (9.6.9)$$

If B is very large, then the above method can be executed by a random sample with a replacement of size b. The α spending function for an overall significance level of 2.5% for the one-sided alternative is given by

$$\alpha(s) = \begin{cases} \dfrac{\alpha}{2} s & \text{if } s < 1, \\ \alpha & \text{if } s = 1. \end{cases}$$

Table 9.6.7 Summary of the Interim Results Based on the Number of Deaths from Arrhythmia or Cardiac Arrest

	Calendar Times for Interim Analyses	
	9/1/88	3/30/89
Placebo	7(576)[a]	9(725)
Active[b]	22(571)	33(730)
Total	29(1147)	42(1455)
Total Expected	425	300
One-sided α increment	0.0009	0.0011
Observed Logrank	−2.82	−3.22
Lower Bound[c]	−3.18	−3.04

[a] Numbers in the parnetheses are the number of patients assigned to the treatment.
[b] Active treatments include encainide and flecainide.
[c] Lower bounds were computed from a random sample with replacement of size $b = 4000$.
Source: Modified from Pawitan and Hallstrom (1990).

In other words, CAST uniformly spent 50% of the overall probability of type I error until the end of the trial and allocated the other half for the final analysis. The primary event of interest for CAST was death from arrhythmia or cardiac arrest which was original expected to be 425 based on a projection of an event rate of 11% over a three-year follow-up period. However, the observed event rate was about 2% for the placebo group in the first year. Therefore the event rate was revised to 300 on the second interim analysis on March 30, 1989. The results of interim analyses and boundaries are given in Table 9.6.7. At the time of the first interim analysis, the observed number of events was 29. Therefore $\alpha(29/425) = 0.0125(29/425) = 0.0009$. The corresponding lower boundary using a random sample with replacement of size 4000 computed from (9.6.9) was −3.18. Since the observed logrank statistic is −2.82 which is greater than −3.18, the trial continued. Similarly the boundary for the second interim analysis based on an observed 42 events and a revised total expected 300 events was −3.04. But the observed logrank test statistic was −3.22 which crossed the boundary. This result coupled with other considerations, the DMC of CAST recommended on April 17, 1989, that the trial be terminated because of the adverse effect of the class IC arrhythmic suppressant agents.

CAST study (Pawitan and Hallstrom, 1990) also applied the stochastic curtailment method (Lan et al., 1982) to compute the conditional power of the trial given the current data for consideration of possible termination of the trial due to lack of benefit if the conditional power is too low. Lan and Wittes (1988) propose the B-value as a tool for monitoring data for the calculation of conditional power under the assumption that the current trend indicated by the observed data at information time s continues. For simplicity we consider the one-sample problem. Let $\theta = \sqrt{N}\mu$, where N is the maximum information that a trial is planned to accumulate and μ is the population mean. The power of rejecting the null hypothesis at the α significance level for a two-sided test conditioning on the current trend at information time s is given as

$$\phi(\Theta) = 1 - \Phi \left\{ \frac{[z(\alpha/2) - B(s)/s]}{\sqrt{1-s}} \right\}$$

$$= 1 - \Phi \left\{ \frac{[z(\alpha/2) - Z(s)/\sqrt{s}]}{\sqrt{1-s}} \right\}, \qquad (9.6.10)$$

where the B-value at information s is defined as

$$B(s) = \frac{Z(s)}{\sqrt{s}}. \qquad (9.6.11)$$

Lan and Wittes (1988) also consider the application of the B-value to the two-sample Z-test, the two-sample Wilcoxon rank sum test, and the logrank test.

9.7 DISCUSSION

The statistical inference for censored data covered in this chapter was restricted to parallel-group designs in which the time to occurrence of a well-defined clinical event is independent from patient to patient. On the other hand, in the two-sequence, two-period crossover design discussed in Chapter 5, each patient with chronic conditions received both treatments during the study. The result was two censored responses, conditioned under different treatments, were obtained from the same patient. Since they were observed from the same patient, the two censored clinical endpoints were correlated.

There are many examples of correlated censored endpoints in clinical trials. For example, France et al. (1991) report a study whose objective is to evaluate the efficacy of the monotherapy of atenolol at 50 mg twice a day with that of a combination therapy of atenolol and nifedipine in the treatment of angina pectoris. Although both drugs are widely used in the treatment of angina, their modes of action are quite different. Atenolol is a beta-blocker, while nifedipine is a calcium-channel blocker. In addition the dose of atenolol selected at 50 mg b.i.d. for this study was near the top of the dose-response curve such that any additional increase in dose was unlikely to improve the drug's efficacy. Therefore it is expected that the combined therapy of nifedipine and atenolol might provide a more effective treatment of angina than atenolol alone. The study started with a four-week placebo run-in period on atenolol, followed by a two-sequence, two-period crossover design with a duration of four weeks for each treatment period. There was a total of 106 patients who completed the study. At the end of each period, a treadmill test was performed for each patient according to the standard Bruce protocol. The clinical endpoints for assessment of efficacy of the treatment included the time to 1 mm ST segment depression, time to pain, and total time of exercise as well as the reasons for early stopping on the treadmill test. These responses were censored if the treadmill test stopped before symptoms occurred.

Liu and Chow (1993) give another example of correlated censored data in the assessment of clinical equivalence of an albuterol metered dose inhaler (MDI) for acute bronchospasm between the test and reference formulations in patients with reversible obstructive airway disease. Because of its intended route of administration, we have negligible plasma levels. As a result clinical endpoints are used for assessment of bioequivalence for MDI products. The clinical endpoints for MDI products recommended by the FDA guidance are derived from the volume of air forced out of the lung within one second (FEV_1) measured at 0, 10, 15, 30, 60, 90, 120, 180, 240, 300, and 360 minutes after dosing. One of the clinical endpoints derived from FEV_1 is the time to the onset of a therapeutic response which is defined as the event that the FEV_1 measurement evaluated within 30 minutes after dosing exceeds 115% of its baseline measurement at time 0. In order to reduce the variability between patients, crossover designs are the designs of choice in evaluations of bioequivalence for MDI products. As a result the censored data, such as the time to the onset of therapeutic response in the same patient, are correlated.

Several methods are available for paired censored clinical endpoints. For example, O'Brien and Fleming (1987) proposed the paired Prentice-Wilcoxon statistic, Huster, Brookmeyer, and Self (1989) suggest parametric models with adjustments by covariates, Holt and Prentice (1974) and Kalbfleisch and Prentice (1980) employ the Cox proportional hazard model to analyze paired censored data. However, these methods fail to take into account the structure of crossover designs. Recently, under assumption of no carryover effect, France et al. (1991) and Liu and Chow (1993) have extended the method of Cox's proportional hazard model for paired censored data proposed by Kalbfleisch and Prentice (1980) to the crossover design.

Basically the method proposed by Kalbfleisch and Prentice (1980) is the sign test generalized to the censored data, which can be derived from the binary logistic regression discussed in Chapter 8. For each sequence within a two-sequence, two-period design, the treatment effect can be estimated as the regression coefficient for the treatment indicator variable according to the method suggested by Kalbfleisch and Prentice (1980). The overall estimated treatment effect is then the average of individual regression coefficients estimated from each sequence. On the other hand, the period effect is the difference in individual regression coefficients between the two sequences. However, under Cox's proportional hazard model, it is not possible to provide a test for the presence of the carryover effect. Liu and Chow (1993) suggest using the hazard ratio as a criterion in the assessment of bioequivalence between MDI products. Jung and Su (1995) propose a nonparametric estimation procedure for the difference or ratio of median failure times for the correlated censored observations from a two-sequence, two-period crossover study. They recommended the use of a 95% confidence interval for the difference or ratio of median failure time as a criterion in the assessment of equivalence between drug products.

As was mentioned above, the method for correlated censored data obtained from crossover designs proposed by France et al., (1991) and Liu and Chow (1993) is an extension of a sign test that may not be powerful under certain alternatives. Feingold and Gillespie (1996) therefore suggest two methods to test the treatment effect. The first method is to transform the data into a score and to analyze the score according to the standard procedure for continuous data under the linear model for a two-sequence, two-period crossover design, assuming compound symmetry in the covariance matrix. They recommended the use of the Gehan score (1965b). For a two-sequence, two-period crossover design, the treatment effect, period effect, and carryover (or sequence) effect can be obtained from within-patient contrasts. The second testing method, based on the censoring pattern in each period and signs of linear contrast, is to classify the result of each patient as uncensored, left-censored, right-censored, or undefined. Feingold and Gillespie then suggest testing the differences between the two sequences with a distribution-free test developed by Gehan (1965a) and Mantel (1967), which is an extension of the two-sample generalized Wilcoxon test for right-censored data. Feingold and Gillespie also select the average quantile as the parameter for estimation of the treatment effect. The average quantile is referred to as the average distance of each survival curve from the origin, computed over the quantile range where all of the survival curves are defined. If all data are uncensored, then the average quantile is the mean. However, their procedure for hypothesis testing is not in accordance with the estimation results. In other words, if one of the proposed tests indicates a statistically significant treatment effect, this does not imply that the confidence interval for that parameter does not include zero.

The group sequential procedures for interim analyses are basically in the context of hypothesis testing which is aimed at pragmatic study objectives concerning which treatment is better. However, most new treatments such as cancer drugs are very expensive, very toxic, or both. As a result only if the degree of the benefit provided by the new treatment exceeds some minimum clinically significant requirement, it will then be considered for the treatment of the intended medical conditions. Therefore an adequate well-controlled trial should be able to provide not only the qualitative evidence whether the experimental treatment is effective but also the quantitative evidence from the unbiased estimation of the size of the effectiveness or safety over the placebo given by the experimental therapy. For a fixed sample design without interim analyses for early termination, it is possible to achieve both qualitative and quantitative goals with respect to the treatment effect. However, with the group sequential procedure the benefit size of the experimental treatment by the maximum likelihood method is usually overestimated because of the choice of stopping rule. Jennison and Turnbull (1990) point out that the sample mean might not be even contained in the final confidence interval. As a result estimation of the size of treatment effect has received a lot of attention. Various estimation procedures have been proposed such as the modified maximum likelihood estimator (MLE), the median unbiased estimator (MUE) and the midpoint of the equal two tailed 90% confidence interval. For more detail, see Cox (1952), Tsiatis et al., (1984), Kim and DeMets (1987), Kim (1989), Chang and O'Brien (1986), Chang et al. (1989), Chang (1989), Hughes and Pocock (1988), and Pocock and Hughes (1989).

The estimation procedures proposed in the above literature require extensive computation. On the other hand, simulation results (Kim 1989; Hughes and Pocock 1989) show that the alpha spending function corresponding to the O'Brien-Fleming group sequential procedure is quite concave and allocates only a very small total nominal significance level to early stages of the interim analyses, and hence the bias, variance, and mean square error of the point estimator following O'Brien-Fleming procedure are also the smallest. Current research has mainly focused on the estimation of the size of the treatment effect for the primary clinical endpoints on which the group sequential procedure is based. However, there are many other secondary efficacy and safety endpoints to be evaluated in the same trial. The impact of early termination of a trial based on the results from primary clinical endpoints on the statistical inference for these secondary clinical endpoints is unclear. In addition group sequential methods and their followed estimation procedures so far are only concentrated on the population average. On the other hand, inference of variability is sometimes also of vital importance for certain classes of drug products and diseases. Research on estimation of variability following early termination is still lacking. Other areas of interest for interim analyses include clinical trials with more than two treatments and bioequivalence assessment. For group sequential procedures for trials with multiple treatments, see Hughes (1993) and Proschan et al. (1994). For the group sequential bioequivalence testing procedure, see Gould (1995b).

CHAPTER 10

Sample Size Determination

10.1 INTRODUCTION

A major consideration of most clinical studies is to determine whether the drug under investigation is effective and safe. During the planning stage of a clinical study, the following questions are of particular interest to the investigators: (1) How many subjects are needed in order to have a desired power (e.g., 80% chance of correctly detecting a clinically meaningful difference? (2) What's the *trade-off* if only a small number of subjects are available for the study due to limited budget and/or some medical considerations? To address these questions, a statistical evaluation for sample size determination/justification is often employed. Sample size determination usually involves the calculation of a required sample size for some desired statistical properties such as precision and power, whereas sample size justification provides statistical justification for a selected sample size which may be small in number because of budget constraints or some medical considerations. In this chapter our emphasis will be placed on sample size determination. The concept can be easily implemented to provide statistical justification for a selected sample size.

For a given study, sample size is usually determined based on some criteria on type I and/or type II errors. For example, we can choose sample size in such a way that there is a desired precision at a fixed confidence level (or fixed type I error). This approach is referred to as precision analysis for sample size determination. The disadvantage of the precision analysis is that it has a small chance of detecting a true difference. As an alternative, a pre-study power analysis is usually conducted to calculate sample size. The concept of power analysis is to select an acceptable type II error for a fixed type I error. In other words, the selected sample size will have a desired power for correctly detecting a clinical/scientific meaningful difference at a fixed type I error rate. In most clinical trials the pre-study power analysis for sample size determination is the most commonly used method for choosing the sample size. Therefore in this chapter we will focus on sample size determination based on power analysis for various situations of clinical trials.

To perform a pre-study power analysis for sample size determination, the power function of an appropriate test for the hypotheses of interest is necessarily characterized. The hypotheses should be established to reflect the study objectives under the study design. In practice, it is not uncommon to observe discrepancies among study objective (hypotheses), study design, statistical analysis (test statistic), and sample size calculation. These inconsistencies often result in (1) wrong test for right hypotheses, (2) right test for wrong hypotheses, (3) wrong test for wrong hypotheses, or (4) right test for right hypotheses with insufficient power. Therefore, before the sample size can be determined, it is suggested that the following be carefully considered: (1) the study objectives or the hypotheses of interest be clearly stated, (2) a valid design with appropriate statistical tests be used, and (3) the sample size be determined based on a test for the hypotheses of interest.

In the next section we will introduce some basic concepts for sample size determination and outline the precision analysis approach and the method of pre-study power and analysis for sample size calculation. The pre-study power analysis for sample size determination will be considered to derive formulas for complex clinical trials with different study objectives. In Sections 10.3 and 10.4 sample size calculations for two samples comparing two treatments and multiple samples comparing more than two treatments are discussed, respectively. The sample size required for complex clinical trials with survival endpoints is outlined in Section 10.5. In Section 10.6 we explore sample size determination for dose-response studies. Sample size determination for crossover design is described in Section 10.7. A brief concluding remark is presented in the last section.

10.2 BASIC CONCEPT

Study Objectives and Hypotheses

In most clinical trials the primary study objective is related to the evaluation of the effectiveness and safety of a drug product. For example, it may be of interest to show that the study drug is effective and safe compared to a placebo for some intended indications. In some cases it may be of interest to show that the study drug is as effective as, superior to, or equivalent to an active control agent or a standard therapy. In practice, hypotheses regarding medical or scientific questions of the study drug are usually formulated based on the primary study objectives. The hypotheses are then evaluated using appropriate statistical tests under a valid study design. To ensure that the test results will have the desired power with a certain degree of accuracy and reliability, a sufficient number of subjects is needed. As a result sample size determination plays a crucial role in the design of clinical trials.

For a given study objective, a corresponding hypothesis can be formulated. As was discussed in Section 2.6, in clinical trials a hypothesis is usually referred to as a postulation, assumption, or statement that is made about the population regarding the effectiveness and safety of a drug under investigation. For example, the statement that there is a direct drug effect is a hypothesis regarding the treatment effect. For testing the hypotheses of interest, a random sample is usually drawn from the targeted population to evaluate hypotheses about the drug product. A statistical test is then performed to determine whether the null hypothesis should be rejected at a prespecified significance level. Based on the test result, conclusion(s) regarding the hypotheses can be drawn.

The selection of hypotheses depends on the study objectives. For example, if we want to demonstrate the effectiveness of a drug product compared to a placebo in terms of an outcome variable, we could consider the following hypotheses:

$$H_0: \mu_T = \mu_P,$$
$$\text{vs.} \quad H_a: \mu_T \neq \mu_P, \tag{10.2.1}$$

where μ_T and μ_P are the mean response of the outcome variable for the test drug and the placebo, respectively. As was discussed before, a typical approach is to show that there is a statistically significant difference between the test drug and the placebo by rejecting the null hypothesis and then demonstrating that there is a high chance of correctly detecting a clinically meaningful difference if such a difference truly exists.

On the other hand, one may want to show that the test drug is as effective as an active agent or a standard therapy. In this case Blackwelder (1982) suggests testing the following hypotheses:

$$H_0: \mu_T - \mu_S, \leq -\delta,$$
$$\text{vs.} \quad H_a: \mu_T - \mu_S, > -\delta, \tag{10.2.2}$$

where μ_S is the mean for a standard therapy and δ is a difference of clinical importance. The concept is to reject the null hypothesis and conclude that the difference between the test drug and the standard therapy is less than a clinically meaningful difference δ and hence that the test drug is as effective as the standard therapy. This study objective is not uncommon in clinical trials especially when the test drug is considered to be less toxic, easier to administer, or less expensive than the established standard therapy.

To show superiority of a test drug over an active control agent or a standard therapy, we may consider the following hypotheses:

$$H_0: \mu_T - \mu_S \leq \delta,$$
$$\text{vs.} \quad H_a: \mu_T - \mu_S > \delta, \tag{10.2.3}$$

The rejection of the above null hypothesis suggests that the difference between

BASIC CONCEPT

the test drug and the standard therapy is greater than a clinically meaningful difference. Therefore we can conclude that the test drug is superior to the standard therapy by rejecting the null hypothesis of (10.2.3).

In practice, unless there is some prior knowledge regarding the test drug, usually we do not know the performance of the test drug as compared to the standard therapy. Therefore hypotheses (10.2.2) and (10.2.3) are not preferred because they have predetermined the performance of the test drug as compared to the standard therapy. As an alternative, the following hypotheses for therapeutic equivalence are usually considered:

$$H_0: |\mu_T - \mu_S| \geq \delta,$$
$$\text{vs.} \quad H_a: |\mu_T - \mu_S| < \delta. \quad (10.2.4)$$

We then conclude that the difference between the test drug and the standard therapy is of no clinical importance if the null hypothesis of (10.2.4) is rejected.

It should be noted that a valid sample size calculation can only be done based on appropriate statistical tests for the hypotheses that can reflect the objectives under a valid study design. It is then suggested that the hypotheses be clearly stated when performing a sample size calculation. Each of the above hypotheses has a different requirement for the sample size in order to achieve a desired power or precision of the corresponding tests.

Type I and Type II Errors

In practice, two kinds of errors occur when testing hypotheses. If the null hypothesis is rejected when it is true, then a type I error has occurred. If the null hypothesis is not rejected when it is false, then a type II error has been made. The probabilities of making type I and type II errors, denoted by α and β, respectively, are given below:

$$\alpha = P\{\text{type I error}\}$$
$$= P\{\text{reject } H_0 \text{ when } H_0 \text{ is true}\},$$

$$\beta = P\{\text{type II error}\}$$
$$= P\{\text{fail to reject } H_0 \text{ when } H_0 \text{ is false}\}.$$

The probability of making a type I error, α, is called the level of significance. The power of the test is defined as the probability of correctly rejecting the null hypothesis when the null hypothesis is false, namely,

$$\text{Power} = 1 - \beta$$
$$= P\{\text{reject } H_0 \text{ when } H_0 \text{ is false}\}.$$

For example, suppose that we want to test the following hypotheses:

H_0: The drug is ineffective;
vs. H_a: The drug is effective.

Then a type I error occurs if we conclude that the drug is effective when in fact it is not. On the other hand, a type II error occurs if we claim that the drug is ineffective when in fact it is effective. In clinical trials none of these errors is desirable. A typical approach is to avoid type I error but at the same time to decrease type II error so that there is a high chance of correctly detecting a drug effect when the drug is indeed effective. Chow and Liu (1992a) illustrate the relationship between type I error and type II error (or power). It appears that α decreases as β increases and α increases as β decreases. The only way of decreasing both α and β is to increase the sample size.

Typically sample size can be determined by controlling either a type I error (or confidence level) or a type II error (or power). In what follows we will introduce two concepts for sample size determination, namely precision analysis based on type I error and power analysis based on type II error.

Precision Analysis

In practice, the maximum probability of commiting a type I error that one can tolerate is usually considered the level of significance. The confidence level, $1 - \alpha$, then reflects the probability or confidence of not rejecting the true null hypothesis. Since the confidence interval approach is equivalent to the method of hypotheses testing, we can determine the sample size required based on the type I error rate using the confidence interval approach. For a $(1 - \alpha)100\%$ confidence interval, the precision of the interval depends on its width. The narrower the interval is, the more precise the inference is. Therefore the precision analysis for sample size determination is to consider the maximum half width of the $(1 - \alpha)100\%$ confidence interval of the unknown parameter that one is willing to accept. Note that the maximum half width of the confidence interval is usually referred to as the *maximum error* of an estimate of the unknown parameter. For example, let $Y_1, Y_2, \ldots, Y_n$ be independent and identically distributed normal random variables with mean μ and variance σ^2. When σ^2 is known, a $(1 - \alpha)100\%$ confidence interval for μ can be obtained as

$$\overline{Y} \pm z(\alpha/2)\, \sigma/\sqrt{n},$$

where $z(\alpha/2)$ is the upper $(\alpha/2)$th quantile of the standard normal distribution. The maximum error, denoted by E in estimating the value of μ that one is willing to accept, is then defined as

BASIC CONCEPT

$$E = |\bar{Y} - \mu| = z(\alpha/2)\,\sigma/\sqrt{n}.$$

Thus the sample size required can be chosen as

$$n = \frac{z(\alpha/2)^2 \sigma^2}{E^2}. \qquad (10.2.5)$$

Note that the maximum error approach for choosing n is to attain a specified precision while estimating μ which is derived only based on the interest of type I error. A nonparametric approach can be obtained by using the following Chebyshev inequality:

$$P\{|\bar{Y} - \mu| \leq E\} \geq 1 - \frac{\sigma^2}{nE^2}$$

and hence

$$n = \frac{\sigma^2}{\alpha E^2} \qquad (10.2.6)$$

Note that the precision analysis for sample size determination is very easy to apply based on either (10.2.5) or (10.2.6). For example, suppose that we wish to have a 95% assurance that the error in the estimated mean is less than 10% of the standard deviation (i.e., 0.1σ). Thus

$$z(\alpha/2)\,\sigma/\sqrt{n} = 0.1\sigma.$$

Hence

$$n = \frac{z(\alpha/2)^2 \sigma^2}{E^2} = \frac{(1.96)^2 \sigma^2}{(0.1\sigma)^2} = 384.2 \approx 385.$$

The above concept can be applied to binary data (or proportions). In addition it can be easily implemented for sample size determination when comparing two treatments. Table 10.2.1 provides a summary for sample size determination based on precision analysis for situations where there are one and two samples, respectively.

Table 10.2.1 Sample Size Determination Based on Precision Analysis

Parameter	Statistic	Confidence Interval	Sample Size
μ	$\overline{Y}$	$\overline{Y} \pm z(\alpha/2) \dfrac{\sigma}{\sqrt{n}}$	$n = \dfrac{z(\alpha/2)^2 \sigma^2}{E^2}$
$\mu_1 - \mu_2$	$\overline{Y}_1 - \overline{Y}_2$	$(\overline{Y}_1 - \overline{Y}_2) \pm z(\alpha/2) \sqrt{\dfrac{\sigma_1^2}{n} + \dfrac{\sigma_2^2}{n}}$	$n = \dfrac{z(\alpha/2)^2 (\sigma_1^2 + \sigma_2^2)}{E^2}$

Power Analysis

Since a type I error is usually considered to be a more important and/or serious error that one would like to avoid, a typical approach in hypothesis testing is to control α at an acceptable level and try to minimize β by choosing an appropriate sample size. In other words, the null hypothesis can be tested at a predetermined level (or nominal level) of significance with a desired power. This concept for determination of the sample size is usually referred to as a *power analysis* for sample size determination.

For determination of the sample size based on a power analysis, the investigator(s) is required to specify the following information: First of all, a significance level has to be selected at which the chance of wrongly concluding that a difference exists when in fact there is no real difference (type I error) can be tolerated. Typically a 5% level of significance is chosen to reflect a 95% confidence regarding the unknown parameter. Second, the desired power has to be selected at which the chance of correctly detecting a difference when the difference truly exists that one wishes to achieve. A conventional choice of power is either 90% or 80%. Third, a clinically meaningful difference must be specified. In most clinical trials the objective is to demonstrate that effectiveness and safety of a drug under study compared to a placebo. Therefore it is important to specify what difference in terms of the primary endpoint is considered clinically or scientifically important. We denote such a difference by Δ. If the investigator will settle for detecting only a large difference, then fewer subjects will be needed. If the difference is relatively small, a larger study group (i.e., a larger number of subjects) will be needed. Finally, the knowledge regarding the standard deviation (i.e., σ) of the primary endpoint considered in the study is required for the sample size determination. A very precise method of measurement (i.e., a small σ) will permit detection of any given difference with a much smaller sample size than would be required with a less precise measurement.

In the following sections we describe some methods for sample size determination based on the power analyses for various situations that are commonly encountered in clinical trials.

10.3 TWO SAMPLES

One-Sample Test for Mean

Suppose that one wishes to test the following hypotheses:

$$H_0: \mu = \mu_0,$$
$$\text{vs.} \quad H_a: \mu > \mu_0, \quad (10.3.1)$$

with a significance level α when the variance σ^2 is known. For a specific alternative hypothesis, say

$$H_a: \mu = \mu_0 + \Delta,$$

where $\Delta > 0$ is a constant, the power of the test is given by

$$1 - \beta = P\{\text{reject } H_0 | H_a \text{ is true}\}$$

$$= P\left\{ \frac{\overline{Y} - (\mu_0 + \Delta)}{\sigma/\sqrt{n}} > z(\alpha) - \frac{\Delta}{\sigma/\sqrt{n}} \,\bigg|\, \mu = \mu_0 + \Delta \right\}.$$

Under the alternative hypothesis that $\mu = \mu_0 + \Delta$, the test statistic

$$\frac{\overline{Y} - (\mu_0 + \Delta)}{\sigma/\sqrt{n}}$$

follows a standard normal variable. Therefore

$$1 - \beta = P\left\{ Z > z(\alpha) - \frac{\Delta\sqrt{n}}{\sigma} \right\},$$

from which we conclude that

$$-z(\beta) = z(\alpha) - \frac{\Delta\sqrt{n}}{\sigma}$$

and hence

$$n = \frac{\sigma^2 [z(\alpha) + z(\beta)]^2}{\Delta^2}. \quad (10.3.2)$$

This result is also true when the alternative is

$$H_a: \mu < \mu_0.$$

Then, for testing a two-sided hypotheses

$$H_0: \mu = \mu_0,$$
$$\text{vs.} \quad H_a: \mu \neq \mu_0,$$

we obtain $1 - \beta$ power for a specified alternative when

$$n = \frac{\sigma^2[z(\alpha/2) + z(\beta)]^2}{\Delta^2} \qquad (10.3.3)$$

For the one-sample test for a proportion, the required sample size can be similarly derived. Let Y be the Bernoulli random variable of interest with the probability of success P and the probability of failure $1 - P$. The objective of the study is to choose between $H_0: P = P_0$ and $H_a: P = P_1$ $(P_1 > P_0)$ based on a sample of size n. The sample proportion

$$p = \frac{1}{n} \sum_{i=1}^{n} Y_i$$

approximately follows a normal distribution with mean P and variance $P(1 - P)/n$, then the required sample size is given by

$$n = \frac{[z(\alpha)\sqrt{P_0(1 - P_0)} + z(\beta)\sqrt{P_1(1 - P_1)}]^2}{(P_1 - P_0)^2}. \qquad (10.3.4)$$

To simplify (10.3.4), many authors (e.g., Lemeshow, Hosmer, and Stewart, 1981; Haseman, 1978) have suggested to consider the following arcsin transformation:

$$A(p) = 2 \arcsin \sqrt{p}$$

before performing the normal test. In this case (10.3.4) becomes

$$n = \frac{[z(\alpha) + z(\beta)]^2}{[A(P_1) - A(P_0)]^2} \qquad (10.3.5)$$

TWO SAMPLES

The above result is also valid for the case where $H_a: P = P_1$ $(P_1 < P_0)$. Hence, for testing a two-sided hypotheses, the required sample size is given by

$$n = \frac{[z(\alpha/2) + z(\beta)]^2}{[A(P_1) - A(P_0)]^2}. \tag{10.3.6}$$

Two-Sample Test for Comparing Means

A similar procedure can be applied to determine the sample size required for achieving a specific power in comparing two treatment means. Let μ_1 and μ_2 denote the means for treatment 1 and treatment 2, respectively. The hypotheses of interest are then given by

$$H_0: \mu_1 = \mu_2,$$
$$\text{vs.} \quad H_a: \mu_1 \neq \mu_2. \tag{10.3.7}$$

Assuming that σ_1^2 and σ_2^2 are known, for a specific alternative hypothesis that $\mu_1 = \mu_2 + \Delta$, the power is given by

$$1 - \beta = P\left\{ \left| \frac{\bar{Y}_1 - \bar{Y}_2}{\sigma_d} \right| > z(\alpha/2) \middle| \mu_1 = \mu_2 + \Delta \right\},$$

where

$$\sigma_d = \sqrt{\frac{\sigma_1^2}{n_1} + \frac{\sigma_2^2}{n_2}}.$$

Therefore

$$\beta = P\left\{ -z(\alpha/2) - \frac{\Delta}{\sigma_d} < \frac{(\bar{Y}_1 - \bar{Y}_2) - \Delta}{\sigma_d} < z(\alpha/2) - \frac{\Delta}{\sigma_d} \middle| \mu_1 = \mu_2 + \Delta \right\}.$$

Under the alternative hypothesis, the statistic

$$\frac{(\bar{Y}_1 - \bar{Y}_2) - \Delta}{\sigma_d}$$

is a standard normal variable. Therefore

$$\beta = P\left\{-z(\alpha/2) - \frac{\Delta}{\sigma_d} < Z < z(\alpha/2) - \frac{\Delta}{\sigma_d}\right\},$$

from which we conclude that

$$-z(\beta) = z(\alpha/2) - \frac{\Delta}{\sigma_d}.$$

If we assume that $n = n_1 = n_2$, then

$$n = \frac{(\sigma_1^2 + \sigma_2^2)[z(\alpha/2) + z(\beta)]^2}{\Delta^2}$$

$$= \frac{2\sigma^2[z(\alpha/2) + z(\beta)]^2}{\Delta^2}, \quad \text{if } \sigma_1^2 = \sigma_2^2. \quad (10.3.8)$$

For the one-sided test, the above expression for the required sample size becomes

$$n = \frac{(\sigma_1^2 + \sigma_2^2)[z(\alpha) + z(\beta)]^2}{\Delta^2}. \quad (10.3.9)$$

Note that when the population variance is unknown, the choice of sample size is not straightforward. For example, in testing the null hypothesis of (10.3.1), when the true value is $\mu = \mu_0 + \Delta$, the statistic

$$\frac{\overline{Y} - (\mu_0 + \Delta)}{s/\sqrt{n}}$$

follows a noncentral t distribution and noncentrality parameter $\delta = \Delta/\sigma$. Tables 10.3.1 and 10.3.2 provide sample sizes for the t test of the mean and the difference between treatments, respectively, for various values of δ.

When the outcome variable is dichotomous (e.g., either improves or does not improve), the outcome variable of interest is the proportion of patients who have the disease rather than the mean of a specified measurement. Let P_1 and P_2 be the proportions of success (e.g., cure or improvement) in the treatment group and the control group, respectively. Then the sample size can be similarly determined based on two-sided test as follows:

TWO SAMPLES

$$n = \frac{[z(\alpha/2)\sqrt{2P(1-P)} + z(\beta)\sqrt{P_1(1-P_1) + P_2(1-P_2)}]^2}{(P_1 - P_2)^2}, \quad (10.3.10)$$

where $P = 1/2(P_1 + P_2)$. An arcsin transformation, as defined earlier, gives

$$n = \frac{[z(\alpha/2) + z(\beta)]^2}{[A(P_1) - A(P_2)]^2}. \quad (10.3.11)$$

Stolley and Strom (1986) indicate that (10.3.7) is useful for prospective cohort studies and retrospective case-control studies as well. For prospective cohort studies, P_1 and P_2 represent the smallest proportion developing the disease in the exposed study group and the proportion expected to develop the disease in the unexposed control group, respectively. The ratio P_1/P_2, which is the incidence rate in the exposed group divided by the incidence rate in the control group, is usually referred to as the relative risk. A relative risk greater than 1.0 indicates that the exposure appears to increase the risk of the outcome. A relative risk less than 1.0 indicates that the exposure appears to decrease the risk of the outcome. A relative risk of 1.0 indicates that there is no association between the exposure and the outcome. For retrospective case-control studies, P_1 and P_2 represent the smallest proportion exposed to the risk factor of interest that one would consider important to detect in the diseased (case) group and the proportion expected to experience the exposure of interest in the undiseased (control) group.

A comprehensive review of the formulas and tables for the determination of sample sizes and power in clinical trials for testing differences in proportions for the two-sample design can be found in Sahai and Khurshid (1996), who provide separate formulas for equal and unequal treatment group sizes, formulas for the calculation of power given the sample size, and complete references for all formulas and tables cited.

Example 10.3.1

Suppose that a pharmaceutical company is interested in conducting a clinical trial to compare two cholesterol lowering agents for treatment of hypercholesterolemic patients. The primary efficacy parameter is a low-density lipidprotein cholesterol (LDL-C). Suppose that a difference of 8% in the percent change of LDL-C is considered a clinically meaningful difference and that the standard deviation is assumed to be 15%. Then, by (10.3.8), at $\alpha = 0.05$, the required sample size for having an 80% power can be obtained as follows:

Table 10.3.1 Sample Size for the t Test of the Mean

One-Sided Test	$\alpha = 0.005$				$\alpha = 0.01$				$\alpha = 0.025$				$\alpha = 0.05$			
Two-Sided Test	$\alpha = 0.01$				$\alpha = 0.02$				$\alpha = 0.05$				$\alpha = 0.1$			
$\beta =$	0.01	0.05	0.1	0.2	0.5	0.01	0.05	0.1	0.2	0.5	0.01	0.05	0.1	0.2	0.5	0.01	0.05	0.1	0.2	0.5

Value of $\Delta = \dfrac{\mu - \mu_0}{\sigma}$

	0.01	0.05	0.1	0.2	0.5	0.01	0.05	0.1	0.2	0.5	0.01	0.05	0.1	0.2	0.5	0.01	0.05	0.1	0.2	0.5
0.15																				122
0.20																				70
0.25					110					139					99					45
0.30				134	78				115	90			119	128	64		122	139	101	32
0.35			125	99	58			109	85	63		109	88	90	45	101	90	97	71	24
0.40		115	97	77	45		101	85	66	47	117	84	68	67	34	80	70	72	52	19
0.45		92	77	62	37	110	81	68	53	37	93	67	54	51	26	65	55	55	40	15
0.50	100	75	63	51	30	90	66	55	43	30	76	54	44	41	21	54	45	44	33	13
0.55	83	63	53	42	26	75	55	46	36	25	63	45	37	34	18	46	38	36	27	13
0.60	71	53	45	36	22	63	47	39	31	21	53	38	32	28	15	39	32	30	22	11
0.65	61	46	39	31	20	55	41	34	27	18	46	33	27	24	13	34	28	26	19	9
0.70	53	40	34	28	17	47	35	30	24	16	40	29	24	21	12	30	24	22	17	9
0.75	47	36	30	25	16	42	31	27	21	14	35	26	21	19	10	27	21	19	15	8
0.80	41	32	27	22	14	37	28	24	19	13	31	22	19	16	9	24	19	17	13	8
0.85	37	29	24	20	13	33	25	21	17	12	28	21	17	15	9	21	17	15	12	7
0.90	34	26	22	18	12	29	23	19	16	11	25	19	16	13	8	19	15	14	11	6
0.95	31	24	20	17	11	27	21	18	14	10	23	17	14	11	7	17	14	11	9	5

	C1	C2	C3	C4	C5	C6	C7	C8	C9	C10	C11	C12	C13	C14	C15	C16	C17	C18	C19	C20
1.00	28																			
1.1	24					25														
1.2	21		19			21	19				21									
1.3	18	22	16			18	16	16			18									
1.4	16	19	14	16		16	14	14			15									
1.5	15	16	13	14		14	13	12			14	16				18				
1.6	13	15	12	12		13	11	11	13		12	13				15				
1.7	12	13	11	11		12	10	10	12		11	12				13				
1.8	12	12	10	10	10	11	10	9	10		10	10	13			11				
1.9	11	11	9	9	9	10	9	9	9		9	9	11			10	13			
2.0	10	10	9	8	8	10	8	8	9		8	8	10			9	11			
2.1	10	10	8	8	8	9	8	7	8		8	7	9	10		8	10	11		
2.2	9	9	8	8	7	8	7	7	7	9	7	7	8	9		8	8	9		
2.3	9	8	7	7	7	8	7	7	7	8	7	7	7	8		7	8	8		
2.4	8	8	7	7	6	8	7	6	7	7	7	6	7	7		7	7	7	8	
2.5	8	7	7	6	6	7	6	6	6	6	6	6	6	7		6	6	7	7	
3.0	7	7	6	6	6	7	6	6	6	6	6	6	6	6	6	6	6	6	6	
3.5	6	6	6	6	6	6	6	6	6	6	6	6	6	6	6	6	6	6	6	
4.0	6	5	5	5	5	5	5	5	5	5	5	5	5	5	5	5	5	5	5	5

Table 10.3.2 Sample Size for the t Test of the Difference Between Two Means

| One-Sided Test | \multicolumn{5}{c|}{$\alpha = 0.005$} | \multicolumn{5}{c|}{$\alpha = 0.01$} | \multicolumn{5}{c|}{$\alpha = 0.025$} | \multicolumn{5}{c|}{$\alpha = 0.05$} |
| Two-Sided Test | \multicolumn{5}{c|}{$\alpha = 0.01$} | \multicolumn{5}{c|}{$\alpha = 0.02$} | \multicolumn{5}{c|}{$\alpha = 0.05$} | \multicolumn{5}{c|}{$\alpha = 0.1$} |

Level of t Test

$\beta =$	0.01	0.05	0.1	0.2	0.5	0.01	0.05	0.1	0.2	0.5	0.01	0.05	0.1	0.2	0.5	0.01	0.05	0.1	0.2	0.5

Value of $\Delta = \dfrac{\mu_1 - \mu_2}{\sigma}$

Δ	0.01	0.05	0.1	0.2	0.5	0.01	0.05	0.1	0.2	0.5	0.01	0.05	0.1	0.2	0.5	0.01	0.05	0.1	0.2	0.5
0.20																				137
0.25															124					88
0.30										123					87					61
0.35					110					90					64				102	45
0.40					85					70				100	50			108	78	35
0.45				118	68				101	55			105	79	39		108	86	62	28
0.50				96	55			106	82	45		106	86	64	32		88	70	51	23
0.55			101	79	46		106	88	68	38		87	71	53	27	112	73	58	42	19
0.60		101	85	67	39		90	74	58	32	104	74	60	45	23	89	61	49	36	16
0.65		87	73	57	34	104	77	64	49	27	88	63	51	39	20	76	52	42	30	14
0.70	100	75	63	50	29	90	66	55	43	24	76	55	44	34	17	66	45	36	26	12
0.75	88	66	55	44	26	79	58	48	38	21	67	48	39	29	15	57	40	32	23	11
0.80	77	58	49	39	23	70	51	43	33	19	59	42	34	26	14	50	35	28	21	10
0.85	69	51	43	35	21	62	46	38	30	17	52	37	31	23	12	45	31	25	18	9
0.90	62	46	39	31	19	55	41	34	27	15	47	34	27	21	11	40	28	22	16	8
0.95	55	42	35	28	17	50	37	31	24	14	42	30	25	19	10	36	25	20	15	7

1.00	50	38	32	26	15	45	33	28	22	13	38	27	23	17	9	33	23	18	14	7
1.1	42	32	27	22	13	38	28	23	19	11	32	23	19	14	8	27	19	15	12	6
1.2	36	27	23	18	11	32	24	20	16	9	27	20	16	12	7	23	16	13	10	5
1.3	31	23	20	16	10	28	21	17	14	8	23	17	14	11	6	20	14	11	9	5
1.4	27	20	17	14	9	24	18	15	12	8	20	15	12	10	6	17	12	10	8	4
1.5	24	18	15	13	8	21	16	14	11	7	18	13	11	9	5	15	11	9	7	4
1.6	21	16	14	11	7	19	14	12	10	6	16	12	10	8	5	14	10	8	6	4
1.7	19	15	13	10	7	17	13	11	9	6	14	11	9	7	4	12	9	7	6	3
1.8	17	13	11	10	6	15	12	10	8	5	13	10	8	6	4	11	8	7	6	
1.9	16	12	11	9	6	14	11	9	8	5	12	9	8	6	4	10	7	6	5	
2.0	14	11	10	8	6	13	10	8	7	5	11	8	7	6	4	9	7	6	5	
2.1	13	10	9	8	5	12	9	8	7	5	10	8	7	5	3	8	6	6	5	
2.2	12	10	8	7	5	11	9	7	6	4	9	7	6	5		8	7	5	5	4
2.3	11	9	8	7	4	10	8	7	6	4	9	7	6	5		8	6	5	5	4
2.4	11	9	8	6	4	10	8	7	6	4	8	6	5	4		7	6	5	4	4
2.5	10	8	7	6	3	9	7	6	5	3	8	6	5	4		6	5	4	4	3
3.0	8	6	6	5		7	6	5	4		6	5	4			5	4	4	3	
3.5	6	5	5	4		6	5	4	4		5	4	4	3		4	4			
4.0	6	5	4	4		5	4	4	3		4	3	3							

$$n = \frac{2\sigma^2[z(\alpha/2) + z(\beta)]^2}{\Delta^2}$$

$$= \frac{2(15)^2[1.96 + 0.842]^2}{(8)^2}$$

$$= 55.2 \approx 56.$$

Therefore a sample size of 56 patients per arm is required to obtain an 80% power for detection of an 8% difference in percent change of LDL-C for the intended clinical study.

Example 10.3.2
A pharmaceutical company is interested in examining the effect of an antidepressant agent in patients with generalized anxiety disorder. A double-blind, two-arm parallel, placebo-controlled randomized trial is planned. To determine the required sample size for achieving an 80% power, the HAM-A score is considered as the primary efficacy variable. It is believed that a difference of 4 in the HAM-A scores between the antidepressant and the placebo is of clinical importance. Assuming that the standard deviation is 7.0 obtained from previous studies, the required sample size can be obtained based on (10.3.8), which is given by

$$n = \frac{2\sigma^2[z(\alpha/2) + z(\beta)]^2}{\Delta^2}$$

$$= \frac{2(7)^2[1.96 + 0.842]^2}{(4)^2}$$

$$= 48.1 \approx 49.$$

Therefore a sample size of 49 patients per arm is required to obtain an 80% power for detection of a difference of 4 in HAM-A scores between the antidepressant agent and the placebo when performing a two-tailed test at $\alpha = 0.05$.

Example 10.3.3
To evaluate the efficacy and safety of an anti-infective agent compared to active control in the treatment of lower respiratory tract infections, a clinical trial is planned. A response rate of 90% for the active control agent is assumed (based on previous studies). In the interest of having an 80% power for detection of a difference of 10% between the treatments is such a difference truly exists, (10.3.10) can be used to calculate the required sample size as follows:

MULTIPLE SAMPLES

$$n = \frac{[z(\alpha/2)\sqrt{2P(1-P)} + z(\beta)\sqrt{P_1(1-P_1) + P_2(1-P_2)}]^2}{(P_1 - P_2)^2}$$

$$= \frac{[(1.96)\sqrt{2(0.85)(0.15)} + 0.842\sqrt{0.9(0.1) + 0.8(0.2)}]^2}{(0.1)^2}$$

$$= 199.02 \approx 200.$$

Therefore a sample size of 200 patients per arm is required to obtain an 80% power for detection of a difference of 10% between treatment response rates.

10.4 MULTIPLE SAMPLES

Sample Size Calculations for Analysis of Variance Models

For one-way analysis of variance with n observations per treatment, the main objective is to test

$$H_0: \tau_1 = \tau_2 = \cdots = \tau_k,$$
vs. H_a: At least one of τ_i's are not zero.

Recall that

$$E(\text{MSA}) = E\left(\frac{\text{SSA}}{k-1}\right) = \sigma^2 + \frac{n}{k-1}\sum_{i=1}^{k}\tau_i^2,$$

$$E(\text{MSE}) = E\left(\frac{\text{SSE}}{k(n-1)}\right) = \sigma^2.$$

Thus, for a given deviation from the null hypothesis H_0, as measured by $n\sum_{i=1}^{k}\tau_i^2/(k-1)$, large value of σ^2 decreases the chance of obtaining a value $F_A = \text{MSA}/\text{MSE}$ that is in the critical region for the test. The sensitivity of the test describes the ability of the procedure to detect differences in the population means, which is measured by the power of the test. The power is interpreted as the probability that the F statistic is in the critical region when the null hypothesis is false and the treatment means differ. Since under the null hypothesis, $F_A = \text{MSA}/\text{MSE}$ follows an F distribution with (ν_1, ν_2) degrees of freedom, where $\nu_1 = k-1$ and $\nu_2 = k(n-1) = N-k$. For the one-way analysis of variance, the power is given by

$$1 - \beta = P\{F_A > f(\alpha, \nu_1, \nu_2) | H_a \text{ is true}\}$$

$$= P\left\{F_A > f(\alpha, \nu_1, \nu_2) \Big| \frac{n}{k-1} \sum_{i=1}^{k} \tau_i^2\right\}. \quad (10.4.1)$$

For given values of $n \sum_{i=1}^{k} \tau_i^2/(k-1)$ and σ^2, the power can be increased by using a large sample size. The problem becomes one of designing the experiment with a value of n so that the power requirements are met. Under the alternative hypothesis that $\sum_{i=1}^{k} \tau_i^2 \neq 0$, F_A follows a noncentral F distribution with a noncentrality parameter δ where

$$\delta^2 = \frac{n \sum_{i=1}^{k} \tau_i^2}{\sigma^2}.$$

Thus (10.4.1) becomes

$$1 - \beta = P\left\{F_A > f(\alpha, \nu_1, \nu_2) \Big| \frac{n}{k-1} \sum_{i=1}^{k} \tau_i^2\right\}$$

$$= P\{F_A > f(\alpha, \nu_1, \nu_2, \delta)\}. \quad (10.4.2)$$

As a result the required sample size per group can be determined because ν_2 and δ are functions of n. As it can be seen from the above expression, there exists no explicit form for the required sample size n. As alternative, we can consider a normal approximation proposed by Laubscher (1960) for solution of n. The idea is to consider the following approximation:

$$Z = \frac{\sqrt{\dfrac{\nu_1(2\nu_2 - 1)F_A}{\nu_2}} - \sqrt{2(\nu_1 + \delta^2) - \dfrac{\nu_1 + 2\delta^2}{\nu_1 + \delta^2}}}{\sqrt{\dfrac{\nu_1 F_A}{\nu_2} + \dfrac{\nu_1 + 2\delta^2}{\nu_1 + \delta^2}}} \quad (10.4.3)$$

which is approximately normally distributed with mean zero and standard deviation one. From (10.4.3) the required sample size can be determined by solving the following equation:

MULTIPLE SAMPLES

$$z(\beta) = \frac{\sqrt{\nu_2[2(\nu_1 + \delta^2)^2 - (\nu_1 + 2\delta^2)]} - \sqrt{\nu_2(\nu_1 + \delta^2)(2\nu_2 - 1)F_A^*}}{\sqrt{\nu_1(\nu_1 + \delta^2)F_A^* + \nu_2(\nu_1 + 2\delta^2)}}, \quad (10.4.4)$$

where $F_A^* = F(\alpha, \nu_1, \nu_2)$.

Note that tables for the solution of (10.4.2) have been constructed by many researchers such as Kastenbaum, Hoel, and Bowman (1970) and Cohen (1977). Charts for solutions of (10.4.2) have also been developed by many authors such as Pearson and Hartley (1951) and Feldt and Mahmound (1958). For example, consider the use of the charts developed by Pearson and Hartley (1951). For convenience sake, define

$$\lambda = \frac{n \sum_{i=1}^{k} \tau_i^2}{2\sigma^2},$$

and let $\phi^2 = 2\lambda/(\nu_1 + 1)$. Then λ and ϕ^2 can be expressed as

$$\lambda = \frac{\nu_1 E(MSA)}{2\sigma^2} - \frac{\nu_1}{2},$$

$$\phi^2 = \frac{\nu_1}{\nu_1 + 1} \frac{E(MSA) - \sigma^2}{\sigma^2}.$$

Table 10.4.1 gives the power of the analysis of variances as a function of ϕ for various values of ν_1, ν_2, and the significance level α. This table can be used to determine the sample size for other fixed effects models such as the randomized complete block model in which λ and ϕ are given. For example, for the randomized complete block model,

$$\lambda = \frac{b \sum_{i=1}^{k} \tau_i^2}{2\sigma^2},$$

$$\phi^2 = \frac{b \sum_{i=1}^{k} \tau_i^2}{k\sigma^2}.$$

Example 10.4.1

To illustrate the use of (10.4.4) for the sample size determination in comparing more than two treatments, consider the following example: Suppose that we are interested in conducting a four-arm parallel group, double-blind, randomized clinical trial to compare four treatments. The comparison will be made based on an F test with a significance level of $\alpha = 0.05$. Assume that the standard error within each group is expected to be $\sigma = 3.5$ and that the clinically important differences for the four treatment groups are given by

Table 10.4.1 Power of the Analysis of Variance Test

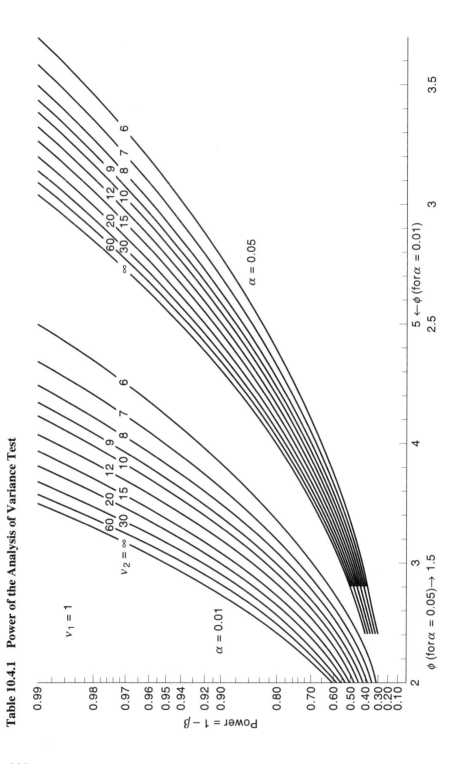

Table 10.4.1 (*Continued*)

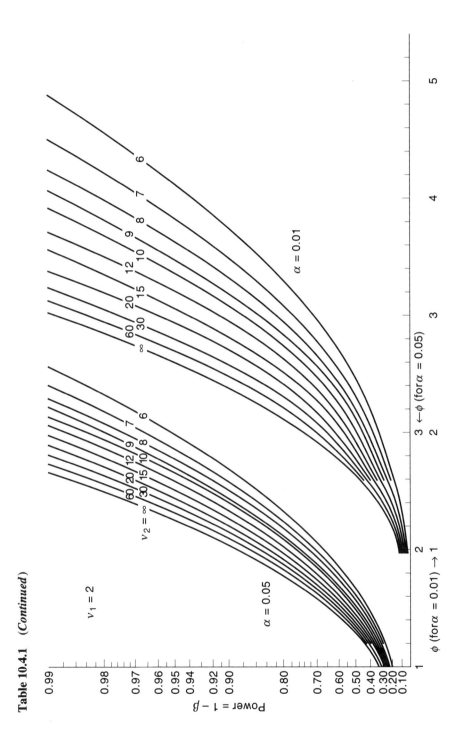

Table 10.4.1 (*Continued*)

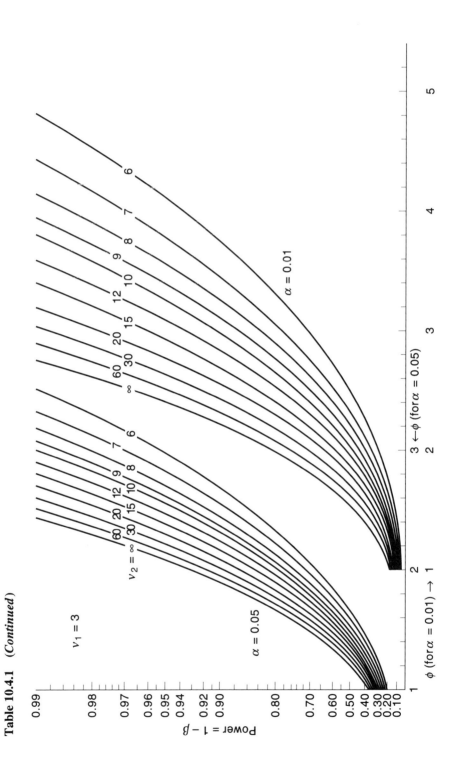

Table 10.4.1 (*Continued*)

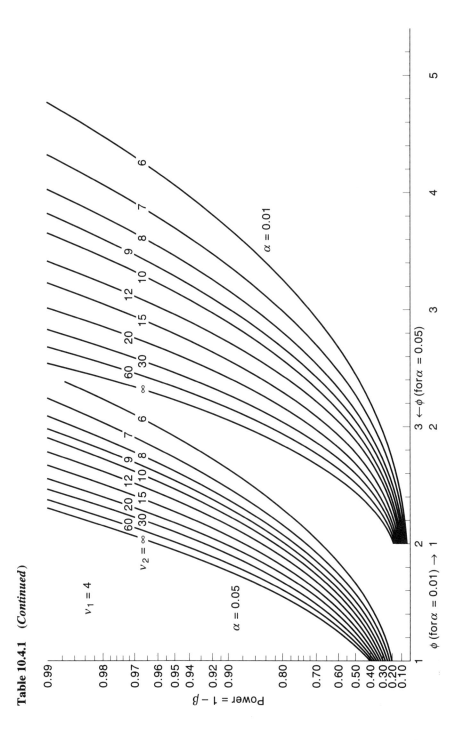

Table 10.4.1 *(Continued)*

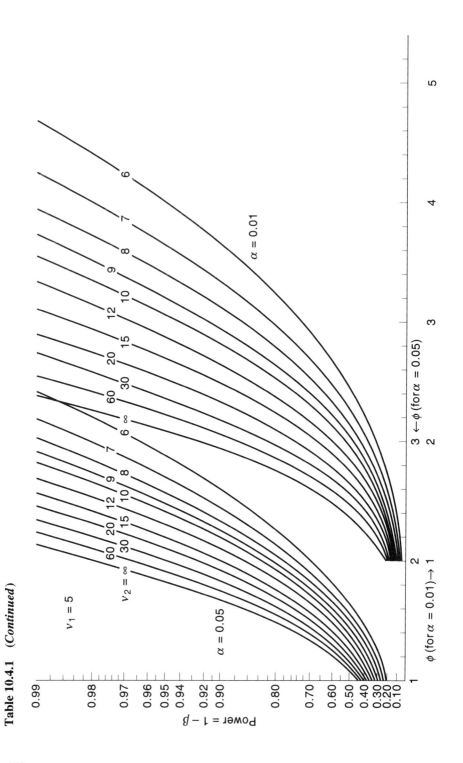

Table 10.4.1 (*Continued*)

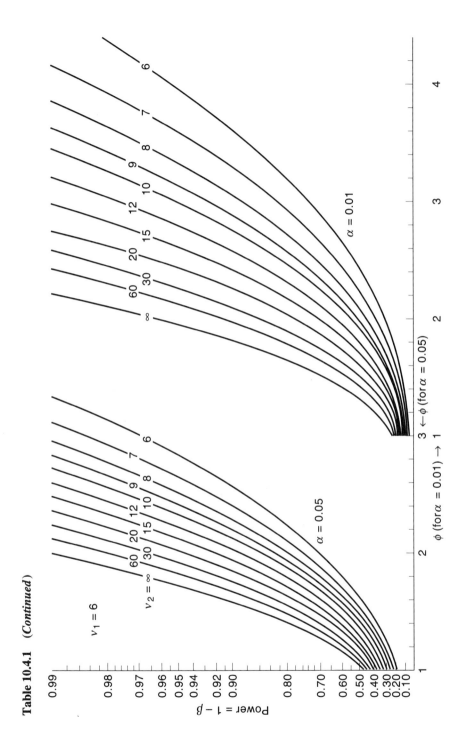

Table 10.4.1 *(Continued)*

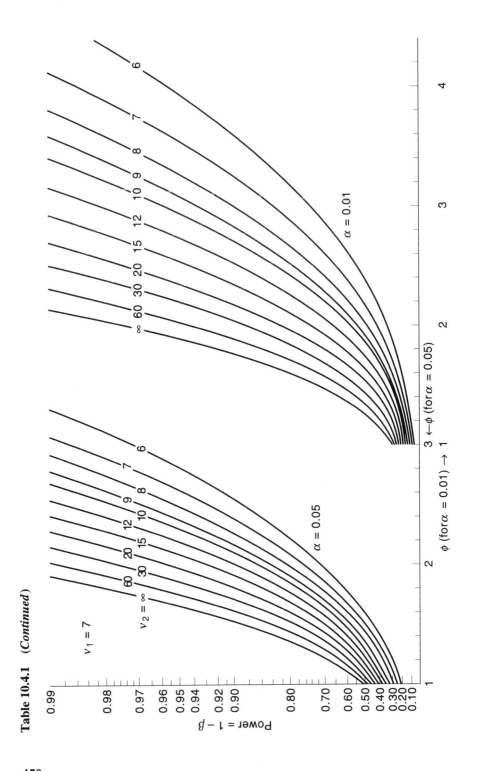

Table 10.4.1 *(Continued)*

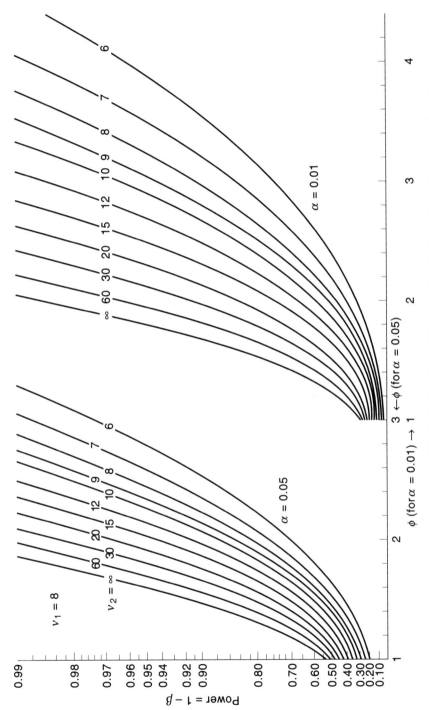

Source: Reproduced from E. S. Pearson and H. O. Hartley, Charts of the power function for analysis-of-variance tests, derived from the noncentral F distribution, *Biometrika*, **38**, 1951, by permission of the editor.

Table 10.4.2 Sample Size Calculation for Multiple Samples

n	ν_1	ν_2	F_A^*	$Z(\beta)$
9	3	32	2.901	0.524
10	3	36	2.866	0.693
11	3	40	2.839	0.853
12	3	44	2.817	1.006
13	3	48	2.798	1.152
14	3	52	2.783	1.293

Note: $Z(\beta)$ are obtained based on (10.4.4).

$$\tau_1 = -0.75, \quad \tau_2 = 3.0, \quad \tau_3 = -0.5, \quad \text{and} \quad \tau_4 = -1.75$$

Thus we have

$$\delta^2 = \frac{n \sum_{i=1}^{k} \tau_i^2}{\sigma^2}$$

$$= \frac{n[(0.75)^2 + (3.0)^2 + (0.5)^2 + (1.75)^2]}{(3.5)^2}$$

$$= 1.092n.$$

The sample size can be determined using (10.4.4). To obtain the required sample size, we apply various n to (10.4.4). The results are summarized in Table 10.4.2. From Table 10.4.2, it can be seen that for $n = 11$ and $F^* = F(0.05, 3, 40) = 2.8387$, equation (10.4.4) yields $z(\beta) \approx 0.853$ which is the closest to $z(0.2) = 0.842$. Therefore $n = 11$ is the required sample size per treatment group. Note that the required sample size $n = 11$ can also be obtained from Table 10.4.1 by specifying λ and ϕ.

Sample Size Calculations for Generalized Linear Models

In many cases test statistics for the null hypothesis of no treatment differences can be derived within the framework of generalized linear models. Self and Mauritsen (1988) propose an approach for sample size and power calculations within the framework of generalized linear models (e.g., McCullagh and Nelder, 1989) that is based on a noncentral chi-square approximation to the distribution of score statistics. In addition Self, Mauritsen, and Ohara (1992) derive a more accurate approach that is based on a noncentral chi-square approximation to the distribution of the likelihood ratio statistic. Their approaches provide a unified tool for sample size calculations for continuous and discrete responses. In general, there exist no explicit formulas to use in determining the sample size for a generalized linear model. Hence the numerical methods are usually required. These methods have been implemented in EGRET SIZ (SERC, 1993). The SIZ of EGRET not only calculates sample size for a given model but also performs a simulation to compare the empirical power with respect to the nominal one. This provides a way to assess the performance of the calculated sample size in practice.

Liu and Liang (1995) propose a method to compute sample size and power with correlated observations. Let Y_{ij} be the jth repeated measurement of a response variable for the ith subject, where $i = 1, \ldots, n$ and $j = 1, \ldots, m$. Then Y_{ij} can be described by the following linear regression model:

$$Y_{ij} = X_{ij}\psi + Z_{ij}\lambda + \epsilon_{ij},$$

where X_{ij} and Z_{ij} represent the design matrix and covariates, respectively, ψ is a $p \times 1$ vector of parameters of interest, and λ is a $q \times 1$ vector of nuisance parameters. It is assumed that $\epsilon_i = (\epsilon_{i1}, \ldots, \epsilon_{im})'$ follows a multivariate normal with mean $\mathbf{0}$ and covariance matrix $\sigma^2 \Sigma$. Under the above model, if we consider the special case where $p = q = 1$ with $Z_{ij} = 1$ for all i and j and $X_{ij} = X_i = 1$ (treatment) or 0 (placebo), then it reduces to a typical two samples problem with repeated measures. In this case the sample size required can be obtained as

$$n = \frac{[z(\alpha/2) + Z(\beta)]^2 \sigma^2 [1 + (m-1)\rho]}{\pi(1-\pi)\psi_1^2 m}, \qquad (10.4.5)$$

where ρ is the within-subject correlation, π is the proportion of subjects in the treatment group, and $\psi_1 = \Delta$ is the difference of clinical importance. The above formula can be applied to the case where two treatments are compared with binary responses with the following modification:

$$n = \frac{[z(\alpha/2) + Z(\beta)]^2 [\pi P_0(1-P_0) + (1-\pi)P_1(1-P_1)][1 + (m-1)\rho]}{m\pi(1-\pi)(P_1-P_2)^2},$$
(10.4.6)

where P_0 and P_1 are response probabilities for the placebo and the treatment groups, respectively.

10.5 SURVIVAL ANALYSIS

For survival analysis in clinical trials, in order to compare the median survival time to some event between two populations, some assumptions are essential (Lachin, 1981). Let $P_i(t)$ denote the probability of surviving to time t for subjects in the ith population, and assume exponential survival with hazard rate λ_i. Then

$$P_i(t) = e^{-\lambda_i t},$$

and the median survival time is

$$M_i = \frac{\ln 2}{\lambda_i}.$$

As indicated by Lachin (1981), if we further assume uniform censoring to time T_0 from the start of the study to termination of follow-up at time T after the start of the study, then

$$\hat{\lambda}_i = \frac{\text{Number of events in sample for population } i}{\text{Total follow-up time for population } i}$$

approximately follows a normal distribution with mean λ_i and variance $\phi(\lambda_i)/n$, where

$$\phi(\lambda_i) = \lambda_i^2 \left[1 - \frac{e^{-\lambda_i(T - T_0)} - e^{-\lambda_i T}}{\lambda_i T_0} \right]^{-1}.$$

Hence the required sample size for testing H_0: $M_1 = M_2$ versus H_a: $M_2 > M_1$ using a normal approximation is given by

$$n = \frac{\left[z(\alpha)\sqrt{2\phi(\bar{\lambda})} + z(\beta)\sqrt{\phi(\lambda_1) + \phi(\lambda_2)} \right]^2}{(\lambda_2 - \lambda_1)^2}. \tag{10.5.1}$$

Sample size determination in clinical trials with survival endpoints is complicated by the fact that the risk of event for patients may not remain constant during the trial. The comparison of survival curves between treatments is often performed based on censored data to determine whether a test treatment can reduce mortality rate or improve survival probability as compared to a placebo or a standard therapy. Let F and G be the failure time distributions in the treatment and placebo groups. Then the hypotheses of interest as given in (9.3.1) are

$$H_0: 1 - F = 1 - G,$$
$$\text{vs.} \quad H_a: 1 - F \neq 1 - G. \tag{10.5.2}$$

SURVIVAL ANALYSIS

In a similar manner the sample size can be determined based on a power analysis of a test statistic for the above hypotheses. For testing (10.5.2), several tests have been proposed in literature. For example, George and Desu (1973) and Rubinstein, Gail, and Santner (1981) discuss tests based on exponential survival curves. Tests based on exponential survival curves, however, are derived under the very restrictive assumption of constant hazard ratios. In practice, the hazard rate, which is often time dependent, can vary even if the effect of therapy is constant over time. As a result the proportional hazards assumption is not valid (Wu, Fisher, and DeMets, 1980; Lachin and Foulkes, 1986). Therefore the usual tests based on exponential models with constant hazard ratios no longer apply.

As an alternative, the logrank test is often considered. Schoenfeld (1981) and Freedman (1982) present methods for sample size calculation based on the asymptotic expectation and variance of the logrank statistic. The conditions under which the formulas for sample size are derived, however, are very restrictive. In this section we will introduce a sample size formula by Lakatos (1986, 1988) which is derived under very general conditions. As was discussed in Section 9.3, the logrank test statistic is a special case of a general class of nonparametric tests for comparing two survival functions. Lakatos (1988) considered expressing the logrank statistic as a member of the Tarone-Ware class of statistics as follows:

$$T = \frac{\sum_{k=1}^{d} w_k \left(X_k - \frac{n_{2k}}{n_{2k} + n_{1k}} \right)}{\sum_{k=1}^{d} w_k^2 \left[\frac{n_{2k} n_{1k}}{(n_{2k} + n_{1k})^2} \right]^{1/2}}, \qquad (10.5.3)$$

where $d = \sum_{k=1}^{K} d_k$ is the total of deaths, X_k is the indicator of the placebo, w_k is the kth Tarone-Ware weight, and n_{jk} are the numbers at risk just before the kth death in the jth treatment. Under a fixed local alternative, Lakatos (1988) indicates that the expectation of (10.5.3) can be approximated by

$$E = \frac{\sum_{i=1}^{N} \sum_{k=1}^{d_i} w_{ik} \left[\frac{\phi_{i_k} \theta_{i_k}}{1 + \phi_{i_k} \theta_{i_k}} - \frac{\phi_{i_k}}{1 + \phi_{i_k}} \right]}{\left[\sum_{i=1}^{N} \sum_{k=1}^{d_i} \frac{w_{ik}^2 \phi_{i_k}}{(1 + \phi_{i_k})^2} \right]^{1/2}}, \qquad (10.5.4)$$

where $N = y(K)$, ϕ_{i_k} is the ratio of patients in the two treatment groups at risk just prior to the kth death in the $y(i)$th interval,

$$\theta_{i_k} = \frac{P_{1i_k}}{P_{2i_k}},$$

where P_{ji_k} is the hazard just prior to the kth death in $y(i)$th interval in the jth treatment, and w_{i_k} is the corresponding Tarone-Ware weight. When $w_{i_k} = 1$ for all i and k, (10.5.3) reduces to the logrank test. Treating this statistic as a normal random variable with mean E and variance 1, we have

$$E = z(\alpha/2) + z(\beta).$$

Assuming that $\phi_{i_k} = \phi_i$ and $w_{i_k} = w_i$ for all k in the $y(i)$th interval and letting $\rho_i = d_i/d$, then (10.5.4) becomes

$$E = \frac{\sqrt{d} \sum_{i=1}^{N} w_i \rho_i \gamma_i}{\left(\sum_{i=1}^{N} w_i^2 \rho_i \eta_i\right)^{1/2}},$$

where

$$\gamma_i = \frac{\phi_i \theta_i}{1 + \phi_i \theta_i} - \frac{\phi_i}{1 + \phi_i}$$

and

$$\eta_i = \frac{\phi_i}{(1 + \phi_i)^2}.$$

This leads to

$$d = \frac{[z(\alpha/2) + z(\beta)]^2 \sum_{i=1}^{N} w_i^2 \rho_i \eta_i}{\left(\sum_{i=1}^{N} w_i \rho_i \eta_i\right)^2}. \tag{10.5.5}$$

Since

$$d = \frac{n(P_1 + P_2)}{2},$$

where P_1 and P_2 are cumulative event rates for treatment and placebo groups, respectively, the required total sample size can be obtained as

$$n = \frac{2d}{P_1 + P_2}. \tag{10.5.6}$$

Yateman and Skene (1992) also propose a method for sample size determination

Example 10.5.1

To illustrate the above methodology for sample size determination, consider the example described in Gail (1985). Suppose that we have a two-year trial with event rates of $1 - \exp(-1) \approx 0.6321$ and $1 - \exp(-\frac{1}{2}) \approx 0.3935$ per year in the placebo and treatment groups, respectively, and that the yearly loss to follow-up and noncompliance rates are 3% and 4%, respectively. Assuming that hazard rate is constant. The quantities ρ_i, η_i, and γi can readily be determined using the Markov model as described in Lakatos (1988) which accounts for lost to follow-up, noncompliance, lag time, and so forth. Table 10.5.1 displays these parameters including ϕ_i and θ_i. By (10.5.4) it can be verified that the number of deaths d is 102. In addition, from the last row of Table 10.5.1, the cumulative event rates over a two-year period are given by 83.7% and 62.7% for the treatment and placebo groups, respectively. Thus the required total sample size for achieving a 90% power at 5% level of significance can be obtained as

$$n = \frac{2(102)}{0.837 + 0.627} = 139.$$

Note that a SAS computer program implementing the Markov models and some variations is given by Lakatos (1986). In addition Shih (1995) illustrate the use of SIZE, a comprehensive computer program for calculating sample size, power, and duration of study in clinical trials with time-dependent rates of event, crossover, and loss to follow-up. As indicated by Shih (1995), SIZE covers a wide range of complexities commonly occurring in clinical trials such as nonproportional hazards, lag in treatment effect, and uncertainties in treatment benefit.

10.6 DOSE-RESPONSE STUDIES

As indicated by Ruberg (1995a, 1995b), the following fundamental questions that dictate design and analysis strategies are necessarily addressed when studying the dose-response relationship of a new drug:

1. Is there any drug effect?
2. What doses exhibit a response different from control?
3. What is the nature of the dose-response relationship?
4. What is the optimal dose?

Table 10.5.1 Parameters for Example 10.5.1

t_i	Control				Experimental							
	L	E	A_C	A_E	L	E	A_E	A_C	γ	η	ρ	θ	ϕ
0.1	0.003	0.095	0.897	0.005	0.003	0.049	0.944	0.004	0.167	0.222	0.098	2.000	1.000
0.2	0.006	0.181	0.804	0.009	0.006	0.095	0.891	0.007	0.166	0.226	0.090	1.986	0.951
0.3	0.008	0.258	0.721	0.013	0.009	0.139	0.842	0.010	0.166	0.230	0.083	1.972	0.905
0.4	0.010	0.327	0.647	0.016	0.011	0.181	0.795	0.013	0.165	0.234	0.076	1.959	0.862
0.5	0.013	0.389	0.580	0.018	0.014	0.221	0.750	0.015	0.164	0.237	0.070	1.945	0.821
0.6	0.014	0.445	0.520	0.020	0.016	0.259	0.708	0.016	0.163	0.240	0.064	1.932	0.782
0.7	0.016	0.496	0.466	0.022	0.018	0.295	0.669	0.017	0.162	0.242	0.059	1.920	0.746
0.8	0.017	0.541	0.418	0.023	0.020	0.330	0.632	0.018	0.160	0.244	0.054	1.907	0.711
0.9	0.019	0.582	0.375	0.024	0.022	0.362	0.596	0.019	0.158	0.246	0.050	1.894	0.679
1.0	0.020	0.619	0.336	0.024	0.024	0.393	0.563	0.019	0.156	0.248	0.046	1.882	0.648
1.1	0.021	0.652	0.302	0.025	0.026	0.423	0.532	0.020	0.154	0.249	0.043	1.870	0.619
1.2	0.022	0.682	0.271	0.025	0.028	0.450	0.502	0.020	0.152	0.249	0.039	1.857	0.592
1.3	0.023	0.709	0.243	0.025	0.029	0.477	0.474	0.020	0.149	0.250	0.036	1.845	0.566
1.4	0.024	0.734	0.218	0.025	0.031	0.502	0.448	0.020	0.147	0.250	0.034	1.833	0.542
1.5	0.025	0.755	0.195	0.025	0.032	0.525	0.423	0.020	0.144	0.250	0.031	1.820	0.519
1.6	0.025	0.775	0.175	0.024	0.033	0.548	0.399	0.019	0.141	0.249	0.029	1.808	0.497
1.7	0.026	0.793	0.157	0.024	0.035	0.569	0.377	0.019	0.138	0.248	0.027	1.796	0.477
1.8	0.026	0.809	0.141	0.023	0.036	0.589	0.356	0.018	0.135	0.247	0.025	1.783	0.457
1.9	0.027	0.824	0.127	0.023	0.037	0.609	0.336	0.018	0.132	0.246	0.023	1.771	0.439
2.0	0.027	0.837	0.114	0.022	0.038	0.627	0.318	0.018	0.129	0.244	0.021	1.758	0.421

DOSE-RESPONSE STUDIES

The first two questions are usually addressed by means of the techniques of analysis of variance, and the last two questions are related to the identification of minimum effective dose (MED). In this section, we will introduce sample size calculations based on the concept of the dose-response study using analysis of variance and statistical test for MED.

Dose-Response Relationship

In a dose-response study suppose that there is a control group and K dose groups. The null hypothesis of interest is then given by

$$H_0: \mu_0 = \mu_1 = \cdots = \mu_K, \qquad (10.6.1)$$

where μ_0 is mean response for the control group and μ_i is mean response for the ith dose group. The rejection of hypothesis (10.6.1) indicates that there is a drug effect. The dose-response relationship can then be examined under appropriate alternative hypotheses. Under a specific alternative hypothesis, the required sample size per dose group can then be obtained. Spriet and Dupin-Spriet (1996) identify the following alternative hypotheses for dose response:

1. $H_a: \mu_0 < \mu_1 < \cdots < \mu_{K-1} < \mu_K$.
2. $H_a: \mu_0 < \cdots < \mu_i = \cdots = \mu_j > \ldots > \mu_K$.
3. $H_a: \mu_0 < \cdots < \mu_i = \cdots = \mu_K$.
4. $H_a: \mu_0 = \mu_1 < \cdots < \mu_K$.
5. $H_a: \mu_0 < \mu_1 < \cdots = \mu_i = \cdots = \mu_K$.
6. $H_a: \mu_0 = \cdots = \mu_i < \cdots < \mu_{K-1} < \mu_K$.
7. $H_a: \mu_0 = \mu_1 < \cdots < \mu_i = \cdots = \mu_K$.
8. $H_a: \mu_0 = \cdots = \mu_i < \cdots < \mu_{K-1} = \mu_K$.

Figure 10.6.1 exhibits dose-response patterns under the above alternative hypotheses. Under these alternative hypotheses the statistical tests can be very complicated, and hence there may exist no close form for the corresponding power functions. As an alternative, Spriet and Dupin-Spriet (1996) depend on simulations to obtain adequate sample sizes for parallel-group dose-response clinical trials. Most recently Fine (1997) has derived a formula for calculating the power of a sample size by the alternative hypothesis on the various linear contrasts of $\mu_j, j = 0, 1, \ldots, K$ (i.e., linear, quadratic, and cubic). An example of an alternative hypothesis on linear contrast is

$$H_a: \sum_{j=0}^{K} c_j \mu_j \neq 0,$$

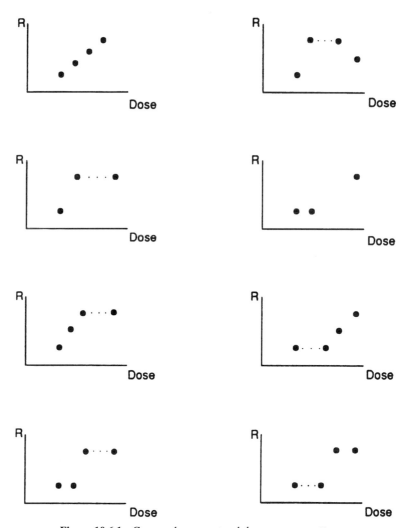

Figure 10.6.1 Commonly encountered dose-response patterns.

where

$$\sum_{j=0}^{K} c_j = 0.$$

The given contrast of means is

$$C = \sum_{j=0}^{K} c_j \mu_j;$$

it can be estimated by the contrast of sample means

$$\hat{C} = \sum_{j=0}^{K} c_j \overline{Y}_j,$$

where $\overline{Y}_j$ is the mean response of n patients receiving the jth dose level. Thus the statistic

$$T = \frac{\sum_{j=0}^{K} c_j \overline{Y}_j}{\sqrt{\hat{\sigma}^2 \sum_{j=0}^{K} c_j^2 / n}},$$

can be used to test the significance of the dose response of contrast C, where $\hat{\sigma}^2$ is the mean square error from the analysis of variance model used to estimate C. Under the alternative hypothesis, the contrast power is given by

$$1 - \beta = \Phi\left[z(\alpha/2) - \frac{\sum_{j=0}^{K} c_j \mu_j}{\sqrt{\sigma^2 \sum_{j=0}^{K} c_j^2 / n}} \right]$$

$$+ 1 - \Phi\left[z(1 - \alpha/2) - \frac{\sum_{j=0}^{K} c_j \mu_j}{\sqrt{\sigma^2 \sum_{j=0}^{K} c_j^2 / n}} \right],$$

where $\Phi(\cdot)$ is the cumulative standard normal distribution function. Hence the per-dose sample size necessary for the $(1 - \beta)100\%$ power to show a statistical significance in the dose response with the contrast is

$$n = \frac{[z(\alpha/2) + z(\beta)]^2 \hat{\sigma}^2 \sum_{j=0}^{K} c_j^2}{\left[\sum_{j=0}^{K} c_j \mu_j \right]^2} \quad (10.6.2)$$

Minimum Effective Dose

For the comparison of the mean dose treatment with the control mean, Williams (1971, 1972) proposes a test that determines the lowest dose level at which there is evidence for a difference from control. Williams considers the alternative hypothesis

$$H_a: \mu_0 = \mu_1 = \cdots = \mu_{i-1} < \mu_i \leq \mu_{i+1} \leq \cdots \leq \mu_K,$$

and proposes the test statistic

$$T_i = \frac{\hat{\mu}_i - \overline{Y}_0}{s\sqrt{(1/n_i) + (1/n_0)}},$$

where s^2 is an unbiased estimate of σ^2 that is independent of $\overline{Y}_i$ and is distributed as $\sigma^2 \chi_\nu^2/\nu$ and $\hat{\mu}_i$ is the maximum likelihood estimate of μ_i, which is given by

$$\hat{\mu}_i = \max_{1 \leq u \leq i} \min_{i \leq v \leq K} \left\{ \frac{\sum_{j=u}^{v} n_j \overline{Y}_j}{\sum_{j=1}^{v} n_j} \right\}.$$

When $n_i = n$ for $i = 0, 1, \ldots, K$, the test statistic can be simplified as

$$T_i = \frac{\hat{\mu}_i - \overline{Y}_0}{s\sqrt{2/n}},$$

which can be approximated by

$$T_i \sim \frac{X_i - Z_0}{s},$$

where

$$X_i = \max_{1 \leq u \leq i} \sum_{j=u}^{i} \frac{Z_j}{i - u + 1}$$

and Z_j follows a standard normal distribution. We then reject the null hypothesis (10.6.1) and conclude that dose i is the minimum effective dose if

$$T_j > t_j(\alpha) \quad \text{for all } j \geq i,$$

where $t_j(\alpha)$ is the upper αth quantile of the distribution of T_j. Note that $t_j(\alpha)$ are given in Tables 10.6.1 through 10.6.4.

Since the power function of the above test is rather complicated, as an alternative, we may consider the following approximation to obtain the required sample size per-dose group:

DOSE-RESPONSE STUDIES

Table 10.6.1 Upper 5% Points $t_k(\alpha)$ of the Distribution T_k

df ν	\multicolumn{10}{c}{k = Number of Dose Levels}
	1	2	3	4	5	6	7	8	9	10
5	2.02	2.14	2.19	2.21	2.22	2.23	2.24	2.24	2.25	2.25
6	1.94	2.06	2.10	2.12	2.13	2.14	2.14	2.15	2.15	2.15
7	1.89	2.00	2.04	2.06	2.07	2.08	2.09	2.09	2.09	2.09
8	1.86	1.96	2.00	2.01	2.02	2.03	2.04	2.04	2.04	2.04
9	1.83	1.93	1.96	1.98	1.99	2.00	2.00	2.01	2.01	2.01
10	1.81	1.91	1.94	1.96	1.97	1.97	1.98	1.98	1.98	1.98
11	1.80	1.89	1.92	1.94	1.94	1.95	1.95	1.96	1.96	1.96
12	1.78	1.87	1.90	1.92	1.93	1.93	1.94	1.94	1.94	1.94
13	1.77	1.86	1.89	1.90	1.91	1.92	1.92	1.93	1.93	1.93
14	1.76	1.85	1.88	1.89	1.90	1.91	1.91	1.91	1.92	1.92
15	1.75	1.84	1.87	1.88	1.89	1.90	1.90	1.90	1.90	1.91
16	1.75	1.83	1.86	1.87	1.88	1.89	1.89	1.89	1.90	1.90
17	1.74	1.82	1.85	1.87	1.87	1.88	1.88	1.89	1.89	1.89
18	1.73	1.82	1.85	1.86	1.87	1.87	1.88	1.88	1.88	1.88
19	1.73	1.81	1.84	1.85	1.86	1.87	1.87	1.87	1.87	1.88
20	1.72	1.81	1.83	1.85	1.86	1.86	1.86	1.87	1.87	1.87
22	1.72	1.80	1.83	1.84	1.85	1.85	1.85	1.86	1.86	1.86
24	1.71	1.79	1.82	1.83	1.84	1.84	1.85	1.85	1.85	1.85
26	1.71	1.79	1.81	1.82	1.83	1.84	1.84	1.84	1.84	1.85
28	1.70	1.78	1.81	1.82	1.83	1.83	1.83	1.84	1.84	1.84
30	1.70	1.78	1.80	1.81	1.82	1.83	1.83	1.83	1.83	1.83
35	1.69	1.77	1.79	1.80	1.81	1.82	1.82	1.82	1.82	1.83
40	1.68	1.76	1.79	1.80	1.80	1.81	1.81	1.81	1.82	1.82
60	1.67	1.75	1.77	1.78	1.79	1.79	1.80	1.80	1.80	1.80
120	1.66	1.73	1.75	1.77	1.77	1.78	1.78	1.78	1.78	1.78
∞	1.645	1.716	1.739	1.750	1.756	1.760	1.763	1.765	1.767	1.768

Source: Williams (1971).

$$1 - \beta = P\{\text{reject } H_0 | \mu_i \geq \mu_0 + \Delta \text{ for some } i\}$$

$$> P\{\text{reject } H_0 | \mu_0 = \mu_1 = \cdots = \mu_{K-1}, \mu_K = \mu_0 + \Delta\}$$

$$\geq P\left\{\frac{\overline{Y}_K - \overline{Y}_0}{\sigma\sqrt{2/n}} > t_K(\alpha/2) | \mu_K = \mu_0 + \Delta\right\}$$

$$= P\left\{Z > t_K(\alpha/2) - \frac{\Delta}{\sigma\sqrt{2/n}}\right\},$$

where Δ is the clinically meaningful difference. If a one-sided alternative is

Table 10.6.2 Upper 2.5% Points $t_k(\alpha)$ of the Distribution T_k

df ν	k = Number of Dose Levels						
	2	3	4	5	6	8	10
5	2.699	2.743	2.766	2.779	2.788	2.799	2.806
6	2.559	2.597	2.617	2.628	2.635	2.645	2.650
7	2.466	2.501	2.518	2.528	2.535	2.543	2.548
8	2.400	2.432	2.448	2.457	2.463	2.470	2.475
9	2.351	2.381	2.395	2.404	2.410	2.416	2.421
10	2.313	2.341	2.355	2.363	2.368	2.375	2.379
11	2.283	2.310	2.323	2.330	2.335	2.342	2.345
12	2.258	2.284	2.297	2.304	2.309	2.315	2.318
13	2.238	2.263	2.275	2.282	2.286	2.292	2.295
14	2.220	2.245	2.256	2.263	2.268	2.273	2.276
15	2.205	2.229	2.241	2.247	2.252	2.257	2.260
16	2.193	2.216	2.227	2.234	2.238	2.243	2.246
17	2.181	2.204	2.215	2.222	2.226	2.231	2.234
18	2.171	2.194	2.205	2.211	2.215	2.220	2.223
19	2.163	2.185	2.195	2.202	2.205	2.210	2.213
20	2.155	2.177	2.187	2.193	2.197	2.202	2.205
22	2.141	2.163	2.173	2.179	2.183	2.187	2.190
24	2.130	2.151	2.161	2.167	2.171	2.175	2.178
26	2.121	2.142	2.151	2.157	2.161	2.165	2.168
28	2.113	2.133	2.143	2.149	2.152	2.156	2.159
30	2.106	2.126	2.136	2.141	2.145	2.149	2.151
35	2.093	2.112	2.122	2.127	2.130	2.134	2.137
40	2.083	2.102	2.111	2.116	2.119	2.123	2.126
60	2.060	2.078	2.087	2.092	2.095	2.099	2.101
120	2.037	2.055	2.063	2.068	2.071	2.074	2.076
∞	2.015	2.032	2.040	2.044	2.047	2.050	2.052

Source: Williams (1972).

desired, $t_K(\alpha)$ should be used. To have a power of $1 - \beta$, the required sample size is obtained by solving

$$1 - \beta = P\left\{Z > t_K(\alpha/2) - \frac{\Delta}{\sigma\sqrt{2/n}}\right\}$$

and hence

$$-z(\beta) = t_K(\alpha/2) - \frac{\Delta}{\sigma\sqrt{2/n}}.$$

Thus we have

Table 10.6.3 Upper 1% Points $t_k(\alpha)$ of the Distribution T_k

df ν	\multicolumn{10}{c}{k = Number of Dose Levels}
	1	2	3	4	5	6	7	8	9	10
5	3.36	3.50	3.55	3.57	3.59	3.60	3.60	3.61	3.61	3.61
6	3.14	3.26	3.29	3.31	3.32	3.33	3.34	3.34	3.34	3.35
7	3.00	3.10	3.13	3.15	3.16	3.16	3.17	3.17	3.17	3.17
8	2.90	2.99	3.01	3.03	3.04	3.04	3.05	3.05	3.05	3.05
9	2.82	2.90	2.93	2.94	2.95	2.95	2.96	2.96	2.96	2.96
10	2.76	2.84	2.86	2.88	2.88	2.89	2.89	2.89	2.90	2.90
11	2.72	2.79	2.81	2.82	2.83	2.83	2.84	2.84	2.84	2.84
12	2.68	2.75	2.77	2.78	2.79	2.79	2.79	2.80	2.80	2.80
13	2.65	2.72	2.74	2.75	2.75	2.76	2.76	2.76	2.76	2.76
14	2.62	2.69	2.71	2.72	2.72	2.73	2.73	2.73	2.73	2.73
15	2.60	2.66	2.68	2.69	2.70	2.70	2.70	2.71	2.71	2.71
16	2.58	2.64	2.66	2.67	2.68	2.68	2.68	2.68	2.68	2.69
17	2.57	2.63	2.64	2.65	2.66	2.66	2.66	2.66	2.67	2.67
18	2.55	2.61	2.63	2.64	2.64	2.64	2.65	2.65	2.65	2.65
19	2.54	2.60	2.61	2.62	2.63	2.63	2.63	2.63	2.63	2.63
20	2.53	2.58	2.60	2.61	2.61	2.62	2.62	2.62	2.62	2.62
22	2.51	2.56	2.58	2.59	2.59	2.59	2.60	2.60	2.60	2.60
24	2.49	2.55	2.56	2.57	2.57	2.57	2.58	2.58	2.58	2.58
26	2.48	2.53	2.55	2.55	2.56	2.56	2.56	2.56	2.56	2.56
28	2.47	2.52	2.53	2.54	2.54	2.55	2.55	2.55	2.55	2.55
30	2.46	2.51	2.52	2.53	2.53	2.54	2.54	2.54	2.54	2.54
35	2.44	2.49	2.50	2.51	2.51	2.51	2.51	2.52	2.52	2.52
40	2.42	2.47	2.48	2.49	2.49	2.50	2.50	2.50	2.50	2.50
60	2.39	2.43	2.45	2.45	2.46	2.46	2.46	2.46	2.46	2.46
120	2.36	2.40	2.41	2.42	2.42	2.42	2.42	2.42	2.42	2.43
∞	2.326	2.366	2.377	2.382	2.385	2.386	2.387	2.388	2.389	2.389

Source: Williams (1971).

$$n = \frac{2\sigma^2[t_K(\alpha/2) + z(\beta)]^2}{\Delta^2}. \qquad (10.6.3)$$

Example 10.6.1

To illustrate sample size calculation based on Williams's test, consider a dose-response study on three doses of an active drug in determining a minimum effective dose. Suppose that the standard deviation of the response variable is 45 and that the clinically meaningful difference is 25. By Table 10.6.2, we have $t_3(0.025) = 2.032$. Then, by (10.6.3), we have

Table 10.6.4 Upper 0.5% Points $t_k(\alpha)$ of the Distribution T_k

df ν	\multicolumn{7}{c}{k = Number of Dose Levels}
	2	3	4	5	6	8	10
5	4.179	4.229	4.255	4.270	4.279	4.292	4.299
6	3.825	3.864	3.883	3.895	3.902	3.912	3.197
7	3.599	3.631	3.647	3.657	3.663	3.670	3.674
8	3.443	3.471	3.484	3.492	3.497	3.504	3.507
9	3.329	3.354	3.366	3.373	3.377	3.383	3.886
10	3.242	3.265	3.275	3.281	3.286	3.290	3.293
11	3.173	3.194	3.204	3.210	3.214	3.218	3.221
12	3.118	3.138	3.147	3.152	3.156	3.160	3.162
13	3.073	3.091	3.100	3.105	3.108	3.112	3.114
14	3.035	3.052	3.060	3.065	3.068	3.072	3.074
15	3.003	3.019	3.027	3.031	3.034	3.037	3.039
16	2.957	2.991	2.998	3.002	3.005	3.008	3.010
17	2.951	2.966	2.973	2.977	2.980	2.938	2.984
18	2.929	2.944	2.951	2.955	2.958	2.960	2.962
19	2.911	2.925	2.932	2.936	2.938	2.941	2.942
20	2.894	2.903	2.915	2.918	2.920	2.923	2.925
22	2.866	2.879	2.885	2.889	2.891	2.893	2.895
24	2.842	2.855	2.861	2.864	2.866	2.869	2.870
26	2.823	2.835	2.841	2.844	2.846	2.848	2.850
28	2.806	2.819	2.824	2.827	2.829	2.831	2.832
30	2.792	2.804	2.809	2.812	2.814	2.816	2.817
35	2.764	2.776	2.781	2.783	2.785	2.787	2.788
40	2.744	2.755	2.759	2.762	2.764	2.765	2.766
60	2.697	2.707	2.711	2.713	2.715	2.716	2.717
120	2.651	2.660	2.664	2.666	2.667	2.669	2.669
∞	2.607	2.615	2.618	2.620	2.621	2.623	2.623

Source: Williams (1972).

$$n = \frac{2(45)^2(2.032 + 0.842)^2}{(25)^2}$$

$$= 54.$$

Thus 54 subjects per treatment group are needed in order to have an 80% power for determining the minimum effective dose for the subjects under study.

10.7 CROSSOVER DESIGNS

Point Hypotheses for Equality

Let Y_{ijk} be the response of the ith subject in the kth sequence at the jth period. Then the following model without consideration of unequal carryover effects can be used to describe a standard two-sequence, two-period crossover design:

$$Y_{ijk} = \mu + S_{ik} + P_j + T_{(j,k)} + e_{ijk}, \qquad (10.7.1)$$

where $i(\text{subject}) = 1, 2, \ldots, n_k$, $j(\text{period})$, $k(\text{sequence}) = 1, 2$. In model (10.7.1), μ is the overall mean, S_{ik} is the random effect of the ith subject in the kth sequence, P_j is the fixed effect of the jth period, $T_{(j,k)}$ is the direct fixed effect of the treatment administered at period j in sequence k, namely

$$T_{(j,k)} = \begin{cases} \text{Placebo} & \text{if } k = j, \\ \text{Test Drug} & \text{if } k \neq j, \end{cases} \quad k = 1, 2, j = 1, 2,$$

and e_{ijk} is the within-subject random error in observing Y_{ijk}. For model (10.7.1) it is assumed that $\{S_{ik}\}$ are independently and identically distributed with mean 0 and variance σ_S^2 and that $\{e_{ijk}\}$ are independently distributed with mean 0 and variance σ^2. $\{S_{ik}\}$ and $\{e_{ijk}\}$ are assumed to be mutually independent. Let us test the following hypotheses:

$$H_0: \mu_T = \mu_P,$$
$$\text{vs.} \quad H_a: \mu_T \neq \mu_P. \qquad (10.7.2)$$

Under model (10.7.1), we can consider period differences for each subject within each sequence which are defined as

$$d_{ik} = \tfrac{1}{2}(Y_{i2k} - Y_{i1k}),$$

where $i = 1, \ldots, n_k$; $k = 1, 2$. Then a test for hypotheses (10.7.2) can be obtained based on a two-sample t statistic as follows:

$$T_d = \frac{\bar{Y}_T - \bar{Y}_P}{\hat{\sigma}_d \sqrt{(1/n_1) + (1/n_2)}},$$

where

$$\overline{Y}_T = \tfrac{1}{2}(\overline{Y}_{.21} + \overline{Y}_{.12}),$$

$$\overline{Y}_P = \tfrac{1}{2}(\overline{Y}_{.11} + \overline{Y}_{.22}),$$

$$\hat{\sigma}_d^2 = \frac{1}{n_1 + n_2 - 2} \sum_{k=1}^{2} \sum_{i=1}^{n_k} (d_{ik} - \overline{d}_{.k})^2,$$

and

$$\overline{Y}_{.jk} = \frac{1}{n_k} \sum_{i=1}^{n_k} Y_{ijk},$$

$$\overline{d}_{.k} = \frac{1}{n_k} \sum_{i=1}^{n_k} d_{ik}.$$

Under the null hypothesis (10.7.2), T_d follows a t distribution with $n_1 + n_2 - 2$ degrees of freedom. We can reject the null hypothesis of (10.7.2) if

$$|T_d| > t(\alpha/2, n_1 + n_2 - 2).$$

Under the alternative hypothesis that $\mu_T = \mu_P + \Delta$, the power of the test T_d can be similarly evaluated. In the interest of balance, we assume that $n_1 = n_2 = n$; that is, each sequence will be allocated the same number of subjects at random. As a result the sample size per sequence for testing the hypotheses of equality (10.7.2) can be determined by the formula

$$n \geq \frac{2\sigma_d^2 [t(\alpha/2, 2n-2) + t(\beta, 2n-2)]^2}{\Delta^2}, \qquad (10.7.3)$$

where σ_d^2 can be estimated from previous studies and Δ is the clinically meaningful difference which we want to detect. If we need to have a power of 80% for detection of a difference of at least 20% of the unknown placebo mean, then (10.7.3) can be simplified as

$$n \geq [t(\alpha/2, 2n-2) + t(\beta, 2n-2)]^2 \left[\frac{CV}{20}\right]^2, \qquad (10.7.4)$$

where

CROSSOVER DESIGNS 469

$$CV = \frac{\sqrt{2}\sigma_d}{\mu_P} \times 100\%.$$

Since $(2n - 2)$ in (10.7.3) and (10.7.4) are unknown, a numerical iterative procedure is required to solve for n.

Example 10.7.1
To illustrate (10.7.3), consider the problem of determining the sample size for a bioequivalence trial that compares a test drug with a placebo under a standard two-sequence, two-period crossover design (Chow and Liu, 1992a). For this trial, the sponsor is interested in having an 80% power for detection of a 20% difference between the test drug and the placebo. Based on the results from previous studies, it is estimated that the reference mean is 82.559 with a CV of 15.66. Thus (10.7.4) can be applied to determine the sample size per sequence. For iterative purposes we first guess that $n = 9$. This gives degrees of freedom $2(n - 2) = 16$, $t(0.025, 16) = 2.12$, and $t(0.2, 16) = 0.865$. By (10.7.4),

$$n = (2.12 + 0.865)^2 \left[\frac{15.66}{20} \right]^2$$

$$= 5.5 \approx 6.$$

We then start with $n = 6$ and repeat the same calculation, which gives $n = 5.9 \approx 6$ which is very close to the previous solution. Therefore we conclude that a total of $N = 2n = 12$ subjects are required to provide an 80% power for detection of a 20% difference of the reference mean at the 5% level of significance.

Interval Hypotheses for Equivalence

As pointed out by Chow and Liu (1992a), the power approach for sample size determination based on the hypothesis of equality (10.7.2) is not statistically valid in assessing *equivalence* between treatments. For the assessment of equivalence between treatments under the standard two-sequence, two-period crossover design, it is suggested that the following interval hypotheses be tested:

$$H_0: \mu_T - \mu_P \leq \theta_L \quad \text{or} \quad \mu_T - \mu_P \geq \theta_U,$$
$$\text{vs.} \quad H_a: \theta_L < \mu_T - \mu_P < \theta_U, \tag{10.7.5}$$

where θ_L and θ_U are some clinically meaningful limits for equivalence. The concept of interval hypotheses is to show equivalence by rejecting the null hypothesis of inequivalence. The above hypotheses can be decomposed into two sets of one-sided hypotheses:

$$H_{01}: \mu_T - \mu_P \leq \theta_L,$$
$$\text{vs.} \quad H_a: \mu_T - \mu_P > \theta_L,$$

and

$$H_{01}: \mu_T - \mu_P \geq \theta_U,$$
$$\text{vs.} \quad H_a: \mu_T - \mu_P < \theta_U.$$

Under model (10.7.1), Schuirmann (1987) proposes two one-sided test procedures for the above two one-sided hypotheses. We can reject the null hypothesis of inequivalence if

$$T_L = \frac{\bar{Y}_T - \bar{Y}_P - \theta_L}{\hat{\sigma}_d \sqrt{(1/n_1) + (1/n_2)}} > t(\alpha, n_1 + n_2 - 2)$$

and

$$T_U = \frac{\bar{Y}_T - \bar{T}_P - \theta_U}{\hat{\sigma}_d \sqrt{(1/n_1) + (1/n_2)}} < -t(\alpha, n_1 + n_2 - 2).$$

Let $\theta = \mu_T - \mu_P$ and $\phi_S(\theta)$ be the power of Schuirmann's two one-sided tests at θ. Assuming that $n_1 = n_2 = n$, the power at $\theta = 0$ is given by

$$1 - \beta = \phi_S(0)$$
$$= P\left\{ \frac{-\Delta}{\hat{\sigma}_d \sqrt{2/n}} + t(\alpha, 2n - 2) < \frac{Y}{\hat{\sigma}_d \sqrt{2/n}} < \frac{\Delta}{\hat{\sigma}_d \sqrt{2/n}} - t(\alpha, 2n - 2) \right\},$$

(10.7.6)

where $Y = \bar{Y}_T - \bar{Y}_P$. Since a central t distribution is symmetric about 0, the lower and upper endpoints of (10.7.6) are also symmetric about 0:

$$\frac{-\Delta}{\hat{\sigma}_d \sqrt{2/n}} + t(\alpha, 2n - 2) = -\left\{ \frac{\Delta}{\hat{\sigma}_d \sqrt{2/n}} - t(\alpha, 2n - 2) \right\}.$$

Therefore $\phi_S(0) \geq 1 - \beta$ implies that

$$\left| \frac{\Delta}{\hat{\sigma}_d\sqrt{2/n}} - t(\alpha, 2n-2) \right| \geq t(\beta/2, 2n-2),$$

or that

$$n(\theta = 0) \geq 2[t(\alpha, 2n-2) + t(\beta/2, 2n-2)]^2 \left[\frac{\hat{\sigma}_d}{\Delta} \right]^2 \tag{10.7.7}$$

If we must have an 80% power for detection of a 20% difference of placebo mean, then (10.7.7) becomes

$$n(\theta = 0) \geq [t(\alpha, 2n-2) + t(\beta/2, 2n-2)]^2 \left[\frac{CV}{20} \right]^2, \tag{10.7.8}$$

We will now consider the case where $\theta \neq 0$. Since the power curves of Schuirmann's two one-sided test procedures are symmetric about zero (Phillips, 1990), we will only consider the case where $0 < \theta = \theta_0 < \Delta$. In this case the statistic

$$\frac{Y - \theta_0}{\hat{\sigma}_d \sqrt{2/n}}$$

has a central t distribution with $2n - 2$ degrees of freedom. The power of Schuirmann's two one-sided test procedures can be evaluated at θ_0, which is given by

$1 - \beta = \phi_S(\theta_0)$

$$= P\left\{ \frac{-\Delta - \theta_0}{\hat{\sigma}_d\sqrt{2/n}} + t(\alpha, 2n-2) < \frac{Y - \theta_0}{\hat{\sigma}_d\sqrt{2/n}} < \frac{\Delta - \theta_0}{\hat{\sigma}_d\sqrt{2/n}} - t(\alpha, 2n-2) \right\}.$$

$$\tag{10.7.9}$$

Note that unlike the case where $\theta = 0$, the lower and upper endpoints of (10.7.9) are not symmetric about 0. Therefore, as indicated by Chow and Liu (1992a), if we choose

$$\frac{\Delta - \theta_0}{\hat{\sigma}_d\sqrt{2/n}} - t(\alpha, 2n-2) = t(\beta/2, 2n-2),$$

then the resultant sample size may be too large to be of practical interest, and the power may be more than we need. As an alternative, Liu and Chow (1992a) consider the inequality for obtaining an approximate formula for n

$$\phi_S(\theta_0) \leq P\left\{\frac{Y - \theta_0}{\hat{\sigma}_d\sqrt{2/n}} < \frac{\Delta - \theta_0}{\hat{\sigma}_d\sqrt{2/n}} - t(\alpha, 2n - 2)\right\}.$$

As a result $\phi_S(\theta_0) \geq 1 - \beta$ gives

$$\frac{\Delta - \theta_0}{\hat{\sigma}_d\sqrt{2/n}} - t(\alpha, 2n - 2) = t(\beta, 2n - 2),$$

or

$$n(\theta_0) \geq 2[t(\alpha, 2n - 2) + t(\beta, 2n - 2)]^2 \left[\frac{\hat{\sigma}_d}{\Delta - \theta_0}\right]^2. \quad (10.7.10)$$

Similarly, if we must have an 80% power for detection of a 20% difference of placebo mean, then (10.7.10) becomes

$$n(\theta_0) \geq [t(\alpha, 2n - 2) + t(\beta, 2n - 2)]^2 \left[\frac{CV}{20 - \theta'_0}\right]^2, \quad (10.7.11)$$

where

$$\theta'_0 = 100 \times \frac{\theta_0}{\mu_P}.$$

Example 10.7.2

We continue the previous example to illustrate the computation of the sample size in achieving an 80% power when $\theta'_0 = 5\%$ and $\Delta = 20\%$. From the previous example, the estimated CV is given by 15.66%. If our initial guess of n is 6, then

$$t(\alpha, 2n - 2) = t(0.05, 10) = 1.812,$$
$$t(\beta, 2n - 2) = t(0.20, 10) = 0.879.$$

Thus

$$n = (1.812 + 0.879)^2 \left[\frac{15.66}{20 - 5}\right]^2$$

$$= 7.9 \approx 8.$$

We then use $n = 8$ as the initial value for the next enumeration. Since $t(0.05, 14) = 1.761$ and $t(0.20, 14) = 0.868$, (10.7.11) gives

$$n = (1.761 + 0.868)^2 \left[\frac{15.66}{20 - 5} \right]^2$$

$$= 7.6 \approx 8.$$

Since the last two enumerations generate the same required sample size per sequence, a total of 16 subjects would be required to achieve an 80% power.

Table 10.7.1 presents the required total sample sizes necessary to achieve either an 80% or a 90% power for θ from 0 to 15% by increments of 5% as well as CVs from 10% to 40% by increments of 2% (Liu and Chow, 1992).

Higher-Order Crossover Designs

As was demonstrated in Chapter 5, the standard two-sequence, two-period crossover design is not useful in the presence of carryover effects. In addition it does not provide independent estimates of intrasubject variabilities. To account for these disadvantages, in practice, it is of interest to consider a higher-order crossover design (Chow and Liu, 1992a, 1992b). A higher-order crossover design is defined as a crossover design in which either the number of periods or the number of sequences is greater than the number of treatments to be compared. The most commonly used higher-order crossover designs are Balaam's design, the two-sequence dual design, and four-period designs with two or four sequences. For a higher-order crossover design, the general model given in (10.7.1) can be used with the appropriate indexes on j and k. Table 10.7.2 summarizes these higher-order crossover designs.

For a given higher-order crossover design, the sample size is similarly determined based on either the point hypotheses for equality or the interval hypotheses for equivalence under model (10.7.1). Let us consider the sample size determined by the interval hypotheses (10.7.5). Let $n_i = n$ be the number of subjects in sequence i of a higher-order crossover design, and let F_ν denote the cumulative distribution function of the t distribution with ν degrees of freedom. Then it can be verified that the power of Schuirmann's two one-sided tests at the α level of significance for the mth design is given by

$$\phi_m(\theta) = F_{\nu_m}\left(\frac{\Delta - \theta}{CV\sqrt{b_m/n}} - t(\alpha, \nu_m) \right)$$

$$- F_{\nu_m}\left(t(\alpha, \nu_m) - \frac{\Delta + \theta}{CV\sqrt{b_m/n}} \right)$$

Table 10.7.1 Sample Sizes for Schuirmann's Two One-Sided Test Procedures at $\Delta = 0.2$ and the 5% Nominal Level in 2×2 Standard Crossover Design

		$\theta = \mu_T - \mu_R$			
Power	CV (%)	0%	5%	10%	15%
80%	10	8	8	16	52
	12	8	10	20	74
	14	10	14	26	100
	16	14	16	34	126
	18	16	20	42	162
	20	20	24	52	200
	22	24	28	62	242
	24	28	34	74	288
	26	32	40	86	336
	28	36	46	100	390
	30	40	52	114	448
	32	46	58	128	508
	34	52	66	146	574
	36	58	74	162	644
	38	64	82	180	716
	40	70	90	200	794
90%	10	10	10	20	70
	12	10	14	28	100
	14	14	18	36	136
	16	16	22	46	178
	18	20	28	58	224
	20	24	32	70	276
	22	28	40	86	334
	24	34	46	100	396
	26	40	54	118	466
	28	44	62	136	540
	30	52	70	156	618
	32	58	80	178	704
	34	66	90	200	794
	36	72	100	224	890
	38	80	112	250	992
	40	90	124	276	1098

Source: Liu and Chow (1992).

for $m = 1$ (Balaam design), 2 (two-sequence dual design), 3 (four-period design with two sequences), and 4 (four-period design with four sequences), where

$$\nu_1 = 4n - 3, \quad \nu_2 = 4n - 4, \quad \nu_3 = 6n - 5, \quad \nu_4 = 12n - 5;$$
$$b_1 = 2, \quad b_2 = \frac{3}{4}, \quad b_3 = \frac{11}{20}, \quad b_4 = \frac{1}{4}.$$

Table 10.7.2 Commonly Used Higher-Order Crossover Designs

Design	Sequence	Period	Design
1	4	2	Balamm design
2	2	3	Two-sequence dual design
3	2	4	Four-period design with two sequences
4	4	4	Four-period design with four sequences

Hence the formula of n required to achieve a $1 - \beta$ power at the α level of significance for the mth design when $\theta = 0$ is given by

$$n \geq b_m[t(\alpha, \nu_m) + t(\beta/2, \nu_m)]^2 \left[\frac{CV}{\Delta}\right]^2, \quad (10.7.12)$$

and if $\theta = \theta_0 > 0$, the approximate formula for n is given by

$$n(\theta_0) \geq b_m[t(\alpha, \nu_m) + t(\beta, \nu_m)]^2 \left[\frac{CV}{\Delta - \theta}\right]^2 \quad (10.7.13)$$

for $m = 1, 2, 3,$ and 4.

Note that Tables 10.7.3 through 10.7.6 give the required total number of subjects N_m for each dosing (m) to achieve either an 80% or a 90% power for θ from 0 to 15% by increments of 5% as well as CVs from 10% to 40% by increments of 2%, where

$$N_1 = 4n, \quad N_2 = 2n, \quad N_3 = 2n, \quad \text{and} \quad N_4 = 4n.$$

10.8 MULTIPLE-STAGE DESIGN IN CANCER TRIALS

In phase II cancer trials, it is undesirable to stop a study early when the treatment appears to be effective but desirable when the treatment seems to be ineffective. For this purpose, a multiple-stage design is often employed to determine whether an experimental treatment holds sufficient promise to warrant further testing (see, e.g., Fleming, 1982; Simon, 1989; Chang et al., 1987; Therneau et al., 1990). The concept of a multiple-stage design is to permit early stopping when a moderately long sequence of initial failures occurs. For example, in the two-stage Simon's design, n_1 patients are treated and the trial terminates if all n_1 are treatment failures. If there are one or more successes in stage 1,

Table 10.7.3 Sample Sizes for Schuirmann's Two One-Sided Test Procedures at $\Delta = 0.2$ and the 5% Nominal Level in Balaam's Design

Power	CV (%)	θ 0%	5%	10%	15%
80%	10	20	24	52	200
	12	28	36	76	288
	14	36	48	100	392
	16	48	60	132	508
	18	60	76	164	644
	20	72	92	200	796
	22	88	108	244	960
	24	104	132	288	1144
	26	120	152	336	1340
	28	136	176	392	1556
	30	156	200	448	1784
	32	180	228	508	2028
	34	200	256	576	2292
	36	224	288	644	2568
	38	252	320	716	2860
	40	276	356	796	3168
90%	10	24	36	72	276
	12	36	48	104	400
	14	48	64	136	540
	16	60	80	180	704
	18	76	104	224	892
	20	92	124	276	1100
	22	108	152	336	1328
	24	128	180	400	1584
	26	152	208	468	1856
	28	172	244	540	2152
	30	200	276	620	2472
	32	224	316	704	2808
	36	284	400	892	3556
	38	316	444	992	3960
	40	352	492	1100	4388

then stage 2 is implemented by including the other n_2 patients. A decision is then made based on the response rate of the $n_1 + n_2$ patients. The drawback of Simon's design is that it does not allow early termination if there is a long run of failures at the start. To overcome this disadvantage, Ensign, et al. (1994) proposed an optimal three-stage design which modifies the Simon's two-stage design. The optimal three-stage design is outlined below.

Let p_0 be the response rate that is not of interest for conducting further studies and p_1 be the response rate of definite interest ($p_1 > p_0$). The optimal three-stage design is implemented by testing the following hypotheses:

Table 10.7.4 Sample Sizes for Schuirmann's Two One-Sided Test Procedures at $\Delta = 0.2$ and the 5% Nominal Level in Two-Sequence Dual Design

Power	CV (%)	θ 0%	5%	10%	15%
80%	10	6	6	12	38
	12	6	8	16	56
	14	8	10	20	74
	16	10	12	26	96
	18	12	16	32	122
	20	14	18	38	150
	22	18	22	46	182
	24	20	26	56	216
	26	24	30	64	252
	28	28	34	74	292
	30	30	38	86	336
	32	34	44	96	382
	34	38	50	108	430
	36	44	56	122	482
	38	48	62	136	538
	40	54	68	150	596
90%	10	6	8	14	54
	12	8	10	20	76
	14	10	14	28	102
	16	12	16	34	134
	18	16	20	44	168
	20	18	24	54	208
	22	22	30	64	250
	24	26	34	76	298
	26	30	40	88	350
	28	34	46	102	404
	30	38	54	118	464
	32	44	60	134	528
	34	48	68	150	596
	36	54	76	168	668
	38	60	84	188	744
	40	66	94	208	824

$$H_0: p \leq p_0 \quad \text{vs.} \quad H_1: p \geq p_1.$$

Rejection of H_0 (or H_1) meant that further (or not further) study of the test treatment should be carried out in phase III. At stage 1, n_1 patients are treated. We would reject H_1 (i.e., the test treatment is not promising) and stop the trial if there is no response. If there are one or more responses, then proceed to stage 2

Table 10.7.5 Sample Sizes for Schuirmann's Two One-Sided Test Procedures at $\Delta = 0.2$ and the 5% Nominal Level in Four-Period Design With Two Sequences

Power	CV (%)	θ			
		0%	5%	10%	15%
80%	10	4	4	8	28
	12	6	6	12	40
	14	6	8	14	54
	16	8	10	18	72
	18	10	12	24	90
	20	12	14	28	110
	22	14	16	34	134
	24	16	18	40	158
	26	18	22	48	186
	28	20	26	54	214
	30	22	28	62	246
	32	26	32	72	280
	34	28	36	80	316
	36	32	40	90	354
	38	36	46	100	394
	40	40	50	110	436
90%	10	4	6	12	40
	12	6	8	16	56
	14	8	10	20	76
	16	10	12	26	98
	18	12	16	32	124
	20	14	18	40	152
	22	16	22	48	184
	24	18	26	56	218
	26	22	30	66	256
	28	24	34	76	296
	30	28	40	86	340
	32	32	44	98	388
	34	36	50	110	438
	36	40	56	124	490
	38	44	62	138	546
	40	50	68	152	604

by including additional n_2 patients. We would reject H_1 and stop the trial if the total number of responses is less than a prespecified number of r_2; otherwise continue to stage 3. At stage 3, n_3, n_3 more patients are treated. We would reject H_1 if the total number of responses is less than a prespecified number of r_3. In this case, we conclude the test treatment is ineffective. On the other hand, if there are more than r_3 responses, we reject H_0 and conclude the test treatment is promising. Based on the above three-stage design, Ensign et al.

Table 10.7.6 Sample Sizes for Schuirmann's Two One-Sided Test Procedures at $\Delta = 0.2$ and the 5% Nominal Level in Four-Period Design With Four Sequences

Power	CV (%)	θ 0%	5%	10%	15%
80%	10	4	4	8	28
	12	4	8	12	40
	14	8	8	16	52
	16	8	8	20	64
	18	8	12	24	84
	20	12	12	28	100
	22	12	16	32	124
	24	16	20	40	144
	26	16	20	44	168
	28	20	24	52	196
	30	20	28	60	224
	32	24	32	64	256
	34	28	36	72	288
	36	32	40	84	324
	38	32	44	92	360
	40	36	48	100	400
90%	10	4	8	12	36
	12	8	8	16	52
	14	8	12	20	68
	16	8	12	24	92
	18	12	16	32	112
	20	12	16	36	140
	22	16	20	44	168
	24	20	24	52	200
	26	20	28	60	236
	28	24	32	68	272
	30	28	36	80	312
	32	32	40	92	352
	34	32	48	100	400
	36	36	52	112	448
	38	40	56	128	496
	40	44	64	140	552

(1994) considered the following to determine the sample size. For each value of n_1 satisfying

$$(1 - p_1)^{n_1} < \beta,$$

where

$$\beta = P \text{ (reject } H_1 | p_1),$$

compute the values of r_2, n_2, r_3, and n_3 that minimize the null expected sample size $EN(p_0)$ subject to the error constraints α and β, where

$$EN(p) = n_1 + n_2\{1 - \beta_1(p)\} + n_3\{1 - \beta_1(p) - \beta_2(p)\},$$

and β_i are the probability of making type II error evaluated at stage i. Ensign et al. (1994) use the value of

$$\beta - (1 - p_1)^{n_1}$$

as the type II error rate in the optimization along with type I error

$$\alpha = P(\text{reject } H_0 | p_0)$$

to obtain r_2, n_2, r_3, and n_3. Repeating this, n_1 can then be chosen to minimize the overall $EN(p_0)$.

Example 10.8.1

Suppose an investigator is interested in planning a clinical trial in patients with a specific carcinoma where standard therapy has a 20% response rate. Suppose further that the test treatment is a combination of the standard therapy and a new agent. The investigator would like to determine whether the test treatment will achieve a response rate of 40%. Since the test treatment may not be effective, it is desirable to warrant early study termination. In this case, the optimal three-stage design described above is useful. We consider the optimal three-stage design for testing

$$H_0: p \leq 0.20 \quad \text{vs.} \quad H_1: p \geq 0.40$$

with $\alpha = \beta = 0.10$ which result in the following sample size allocation at each stage:

$$(0/8, 3/16, 11/42).$$

That is, at stage 1, 8 patients are to be tested. We would terminate the trial if no response is observed in the eight patients. If there are one or more responses, we continue to the next stage. At the second stage, eight more patients are treated. We stop the trial if less than three responses are observed in the sixteen patients; otherwise continue to stage 3. At stage 3, 26 more patients are treated. We conclude the test treatment is effective if there are more than 11 responses in the 42 patients.

10.9 DISCUSSION

For convenience sake, we have considered $n_i = n$ for all i, where i can be the ith treatment group in parallel designs or the ith sequence in crossover designs. In practice, n_i are not necessarily the same for all i. For example, when conducting a placebo control clinical trial with very ill patients or patients with severe or life-threatening diseases, it is not ethical to put too many patients in the placebo arm. Then the investigator must put fewer patients in the placebo (if the placebo arm is considered necessary to demonstrate the effectiveness and safety of the drug under investigation). A typical ratio of patient allocation for situations of this kind is $1:2$; that is, each patient has a one-third chance of being assigned to the placebo group and a two-third chance of receiving the active drug. For different ratios of patient allocation, the sample size formulas discussed in this chapter can be directly applied with appropriate modifications of the corresponding degrees of freedom.

It should be noted that the sample size obtained based on the formulas in this chapter is the number of *evaluable* patients required in order to achieve a desired power. In practice, we may have to enroll more patients in order to obtain the required evaluable patients due to potential dropout. Therefore, during the stage of planning, a sample size that can account for the potential dropout is usually selected. For example, if the sample size required for an intended clinical trial is n and the potential dropout rate is p, then we need to enroll $n/(1-p)$ patients in order to obtain n evaluable patient at the completion of the trial. It should also be noted that the investigator may have to screen more patients in order to obtain $n/(1-p)$ *qualified* patients at the entry of the study based on inclusion/exclusion criteria of the trial.

In many clinical trials, multiple comparisons may be performed. In the interest of controlling the overall type I error rate at the α level an adjustment for multiple comparisons such as the Bonferroni adjustment is necessary. The formulas for sample size determination discussed in this chapter can still be applied by simply replacing the α level with an adjusted α level.

Fleiss (1986a) points out that the required sample size may be reduced if the response variable can be described by a covariate. Let n be the required sample size per group when the design does not call for the experimental control of a prognostic factor. Also let n^* be the required sample size for the study with the factor controlled. The relative efficiency (RE) between the two designs is defined as

$$RE = \frac{n}{n^*}.$$

As indicated by Fleiss (1986a), if the correlation between the prognostic factor (covariate) and the response variable is r. Then RE can be expressed as

$$RE = \frac{100}{1-r^2}.$$

Hence we have

$$n^* = n(1-r^2).$$

As a result the required sample size per group can be reduced if the correlation exists. For example, a correlation of $r = 0.32$ could result in a 10% reduction in the sample size.

In most clinical trials, although the primary objectives are usually to evaluate the effectiveness and safety of the test drug under study, the assessment of drug safety has not received the same level of attention as the assessment of efficacy. As a result sample size calculations are usually performed based on a pre-study power analysis for the primary efficacy variable. If the sample size is based on a primary safety variable such as the adverse event rate, a large sample size may be required especially when the incidence rate is low. For example, if the incidence rate is one per 10,000, then we will need to include 10,000 in order to observe one incidence. O'Neill (1988) indicates that the magnitude of rates that can feasibly be studied in most clinical trials is about 0.01 and higher. However, observational cohort studies usually can assess rates on the order of 0.001 and higher. O'Neill (1988) also indicates that it is informative to examine the sample sizes that are needed to estimate a rate or to detect or estimate differences of specified amounts between the rates for two different treatment groups.

CHAPTER 11

Issues in Efficacy Evaluation

11.1 INTRODUCTION

As was discussed in Chapter 1, the characteristics of an adequate, well-controlled clinical trial include a study protocol with a valid statistical design, adequate controls, appropriate randomization and blinding procedures, the choice of sensitive efficacy and safety clinical endpoints, a strict adherence to the study protocol during the conduct of the trial, and a sound statistical analysis. These components are crucial for providing a scientific and unbiased assessment of the effectiveness and safety of a drug product. After the completion of a study, it is extremely important to summarize the clinical results and provide a valid scientific interpretation.

To assist the sponsors in the preparation of final clinical reports for regulatory submission and review, most regulatory agencies have developed guidelines for the format and content of a clinical report. In the United States, before the mid-1980s, clinical study reports were typically prepared by the study's medical monitor using the study's statistical reports. As a result, for each clinical study, two reports were submitted to the FDA for review. The clinical report was reviewed by the respective therapeutic division and the statistical report was reviewed by the biometrics division. The disadvantage of this arrangement is that the clinical reports may not provide the medical reviewers with the sufficient information for an adequate and correct interpretation of the results of statistical analyses. On the other hand, statistical reports often lack important clinical interpretations for the statistical reviewers to appreciate and understand the clinical significance and the magnitude of the study findings. To overcome this disadvantage, the FDA made a revolutionary change in the reporting of a clinical trial by publishing the *Guideline for the Format and Content of the Clinical and Statistical Sections of an Application* in July 1988. This guideline specifies that two separate clinical and statistical reports must be combined into a full integrated report. The full integrated report cannot be derived by simply attaching a separate statistical report to the clinical report. The advantage of this revolutionary change is that it forces both medical and statistical reviewers to review the relevant material and information in a single report.

In 1994 the Committee for Proprietary Material Products (CPMP) Working Party on Efficacy on Medicinal Products of the European Community issued a similar guideline entitled *An Note for Guidance on Biostatistical Methodology in Clinical Trials in Applications for Marketing Authorizations for Medical Products* (CPMP, 1995). At the same time the International Conference on Harmonization (ICH) signed off its *Step 4* final draft of *the Structure and Contents of Clinical Study Reports*. The ICH recommended its adoption to the three regulatory agencies. Technically speaking, clinical and statistical reports must now include clinical data listing, statistical methods, and results of statistical analysis in addition to documentation on the conduct of the clinical trial. For an integrated clinical and statistical report, the FDA guidelines suggest that tables and figures be incorporated into the main text of the report or placed at the end of the text. The study protocol, investigator information, related publications, patient data listings, and technical statistical details such as derivations, computations, analyses, and computer output must be provided in the appendixes of the report.

In clinical trials the observed clinical data can basically be summarized at three different levels. The first level concerns patient data listings which include death, discontinued patients, protocol deviations, patients excluded from the efficacy analysis, listings of serious adverse events, laboratory abnormal values, individual demographic, efficacy and adverse events listings, and laboratory measurements. The FDA also requires that the case report tabulations be included at this level. The second level concerns statistical tables and figures provided at the end but not in the text of the report. These statistical tables and figures are derived from the results of the designated statistical methods stated in the protocol. The third level concerns the summary tables and figures in the text which are condensed from the summary statistical tables and figures. Cross-references must be provided to summary tables and figures in the text, statistical tables and figures at the end of the study, and individual patient data listings in the appendixes of the study.

For efficacy evaluation of the drug product under investigation, formal statistical inferential procedures are usually performed to establish the benefit of the treatment based on the efficacy data. However, many clinical and statistical issues may be raised during the analysis of efficacy data. These issues need to be addressed before a fair and unbiased assessment of efficacy can be reached. The FDA guidelines on the format and content for the full integrated clinical and statistical report and the ICH guidelines on the structure and contents of clinical study reports require the following issues be addressed in the final reports: (1) baseline comparability, (2) analyses of the intention-to-treat dataset versus evaluable dataset, (3) adjustments of covariates, (4) multicenter trials, (5) subgroups analysis, (6) multiple endpoints, (7) interim analysis and data monitoring, (8) active control studies, and (9) handling of dropouts or missing data. Note that the issues of active control studies and equivalence trials were previously addressed in Sections 6.2 and 6.5, respectively. In addition the issue of group sequential procedures for interim analyses was discussed in Chapter 9. In this chapter, we will focus on some remaining issues.

In Section 11.2, we will discuss baseline comparability between treatment groups with respect to patient characteristics. The concept of the intention-to-treat dataset is elucidated and compared with the evaluable dataset in Section 11.3. Statistical methods for adjustment of treatment effects by covariates are discussed in Section 11.4. Some statistical issues regarding multicenter studies, multiple comparisons such as subgroup analyses, and multiple endpoints are addressed in Sections 11.5 and 11.6. Various aspects with respect to data monitoring are provided in Section 11.7. In Section 11.8 some final remarks are given.

11.2 BASELINE COMPARISON

Baseline measurements are those collected during the baseline periods as defined in the study protocol. Baseline usually refers to as measurements obtained at randomization and prior to treatment. Sometimes measurements obtained at screening are used as baselines. Also, baselines are not always restricted to pretreatment measurements. For example, for complicated designs such as the enrichment design for the assessment of Tacrine in the treatment of Alzheimer's disease and CAST for arrhythmia, there may have several phases with the primary clinical endpoints evaluated at two successive phases. Then the measurements that would be used as the baseline for the entire trial or for a particular phase of the study should be precisely stated in the study protocol and in the final clinical report.

Basically the objectives for the analysis of baseline data are threefold. First, the analysis of baseline data provides a description of patient characteristics of the target population to which statistical inference is made, along with useful information on whether the patients enrolled in the study are representative of a targeted population according to the inclusion and exclusion criteria of the trial. Second, since baseline data measure the initial patient disease status, they can serve as reference values for the assessment of the primary efficacy and safety clinical endpoints evaluated after the administration of the active treatment. Finally, the comparability between treatment groups can be assessed based on baseline data to determine potential covariates for statistical evaluations of treatment effects.

In clinical trials the baseline data usually consist of demographic data such as age, gender, or race; initial disease status as evaluated by the primary efficacy, safety endpoints, and other relevant data; and medical history. The ICH *Guideline on Structure and Contents of Clinical Study Reports* requires that baseline data on demographic variables and some disease factors be collected and presented. These disease factors include (1) specific entry criteria, duration, stage and severity of disease and other clinical classifications and subgroupings in common usage or of known prognostic significance, (2) baseline values for critical measurements carried out during the study or identified as important indicators of prognosis or response to therapy, (3) concomitant illness at trial initiation, such as renal disease, diabetes, and heart failure, (4) relevant previous illness, (5) relevant previous treatment for illness treated in the study, (6) concomitant treatment maintained, (7) other factors that might affect response to therapy (i.e., weight, renin status, and antibody level), (8) other possibly relevant variables (e.g., smoking, alcohol intake, and special diets), and for women, menstrual status and date of last menstrual period if pertinent for the study.

Note that in most clinical trials it is tempting to collect more baseline information than needed. It is suggested that we only focus on the baseline or prognostic variables that might affect the response to the treatment.

The effectiveness and safety of the treatment is usually assessed by the change from the baseline of some primary clinical endpoints. The change can be either the absolute change from the baseline, which is defined as the difference between the post-treatment value and the baseline value, or the relative change such as the percent change from the baseline, which is the absolute change from the baseline divided by the baseline value multiplied by 100. The change from the baseline, which can be negative or positive, is symmetric about zero if there is no treatment effect. The percent change from the baseline is in fact the ratio of the post-treatment value to the baseline value minus 1. Thus the range of the percent change from the baseline is from -100 to infinity. As a result the percent change from the baseline is not symmetric about 0 nor 1. If the post-treatment and the baseline values are positively correlated, then the variability of the change from the baseline will be smaller than that of the raw value. Therefore the change from the baseline measures the alteration caused by the treatment. In addition it may provide a more precise statistical inference of the treatment effects because of a possible reduction of variability. If the distribution of a clinical endpoint is approximately normal, then the distribution of the change from the baseline for this endpoint is also normal. However, the distribution of the percent change from the baseline is not normal; rather it is a Cauchy distribution whose moments such as mean and variance do not exist. As a result no inference about the mean or variance can be made. In this case it is recommended that nonparametric methods be employed to obtain inference of the treatment effect based on the median.

As was indicated above, the baseline data are often used as reference values for assessing changes in disease status after administration of the treatment. In practice, multiple baseline measurements are often obtained not only to assess variability but also to evaluate the stability of the disease status before the administration of active treatment. However, the reasons for multiple baseline measurements are manifold (Carey et al., 1984; Frick et al., 1987; Manninen et al., 1988). For example, some investigators feel that due to various causes, a single baseline measurement might not provide reliable assessments for some critical disease characteristics. On the other hand, in antiarrhythmia trials, multiple baseline measurements are evaluated during the washout period to ensure that patients are free of previous antiarrhythmia medications. As a result some of multiple baseline measurements evaluated at the end of the washout period can provide an actual description of the initial disease status for each patient. In addition the FDA guideline for the studies of benign prostatic hyperplasia (Boyarsky and Paulson, 1977) suggests that baseline measurements be collected in a placebo run-in period of at least 28 days. The purpose in collecting multiple baseline measurements during the placebo run-in period is to assess the placebo effect by some important clinical endpoints and to evaluate the stability of the disease status.

However, when there are more than one baseline measurement for a particular endpoint, the question arises *whether one of the measurements or the average of all or part of the measurements should be used as a single baseline value.* For example, in antihypertensive trials it is customarily to use the average of two or three measurements of systolic and diastolic blood pressures taken within the prespecified window of the baseline period as the baseline value without formal statistical justification. The question regarding which measurements should be used as baseline is subjective to clinical judgment, which should be addressed either in the study protocol or prior to the initiation of the trial based on the information from previous or related trials and pilot studies. In some cases, however, it may be of interest to establish the stability of the disease state or to evaluate the placebo effect through the placebo run-in period. In this case multivariate statistical procedures are useful. In practice, for multiple baseline measurements, it is suggested that valid statistical methods be used to justify a single baseline, such as the average, by combining baseline measurements.

Let Y_{hij} be the response of a clinical endpoint evaluated at the jth time point during the baseline period for the ith subject receiving the hth treatment; $h = 1, \ldots, H, i = 1, \ldots, n_i, j = 1, \ldots, p$, and $n_i > p+1$. Since baseline measurements are obtained prior to treatments, it is not anticipated that there are differences among treatment groups (due to the treatment). The following model is commonly used for investigation of the time effect of baseline measurements by suppressing the subscript h for treatment:

$$Y_{ij} = \alpha_j + e_{ij}, \qquad j = 1, \ldots, p, \; i = 1, \ldots, N, \qquad (11.2.1)$$

where α_j is the average of the clinical endpoint at time point j, e_{ij} is the random error associated with Y_{ij}, and $N = n_1 + \ldots + n_H$;

Let the $p \times 1$ vector $\mathbf{Y}_i = (Y_{i1}, \ldots, Y_{ip})$ be the p measurements of the clinical endpoint measured at p time points during the baseline period. Also let $\mathbf{e}_i = (e_{i1}, \ldots, e_{ip})$ be the corresponding error vector. Assume that $\mathbf{e}_i$ follow a multivariate normal distribution with mean vector 0 and a nonsingular covariance matrix Σ. Then the Hotelling T^2 can be used to investigate the trend and placebo effect for multiple baselines (Morrison, 1976). Let $\mathbf{C}$ be a $(p-1) \times p$ matrix with the jth row vector $\mathbf{c}'_j$ of orthogonal contrast such that

$$\mathbf{1}_p'\mathbf{c}_j = 0 \quad \text{and} \quad \mathbf{c}_j'\mathbf{c}_{j'} = 0 \quad \text{for } j \neq j',$$

where $\mathbf{1}_p$ is $p \times 1$ column of 1. Then the hypothesis of no overall trend effect due to time is rejected at the αth level of significance if the test statistic

$$T^2 = \frac{N-p+1}{(n-1)(p-1)} N\overline{\mathbf{Y}}'\mathbf{C}'(\mathbf{CSC}')^{-1}\mathbf{C}\overline{\mathbf{Y}}$$
$$> F(\alpha, p-1, N-p+1) \tag{11.2.2}$$

where $\overline{\mathbf{Y}}$ is the sample means of $\mathbf{Y}_1, \ldots, \mathbf{Y}_N$, $\mathbf{S}$ is the sample covariance matrix from $\mathbf{Y}_1, \ldots, \mathbf{Y}_N$ with $N-1$ degrees of freedom, and $F(\alpha, p-1, N-p+1)$ is αth upper quantile of an F distribution with $p-1$ and $N-p+1$ degrees of freedom.

Since the Helmert transformation compares a measurement of the baseline value at a particular time point to the average of baseline values at subsequent time points, it can be used to evaluate the placebo effect and determine the time point at which the baseline stabilizes (Searle, 1971). For the Helmert matrix, the jth row of $\mathbf{C}$ is constructed such that the first $j-1$ elements are 0, the jth element is 1, and rest of the elements are $1/(p-j)$, $j = 1, \ldots, p-1$. If the time point where the baselines stabilize can be determined by the Helmert transformation, it is not only possible to estimate the placebo effect but also to combine the stabilized multiple baseline measurements into a single baseline. Note that if there is no time effect for the stabilized baseline measurements, model (11.2.1) can be rewritten as

$$Y_{ij} = \alpha + e_{ij}, \quad j = 1, \ldots, p, i = 1, \ldots, N. \tag{11.2.3}$$

Under model (11.2.3), $\mathbf{Y}_1$ follows a multivariate normal distribution with mean vector $\alpha \mathbf{1}_p$ and covariance matrix Σ. The criterion used to summarize the stabilized multiple baseline measurements of a clinical endpoint is to find the best linear unbiased estimate of α. In addition, since $Y_{i1}, \ldots, Y_{ip}$ represent p unbiased estimators of a scalar parameter α and covariance matrix Σ, the best linear unbiased estimator for α based on $Y_{i1}, \ldots, Y_{ip}$ by the generalized least squares procedure (GLS), under model (11.2.3), is given by (O'Brien, 1984)

$$X_i = \frac{\mathbf{1}_p'\Sigma^{-1}\mathbf{Y}_i}{\mathbf{1}_p'\Sigma^{-1}\mathbf{1}_p} \tag{11.2.4}$$

with variance

$$V_X = (\mathbf{1}_p'\Sigma^{-1}\mathbf{1}_p)^{-1}.$$

Some investigators (e.g., Carey et al., 1984; Frick et al., 1987) suggest the use of a simple average to combine with stabilized multiple baseline measurements, which is given by

$$Z_i = \frac{\mathbf{1}'_p \mathbf{Y}_i}{\mathbf{1}'_p \mathbf{1}_p}, \qquad (11.2.5)$$

with variance

$$V_Z = \frac{\mathbf{1}'_p \Sigma^{-1} \mathbf{1}_p}{p^2}.$$

Although the covariance matrix Σ is usually unknown, it can be estimated by its consistent sample estimator for large samples. Let $\mathbf{S}_1, \ldots, \mathbf{S}_H$ be the within-group sample covariance matrices from H treatment groups. Then an unbiased consistent estimator of Σ can be obtained as

$$\mathbf{S}_p = \frac{1}{N-H} [(n_1 - 1)\mathbf{S}_1 + \ldots + (n_H - 1)\mathbf{S}_H].$$

As a result the estimated generalized least square (EGLS) estimator of α is given by

$$x_i = \frac{\mathbf{1}'_p \mathbf{S}_p^{-1} \mathbf{Y}_i}{\mathbf{1}'_p \mathbf{S}_p^{-1} \mathbf{1}_p}, \qquad (11.2.6)$$

where x_i is asymptotically unbiased and follows an asymptotic distribution such as that of (11.2.4). Unbiased consistent estimators of variances of x_i and Z_i are given by, respectively,

$$v_x = \frac{N-H}{N-p-1} (\mathbf{1}'_p \mathbf{S}_p^{-1} \mathbf{1}_p)^{-1}$$

and

$$v_Z = \frac{\mathbf{1}'_p \mathbf{S}_p^{-1} \mathbf{1}_p}{p^2}.$$

Example 11.2.1

A parallel two-group randomized phase II study with 22 patients per treatment group was conducted to investigate the efficacy and safety of a newly devel-

Table 11.2.1 Daily Mean Heart Rate (Beats per Minute) and the Pooled Within-Group Covariance Matrix

	Day 1	Day 2	Day 3	Day 4
Placebo				
$N = 22$	74.2	73.9	74.1	73.6
Treatment				
$N = 22$	75.0	73.5	73.1	73.8
Covariance				
Matrix ($df = 42$)				
Day 1	140.2	70.2	66.3	88.2
Day 2		110.8	45.6	47.5
Day 3			73.3	59.3
Day 4				92.9

oped pharmaceutical entity, compared to a placebo, in suppressing ventricular arrhythmia. One important safety endpoint in this trial was the supine heart rate (beats/minutes). Baseline measurements were obtained daily prior to the morning (placebo) dose during a four-day placebo run-in period. The sample means and pooled within-group sample covariance matrix are given in Table 11.2.1.

To detect the placebo effect and determine the time point where the mean heart rate stabilizes, a preliminary analysis can be performed by the following 3×4 Helmert matrix:

$$\mathbf{C} = \begin{pmatrix} 1 & -\frac{1}{3} & -\frac{1}{3} & -\frac{1}{3} \\ 0 & 1 & -\frac{1}{2} & -\frac{1}{2} \\ 0 & 0 & 1 & -1 \end{pmatrix}.$$

The corresponding T^2 is 0.23 with a p-value of 0.87 (3 and 41 d.f.). Since we failed to reject the null hypothesis at the α level of significance, we can conclude that there is no placebo effect, and we do not pursue individual hypotheses about the time point where the heart rate stabilizes. Similar results for time trends can be tested based on the following linear contrasts:

$$\mathbf{C} = \begin{pmatrix} -3 & -1 & 1 & 3 \\ 1 & -1 & -1 & 3 \\ -1 & 3 & -3 & 1 \end{pmatrix}.$$

Since there is no placebo effect during the placebo run-in period, an attempt is made to construct a combined baseline heart rate for each patient from the four baseline measurements. The estimated generalized least square estimate of the combined baseline heart rate is given by

$$x_i = -0.1764 Y_{i1} + 0.3027 Y_{i2} + 0.5478 Y_{i3} + 0.3259 Y_{i4}.$$

The simple average of the four baseline heart rates, Z_i, uses equal weights for the combined baseline, while the EGLS procedures uses weights proportional to the row totals of the inverse of the pooled within-group sample covariance matrix. The estimated variances of x_i and Z_i are 66.32 and 73.21. Therefore, with respect to this dataset, the estimated variance of the combined baseline heart rate by the EGLS procedure is reduced about 9% compared to that of the simple average.

11.3 INTENTION-TO-TREAT AND EFFICACY ANALYSIS

In clinical trial development, despite a thoughtful and well-written study protocol, deviations from the protocol may be encountered over the course of a trial. For example, some patients randomly assigned to receive one of the treatments may be found not meeting the inclusion and/or exclusion criteria after the completion of the trial due to the reason that some laboratory evaluations (on which inclusion criteria are based) are not available at the time of randomization and drug dispensation. In some situations patients might receive wrong treatment due to a mix-up in randomization schedule. Some patients who received the assigned treatment might be switched to receive the other treatment in an emergency or after deterioration of the patient's disease status. In addition, for every clinical trial, it is likely that patients will withdraw from the study prematurely before the completion of the trial due to various reasons. For patients who complete the study, they might miss some scheduled visits. Another example is patient compliance to the treatment regimen. As a result a legitimate question is which patients should be included in the analysis for a valid and unbiased assessment of the efficacy and safety of the treatment.

Before we address the question, it is helpful to review the concept of randomization discussed in Chapter 4. Recall that randomization does not guarantee equal distributions of baseline variables such as demographic and patient characteristics among treatment groups. It, however, does provide an unbiased comparison between treatment groups. Cornell (1990) pointed out that randomization can avoid bias not only with respect to clinical endpoints, which can be readily identified, measured, and controlled in advance, but also with respect to those endpoints not measured and perhaps not yet known to be influential. In addition, as was indicated in Chapter 4, the random assignment of patients to treatment groups is the key to a valid statistical inference. These two crucial points lead to the concept of the intention-to-treat (ITT) analysis. The ITT analysis is usually referred to as the primary analysis for evaluation of efficacy and safety in clinical trials based on all available data obtained from all randomized patients, even though they never receive the treatment or they receive the treatment different from what they are supposed to.

Sackett et al. (1991) and Spilker (1991) report on a clinical trial that compares surgical therapy with medical therapy in treatment of patients with bilateral carotid stenosis. This study enrolled 167 patients. Among these patients,

94 were randomized to the surgical group and 73 were randomly assigned the medical therapy. Sixteen patients who either had stroke or died during initial hospitalization were not included in the analysis. Fifteen of the 16 excluded patients were randomized to the surgical group. Consequently a statistically significant number of patients in surgical therapy was excluded in the evaluable analysis (p-value = 0.0011 based on a two-sided Fisher's exact test). The rate of occurrence of subsequent transient ischemic attack, stroke, or death for the evaluable analysis is 54.4% in the surgical therapy as compared to that of 73.6% in the medical therapy. This indicates that there is a statistically significant reduction of risk 27% in the surgical therapy (p-value = 0.018 based on two one-sided Fisher's exact test). However, for the intention-to-treat analysis, which includes all randomized patients, the reduction of risk is estimated as low as 16%, which is not statistically significant at the 5% level of significance (p-value = 0.10 based on the Fisher's two-sided test). These results are summarized in Table 11.3.1. The statistically significant reduction of risk in subsequent transient ischemic attack, stroke, or death of the surgical therapy observed from the evaluable analysis is a typical example of a biased inference and may be misleading because of the exclusion of patients from analysis.

In general, the ITT analysis is a conservative approach with better capability of providing an unbiased inference about the treatment effect. The ITT analysis can reflect real clinical practice better than any other analyses. In order to provide an unbiased and valid inference based on the ITT analysis, the study proto-

Table 11.3.1 Summary of Surgical Versus Medical Therapy in Treatment of Bilateral Carotid Stenosis

	Therapy		
	Surgical	Medical	p-Value
I. Patients excluded			
N	94	73	0.0011
Number excluded	15 (16.0%)	1 (1.4%)	
II. Evaluable analysis			
N	79	72	0.018
TIA, stroke or death	43 (54.4%)	53 (73.6%)	
II. Intention-to-treat analysis			
N	94	73	0.10
TIA, stroke or death	58 (61.7%)	54 (74.8%)	

Note: TIA: Transient ischemic attack; p-values are obtained from the two-tailed Fisher's exact test.

Source: Adapted from Sackett et al. (1991) and Spilker (1991).

col should include dropout rates, rates of poor compliance, patterns of missing values, and other related variables as response variables. These response variables should be compared between the treatment groups. In addition interactions between treatment groups and noncompleters (or completers) should also be examined with respect to demographic variables and baseline disease characteristics. The CPMP Working Party on Efficacy of Medicinal Products suggests that at least demographic and baseline data on the disease status in the wider population screened for entry, or enrolled into a screening phase of the study prior to randomization, should be summarized to provide information about the numbers and characteristics of excluded patients, both eligible and ineligible, together with their reasons for exclusion, in order to guide assessment of the potential practical impact on the study results (European Commission, 1994).

For the ITT analysis, serious bias can be introduced if no data are collected for the patients who discontinued the study prematurely. Therefore it is recommended that the primary clinical endpoints of the trial be evaluated at the time of withdrawal. In many clinical trials the method of the last observation carried forward (LOCF) is applied to patients who withdrew prematurely from the study with no additional follow-up data after discontinuation. This approach, however, is biased if the withdrawal of patients is treatment related. In principle, the ITT analysis is performed according to the random allocation of treatment regardless of the real treatment that patients actually receive. If the proportion of mix-up in treatment assignments exceeds a specified level, the ITT analysis, nor any other analysis, will no longer be valid for the interpretation of the results. This is the fatal consequence of poor conduct of randomization codes, packaging, and dispensation of the study drugs. If this situation does occur, the trial has little value in providing clinical evidence for the efficacy and safety of the study drug.

In addition to the ITT analysis, many subsets of the ITT dataset may be constructed for efficacy analysis. These subsets may include (1) all patients with any efficacy observations or with a certain minimum number of observations, (2) only patients completing the study, (3) all patients with an observation during a particular time window, and (4) only patients with a specified degree of compliance. Analyses based on these subsets are referred to as *evaluable analyses*. The patient population for the evaluable analysis is called the *per-protocol* or *efficacy* patient population. The criteria for inclusion of patients in the evaluable analysis should be specified in the study protocol in advance. It should not be established after the data are collected and randomization codes are unblinded. The criteria for the evaluable analysis usually include the following characteristics:

1. Satisfaction of a prespecified minimal length of exposure to the treatment of the investigated therapy.
2. Provision of data on primary clinical endpoints at prespecific and relevant scheduled time points.

3. Satisfactory compliance with treatment.

4. No major protocol violations or deviations including the inclusion or exclusion criteria, incorrect randomization, concomitant medications.

Note that the above criteria should be specified in the study protocol. The definition of protocol violation should also be specified in the study protocol prior to the initiation of the trial. In addition the nature and reasons for protocol violation should be reviewed, described, decided, and documented by blinded data coordinators, medical monitor and other personnel before the randomization codes are broken and the database is locked for analysis. The evaluable analysis is usually considered the preferred analysis by the sponsor because, in general, it maximizes the probability of showing the efficacy of the drug under study. The evaluable analysis, however, is much more vulnerable to bias than the ITT analysis because of inherent subjectivity for exclusion of patients from the evaluable analysis as demonstrated by the example given in Table 11.2.1. For clinical trials with comparative controls or placebos, the ITT analysis will give a less optimistic estimate for the efficacy, since the inclusion of noncompleters will generally dilute the treatment effect. However, if the objective of the trial is to show therapeutic equivalence between treatment groups, the ITT analysis is still less biased than the evaluable analysis, it is no longer conservative in the context of equivalence trials. The results obtained from the ITT analysis should be interpreted with extreme caution.

From the above discussion, it is helpful to provide a disposition of all patients who entered the study so that the patients in the datasets for the ITT and evaluable analyses can be clearly defined. The principle is to account for every single patient participated in the study. Both the FDA and ICH guidelines on clinical reports require that the numbers of patients who are randomized, and complete each phase of the study be provided, as well as the reasons for all post-randomization discontinuation. Tables 11.3.2 and 11.3.3 give two examples of the summary data on patient disposition as provided in the ICH *Guideline on the Structure and Content of Clinical Study Reports* (ICH, 1995). The FDA guideline also suggests, as illustrated in Table 11.3.4, that there be provided summaries of the number of the patients who entered and completed each phase of the study, or each week/month of the study. In addition a listing should be provided as shown in Table 11.3.5, by center and treatment group, for the patients who discontinued from the trial after enrollment, with the information on patient identifier, demographic characteristics, the specific reasons for discontinuation, the treatment and dose level, and the duration of the treatment before discontinuation. After the status of each patient during the trial is accounted for, we need to decide on the dataset for efficacy, in particular, which patients should be included in the efficacy analysis. These information should be provided in the clinical report. Table 11.3.6 gives a listing of the patients and visits excluded from an efficacy analysis. The

Table 11.3.2 Example of Tables for Disposition of Patients

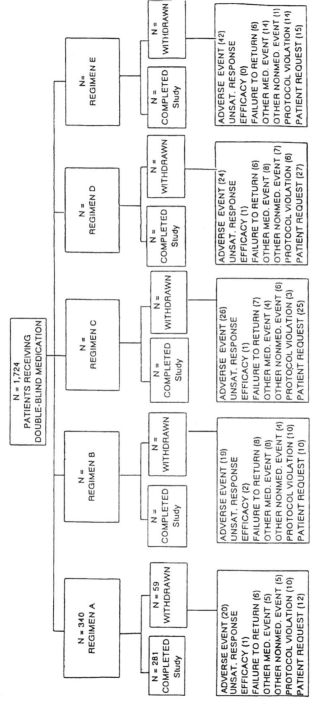

Source: Guideline on Structure and Contents of Clinical Study Reports (ICH, 1995).

Table 11.3.3 Example of Tables for Disposition of Patients

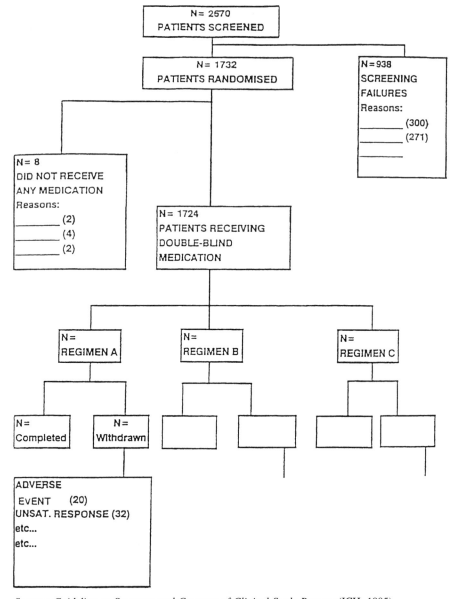

Source: Guideline on Structure and Contents of Clinical Study Reports (ICH, 1995).

FDA guidelines also require the investigator to provide a summary table of the number of patients excluded from the efficacy analysis by reason and week or phase (see Table 11.3.7). The primary efficacy analysis should be based on all randomized patients according to their random assignments of treatment. If

Table 11.3.4 Disposition of Patients

	STUDY # (Data Set Identification) Number of Patients Completing Each Period of Study				
	Randomized	Treated	Week 1	Week 2	Week 4
Test drug	#	#(%)			
Active control					
Placebo	———	———	———	———	———
Total	———	———	———	———	———
Comparability test (p-value)		———	———	———	———

Source: U.S. FDA *Guideline for the Format and Content of the Clinical and Statistical Sections of an Application* (1988).

the preferred efficacy analysis by the sponsor is based on a reduced subset of the intention-to-treat dataset, any discrepancies and inconsistencies between the two analyses should be explained and clarified in the report.

11.4 ADJUSTMENT FOR COVARIATES

For assessment of the efficacy and safety of a drug product, it is not uncommon that primary clinical endpoints are affected by some factors such as the demographic variables of age, gender, and race and/or patient characteristics such as disease severity, concomitant medications, and medical history. These factors are referred to as covariates and are also known as confounding factors, prognostic factors, or risk factors. In clinical trials we might identify some covariates of an impact on the clinical outcomes so that these covariates can be measured during the trial. In practice, it is common not to collect information on covariates that may be influential and yet unknown at the planning stage of the trial. If patients are randomly assigned to receive treatments, as indicated in Chapter 4, the estimated treatment effect is asymptotically free of the accidental bias induced by omission of one or more covariates. In other words, a randomized trial is asymptotically free of covariate imbalance, even for unknown and unmeasured covariates. Covariate balance can typically be reached when the sample size tends to infinity. In practice, however, the sample size is finite; consequently the estimated treatment effect is biased if there is an imbalance in one or more covariates.

As an example, let us consider a antihypertensive trial that compares a test drug against a placebo. Let Y_{ij} and X_{ij} be the diastolic blood pressure (mm Hg) measured at the end of the study and its corresponding baseline value for

Table 11.3.5 Listing of Patients Who Discontinued Therapy

Center: Study #
          Data Set Identification

Treatment	Patient #	Sex	Age	Last Visit	Duration	Dose	Concomitant Medication	Reason for Discontinuation
Test drug/ Investigational product								Adverse event* ... Therapy failure
Treatment	Patient #	Sex	Age	Last Visit	Duration	Dose	Concomitant Medication	Reason for Discontinuation
Active control/ Compactor								
Treatment	Patient #	Sex	Age	Last Visit	Duration	Dose	Concomitant Medication	Reason for Discontinuation
Placebo								

(Repeat for other centers)

Source: The U.S. FDA *Guideline for the Format and Contents of the Clinical and Statistical Sections of an Application* (1988).
*The specific reaction leading to discontinuation.

ADJUSTMENT FOR COVARIATES

Table 11.3.6 Listing of Patients and Observations Excluded from Efficacy Analysis

| | STUDY # | | | | | |
	(Data Set Identification)					
Center:						
Treatment	Patient #	Sex	Age	Observation	Excluded	Reason(s)
Test Drug/ Investigational Product						
Treatment	Patient #	Sex	Age	Observation	Excluded	Reason(s)
Active Control/ Comparator						
Treatment	Patient #	Sex	Age	Observation	Excluded	Reason(s)
Placebo						
(Repeat for other centers)						
Reference Tables						
Summary:						

Source: U.S. FDA *Guideline for the Format and Content of the Clinical and Statistical Sections of an Application* (1988).

jth patient and the ith treatment; $j = 1, \ldots, n_i$, $i = T, P$. The analysis of covariance model given in (7.5.1) is

$$Y_{ij} = \mu + \tau_i + \beta(X_{ij} - \overline{X}_{..}) + \epsilon_{ij}, \qquad (11.4.1)$$

where $\overline{X}_{..}$ is the overall average of the baseline values. The parameter of interest is the treatment effect, which is given by

$$\theta = \tau_T - \tau_P.$$

We denote $\overline{Y}_{i.}$ and $\overline{X}_{i.}$ as the average of the diastolic blood pressures and the corresponding baseline values of the ith treatment, $i = T, P$. If the baseline diastolic blood pressure is ignored, then an intuitive estimate of the treatment effect is the difference in the unadjusted treatment average between the test drug and the placebo, namely

$$\hat{\theta}^* = \overline{Y}_{T.} - \overline{T}_{P.} \qquad (11.4.2)$$

Table 11.3.7 Number of Patients Excluded from Efficacy Analysis

| | STUDY # | | | |
	(Data Set Identification)			
Test Drug	$N =$			
		Week		
Reason	1	2	4	8
Total				

Note: Similar tables should be prepared for the other treatment groups.
Source: U.S. FDA *Guideline for the Format and Content of the Clinical and Statistical Sections of an Application* (1988).

The expected value of $\hat{\theta}^*$ is

$$E(\hat{\theta}^*) = \tau_T - \tau_P + \beta(\overline{X}_{T.} - \overline{X}_{P.}). \qquad (11.4.3)$$

From (11.4.3) we note that in addition to the true treatment effect, $E(\hat{\theta}^*)$ contains a covariate. Therefore, unless the baseline diastolic blood pressure is balanced between the test and the placebo groups, namely $\overline{X}_{T.} = \overline{X}_{P.}$, the estimated treatment effect (11.4.2) is biased. We define

$$\overline{Y}_i^* = \overline{Y}_{i.} - \hat{\beta}(\overline{X}_{i.} - \overline{X}_{..}) \qquad (11.4.4)$$

as the adjusted treatment average for the ith treatment, $i = T, P$, where $\hat{\beta}$ is the least squares estimate (LSE) of the regression coefficient of the diastolic blood pressure at the end of the study on the corresponding baseline value. Since the LSE of β is unbiased, the expected value of the adjusted treatment average is given by

$$\begin{aligned} E(\overline{Y}_i^*) &= E[\overline{Y}_{i.} - \hat{\beta}(\overline{X}_{i.} - \overline{X}_{..})] \\ &= \mu + \tau_i + \beta(\overline{X}_{i.} - \overline{X}_{..}) - \beta(\overline{X}_{i.} - \overline{X}_{..}) \\ &= \mu + \tau_i. \end{aligned}$$

Let $\hat{\theta}$ denote the difference in adjusted treatment average between the test drug and the placebo. The expected value of $\hat{\theta}$ is given by

$$E(\hat{\theta}) = E(\overline{Y}_T^* - \overline{Y}_P^*)$$
$$= \tau_T - \tau_P.$$

Since the expected value of $\hat{\theta}$ consists of only the treatment effect which is independent of the covariate, it is an unbiased estimator of the unknown but true treatment effect. As discussed above, if the covariates are not balanced, then the difference in the simple average between treatment groups will be biased for estimation of the treatment effect. Hence the covariates must be included in the statistical model for an unbiased estimate of the treatment effect.

In the case where covariates are not balanced between the treatment groups, to obtain a valid inference of the treatment effect, it is necessary to adjust for covariates that are statistically significantly correlated with the clinical endpoints. As was mentioned above, if a covariate is balanced, then the difference in the simple treatment of averages is an unbiased estimate for the treatment effect. This is equivalent to assuming the following one-way analysis of variance model:

$$Y_{ij} = \mu + \tau_i + \epsilon_{ij}^*. \tag{11.4.5}$$

A comparison between models (11.4.1) and (11.4.5) reveals that the analysis of covariance is a combination of analysis of variance and regression analysis techniques. As a result the error term ϵ_{ij}^* in model (11.4.5) not only includes the pure error term ϵ_{ij} but also the part of the regression of the response endpoint on the covariate. If a statistically significant correlation exists between the response endpoint and the covariate, then a significant portion of variability of the response endpoint can be explained by the covariate. As a result the residual sum of squares obtained under model (11.4.1) is much smaller than that under model (11.4.5) without the inclusion of the covariate. Although both models (11.4.1) and (11.4.5) yield unbiased estimators for the treatment effect, the precision of the estimated treatment effect under model (11.4.1) is better because the covariate helps to remove the variability of the response endpoint. In summary, an adjustment for covariates not only provides unbiased statistical inference but also increases precision of the statistical inference.

Note that the model (11.4.1) with covariates in statistical inference for estimation of the treatment effect assumes a common slope for both the test and placebo groups. The treatment effect at a particular value of the covariate is then the distance between the two lines at that value. Under the assumption of a common slope, the regression lines for the two groups in their relationship to the response endpoint and the covariate, which are parallel to each other, are shown in Figure 11.4.1. Since the distance between the two lines is the same for the entire range of the covariate, a common treatment effect can be estimated.

If there is an interaction between treatment and covariate, then the two regression lines will not have the same slope. In this case model (11.4.1) can be modified as

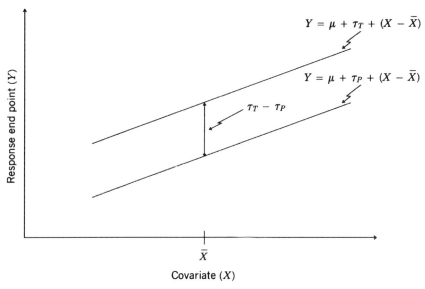

Figure 11.4.1 Adjustment for covariate in estimation of treatment effect. Case I: Common slope.

$$Y_{ij} = \mu + \tau_i + \beta_i(X_{ij} - \overline{X}_{..}) + \epsilon_{ij}. \qquad (11.4.6)$$

It follows that the two regression lines are not parallel any more. This implies that the treatment effect is different at different values of the covariate. Although an unbiased estimate can still be obtained at a particular value of the covariate, it is not possible to unbiasedly estimate the common treatment effect over the entire range of the covariate. Figure 11.4.2 depicts a situation where the treatment effect has the same direction for the possible range of the covariate. The treatment effect increases as the value of the covariate increases. This might indicate that test drug is more efficacious for those patients with a large covariate value. On the other hand, the treatment effect might change signs at the different ranges of the covariate as shown in Figure 11.4.3. In other words, the test drug is worse than the placebo for patients with low covariate values. If interaction between the treatment and a covariate exists, a stratified analysis might be performed by dividing the entire range of covariates into several strata so that the treatment effect is homogeneous. This subgroup analysis will be discussed later in the chapter.

Another key assumption in the adjustment of covariates of the treatment effect under model (11.4.1) is that the treatment should not affect the covariates. This assumption is easily satisfied by subject-specific covariates such as demographic variables and baseline disease characteristics that are measured only once prior to the initiation of the treatment and are time independent. Certain other covariates may be measured at every post-treatment visit when the pri-

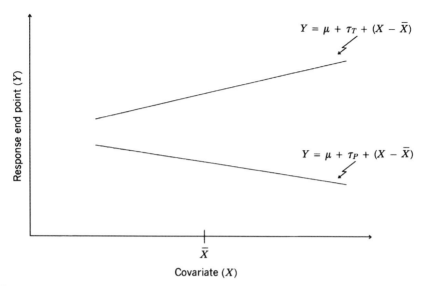

Figure 11.4.2 Adjustment for covariate in estimation of treatment effect. Case II: Different slopes, same direction but different magnitude.

mary clinical endpoints are evaluated. Typical examples of such covariates are heart rates and cholesterol levels in antihypertensive trials or in a prevention trial of cardiovascular events, and CD4 in AIDS trial. Since these covariates are likely to be influenced by the treatment, interpretation of the estimated treatment

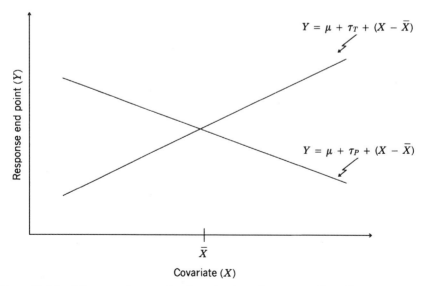

Figure 11.4.3 Adjustment for covariate in estimation of treatment effect. Case III: Different slopes, different direction with different magnitude.

effect after adjustment for the time-dependent covariates must be made with extreme caution. As a result the CPMP Working Party on Efficacy of Medicinal Products advises investigators not to adjust the primary analysis for covariates measured after randomization (European Commission, 1994). In addition it is recommended that statistical methods that account for covariate imbalance between treatment groups be used to improve the precision of the inference.

Model (11.4.1) is a model for the adjustment of covariates with continuous endpoints. The blocked Wilcoxon rank sum test (Lehmann, 1975) can be used for categorical covariates when the normality assumption of continuous endpoints are questionable. The logistic regression or generalized estimation equations (GEE) discussed in Chapter 8 can be employed for the analysis of categorical endpoints. The adjustment for covariates with censored data can be handled through Cox's proportional hazard model, as described in Chapter 9. Since there are many possible covariates and statistical methods for adjustment, the FDA and ICH guidelines for clinical reports require that the selection of and adjustment for any covariate should be explained in the reports. In addition methods for adjustment, results of analyses, and supportive information should be fully documented along with other statistical methods. If the covariates or the methods for adjustment are different from those planned in the protocol, the results of the planned analyses should be presented and the discrepancies should be explained in detail. Examples of how the adjustment of covariates results should be presented are given in the FDA guideline, these examples are reproduced in Tables 11.4.1 and 11.4.2.

Table 11.4.1 Summary of Results for Prognostic Factors for Analysis of Time to Treatment Failure

Prognostic Factor	Study 1[b]	Study 2
Baseline performance status	<0.01	<0.01
Age	0.02	0.29
Disease-free interval	0.08	0.48
Adjuvant chemotherapy	0.53	0.01
Race	0.96	0.12
Estimated coefficient for treatment, beta[a]	−0.12	−0.27
Standard error of beta	0.12	0.13
p-value of treatment effect	0.30	0.03
e^{BETA} = estimated ratio of hazard rates (active:test)[a]	0.89	0.76
(95% confidence interval)	(0.70, 1.12)	(0.59, 0.99)

[a] Adjusted for all prognostic factors whose p-value was <0.20 (significance at 0.20 level should not be taken as an FDA policy).
[b] p-Value from final Cox regression model.
Source: U.S. FDA *Guideline for the Format and Content of the Clinical and Statistical Sections of an Application* (1988).

Table 11.4.2 Effect of Prognostic Factors of the Comparison of Treatments Before and After Adjusting for Prognostic Factors

Efficacy Variable	Study 1	Study 2
Time to treatment failure		
Estimated hazard ratio (T:A)		
Unadjusted	0.90	0.81
(95% confidence limits)	(0.72, 1.14)	(0.63, 1.04)
p-value	0.39	0.09
Estimated hazard ratio (T:A)		
Adjusted	0.89	0.76
(95% confidence limits)	(0.70, 1.12)	(0.59, 0.99)
p-value	0.30	0.03
Factors adjusted	PS, AGE, INT[a]	PS, ADJ, RACE[a]

Note: T = test drug, A = active control.
[a]PS = Baseline performance status, AGE = age, INT = disease-free interval, ADJ = adjuvant chemotherapy, RACE = race.
Source: U.S. FDA *Guideline for the Format and Content of the Clinical and Statistical Sections of an Application* (1988).

11.5 MULTICENTER TRIALS

A multicenter trial is a single study that is conducted simultaneously at more than one site (or center) according to a common protocol. A multicenter trial is not equivalent to separate *single-site* trials. Hence the data collected from different centers of a multicenter trial are intended to be analyzed as a whole. The history of multicenter trials can be traced back to early 1960s when Sir Bradford Hill designed his classical multicenter controlled trial of antihistamine, cortisone, and streptomycin (Hill, 1962).

There are many reasons for conducting multicenter trials. The most important rationale is probably that the framework of multicenter trials provides an efficient means of accruing sufficient numbers of patients in order to achieve a desired power within a predetermined time frame. If a trial is conducted at a single site, the resulting estimate of the treatment effect might be uniquely related to some characteristics, known or unknown, of the center. Thus the generalization of trial results can be very restricted. Multicenter trials provide a basis for a broad generalization of the trial results because the patients are recruited from a wider spectrum of the targeted population. In addition the study drug is administrated under a broader clinical setting and practice, with a large number of investigators who will provide a more comprehensive clinical judgment concerning the value of the study drug.

Although a multicenter trial possesses the characteristics that (1) it is conducted under a single protocol, (2) it follows a set of pre-specified schedules

Table 11.5.1 Example for the Distribution of Patients by Center and Treatment—I

Center	Placebo	Low	Medium	High	Total
San Diego	16	19	17	18	70
Phoenix	15	17	18	14	64
Houston	15	18	20	16	69
Minneapolis	1	0	2	0	3
Total	47	54	57	48	206

and evaluation procedures, and (3) the data are processed under a centralized management system, patients recruited at different centers may be inherently subtly different due to the fact that (1) individual investigator's clinical practice and standard operating procedures might be consistent but not completely the same and (2) equipment might be differently calibrated. These issues and other known and unknown reasons may cause variation among centers. In addition the number of patients recruited at each center is in fact a random number that cannot be controlled as desired. For example, the protocol might call for at least 20 patients at each center. It might turn out that distribution of the patient is quite heterogeneous across centers. Table 11.5.1 provides a tabulation of the number of patients for the example given in Section 2.3. This example presents the situation where a phase IIB multicenter trial with four treatment groups was conducted at a small number of centers. A large number of patients, namely more than 60, were randomized at three of the four centers. However, the other center only recruited three patients. As a result at this center no patients were even randomized to the low- and high-dose groups. Table 11.5.2 presents another situation where a phase IIB multicenter trial with four treatment groups was conducted at a large number of centers. Although it was anticipated that 20 patients were recruited at each center, only 9 of the 25 centers reached the goal which constitutes 60% of 428 randomized patients. On the other hand, fewer than 15 patients were randomized at other 9 centers with a total of 97 patients (or 23% of 428 randomized patients). In addition two centers only randomized no more than three patients. For both examples, since at some centers some treatment groups did not have randomized patients, the result/interpretation of statistical analyses were not reliable.

For the analysis of a multicenter trial, Lewis (1995) posed the following questions which are helpful for the analysis of clinical data in a regulatory context:

1. Are some of the centers too small for reliable separate interpretations of the results?
2. Are some of the centers so big that they dominate the results?
3. Do the results at one or more centers look out of line with the others, even if not significantly so? Can the reason for this be established?

Table 11.5.2 Example for the Distribution of Patients by Center and Treatment—II

Center	Placebo	100 mg	200 mg	300 mg	Total
1	5	5	5	6	21
2	3	4	3	4	14
3	6	6	6	6	24
4	3	2	2	3	10
5	5	5	5	5	20
6	4	3	4	3	14
7	9	9	9	9	36
8	7	6	7	6	26
9	9	9	9	9	36
10	6	6	5	6	23
11	2	2	3	3	10
12	1	0	1	1	3
13	5	5	4	5	19
14	9	9	8	8	34
15	5	5	4	5	19
16	2	2	2	2	8
17	4	5	4	5	18
19	4	5	5	5	19
20	4	3	3	3	13
21	9	9	9	9	36
22	2	4	4	3	13
24	2	3	3	2	10
25	0	1	1	0	2
Total	106	107	106	109	428

4. Do any of the centers show a trend in the *wrong* direction? Can this be explained? How can we restrict the use of a drug to the appropriate patients if we do not understand such trends?
5. If a treatment-by-center interaction is detected, is the trial validated?

Both the FDA and ICH guidelines require statistical tests for homogeneity across centers in order to detect possible quantitative or qualitative treatment-by-center interaction. For example, let Y_{ijk} be the reduction in diastolic blood pressure (mm Hg) observed from patient k receiving treatment i at center j in a clinical trial that evaluates the efficacy and safety of a test drug with a placebo control at J centers in treatment of patients with mild to moderate hypertension; $k = 1, \ldots, n_{ij}, j = 1, \ldots, J, i = T, P$. It is assumed that Y_{ijk} approximately follows a normal distribution with population average μ_{ij} and variance σ^2. Here we use a normal continuous endpoint. Similar concepts can be applied to nonnormal continuous data, categorical data, or censored data from a multicenter trial. Descriptive statistics for site j are summarized in Table 11.5.3. Traditionally data from multicenter trials can be analyzed by the two-way analysis of

Table 11.5.3 Descriptive Statistics for Site j of a Multicenter Trial

Statistics	Treatment		Difference
	Placebo	Test Drug	
N	n_{Pj}	n_{Tj}	
Mean	$\overline{Y}_{Pj.}$	$\overline{Y}_{Tj.}$	$d_j = \overline{Y}_{Pj.} - \overline{Y}_{Tj.}$
Standard deviation	s_{Pj}	s_{Tj}	s_j
Confidence interval	CI_{Pj}	CI_{Tj}	CI_j

$s_j^2 = [(n_{Pj} - 1)s_{Pj}^2 + (n_{Tj} - 1)s_{Tj}^2]/(n_{Pj} + n_{Tj} - 2)$
$CI_{ij} = \overline{Y}_{ij.} + (s_{ij}/\sqrt{n_{ij}})t(\alpha/2, n_{ij} - 1), i = T, P$
$CI_j = (\overline{Y}_{Tj.} - \overline{Y}_{Pj.}) + (s_j/\sqrt{w_j})t(\alpha/2, n_{Pj} + n_{Tj} - 2); w_j = (1/n_{Pj}) + (1/n_{Tj})$

variance model with or without interaction, as discussed in Section 7.4. The model with interaction is given as

$$Y_{ijk} = \mu_{ij} + \epsilon_{ijk}$$
$$= \mu + \tau_i + \beta_j + (\tau\beta)_{ij} + \epsilon_{ijk} \qquad (11.5.1)$$

The analysis of variance model without interaction can be obtained from (11.5.1) by excluding the interaction term $(\tau\beta)_{ij}$

$$Y_{ijk} = \mu_{ij} + \epsilon_{ijk}$$
$$= \mu + \tau_i + \beta_j + \epsilon_{ijk}, \qquad (11.5.2)$$

where ϵ_{ijk} is assumed to follow a normal distribution with mean 0 and variance σ^2.

At center j the treatment effect between the test drug and the placebo can be expressed as

$$\delta_j = \mu_{Tj} - \mu_{Pj}. \qquad (11.5.3)$$

An unbiased estimator for δ_j can be obtained as

$$d_j = \overline{Y}_{Tj.} - \overline{Y}_{Pj.}, \qquad (11.5.4)$$

with an estimate of its variance given by

$$v(d_j) = s_j^2$$
$$= \frac{(n_{Pj} - 1)s_{Pj}^2 + (n_{Tj} - 1)s_{Tj}^2}{n_{Pj} + n_{Tj} - 2}, \quad j = 1, \ldots, J. \qquad (11.5.5)$$

As a result, at site j, the null hypothesis of no treatment effect is rejected at the α level of significance if

$$|t_j| = \left| \frac{d_j}{s_j/\sqrt{w_j}} \right| > t(\alpha/2, n_{Pj} + n_{Tj} - 2), \qquad j = 1, \ldots, J, \qquad (11.5.6)$$

or the $(1 - \alpha)100\%$ confidence interval for δ_j does not contain zero.

The objective for the analysis of clinical data from a multicenter trial is twofold. It is not only to investigate whether a consistent treatment effect can be observed across centers but also to provide an estimate of the overall treatment effect. These objectives, however, depend on (1) whether the center should be considered a fixed or a random factor, (2) what is the appropriate definition of the overall treatment effect, and (3) whether one should estimate the overall treatment in a manner similar to a stratified analysis. Table 11.5.2 illustrates the possibility that a multicenter trial might have a large number of centers, each with a relatively small number of patients. In addition it is reasonable to consider the centers participating in a multicenter trial as a representative sample randomly selected from a population of centers. Chakravorti and Grizzle (1975) recommend that the center effect be considered as random rather than fixed. Under the assumption that the center effect is random, model (11.5.1) becomes a mixed effect model. Under this model, Fleiss (1986b) indicates that the inference for the difference between the treatment averages can be made by a linear combination of the pooled variance s^2 and the interaction mean square MS(AC), where

$$s^2 = \frac{\sum \sum (n_{ij} - 1) s_{ij}^2}{\sum \sum (n_{ij} - 1)}, \qquad (11.5.7)$$

$$\text{MS}(AC) = \frac{1}{J - 1} \sum \left(\frac{1}{w_j} \right) (d_j - \bar{d}^*)^2, \qquad (11.5.8)$$

and

$$\bar{d}^* = \frac{\sum (1/w_j) d_j}{\sum (1/w_j)}.$$

Except for the balanced situation where the equal number of responses for the clinical endpoints observed for each treatment-by-center combination is proportional to its marginal (i.e., $n_{ij} = n_{i.} n_{.j}/n$ for all i, j), statistical methods can be very complicated in the analysis of mixed effects from a multicenter trial, and only approximate results are available (Searle, 1972; Fleiss, 1986b; Chkravorti and Grizzle, 1975; Mielke and McHugh, 1965). Hence, the analysis of the data from a multicenter trial under the assumption of a mixed effects model is unnecessarily perplexing.

In practice, a center is often selected based on the criteria that (1) it can recruit a minimum number of patients from the targeted population as specified by the sponsor, (2) the investigators at the center have expertise and experience in the treatment of the disease, and (3) the center has special equipment or facilities required for the study. As a result the selection of centers in multicenter trials is not a random process but a much deliberated one. Fleiss (1986b) and Goldberg and Koury (1990) suggest that the center effects thus should not be considered random when performing statistical analyses. Note that the primary interest of a multicenter study is not for the comparison between centers. Hence the centers should not be considered as a designed factor rather than a classification factor.

For assessment of the treatment effect in multicenter trials, it is suggested that a parameter be used for the overall treatment effect, which is the simple average over the treatment effects of J individual centers:

$$\delta = \frac{1}{J} \sum_{j=1}^{J} \delta_j$$

$$= \frac{1}{J} \sum_{j=1}^{J} (\mu_{Tj} - \mu_{Pj}). \qquad (11.5.9)$$

Denote $\bar{d}$ as the simple average of d_j:

$$\bar{d} = \frac{1}{J} \sum_{j=1}^{J} d_j$$

$$= \frac{1}{J} \sum_{j=1}^{J} (\bar{Y}_{Tj.} - \bar{Y}_{Pj.}). \qquad (11.5.10)$$

Then $\bar{d}$ is an unbiased estimator for δ with estimated variance

$$v(\bar{d}) = \frac{s^2 \sum w_j}{J^2}. \qquad (11.5.11)$$

The null hypothesis that $H_0: \delta = 0$ is rejected with respect to a two-sided alternative at the α level of significance if

$$|t| = \left| \frac{\bar{d}}{s\sqrt{\sum w_j}/J} \right| > t(\alpha/2, \sum \sum (n_{ij} - 1)). \qquad (11.5.12)$$

The $(1 - \alpha)100\%$ confidence interval for δ is given by

$$\bar{d} \pm (s\sqrt{\Sigma w_j}/J)t(\alpha/2, \sum\sum(n_{ij}-1)). \tag{11.5.13}$$

Note that an unbiased estimator for δ exists and that the test based on (11.5.12) is theoretically correct regardless of the sample sizes and presence of treatment-by-center interaction. When there is no interaction, the cell-means model (11.5.1) reduces to the main-effect model (11.5.2), which assumes that the treatment and center effects are additive. It follows that the treatment effect δ_j at center j is given by

$$\begin{aligned}\delta_j &= \mu_{Tj} - \mu_{Pj} \\ &= (\mu + \tau_T + \beta_j) - (\mu + \tau_P + \beta_j) \\ &= \tau_T - \tau_P, \quad j = 1, \ldots, J.\end{aligned}$$

From the above it can be seen that due to the additivity of the main-effects model, the individual treatment effect within center j does not involve the center effects and is a constant across all centers. In addition, since δ in (11.5.9) is defined as the simple average over the treatment effects of J centers, under the main-effects model the overall treatment effect is also equal to $\tau_T - \tau_P$. Hence the overall treatment effect is a valid parameter for evaluation of treatment effects under both the cell-means model and the main-effects model.

When there is no treatment-by-center interaction, the minimum variance unbiased estimator (MVUE) of $\tau_T - \tau_P$ can be obtained as

$$\bar{d}^* = \frac{\sum (d_j/w_j)}{\sum (1/w_j)}, \tag{11.5.14}$$

with the estimated variance given by

$$v(\bar{d}^*) = \frac{s^2}{\sum (1/w_j)}. \tag{11.5.15}$$

The hypothesis testing can be performed and confidence interval for $\tau_T - \tau_P$ can be constructed through the t statistic as described above.

The estimator for the overall treatment effect $\bar{d}^*$ is the same as the combined estimator obtained from a post-treatment stratified analysis by considering each center as a stratum. The ratio $(1/w_j)/\sum(1/w_j)$ represents the information on the proportion of eligible patients in the targeted population who are from center j. As a result $\bar{d}^*$ is still a consistent estimator in presence of treatment-by-center interaction which is close to the difference between the two treatment averages when the sample size is sufficiently large. Fleiss (1986b) points out that this ratio also reflects the ability and efficiency of patient recruitment for center j.

The null hypothesis of no treatment-by-center interaction is rejected at the α level of significance if

$$F_I = \frac{MS(AC)}{s^2} > F(\alpha, J-1, \sum\sum(n_{ij}-1)), \quad (11.5.16)$$

where $F(\alpha, J-1, \sum\sum(n_{ij}-1))$ is the αth upper quantile of a central F distribution with $J-1$ and $\sum\sum(n_{ij}-1)$ degrees of freedom. If the null hypothesis of no treatment-by-center interaction is rejected, it is helpful to determine whether the interaction is of quantitative or qualitative type. A quantitative interaction implies that the heterogeneity of the treatment effect across centers is due to the magnitude rather than the direction. On the other hand, a qualitative interaction indicates that the treatment effect is not only heterogeneous in magnitude but also changes direction from center to center as shown in Figure 2.4.5. Define

$$Q^- = \frac{d_j}{s_j^2} I[d_j > 0]$$

and

$$Q^+ = \frac{d_j}{s_j^2} I[d_j < 0] \quad (11.5.17)$$

Gail and Simon (1985) propose rejecting the null hypothesis of no qualitative interaction if

$$\min(Q^-, Q^+) > c, \quad (11.5.18)$$

where c is the critical value provided in Table 1 of Simon and Gail (1985).

In multicenter trials, although the sample size is selected to achieve a desired power for detection of the overall treatment effect, it is rarely large enough to identify the treatment-by-center interaction with adequate power. Fleiss (1986b) recommend the use of a 10% level of significance for the detection of treatment-by-center interaction. If the null hypothesis is rejected, then $\bar{d}$ should be used for the inference of the overall treatment effect because it is an unbiased estimator for δ. Note that both $\bar{d}$ and $\bar{d}^*$ are unbiased for δ^* in the absence of treatment-by-center interaction. Since $\bar{d}^*$ is the MVUE for δ^*, there are more degrees of freedom for estimation of σ^2. Hence, if we fail to reject the null hypothesis of no treatment-by-center interaction, then δ^* and σ^2 should be estimated from the reduced main-effects model. However, if the number of centers is relatively small and the treatment-by-center interaction is not significant, both $\bar{d}$ and $\bar{d}^*$ will be close.

If all centers enroll very small numbers of patients, Goldberg and Koury

(1990) suggest ignoring the centers in the analysis. For the unbalanced situations of patient enrollment among centers as presented in Tables 11.5.1 and 11.5.2, since no patients were randomized to treatment groups at some centers, the results and their interpretation are consistent if we ignore the center effect in the analysis. However, one should always follow the rule that *if randomized then analyzed*. One should never omit any centers from the analysis of multicenter trials. One way to resolve the situation, as illustrated in Tables 11.5.1 and 11.5.2, is to randomly assign the patients at those centers to empty cells at other centers before the analysis of data.

Note that the overall treatment effect defined here is the simple average of center-specific treatment effects. Hence, if the qualitative treatment-by-effect interaction does exist, the overall treatment effect may be null. Although the statistical inference provided by $\overline{d}$ is theoretically correct, both FDA and ICH guidelines require that not only individual center results be presented but also any extreme or opposite results among centers be noted and discussed. A graphical presentation of individual center results is provided in Figure 11.5.1. In addition all data, including demographic, baseline, post-baseline data, and efficacy data, should be presented by center.

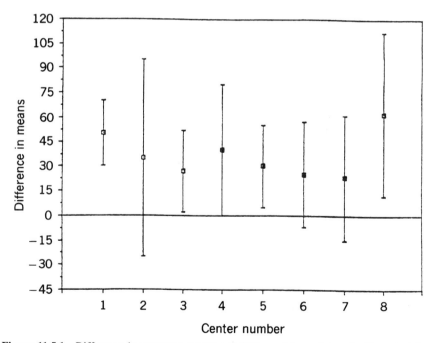

Figure 11.5.1 Difference in treatment means and 95% confidence intervals. Mean test drug change from the baseline minus the mean placebo change from the baseline. (Source: U.S. FDA *Guideline for the Format and Content of the Clinical and Statistical Sections of an Application*, 1988).

11.6 MULTIPLICITY

Lepor et al. (1996) report the results of a double-blind, randomized multicenter clinical trial that evaluated the efficacy and safety of terazosin (10 mg daily), an α_1-adrenergic-antagonist, finasteride, a 5α-reductase inhibitor (5 mg daily) or both with a placebo control in equal allocation in 1229 men with benign prostatic hyperplasia. The primary efficacy endpoints of this trial are the American Urological Association (AUA) symptom score (Barry et al., 1992) and the maximum uroflow rate. These endpoints were evaluated twice during the four-week placebo run-in period and at 2, 4, 13, 26, 39, and 52 weeks of therapy. The primary comparisons of interest included pairwise comparisons among the active drugs and combination therapy, while the secondary comparisons consisted of a pairwise comparison of the active drugs and combination therapy with the placebo. The results for the primary efficacy endpoints presented in Lepor et al. (1996) were obtained by performing analyses of covariance with repeated measurements based on the intention-to-treat population.

One of the objectives of the trial is to determine the time when the treatments reach therapeutic effects. Therefore comparisons among treatment groups were performed at each scheduled postrandomization visits at 2, 4, 13, 26, 39, and 52 weeks. In addition to the original observations of the primary endpoints, the change from baseline can also be employed to characterize the change after treatment for each patient. It may be of interest to see whether the treatment effects are homogeneous across race, age, and baseline disease severity. Therefore some subgroup analyses can be performed such as for caucasians and for noncaucasians patients, for patients below or at least 65 years of age, for patients with the baseline AUA symptom score below 16 or at least 16, or for patients with the maximum uroflow rate below 10 or at least 10 ml/s. As a result as illustrated in Table 11.6.1 the number of the total comparisons for the primary efficacy endpoints can be as large as 1344. If there is no difference among the four treatment groups and each of the 1344 comparisons are performed at the 5% level of significance, we can expect 67 statistically significant comparisons with reported p-values smaller than 0.05. The probability of observing at least one statistically significant difference among 1344 comparisons could be as large as 1 under the assumption that all 1344 comparisons are statistically independent. The number of p-values does not include those from the center-specific treatment comparisons and from other types of comparisons such as treatment-by-center interaction.

Although the above example is a bit exaggerated, it does point out that the multiplicity in multicenter clinical trials is an important issue that has an impact on statistical inference of the overall treatment effect. In practice, however, it is almost impossible to characterize a particular disease by a single efficacy measure due to (1) the multifaceted nature of the disease, (2) lack of understanding of the disease, and (3) lack of consensus on the characterization of the disease. Therefore multiple efficacy endpoints are often considered to evaluate the effectiveness of test drugs in treatment of most diseases such as AIDS, asthma,

Table 11.6.1 Summary of Possible Number of Comparisons

Item	Number of Comparisons
Pairwise comparison	
Primary	3
Secondary	3
Visit	7
Primary end point	2
Response	2
Race	2
Baseline severity of disease	
AUA symptom score	2
Maximum uroflow rate	2
	1344

benign prostatic hyperplasia, arthritis, postmenopausal osteoporosis, and ventricular tachycardia. Some of these endpoints are objective histological or physiological measurements such as the maximum uroflow rate for benign prostatic hyperplasia or pulmonary function FEV_1 (forced expiratory volume in one second) for asthma. Other may include the symptoms or subjective judgment of the well-being of the patients improved by the treatments such as the AUA symptom scores for benign prostatic hyperplasia, asthma-specific symptom score for asthma, or the Greene climacteric scale for postmenopausal osteoporosis (Greene and Hart, 1987). Hence one type of multiplicity in statistical inference for clinical trials results from the source of multiple endpoints.

On the other hand, a clinical trial may be conducted to compare several drugs of different classes for the same indication. For example, the study by Lepor et al. (1996) compares two monotherapies of terazosin, finasteride with the combination therapy, and a placebo control for treatment of patients with benign prostatic hyperplasia. Some other trials might be intended for the investigation of a dose-response relationship of the test drug. For example, Gormley et al. (1992) evaluate the efficacy and safety of 1 and 5 mg of finasteride with a placebo control. This type of multiplicity is inherited from the fact that the number of treatment groups evaluated in a clinical trial is greater than 2. Other types of multiplicity are caused by subgroup analyses. Examples include a trial reported by the National Institute for Neurological Disorders and Stroke rt-PA Study Group (1995) in which stratified analyses were performed according to the time from the onset of stroke to the start of treatment (0–90 or 91–180 minutes). In addition the BHAT (1982) and CAST (1989) studies were terminated early because of overwhelming evidence of either beneficial efficacy or serious safety concern before the scheduled conclusion of the trials by the technique of repeated interim analyses. In summary, multiplicity in clinical trials can be classified as repeated interim analyses, multiple comparisons, multiple endpoints, and subgroup analyses. Since the causes of these multiplicities are different, special attention must be paid to (1) the formulation of statistical hypotheses based on the objectives of the trial, (2) the proper control of experimentwise false positive rates in subsequent analyses of the data, and (3) the interpretation of the results. Since repeated interim analyses have been discussed in Chapter 9, in what follows we will only address the remaining types of multiplicities.

Multiple Comparisons

Multiple comparisons are referred to as the comparisons among more than two treatments. If there is no structure among the treatment groups, then Bonferroni's technique discussed in Section 7.4 is appropriate. The concept of Bonferroni's technique is to adjust the p-values for control of experimentwise type I error rate α for pairwise comparisons. The method is useful when the number of treatments is small. In addition Bonferroni's method does not require that the structure of the correlation among comparisons be specified nor that the number of patients in each treatment group be equal. However, when the number of treatment groups increases, Bonferroni's adjustment for p-values becomes very conservative and may lack adequate power for the alternative in which most or all efficacy endpoints are improved. To overcome this drawback, many modified Bonferroni procedures have been proposed. For illustration purposes, in what follows we will only introduce modified procedures proposed by Holm (1979) and Hochberg (1988).

Holm (1979) proposes that Bonferroni's procedure be modified as follows: Let $p_{(1)} \leq p_{(2)} \leq \ldots \leq p_{(m)}$ be the ordered values of p-values $p_1, p_2, \ldots, p_m$ obtained from testing the following hypotheses:

$$H_{k0}: \mu_i - \mu_j = 0$$
$$\text{vs.} \quad H_{ka}: \mu_i - \mu_j \neq 0, i \neq j; k = 1, \ldots, m. \quad (11.6.1)$$

Holm's procedure starts with the ordered hypothesis $H_{(1)0}$, which is the hypothesis with the smallest p-value, $p_{(1)}$:

Step 1. Stop the procedure and conclude that all pairwise comparisons are not statistically significant at the α level of significance if $mp_{(1)} > \alpha$. Otherwise, reject the null hypothesis $H_{(1)0}$ and continue to step 2.

Step 2. Stop the procedure and conclude that the remaining $(m-1)$ pairwise comparisons are not statistically significant at the α level of significance if $(m-1)p_{(2)} > \alpha$. Otherwise, reject the null hypothesis $H_{(2)0}$ and continue to the next step.

⋮

Step k. Stop the procedure and conclude that the rest of the $(m - k + 1)$ pairwise comparisons are not statistically significant at the α level of significance if $(m - k + 1)p_{(k)} > \alpha$. Otherwise, reject the null hypothesis $H_{(k)0}$ and continue to the next step. Repeat step k until it stops.

The Holm's Bonferroni's procedure is the same as the original Bonferroni's procedure for testing $H_{(1)0}$ at which the p-value is adjusted down by dividing the nominal level of significance by the number of total comparisons. But for the rest of $(m-1)$ hypotheses, Holm's procedure is much sharper than Bonferroni's procedure, since it only requires that the p-values be adjusted according to the number of remaining pairwise comparisons. Because Holm's procedure requires the smallest p-values to be smaller than α/m, Hochberg (1988) suggests another version of Bonferroni's procedure which is even sharper than Holm's procedure. Hochberg's procedure begins with the ordered hypothesis $H_{(m)0}$ corresponding to the largest p-value $p_{(m)}$.

Step 1. Stop the procedure and conclude that all m pairwise comparisons are statistically significant at the α level of significance if $p_{(m)} \leq \alpha$. Otherwise, do not reject the null hypothesis of $H_{(m)0}$ and continue to step 2.

Step 2. Stop the procedure and conclude that the remaining $m-1$ pairwise comparisons are statistically significant at the α level of significance if $2p_{(m-1)} \leq \alpha$. Otherwise, do not reject the null hypothesis of $H_{(m-1)0}$ and continue to the next step.

$\vdots$

Step k. Stop the procedure and conclude that all $(m-k+1)$ pairwise comparisons are statistically significant at the α level of significance if $(m-k+1)p_{(m-k+1)} \leq \alpha$. Otherwise, do not reject the null hypothesis of $H_{(m-k+1)0}$ and continue to the next step. Repeat step k until it stops.

Hochberg's procedure is more powerful than Bonferroni's procedure or Holm's procedure because it only requires that the largest p-value be smaller than α to declare one statistically significant comparison.

Sometimes, clinical trials are planned to compare the effectiveness and safety of K doses of a test drug versus a placebo group to determine the minimum effective and maximum tolerable dose. As a result, in addition to characterization of the dose-response relationship, pairwise comparisons between each dose and the placebo are of clinical interest. In this case the hypothesis of interest can be formulated as follows:

$$H_{k0}: \mu_i - \mu_0 = 0,$$
$$\text{vs.} \quad H_{ka}: \mu_i - \mu_0 \neq 0, \quad i = 1, \cdots, K. \quad (11.6.2)$$

Dunnett's procedure for comparison with a control can be directly applied (Dunnett, 1955). Let $\overline{Y}_i - \overline{Y}_0$ be the observed sample mean difference between dose i and the placebo, and let $v(\overline{Y}_i - \overline{Y}_0)$ denote the estimated variance of $\overline{Y}_i - \overline{Y}_0$, $i = 1, \ldots, K$. Then hypothesis (11.6.2) is rejected at the α level of significance if the $(1-\alpha)100\%$ simultaneous confidence interval for $\mu_i - \mu_0$,

$$\overline{Y}_i - \overline{Y}_0 \pm \sqrt{v(\overline{Y}_i - \overline{Y}_0)} t(\alpha, K, dfE, \rho_{ij}) \qquad (11.6.3)$$

does not contain zero, where $t(\alpha, K, dfE, \rho_{ij})$ is the critical value of the two-sided comparison for comparing K treatments with a control as given in Hochberg and Tamhane (1987), dfE is the error degrees of freedom from the appropriate analysis of variance table, and ρ_{ij} is the correlation between $\overline{Y}_i - \overline{Y}_0$ and $\overline{Y}_j - \overline{Y}_0$ which is given by

$$\rho_{ij} = \sqrt{\frac{n_i n_j}{(n_i + n_K)(n_j + n_K)}}.$$

Dunnett's procedure for the one-sided alternative is also available (1955). In addition step-down and step-up versions of Dunnett's procedure for comparing treatments with a control are proposed by Dunnett and Tamhane (1991, 1992) for various multiple comparisons of treatments with a control. An overview of multiple comparisons in clinical trials can be found in Dunnett and Goldsmith (1995).

Multiple Endpoints

As mentioned earlier, the efficacy of a test drug in treatment of a certain disease can be characterized through multiple clinical endpoints. Capizzi and Zhang (1996) classify the clinical endpoints into primary, secondary, and tertiary categories whose criteria are modified and given below:

Primary endpoints should satisfy the following criteria:

- Should be of biological and/or clinical importance.
- Should form the basis of the objectives of the trial.
- Should not be highly correlated.
- Should have sufficient power for the statistical hypotheses formulated from the objectives of the trial.
- Should be relatively few (e.g., at most 4).

For secondary endpoints, typical criteria include whether the endpoints are (1) biologically and/or clinically important, but with less adequate statistical power, (2) potentially important, but highly correlated with primary endpoints, and (3) address other important, but ancillary objectives. The criteria of tertiary endpoints depend on whether (1) they are exploratory endpoints and (2) they are not of major importance.

Since the sample size of a clinical trial is usually selected to provide a sufficient power for detection of a difference in some or all primary clinical endpoints, we focus on the issue of false positive and false negative rates caused by multiple primary endpoints. It is, however, very important to understand the objective of the trial and to tailor and formulate the corresponding statistical hypotheses to the specific objectives in terms of multiple endpoints. Consider a randomized, parallel group, double-blind trial that evaluates the efficacy of a test drug versus a placebo control through a set of K primary endpoints. Let Y_{ijk} denote the kth endpoint for the jth subject in treatment group i; $k = 1, \ldots, K$, $j = 1, \ldots, n_i$, $i = T, P$. Define the population average for the kth endpoints of group i as

$$E(Y_{ijk}) = \mu_{ik}, \quad k = 1, \ldots, K, i = T, P.$$

The following statistical hypothesis formulates the clinical objectives of the trial. We declare that the test drug is efficacious if the test drug demonstrates a superior efficacy in any one of the K primary endpoints under the assumption that large values are better than small values:

$$H_0: \mu_{Tk} = \mu_{Pk} \quad \text{for all } k,$$
$$\text{vs.} \quad H_a: \mu_{Tk} > \mu_{Pk} \quad \text{for at least one } k = 1, \ldots, K. \quad (11.6.4)$$

Let $\bar{d}_k = \bar{Y}_{Tk} - \bar{Y}_{Pk}$ be the observed difference in sample mean of endpoint k between the test drug and the placebo. Also let $v(\bar{Y}_{Tk} - \bar{Y}_{Pk})$ be the estimated variance of $\bar{Y}_{Tk} - \bar{Y}_{Pk}$, $k = 1, \ldots, K$. Denote p_k as the observed p-value based on the test statistics corresponding to hypothesis (11.6.4):

$$t_k = \frac{\bar{Y}_{Tk} - \bar{Y}_{Pk}}{\sqrt{v(\bar{Y}_{Tk} - \bar{Y}_{Pk})}}, \quad k = 1, \ldots, K. \quad (11.6.5)$$

Then Bonferroni's adjustment of the p-value described above can be directly applied to control the false positive rate based on $p_1, \ldots, p_k$. Table 11.6.2 provides the Bonferroni correction and the true nominal significance level to maintain the overall false positive rate of 5% for various correlation coefficient when the number of endpoints is 4. In general, if the number of primary endpoints is small and the correlation between them is less than 0.5, Bonferroni's correction of p-value works reasonably well. However, when the number of endpoints increases and their correlations become large, then Bonferroni's adjustment will be too conservative and lack the power to detect any treatment effects in the primary endpoints. The criteria for primary endpoints specify that the number of primary endpoints should be at most 4 and that the correlation should not exceed 0.5. Both Bonferroni's technique and its modification proposed by Hochberg (1988) are appropriate methods for adjustment of p-values for hypotheses of (11.6.4).

Note that Bonferroni's method achieves its greatest power when the true treatment effect exists in only one of the K endpoints. However, some early phase II trials may have a large number of clinical endpoints that are highly

Table 11.6.2 Bonferroni's Correction and the True Nominal Significance Level to Maintain the Overall False Positive Rate of 5% for Various Correlation Coefficients the Number of End Points is 4

Bonferroni's Correction	Correlation Coefficient										
	0	0.1	0.2	0.3	0.4	0.5	0.6	0.7	0.8	0.9	1.0
	α level (%)										
1.25	1.27	1.30	1.30	1.38	1.40	1.54	1.70	1.84	2.20	2.69	5.0

correlated one another. In phase III studies it is useful to obtain a single composite index from multiple endpoints that provides a summary of the treatment's efficacy. Several methods for the construction of a composite index have been proposed by O'Brien (1984), Pocock et al. (1987), and Tang et al. (1989).

Assume that Y_{ijk} defined above have been standardized (i.e., subtracting the overall mean from each observation and dividing by the within-group sample standard deviation). Let $\bar{\mathbf{d}}$ be the vector of the observed sample mean differences of the K endpoints and $\mathbf{S}$ denotes the estimated within-group correlation matrix obtained under the assumption of the common covariance matrix of the K endpoints for both treatment groups. Then the global test statistic under the normality assumption proposed by O'Brien (1984) is an estimated generalized least squares (EGLS) method, which is given by

$$T_c = \frac{\mathbf{1}'\mathbf{S}^{-1}\bar{\mathbf{d}}}{\sqrt{w(\mathbf{1}'\mathbf{S}^{-1}\mathbf{1})}} \quad (11.6.6)$$

where $\mathbf{1}$ is the $k \times 1$ vector of 1 and $w = (1/n_T) + (1/n_P)$. The null hypothesis is rejected at the α level of significance if

$$T_c > t(\alpha, N - 2K),$$

where $t(\alpha, N - 2K)$ is the αth upper quantile of a central t distribution with $N - 2K$ degrees of freedom. The denominator of the global test statistic in T_c is a linear combination of the K observed sample mean differences with weights equal to the column sums of $\mathbf{S}^{-1}$. As a result the endpoints that are highly correlated with other endpoints receive small weights.

When the normality assumption for the distribution of multiple endpoints is in doubt, a nonparametric procedure (O'Brien, 1984) is also available for hypothesis (11.6.4). The nonparametric method is a rank-sum type test that starts ranking Y_{ijk} among all observations of endpoints k in the combined sample. Let us denote R_{ij} as the sum of the ranks assigned to the jth patient in group $i, j = 1, \ldots, n_i$, $i = T, P$. We can then apply the Wilcoxon rank-sum test to R_{ij} for testing hypothesis (11.6.4).

Pocock et al. (1987) extends O'Brien's procedure to combine multiple binary clinical endpoints which are commonly used to evaluate the efficacy of patients' response to the treatment. The test statistic of their procedure for combining multiple binary endpoints is to replace the elements in vector $\bar{\mathbf{d}}$ and the estimated correlation matrix $\mathbf{S}$ by $p_{Tk} - p_{Pk}$ and $s_{kk'}$, $1 \leq k \neq k' \leq K$, respectively, where p_{Tk} and p_{Pk} are the proportions of patients responding to the test and placebo groups, respectively, and

$$s_{kk'} = \frac{p_{kk'} - p_k p_{k'}}{\sqrt{p_k p_{k'}(1 - p_k)(1 - p_{k'})}}, \tag{11.6.7}$$

where $p_{kk'}$ is the proportion of patients with responses in both endpoints k and k', and p_k are the proportion of the patients in both groups whose response was defined in (8.3.10). If the sample size is moderately large, then the global test statistic T_c follows a standard normal distribution.

Pocock et al. (1987) propose an extension of the O'Brien EGLS method that combines censored and binary endpoints as is common in cancer chemotherapy trials. For example, let endpoint k be the censored survival time and endpoint k' be the tumor response. The test statistic is obtained by substituting the elements of $\bar{\mathbf{d}}$ and $\mathbf{S}$ in test statistic T_c by z_k and $s_{kk'}$, respectively, where z_k is the square root of the logrank statistic X_{LR} defined in (9.3.5), and

$$s_{kk'} = \frac{R_d - \sum p(t)}{N\sqrt{v_k p_k (1 - p_k)}}, \tag{11.6.8}$$

where R_d is the total number of the patients who responded to the treatment but nevertheless died, v_k given by the denominator of the logrank statistics in (9.3.5), p_k is defined above in (11.6.7), $p(t)$ is the proportion of patients who responded among those who were at risk at time t, and the summation is over all death times.

The null hypothesis (11.6.4) assumes that all K endpoints are equally important. As a result the weights for the test statistics based on the EGLS method given above are only based on the pairwise correlations between endpoints. However, if the relative importance of endpoints can be specified a priori in the protocol with known weights c_k, $k = 1, \ldots, K$, then the global test statistic given in (11.6.6) can be easily modified to accommodate the unequal priorities among the K endpoints as follows:

$$T_c = \frac{\mathbf{1}'(\mathbf{CSC})^{-1}\bar{\mathbf{d}}}{\sqrt{w[\mathbf{1}'(\mathbf{CSC})^{-1}\mathbf{1}]}}, \tag{11.6.9}$$

where $\mathbf{C}$ is a $k \times k$ diagonal matrix with diagonal elements being c_k.

Table 11.6.3 False Positive Rate (%) for the IUT Procedure Performed at the Significance Level When the Number of End Points is 4

Correlation Coefficient					
0	0.2	0.4	0.6	0.8	1.0
<0.01	0.02	0.14	0.5	1.3	5

Source: Capizzi and Zhang (1996).

The objective of clinical trials for a certain class of drug products with respect to some diseases may be more stringent than that for the formulation of hypothesis (11.6.4), and the test drug may be claimed to be efficacious only if it demonstrates a statistically significantly superior effect in all of the K primary endpoints. In other words, the corresponding statistical hypothesis is then expressed as

$$H_0: \mu_{Tk} = \mu_{Pk} \text{ for at least one } k,$$
$$\text{vs. } H_a: \mu_{Tk} > \mu_{Pk} \text{ for all } k, k = 1, \ldots, K. \quad (11.6.10)$$

The null hypothesis (11.6.4) is that the treatment difference between the test drug and placebo in all K endpoints are zero versus the alternative that the test drug is superior in at least one of the K endpoints. One the other hand, the null hypothesis (11.6.10) indicates that the treatment difference between the test drug and placebo is zero in any one of the K endpoints while the alternative hypothesis of (11.6.10) requires that the test drug be effective in all K endpoints. Hence the intersection-union test (IUT) proposed by Berger (1982) can be applied to test hypothesis (11.6.10) which is rejected at the α level of significance if and only if each of the K individual null hypotheses H_{k0} is rejected at the α significance level, $k = 1, \ldots, K$. Although the size of the IUT procedure is α (i.e., an α size test), it is very conservative as demonstrated in Table 11.6.3. Using a simulation, Capizzi and Zhang (1996) provide the nominal significance level at which each individual hypothesis is tested to achieve an experimentwise false positive rate of 5%. The simulation result is reproduced in Table 11.6.4 for four endpoints. For example, if the correlation among four endpoints is equal to 0.4, then for each endpoint the hypothesis must be tested at a nominal level of 0.289 to maintain an experimentwise type I error rate of 0.05. However, this approach with a nominal level 0.289 for each of four tests does not have a size of 5%. Suppose that there is a ranking among the K endpoints. The individual hypotheses for the important primary endpoints are tested at the 5% level as usual. However, for other endpoints with less relevance, individual tests can be performed at a higher nominal significance level to improve the power of the IUT procedure. But the resulting IUT procedure does not have an experimentwise false positive rate of 5%. In general, if the hypothesis for each of the K endpoints is tested at $\alpha_k, k = 1, \ldots, K$, then the experimentwise false positive

Table 11.6.4 Nominal Significance Level for the IUT Procedure to Maintain an Experimentwise False Positive Rate of 5% When the Number of Endpoints is 4

		Correlation Coefficient			
0	0.2	0.4	0.6	0.8	1.0
47.3	37.6	28.9	20.9	13.6	5

Source: Capizzi and Zhang (1996).

rate (i.e., size) of the IUT procedure for hypothesis (11.6.10) is the maximum of $\alpha_1, \ldots, \alpha_K$ (Berger and Hsu, 1996).

Hypothesis (11.6.4) is somewhat too liberal in the sense that the declaration of the test drug's effectiveness only requires a demonstration of superior efficacy at one of the K endpoints. In contrast is the strict requirement of superior efficacy at all K endpoints demanded by hypothesis (11.6.10). As a result Capizzi and Zhang (1996) propose a hybrid hypothesis in which the test drug is claimed to be effective if superior efficacy is demonstrated at M of K endpoints. The corresponding statistical hypotheses are given as

$$H_0: \mu_{Tk} = \mu_{Pk} \quad \text{for at most } K-M \text{ endpoints,}$$
$$\text{vs.} \quad H_a: \mu_{Tk} > \mu_{Pk} \quad \text{for at least } M \text{ of } K \text{ endpoints,} \quad (11.6.11)$$

$k = 1, \ldots, K$. Note that hypothesis (11.6.10), which requires a demonstration of superior efficacy at each of the primary endpoints, is often considered the hypothesis of choice for declaring the effectiveness of a drug. For example, finasteride at 5 mg was approved by U.S. FDA in treatment of patients with benign prostatic hyperplasia because its effectiveness was demonstrated at all three primary endpoints such as a statistically significant decrease in total urinary-symptom scores (p-value < 0.001), a statistically significant increase of 1.6 ml/s in maximal urinary-flow rate (p-value < 0.001), and a statistically significantly 19% reduction in prostatic volume (p-value < 0.001). On the other hand, hypotheses (11.6.4) and (11.6.11) may be appropriate for clinical trials during phase II development because the primary endpoints may not be fully understood for some diseases or the effects of the drug on different aspects of the disease are still under investigation.

Subgroup Analysis

Although the primary objective of clinical trials is to provide a valid and unbiased inference of the treatment effect for the disease under study, it is of great interest to observe whether the treatment effect is consistent across some demographic factors such as age, gender, race, baseline disease severity, some prognostic factors, or previous medical conditions and concomitant medications. As a result subgroup analyses are performed within strata according to some stratification or classification factors. The primary goal of these subgroup analyses is not for a definitive statistical inference of the treatment effect for each subgroup but rather for an exploratory identification of unusual or unexpected results.

The difference between subgroup analyses and multiple comparisons and multiple endpoints is that patients in subgroups stratified by different values of a covariate are different and the test statistics obtained from different subgroups are statistically independent. For example, consider a randomized double-blind, parallel-group clinical trial for the investigation of a new antihypertensive agent's effectiveness versus a placebo. The primary endpoint is the change from the baseline in diastolic blood pressure (mm Hg). Suppose that it is of interest to investigate whether the reduction in diastolic blood pressure is consistent across age groups. Therefore patients are divided into groups according to age groups of 18–30, 31–50, 51–70, and over 70. The two-sample unpaired t-statistic can be computed within each age group to test whether the new antihypertensive agent is efficacious in each of the four age groups. Since patients in each age group are different and these age subgroups are mutually exclusive, the four unpaired t-statistics are independent of each other. As a result the adjustment of p-values is straightforward. Let α be the desired experimentwise false positive rate for a total of K subgroups. Then the nominal significance level α^* of the test performed for each subgroup is given as

$$\alpha^* = 1 - (1 - \alpha)^{1/K}. \qquad (11.6.12)$$

Since the purpose of subgroup analyses includes implicit future hypotheses, the experimentwise false positive rate must be adequately controlled and conservative methods such as Bonferroni's correction of p-values are often used. However, the ICH guidelines on structure and contents of clinical study reports require that *where there is a prior hypothesis of a differential effect in a particular subgroup, this hypothesis and its assessment should be part of the planned statistical analysis*. Even so, it is recommended that the clinical reports should state that their subgroup analyses are exploratory. On the other hand, CPMP also issues a note for *Guidance on Statistical Methodology in Clinical Trials in Applications for Marketing Authorizations for Medicinal Products* which expresses a different view on subgroup analyses (CPMP, 1994). The note emphasizes that the model approaches such as analysis of covariance, logistic regression, or Cox's proportional hazard model should include covariates and treatment-by-covariate interactions in order to obtain an overall treatment effect. The note does not recommend the multiple separate analyses within strata defined by the covariates. Under the assumption of no treatment-by-covariate interaction, $\bar{d}^*$ given in (11.5.14) is the minimum variance unbiased estimator for the overall treatment effect.

11.7 DATA MONITORING

Definition and Objectives

Data monitoring is an active process involving a completely blinded review of clinical data while a trial is in progress (PMA, 1993). O'Neill (1993) provides some viewpoints from a regulatory perspective. He suggests that data monitoring should include the following:

1. Assessments of the quality and relevance of the data and the extent to which the protocol is being followed.
2. Arrangements for data processing, auditing, cleaning, and report generation.
3. Compilation and an assessment of safety data.
4. Compilation and an assessment of efficacy data.
5. Points in time for decision making on whether to continue or stop a trial based on observed results.

Data monitoring is multifaceted process that can vary from stage to stage during the clinical development of a drug product. For example, during phases I and II clinical studies are mainly of an exploratory nature, whereas during phase III they take a confirmatory form. In addition the purpose might differ depending on the sponsors such as the pharmaceutical industry and the U.S. National Institutes of Health. However, if a clinical trial demonstrates overwhelming evidence of beneficial efficacy or unexpected harmful adverse effects for the test drug, then it is recommended that the trial be terminated before the scheduled completion of the trial. Data monitoring and interim analyses are commonly employed for clinical trials on the treatment of life-threatening diseases or severely debilitating illnesses with long-term follow-up and endpoints such as mortality or irreversible morbidity. The decision for early termination of such clinical trials is crucial and cannot be made simply based on an application of one of many statistical procedures for interim analyses discussed in Chapter 9. Data monitoring and interim analysis are extremely complex processes for which there exist no definitive rules, and they are critical to good clinical practice. Canner (1981) points out in decision making in clinical trials, no single statistical decision rule or procedure can take the place of well-reasoned considerations of all data by a group of concerned, competent, and experienced persons with a wide range of scientific backgrounds and points of view. In particular, the decision for early termination of a clinical trial should be based on such considerations as baseline comparability, unbiased evaluation, compliance, internal consistency with other endpoints and subgroup analyses, external consistency, benefit and risk, length of follow-up, public impact, repeating testing, and multiple comparisons.

Regulatory Concerns

Improper and sloppy data monitoring and interim analyses can potentially alter the conduct of a clinical trial. Consequently serious bias may be introduced in the treatment effects. Both the FDA (FDA, 1988) and ICH guidelines (ICH, 1995) on the format, structure, and contents of a clinical report explicitly state that:

> The process of examining and analyzing data accumulating in a clinical trial, either formally or informally, can introduce bias and/or increase type I error. Therefore, all interim analyses, formal or informal, pre-planned or ad hoc, by any study participant, sponsor staff member, or data monitoring group should be described in full, even if the treatment groups were not identified. The need for statistical adjustment because of such analyses should be addressed. Any operating instructions or procedures used for such analyses should be described. The minutes of meetings of any data monitoring group and any data reports reviewed at those meetings, particularly a meeting that led to a change in the protocol or early termination of the study, may be helpful and should be provided. Data monitoring without code-breaking should be also described, even if this kind of monitoring is considered to cause no increase in type I error.

In addition, as indicated by O'Neill (1993), the FDA classifies the issues of data monitoring in clinical trials as planning, reporting, operation, and documentation of the trials. These issues include (1) unreported interim analyses, (2) planned or unplanned interim access to unblinded comparative study results, (3) failure of assessment of impact of unplanned interim analyses on study results, (4) bias on the future conduct of the trial caused by unblinded assess to study results, (5) the recognition of all relevant parties of the regulatory implications of early termination of trials, (6) development of efficient, effective communication and information flow between the data-monitoring committee and the regulatory authority, (7) appropriate evaluation of exploratory trials, (8) planning trials not to stop early for efficacy reasons alone but to balance the need for safety data on longer-term exposure with short-term follow-up of early efficacy results, (9) establishment of policies regarding access to ongoing data, access to unblinded data, and participation in the decision-making chain.

The above regulatory issues on data monitoring and interim analyses are concerned with a potential bias in estimating the treatment effect, an inflation of a false positive rate, and documentation of the processes. Since data monitoring and interim analyses can affect the subsequent conduct of the trial, a bias may be introduced that may not be measurable or quantifiable. As a result the generalizability of the trial results (or inference) is placed in serious jeopardy. In addition the integrity and credibility of the trial becomes seriously affected. Therefore it is important to make any attempts/efforts to eliminate both known and unknown biases. As discussed earlier in this book, one way to avoid a potential bias is to maintain blindness throughout the study. To achieve this objective, standard operating procedures should be followed in the selection of plans, methods, data management, documentation of data monitoring, planned and unplanned interim analyses, and dissemination of the results.

In general, phase IIB, phase III, and phase IV trials must be triple-blinded in the sense that patients, investigators, and the clinical project team responsible for the development are unaware of any individual treatment assignments and treatment results until all data of the trial have been received, entered, edited, and verified and the database locked for analysis. According to the definition given by the PMA, data monitoring is to be used for all trials with and without planned interim analyses. In principal, all parties except for the project statistician, data coordinator, and programmer are kept blinded until after the database is locked and the statistical analyses completed, reviewed, and approved. When an interim analysis is required, then only a statistician who is not involved in the trial is authorized to merge the file of randomization codes of the individual treatment assignments with the dataset of endpoints by patient number required for the interim analysis. The file of randomization codes must be kept in a classified and secure area in a database management system that is separate from the rest of the trial data and can only be accessed by the authorized personnel. The extent of data included in the dataset for the interim analysis should be (1) specified in the protocol and (2) minimal in order to satisfy the objectives of the interim analysis and information for any subsequent decision-making process leading to termination of the trial. On the other hand, the results of the interim analysis could be presented as an overall summary with proper treatment identification for a meaningful interpretation but excluding individual patient listings. These and any results of interim analyses should be accessible only to a few designated people such as the members of an independent data monitoring committee. A standard operating procedure must be established to maintain the blindness of the study.

Early and Late Stages of Drug Development

As was mentioned earlier, the objectives of data monitoring depend on the stage of drug development. In the early stages the trials are often exploratory in nature. As a result the objectives of data monitoring and interim analyses are in the context of trial management such as verification of design assumptions, detection of possible design flaws, monitoring for unexpected side effects, and generation of hypotheses for future use. Although general principles of data monitoring can be applied to clinical trials at early phases of drug development, their protocols usually do not specify the objectives, procedures, frequencies, and methods of interim analyses. In addition, since these trials always proceed to completion regardless of any interim analyses, the p-values obtained from interim analyses remain generally unadjusted. Consequently they should be considered as descriptive rather than inferential. Moreover, since clinical trials at an early stage are focused on generating hypotheses to be confirmed at later stages, decisions made for future trials are sometimes based on incomplete information. Accordingly, these decisions may be misleading due to (1) unadjusted p-values of interim analyses and (2) small sample sizes of these studies. It is suggested that data monitoring and interim analyses for these early trials be planned with respect to dates of interim analyses, numbers of patients, summary statistics, and any other adequately documented decisions.

Clinical trials conducted in the late stage of clinical development are confirmatory, adequate well-controlled studies to provide substantial evidence of effectiveness and safety for approval. On the other hand, phase IV studies and mega trials such as physician health study, ISIS series trials, or GUSTO study are conducted to evaluate the drugs or compare different agents in various combinations for the same indication in a much larger targeted population. There trials are usually very large and involve more patients and take much longer to complete with huge human and financial investment. Therefore, for ethical and scientific reasons interim analyses are planned for these trials, in particular, with survival and irreversible morbidity such as stroke or myocardial infarction as primary endpoints. However, early termination of a trial conducted in non-life-threatening diseases solely due to efficacy is not common. The guidelines for data monitoring and interim analyses for large confirmatory trials include:

1. Group sequential methods introduced in Chapter 9 should be described in the protocol with respect to the required primary endpoints, frequency, boundaries, and references of the employed methods for interim analysis and decision rules derived partly from the results of interm analyses.
2. An independent data monitoring committee with internal or external members should be established to monitor the data and to review the results of interim analyses.
3. Principles of adequate documentation with respect to the results of interim analyses and minutes of data monitoring committee are essential for good practice of data monitoring.
4. Blindness on dissemination of the results of interim analyses and deliberation of data monitoring committee should be strictly enforced and tightly controlled.

Administrative Interim Analyses

From a functional point of view, interim analyses by the data-monitoring committees can be classified as *formal* and *administrative analyses* (PMA, 1993; Williams et al., 1993). The aim of a formal interim analysis is to decide on early termination of a planned study if there is compelling evidence of beneficial effectiveness or harmful side effects. The administrative interim analysis is relevant in pharmaceutical cases and is conducted for the reasons external to a trial at the request of regulatory agencies or upper management. Such an administrative analysis rarely concerns early termination, and it is usually performed at an early stage of the trial in order to verify the design assumptions, check the data entry, editing, and review processes, and compare the simple baseline summary statistics used in the enrollment of patients. Nevertheless, sometimes trials are terminated this early due to major design flaws, an unusual placebo effect or unexpected toxicity, and other reasons relating to recruitment problems, budget considerations, administrative cutoffs, company merges, and contracting the study out to a contract research organization (CRO). For these reasons standard operating procedures for early termination based on an administrative analysis must be established. The standard operating procedures should identify the individual responsible for authorizing an administrative interim analysis and outline a clear paper trail for the reasons and actions as a consequence of such an analysis. It is important to recognize that the potential danger of any administrative interim analysis lies in introduction of bias and inconclusive evidence of trend because of lack of power.

Data Monitoring Committee

An independent data monitoring committee (DMC) should be established for any confirmatory trials with planned interim analysis, in particular, for the trials conducted in life-threatening diseases or severely debilitating ailments. The model of data monitoring committee formulated by the U.S. National Institutes of Health (NIH model) is usually adopted for the trials sponsored by government. The model of the NIH clinical trials consists of sponsor, steering committee, center or principal investigators, data coordination and statistical analysis center, central laboratories, and data monitoring committee. The members of the NIH DMC may include those with the disciplines in clinical, laboratory, epidemiology, biostatistics, data management and ethics. In order to be independent, the members of DMC are not involved in the actual conduct of the trial and the members and their family should have no conflict of interest, such as, no financial holdings in companies. In addition, the members should not discuss nor disseminate the results of the interim analyses outside DMC. A member representing the sponsor is usually included in DMC as a nonvoting member and is allowed to attend the open session of DMC meetings only.

The responsibilities of DMC are to monitor the safety and ethical aspects of the trial with respect to the patients, investigators, sponsors, and the regulatory authorities in descending order of priority. DMC achieves these responsibilities by review of protocols; interim review of study progress and the quality of trial conduct, monitoring safety data and possible efficacy and benefit, and recommendation of early termination of trials and dissemination of the primary results. Documentation is also a crucial part of functions of DMC which include protocol, operational manuals for key data, decision processes, and different situations that might be encountered during the conduct of the trial and their possible resolutions, and interim data reports and minutes of the meetings.

The DMC should be prepared and ready from the start and should concentrate on the primary efficacy and safety endpoint but not on individual case reports. Therefore, an on-line data management and analysis system is essential for minimization of delay in data entry and event verification so that any decision made by DMC is on the currently available data. As a result, sometimes, an independent data coordination and statistical analysis center is usually established for trials sponsored by government. The data coordination and statistical analysis center should be independent of the sponsor and responsible for case report design, on-line data management system for data entry, editing and verification, and for the quality control of the conduct of the trial through training, certification, tracking of case report forms and reports, and design and maintenance of on-line analysis system for interim and final analyses. In order to provide a good practice of data monitoring and to work problematically, the data coordination and statistical analysis center is required to function and interact closely with the sponsor, clinics, steering committee, and most importantly, the data monitoring committee.

The philosophy for decision-making process taken by DMC should be *be ahead of time, be prepared*, and *be problematic*. Various formats of DMC meetings can be set up for different objectives. The first one is an open session in which the representative from the sponsor might attend. The objective of the open session is to blindly review the progress, conduct, recruiting, and logistic issues of the trial. The second format is a closed session at which the members of DMC review the summary efficacy and safety results from interim analyses by treatment group. The real treatments or dummy treatment codes such as A versus B could be used for review of interim results. However, the sponsor is not allowed to attend the close session. The last format is the executive session at which the decision of early termination will be deliberated and made. Sometimes, a quorum is set up for any decision about trial termination. This quorum usually include the statistician of DMC.

The DMC might report to different parties such as the sponsor, the study chair, or the executive or steering committee depending on the model and organization of the trial. For example, the Cooperative North Scandinavian Enalapril Survival Study II (CONSENSUS II) was a Scandinavian study sponsored by a U.S. pharmaceutical company with a planned sample size of 4500 patients to compare enalapril, an angiotensin converting enzyme (ACE) inhibitor, to a placebo in the treatment of acute myocardial infarction (Williams et al., 1993). This trial had a steering committee of 12 members, including a nonvoting sponsor's representative and a DMC consisting of three clinicians and one statistician from the academe and a nonvoting sponsor's statistician. The responsibility of the DMC for this trial was to review unblinded analyses and to make recommendations to the steering committee. The sponsor's statistician at the DMC, who was the unblinded sponsor's employee for this trial, provided analyses to the DMC.

As was mentioned earlier, interim analyses are an established part of the pivotal phase III trials conducted by pharmaceutical companies with mortality or irreversible morbidity as the primary endpoints. The method of the DMC can vary according to external involvement of the trial. For certain trials with non-life-threatening diseases, a complete in-house DMC may be set up for data monitoring and interim analyses. The next level of data monitoring and interim analyses is an independent external DMC but the analyses must be performed blinded internally, such as in CONSENSUS II. Sometimes a trial may have an external DMC with data processing performed internally but with the interim analyses done externally. The sponsor of the trial is allowed to attend the open sessions of the DMC meetings. This structure and function is well accepted. Under this structure the sponsor must have sophisticated computer and data management systems and sufficient resources to process a large volume of data. Data are monitored blindly and independently, and interim analyses are performed independently too. However, an independent external DMC may have both data processing and analyses performed externally. The most extreme method of the DMC is a *complete hands-off* approach with the external data processing and analyses done with no sponsor's representative. A typical example is the GUSTO trial.

Regulatory authorities such as the FDA and ICH have expressed their interest in having access to clinical trial results, including raw data, methods for analyses, detailed assessments of a study's conduct and quality, compliance with protocol, and other trial documentations. Although the regulatory authority is aware of the progress of a trial, it should not interfere with the decision made regarding the progress or termination of the trial. That is to say, the authority should not routinely participate in the DMC meeting and should not be a voting member of DMC. The involvement of the regulatory authority with the DMC should not go beyond communication and regulatory authority's representative's attendance at open sessions so that blinding of the trials can be maintained. However, regulatory agencies should be informed of the DMC decisions as soon as possible. O'Neill (1993) indicates that this model has evolved from involvement in AIDS studies. The model has worked well for various NIH trials especially for the ATCG program of the U.S. National Institute of Allergy and Infectious Disease.

11.8 DISCUSSION

The statistical analysis of efficacy data from clinical trials involves many complicated issues. The problems of missing values and dropouts add yet another dimension to this complex analysis. There are two types of missing values (Diggle, Liang, and Zeger, 1994). The first concerns missing values when patients withdraw from a trial at any time before its planned completion. As a result the data scheduled to be collected beyond the patient's dropout time are missing. We refer to the missing values due to reasons other than dropouts as intermittent missing values.

There are many causes of dropouts and missing values. Dropouts can occur because of the duration of study, the nature of the disease, the efficacy and adverse effects of the drug under study, intercurrent illness, accidents, patient refusal or moving, and any number administrative reasons. Some of these causes are treatment-related and some are not. Based on these causes, the mechanism of missing values can generally be grouped into three types (Little and Rubin, 1987). If the causes of missing values are independent of the observed responses and of the responses which would have been available had they not been missing, then the missing values are said to be *completely random*. On the other hand, if the causes for missing values depend on the observed responses but are independent of the scheduled but unobserved responses, then the missing values are said to be *random*. The missing values are said to be *informative* if the causes for missing values depend on the scheduled but unobserved measurements.

If the missing mechanism is either completely random or random, then the statistical inference derived from the likelihood approaches based on patients who complete the study is still valid. However, this inference is less efficient under the completely random or random missing mechanism (Diggle, Liang, and Zeger, 1994). If the missing values are informative, then the inference based on the completers would be biased. As a result the FDA and ICH guidelines on clinical reports both indicate that despite the difficulty, the possible effects of dropouts and missing values on magnitude and direction of bias must be explored as fully as possible. That is to say, before any analyses of efficacy endpoints, at various intervals during the study the frequency, reasons, and time to dropouts and missing values should also be compared between treatment groups. Their impact on the trial's efficacy has to be fully explored and understood, and only after the missing mechanism has been identified can appropriate statistical methods be employed for the analysis (Diggle, 1989; Ridout, 1991). Although some procedures have been proposed (Diggle and Kenward, 1994), there exists no satisfactory well-developed methodology to account for missing values or intermittent missing values. Therefore the conservative strategy seems to be to continue to collect the data on the primary endpoints after a patient withdraws from the study and then to analyze the efficacy endpoints based on the intention-to-treat principle with all available data from all randomized patients.

In clinical trials the sample size is determined by a clinically meaningful difference and information on the variability of the primary endpoint. Since the natural history of the disease is usually not known or the test drug under investigation is a new class of drug, the estimate of variability for the primary endpoint for sample size determination may not be adequate. As a result the planned sample size may need to be adjusted in the middle of the trial if the observed variance of the accumulated responses on the primary endpoint is very different from that used at the planning stage. Procedures have been proposed for adjusting the sample size (or re-estimation) during the course of the trial without unblinding and altering the significance level (Gould, 1992, 1995a; Gould and Shih, 1992).

For example, let us consider a randomized trial with two parallel groups comparing a test drug and a placebo. Suppose that the distribution of the response of the primary endpoint is (or is approximately) normal. Then the total sample size for a two-sided alternative hypothesis can be determined using the following formula as given in Example 10.3.2:

DISCUSSION

$$N = \frac{4\sigma^2[z(\alpha/2) + z(\beta)]^2}{\Delta^2}. \tag{11.8.1}$$

In general, σ^2, the within-group variance, is unknown and must be estimated based on previous studies. Let σ^{*2} be the within-group variance specified for sample size determination at the planning stage of the trial. At the initiation of the trial, we expect the observed variability to be smaller than σ^{*2} so that the trial will have sufficient power to detect the designated clinical meaningful difference. However, if the variance turns out to be much larger than σ^{*2}, we will need to re-estimate the sample size without breaking the randomization codes. If the true within-group variance is in fact σ'^2, then the sample size to be adjusted to achieve $(1 - \beta)100\%$ power at the α level of significance for a two-tailed alternative is given by

$$N' = N \frac{\sigma'^2}{\sigma^{*2}}, \tag{11.8.2}$$

where N is the planned sample size calculated from σ^{*2}.

However, σ'^2 in (11.8.2) is unknown and must be estimated from the accumulated data available from a total n of N patients. One simple approach to estimate σ'^2 is based on the sample variance calculated from the n responses which is given by

$$(n-1)s^2 = \sum\sum (Y_{ij} - \overline{Y}_{..})^2, \tag{11.8.3}$$

where Y_{ij} is the jth observation in group i and $\overline{Y}_{..}$ is the overall sample mean, $j = 1, \ldots, n_i$, $i = T, P$, and $n = n_T + n_P$. If n is large enough for the mean difference between groups to provide a reasonable approximation to Δ, then it follows that σ'^2 can be estimated by (Gould, 1995a)

$$\sigma'^2 = \frac{n-1}{n-2}\left(s^2 - \frac{\Delta^2}{4}\right). \tag{11.8.4}$$

Note that the estimation of within-group variance σ'^2 does not require the knowledge of treatment assignment, and hence the blindness of the trial is maintained. However, since this approach does depend on the mean difference, which is not calculated and is unknown, is a reasonable estimate for Δ. The other procedure for estimating σ'^2 without a value for Δ is the EM algorithm (Gould and Shih, 1992; Gould, 1995a). Suppose that n observations, Y_i, $i = 1, \ldots, n$, on a primary endpoint have been obtained from n patients. The treatment assignments for these patients are unknown, and Y_i are the observations of patient i that can be randomly allocated to either of the two groups. Gould and Shih (1992) and Gould (1995a) assume that the treatment assignments are *missing at random*. Define π_i as the treatment indicator

$$\pi_i = \begin{cases} 1 & \text{if the treatment is the test drug,} \\ 0 & \text{if the treatment is placebo.} \end{cases} \quad (11.8.5)$$

The E step is, after substitution of the current estimates of μ_T, μ_P, and σ, to obtain the provisional values of the expectation of π_i (i.e., the conditional probability that patient i is assigned to the test drug given Y_i), which is given by

$$P\{\pi_i = 1 | Y_i\} = \frac{1}{1 + \exp[(\mu_T - \mu_P)(\mu_T + \mu_P - 2Y_i)/2\sigma^2]}, \quad (11.8.6)$$

where μ_T and μ_P are the population mean of the test drug and the placebo, respectively. The M step involves the maximum likelihood estimates of μ_T, μ_P, and σ after updating π_i by their provisional values obtained from (11.8.6) in the log-likelihood function of the interim observations, which is given by

$$1 = n \log \sigma + \frac{\sum [\pi_i(Y_i - \mu_T)^2 + (1 - \pi_i)(Y_i - \mu_P)^2]}{2\sigma^2}. \quad (11.8.7)$$

The E and M steps are iterated until the values converge. Gould and Shih (1992) and Gould (1995a) indicate that this procedure can estimate within-group variance quite satisfactorily, but fail to provide a reliable estimate of $\mu_T - \mu_P$. As a result the sample size can be adjusted without knowledge of treatment allocation. For sample size re-estimation with respect to binary clinical endpoints, see Gould (1992, 1995a).

From the above procedures, the sample size adjustment can be performed during the study without unblinding the treatment allocations and knowledge of any treatment differences. Therefore it is not necessary to adjust the significance level. However, the magnitude of adjustment of the planned sample size for the phase III pivotal trials may be relatively significant. This is a consequence of the variability between observations from phase II and from the current phase III trials. One has to find the possible causes for any within-group variances such as due to different patient populations between phase II and phase III trials. Although the technique does not require adjustment of the *p*-value and does maintain blinding, sample size re-estimation is rather seldom performed because of concerns for the integrity of a trial (Williams et al., 1993).

CHAPTER 12

Safety Assessment

12.1 INTRODUCTION

As was indicated in Chapter 1, the safety of marketed drugs did not become a public concern until the Elixir Sulfanilamide disaster in late 1930s which led to the Federal Food, Drug and Cosmetic Act (FD&C Act). The FD&C Act requires the pharmaceutical companies to submit all reports of investigations on the safety of new drugs. This requirement was subsequently strengthened by the Kefauver-Harris Drug Amendments to the FD&C Act. These federal regulations have influenced the quantity and quality of safety information on drugs on the market. Consequently throughout the development and marketing of a new drug, there are involved stages of government safety assessments. O'Neill (1988) classifies these stages as pre-marketing, post-marketing, and drug labeling. O'Neill indicates that the assessment of the safety of a new drug begins in the pre-approval stage with pre-clinical animal studies and with early phase I studies that examine the absorption, excretion, dose ranging, tolerance, and other pharmacokinetic performance of the drug in humans. This continues with phase II and phase III of clinical development. During the post-approval marketing, much broader patient populations become involved. The safety information may be obtained from voluntary reports, monitoring system, uncontrolled patient follow-up, and formal epidemiological studies.

For the pre-marketing safety assessment, the FDA requires that a summarization and analysis of safety information of a new drug be included in the NDA submission. In particular, Section 314 of CFR [314.50 (d)(5)(ii)] indicates that an integrated summary of all available information on the safety of the drug product be submitted, including pertinent animal data, demonstrated or potential adverse effects of the drug, clinically significant drug-to-drug interactions, and other safety considerations such as data from the epidemiological studies of a related drug. The applicant has to date periodically its pending application with new safety information learned about the drug that may reasonably affect the statement of contraindication, warnings, precautions, and adverse reactions in the draft labeling. After the drug is approved, Section 21

CFR 314.80(b) also requires that the sponsor promptly review all adverse drug experience information obtained or otherwise received by the sponsor from any source, foreign or domestic, including information derived from commercial marketing experience, post-marketing clinical investigations, post-marketing epidemiological/surveillance studies, reports in scientific literature, and unpublished papers. Section 21 CFR 314.80(c)(1)(ii) also requires that the sponsor periodically review the frequency of adverse drug experience reports which are both serious and expected (in the labeling) and report any significant increase in frequency that might suggest a drug-related incidence higher than previously observed or expected. The safety update report should be submitted (1) four months after the initial submission, (2) following receipt of an approvable letter, and (3) at the times requested by the FDA. For drug labeling, Section 21 CFR 201.57(e) on warnings states when the label should describe serious adverse reactions and potential safety hazards. In addition Section 21 CFR 201.57(g)(2) requires that the frequency of the serious adverse reactions be expressed under the section of *Adverse Reaction* of the labeling, and if known, the approximate mortality and morbidity rates for patients sustaining the reaction.

In general, the assessment of drug safety in clinical trials has not received the same level of attention as the assessment of efficacy. For example, in most clinical trials, sample size is determined to achieve a desired power for detection of a clinically meaningful difference in the primary efficacy variable rather than safety parameters. In addition, unlike the assessment of efficacy, the hypotheses for safety assessment are usually much less well-defined. As a result statistical methods for the assessment of safety are limited. In practice, descriptive statistics for safety data obtained from clinical trials, both controlled and uncontrolled, are often used to (1) summarize rates of occurrence of adverse events in exposed groups and (2) examine any patterns or trends for subgroups of patients experiencing differential rates of adverse events. In practice, safety data obtained in clinical trials are often summarized in terms of rates or relative risks of certain events. Therefore it is important to develop a sound statistical methodology that provides an accurate and reliable assessment of drug safety. In addition to safety data of rates and relative risks of certain events, most clinical trials also contain a large battery of routine laboratory measures as part of the safety evaluation of the drug under investigation. Since laboratory data are obtained from analytical methods, they are often the least biased and the most precise data collected in clinical trials. Although laboratory data can provide information on systemic toxicity, these data are often underutilized for evaluation of drug safety.

In this chapter our goal is to describe approaches for assessing the safety of a drug product in terms of rates or relative risks of adverse events and laboratory data during its clinical development and marketing stages. In the next section the toxicity (or risk) of the exposure to a drug under study is briefly described including the incidence rate of an event and laboratory tests. In Section 12.3 we provide definitions for adverse drug reaction, adverse event, and serious adverse events adopted by the FDA and ICH. Also included in the section are the coding, filing, and reporting of observed adverse events according to some acceptable dictionaries such as COSTART and WHOART or MEDDRA. Statistical methods for the assessment of safety in terms of adverse events are reviewed in Section 12.4. Analyses of laboratory data for the assessment of safety are discussed in Section 12.5. Some remarks and discussions are given in the last section.

12.2 EXTENT OF EXPOSURE

As indicated in *Federal Register* (vol. 61, no. 138, 1996), the evaluation of safety-related data can be considered at three levels: (1) the extent of exposure, (2) the more common adverse events and laboratory test changes, and (3) serious adverse events and other significant adverse events. In this section we will discuss the extent of exposure of a test drug or an investigational drug product. The extent of exposure can be used to determine the degree to which safety can be assessed from the study. The extent of exposure to a test drug or an investigational product is usually characterized according to the number of patients exposed, the duration of exposure, and the dose to which patients were exposed. Duration of exposure to any dose is usually expressed as a median or mean. In practice, it is helpful to describe the number of patients exposed for specified periods of time. In addition it is suggested that the number of patients exposed to the test drug for various durations be broken down by age, sex, racial subgroups, and any other pertinent subgroups (e.g., disease, disease severity, or concurrent illness) in order to get a good profile of the exposure effect on different populations. Another measure of the extent of exposure is the number of patients exposed to specified daily dose levels. The daily dose levels used could be the maximum dose for each patient, the dose with longest exposure for each patient, or the mean daily dose. Similarly it is suggested that the number of patients exposed to various doses be broken down further by age, sex, racial, and any other pertinent subgroups to examine the profile of the extent of exposure for dose. In some cases a cumulative dose may be pertinent. In practice, it is often useful to provide combined dose-duration information such as the numbers exposed for a given duration to the most common dose, the highest dose, or the maximum recommended dose.

Risk of Exposure

In clinical trials the toxicity or risk of exposure to a drug can generally described as a function of the exposure to the drug:

$$\text{Toxicity} = f(\text{exposure}),$$

where the exposure to the drug depends on the dose and time of exposure. If we assume a constant dose, then the toxicity is a function of the time of exposure.

The toxicity or risk of exposure is usually measured by some parameters such as occurrence, number, duration, and time pattern of an event. The event may be an absorbing (irreversible) event (e.g., death), a recurring event with negligible duration (e.g., seizure), or a recurring event with a duration (e.g., migraine headache episodes). Therefore all of the parameters such as occurrence, number of events, and duration are not meaningful for all events. Meaningful parameters are rather the probability of occurrence, the number of events, and the percentage of time affected for an absorbing event, a recurring event, and a recurring event with a duration, respectively.

For quantification of the exposure risk of patients to a drug, the most commonly used measure is a crude incidence rate, which is defined as

$$CR = \frac{\text{Number of patients with event}}{\text{Number of patients at risk}},$$

where the *patients at risk* are all patients enrolled and treated. Under the assumption that each patient has the same probability of experiencing the event, CR is a binomial estimate. In practice, the binomial holds only if each patient undergoes one exposure unit. Tremmel (1996) points out that even by this assumption, the incidence rate may be misleading in long-term clinical trials because (1) it does not take exposure time into account and (2) every patient experiencing the event must disregard parameters of severity such as duration and the number of events. Further, in cases where there are early terminations, the full initial patient sample used as the denominator will lead to a downward bias. Thus Tremmel (1996) suggests that CR be used only when each patient receives one exposure unit such as in phase I trials. For short-term trials the use of CR is also justifiable if all patients experience about the same amount of exposure.

As an alternative to the CR estimate, Tremmel (1996) suggests the following two basic techniques which can be used to quantify the exposure risk of patients to a drug by controlling for the exposure to the drug. The first is to divide the number of events by total exposure by implicitly assuming a constant hazard. In other words, we calculate the number of events per time units such as the number of deaths per patient year. The other is to consider stratification by exposure time by forming time intervals. In other words, we consider estimates of the number of events per time interval assuming a constant hazard function.

Absorbing Events

For describing the exposure risk for absorbing events, Tremmel (1996) suggests using the number of events per patient-time unit, which is defined as

$$h = \frac{\text{Number of positive patient-time units}}{\text{Number of all patient-time units}}.$$

EXTENT OF EXPOSURE

The basic assumption is that one unit of patient-time is equivalent to and independent of another unit of patient-time. In other words, each patient-time unit is a binomial observation with the probability of success $P = h$. If the absorbing event is death, Rothman (1986) estimates the risk of exposure by means of the number of deaths per time unit of exposure:

$$h = \frac{\text{Number of deaths}}{\text{Number of total exposure}},$$

where h estimates the constant hazard. This is a typical example of constant hazard, also known as death per patient year (DPPY). In practice, since the hazard may vary over time, we need to allow for a nonconstant hazard, such as a decreasing hazard function (O'Neill, 1988). For nonconstant hazards, h is a function of time (known as hazard function). As indicated by Salsburg (1993), hazard functions provide useful information that helps us not only to assess drug causality but also to identify periods of increased risks. In addition the corresponding survival rates

$$S(t) = \prod [1 - h(t)]$$

provide probability statements of occurrence similar to the crude incidence rate which are unbiased because they are based on the corrected denominators (O'Neill, 1988). Besides, they are adjusted for exposure by being functions of exposure time.

As was pointed by Tremmel (1996), event per patient-time unit such as death per patient year is a meaningful parameter for the assessment of exposure risk if the event can only occur once.

Recurring Events with Negligible Duration

Recurring events with negligible duration are referred to as events with relevant recurrences such as reinfections, seizures, or sensitivity reactions. Similar to the absorbing events with constant hazard, one may consider the following for recurring events:

$$h = \frac{\text{Number of positive patient-time units}}{\text{Number of all patient-time units at risk}}.$$

Thus h is the risk of getting the event for the first time or again. h provides a probability estimate under the assumption that each patient-time unit at risk is a binomial observation with the probability of success $P = h$.

For nonconstant hazards, since the recurring events may be dependent on individual subjects, the quantification of the exposure risk is more complicated.

Basically there are two approaches with so-called subject-induced dependencies. For models that ignore subject-induced dependencies, the hazard function $h(t)$ can be estimated within different time intervals by the number of events in that interval divided by the total patient-exposure in that interval. The expected event counts for an individual at risk can be derived by integrating the hazard function for the risk interval. Confidence intervals can also be constructed (Andersen et al., 1993). Based on the same idea, Andersen and Gill (1982) considered Cox's proportional hazards model with the hazard being a function of time and a subject-related risk score to allow for recurring events. To account for the number of preceding events, Andersen and Gill (1982) further considered including the number of preceding events as a time-dependent covariate in the model. As a result the estimation of the expected event counts becomes more difficult and depends on the stochastic structure of the future development of the covariates (Andersen et al., 1993).

To account for subject-induced dependencies, we can consider either a normal model for count data proposed by Hoover (1996) or a random-effects model (or frailty model). For the normal model for count data, the idea is to form time intervals with complete exposures by deleting data that are censored in the interval and then, for each subject and interval, determine the event count. We next estimate the mean counts, variances, and covariances of counts between time intervals, assuming that the matrix of counts is from a multivariate normal distribution. We apply the central limit theorem to justify the estimates, which we finally use to derive point estimates and confidence intervals for event counts over time. The purpose of the random effects model is to examine whether there is a subject effect over a potentially relevant variable such as exposure time. The subject effect is defined as the observed count over time for each subject.

Recurring Events with Nonnegligible Duration

To quantify the risk of exposure of recurring events with nonnegligible duration, the most commonly employed approach is to calculate either the prevalence rate or the incidence rate at a given time. The prevalence rate is the risk of *having* an event at a given time, which is defined as

$$h_P = \frac{\text{Number of patient-time units affected}}{\text{Number of patient-time units exposed}}.$$

As can seen from the above definition, the prevalence rate includes patients in both the numerator and denominator throughout the duration of an event. As a result the prevalence rate is the proportion of a population that is affected by disease at a given point in time. Continued suffering increases the number of patient-time units affected. On the other hand, the incidence rate is the risk of *getting* an event at a given time which is defined as

EXTENT OF EXPOSURE

Table 12.2.1 Recurring Events with Nonnegligible Duration

Patient	1	2	3	4	5	6	7	8	9	10
1	—	—	—	—	×	×	×	×	—	—
2	—	—	—	—	—	×	—	—	—	—
3	—	—	—	—	—	—	—	—	—	—
4	—	—	—	×	×	—	—	×	×	×

Source: Tremmel (1996).
Note: × indicates affected.

$$h_I = \frac{\text{Number of incidences}}{\text{Number of patient-time units at risk}}.$$

The differences between the prevalence rate and the incidence rate are that (1) the continued suffering does not count in numerator and that (2) patient-time units of continued suffering are not at risk of getting the event and hence are removed from the denominator. Let us consider the example listed in Table 12.2.1 which was given in Tremmel (1996). In the table the total patient-time units is given by $10(4) = 40$ and the patient-time units affected is 10 (the number of crosses). Therefore the prevalence rate is given by

$$h_P = \frac{10}{40} = 0.1.$$

On the other hand, the patient-time at risk is $40 - 6 = 34$ and the incidences is 4. Thus the incidence rate is

$$h_I = \frac{10 - 6}{40 - 6} = \frac{4}{34} = 0.12.$$

Note that the prevalence rate can be estimated by the incidence rate if it is multiplied by the average duration of the disease (Rothman, 1986).

Laboratory Data

Another approach that is commonly employed to investigate the extent of exposure to a drug in clinical trials is to perform routine laboratory tests. In practice, many clinical studies contain a large battery of routine laboratory measurements as part of the safety evaluation of the drug under investigation. Compared to the rates or relative risks of certain events observed from clinical trials, laboratory data are often the least biased and the most precise data because they are obtained by analytical methods. The laboratory data not only provide reliable information on system toxicity but also valuable information for evaluating the efficacy and safety of the drug under study.

In clinical trials, routine laboratory tests are usually performed not only to screen patients for inclusion in trials prior to randomization but also to protect patients during trials. In many clinical trials, laboratory tests are considered as the primary efficacy variables and hence are used for assessment of drug effects. For example, laboratory tests for cholesterol and triglycerides are often used to assist the diagnosis of patients with coronary heart disease. Cholesterol measurements includes total cholesterol (TOTAL-C), high-density lipoprotein cholesterol (HDL-C), and low-density lipoprotein cholesterol (LDL-C). A high value of LDL-C, a low value of HDL-C, or a high value of triglycerides combined with a low value of HDL-C constitutes a high-risk factor of coronary heart disease. For a normal subject, its total cholesterol, HDL-C, LDL-C, and triglycerides should be within the ranges of 130–200 md/dL, 35–65 mg/dL, <130 mg/dL, and <250 mg/dL, respectively. The relationship among TOTAL-C, HDL-C, LDL-C, and triglycerides can be expressed by the well-known Fieldewald equation as follows:

$$\text{LDL-C} = \text{TOTAL-C} - [\text{HDL-C} + 0.2(\text{triglycerides})].$$

For another example, consider a laboratory test for glucose for patients with diabetes mellitus. According to the National Diabetes Data Group, a subject with glucose >140 mg/dL or two-hour post cibum (2-hr PC) glucose >200 mg/dL would be diagnosed as a patient with diabetes mellitus. If 2-hr PC glucose is between 140 and 200 mg/dL, then the subject is considered to have impaired glucose tolerance. In general, it is estimated that approximately 1–5% of the patients with impaired glucose tolerance will have diabetes mellitus.

In clinical trials any significant change in laboratory measurements could be an indication of potential toxicity or an exposure risk of the drug under study. A careful assessment of such changes is necessary. In general, different laboratory tests can be performed to meet different study objectives of clinical trials with different indications and patient populations. On average, about 20 to 35 laboratory tests are typically performed in a clinical trials. Tables 12.2.2 through 12.2.4 list some commonly performed laboratory tests for hematology, clinical chemistry, and urinalysis and their corresponding reference ranges (or normal ranges) for normalities (if available).

12.3 CODING OF ADVERSE EVENTS

Adverse Drug Reaction and Adverse Event

In clinical trials the extent of exposure of patients to the test drug under study is usually measured by means of adverse drug reactions (ADR) or adverse events (AE) which are then used for a safety evaluation of the drug. Neither the FDA nor the European regulatory authorities provide a definition of ADR or AE. Therefore the definition of ADR or AE varies among pharmaceutical compa-

Table 12.2.2 Laboratory Tests for Hematology

Laboratory Test	Normal Range
WBC (white blood cell count)	$3.6-9.8 \times 10^3/\mu L$
RBC (red blood cell count)	Male: $4.2-6.2 \times 10^6/\mu L$
	Female: $3.7-5.5 \times 10^6/\mu L$
Hemoglobin	Male: 12.9–17.9 g/dL
	Female: 11.0–15.6 g/dL
Hematocrit	Male: 38–53%
	Female: 33–47%
WBC classification	
Band neutrophil	0–3% or 0–5%
Segmented neutrophil	45–70%
Lymphocyte	25–40% or $2.4 \pm 0.8 \times 10^3/\mu L$
Monocyte	2–8%
Eosinophil	1–3% or $70-400/\mu L$
Basophil	0–0.5%
MCV (mean corpuscular volume)	82–98 fL (i.e., 10^{-15}L)
MCH (mean corpuscular hemoglobin)	27–32 pg (i.e., 10^{-12}g)
MCHC (mean corpuscular hemoglobin concentration)	31–36%
Platelet count	$120-400 \times 10^3/\mu L$
RDW (red cell distribution width)	11.5–14.5
MPV (mean platelet volume)	9.8 ± 1.2 fL
Reticulocyte count	0.5–1.5%
ESR (erythrocyte sedimentation rate)	Male: <10 mm/hr
	Female: <20 mm/hr
Bleeding time	3–10 min
Clotting time	8–10 min
PT (prothrombin time)	10–13 s
APTT (activated partial thromboplastin time)	26–36 s
G-6-PD (glucose-6-phosphatase dehydrogenase)	4.10–7.90 IU/g Hb
Bilirubin	0.3–1.0 mg/dL
Fibrinogen	200–400 mg/dL
FDP (fibrinogen degradation product)	<10 µg/mL
Total eosinophil count	$70-400/mm^3$

nies and research organizations. For example, the ICH *Guideline for Clinical Safety Data Management: Definitions and Standards for Expedited Reporting* defines an adverse drug reaction (ADR) as a response to a drug that is noxious and unintended, and that occurs at any dose used or tested in humans for prophylaxis, diagnosis, or therapy of disease, or for the modification of physiological function (ICH, 1994). More recently the ICH GCP guideline defines an adverse event as any unintended and unfavorable sign (e.g., including an abnormal laboratory finding), symptom, or disease temporally associated with the used of a medical product, whether or not it considered related to the product

Table 12.2.3 **Laboratory Tests for Clinical Chemistry**

Laboratory Test	Normal Range
Liver function tests	
ALP (alkaline phosphatase)	65–272 IU/L
AST/SGOT (serum glutamic oxaloacetin transaminase)	15–35 IU/L
ALT/SGPT (serum glutamic pyruvate transaminase)	3–30 IU/L or 8–45 IU/L
γ-GT (gamma glutamyl transferase)	5–40 IU/L
Bilirubin	0.3–1.0 mg/dL
LDH (lactic acid dehydrogenase)	150–400 IU/dL
Total protein	6.6–8.1 gm/dL
Albumin	3.9–5.1 gm/dL
Globulin	2.3–3.5 gm/dL
Renal function tests	
BUN (blood urea nitrogen)	5–20 mg/dL
Creatinine	0.7–1.5 mg/dL
Creatinine clearance test	Male: 62–108 ml/min
	Female: 57–78 ml/min
Electrolytes	
Sodium (Na^+)	135–140 mmol/L
Potassium (K^+)	3.5–5.0 mmol/L
Chloride (Cl^-)	98–108 mmol/L
Calcium (Ca^{2+})	2.1–2.6 mmol/L
Phosphorus	2.5–4.5 mg/dL
Magnesium (Mg^{2+})	1.9–2.5 mg/dL
Uric acid	Male: 3.5–7.9 mg/dL
	Female: 2.6–6.0 mg/dL
CPK (creatinine phosphokinase)	37–289 IU/L
Aldolase	1.7–4.9 units/L
Amylase	Serum: 30–200 IU/L
	Urine: 4–30 IU/2 h
Lipase	<200 units/L
Cholesterol	
Total cholesterol	130–200 mg/dL
HDL-cholesterol	35–65 mg/dL
LDL-cholesterol	<130 mg/dL
Apo A-1 (apolipoprotein A-1)	Male: 66–151 mg/dL
	Female: 75–170 mg/dL
Apo B (apolipoprotein B)	Male: 49–124 mg/dL
	Female: 26–119 mg/dL
Triglycerides	<250 mg/dL
Glucose	
AC glucose	70–110 mg/dL
30 PC glucose	90–160 mg/dL
1-hr PC glucose	90–160 mg/dL

Table 12.2.3 (*Continued*)

Laboratory Test	Normal Range
2-hr PC glucose	75–125 mg/dL
3-hr PC glucose	70–110 mg/dL
HbA$_{1c}$ (glycosylated hemoglobin)	4–7%
Serum iron	Male: 89–200 μg/dL
	Female: 70–180 μg/dL
Ferritin	Male: 27–300 ng/ml
	Female: 10–130 ng/ml
Acid P-tase (acid phosphatase)	Male: 4.7 IU/L
	Female: <3.7 IU/L
Protein electrophoresis	
Total protein	5.9–8.0 g/dL
Albumin	4.0–5.5 g/dL
Alpha-1 globulin	0.15–0.25 g/dL
Alpha-2 globulin	0.43–0.75 g/dL
Beta globulin	0.50–1.00 g/dL
Gamma globulin	0.60–1.30 g/dL
Lipoprotein electrophoresic	
Pre-beta	20 ± 6%
Beta	50 ± 5%
Alpha	36 ± 7%
Hemoglobin electrophoresis	
Hb A	97%
Hb A$_2$	1.5–3.5%
Hb F	<2.0%
Hb C	0
Hb S	0
Osmolality	280–295 mOsm/kg

(ICH, 1996). Nickas (1995) classifies AE data into three categories: (1) known ADRs, (2) AEs where a causal relationship is uncertain, a possible ADR, and (3) AEs that are considered unrelated to study drug. Northington (1996) provides a general definition which considers an AE any *negative event* that a patient/subject experiences during the course of a clinical trial. The term negative event is rather broad and allows too flexible an interpretation. As a result, although this definition may increase the number of events reported, it may also increase the chance of capturing all potentially important events. More specifically, Northington (1996) indicates that an AE may be defined as any unfavorable change in the structure (signs), function (symptoms), or chemistry (laboratory data) of the body temporally associated with participation in the clinical trials, irrespective of the believed relationship to the study drug. This specific definition would include intercurrent illness or injuries, clinically significant results from laboratory tests or other medical procedures, and clinically significant findings uncovered during a physical examination.

Table 12.2.4 Laboratory Tests for Urinalysis

Laboratory Test	Normal Range
Dipstick tests	
pH	4.6–8.0
Protein	<8 mg/dL
Glucose	
Ketone	
Occult blood	
Urobilinogen	0.1–1.0 EU/dL
Leukocyte esterase	
Nitrite	$<10^5$ colony/ml
Sediment	
RBC	<5/HPF
WBC	<5/HPF
Epithelial cells	0
Casts	0/LPF
Crystal	
Microorganisms	
Parasites	
Spermatozoa	
Specific gravity	1.016–1.022
Gram stain	
Bence-Jones protein	
Paraguat test	
Porphobilinogen	
Myoglobin	
Pregnancy test	
Fractional urinalysis	

Since there exists no universal definition of an AE, Scherer and Wiltse (1996) emphasize that the definition of an AE should encompass the concepts of (1) any undesirable experience (2) that occurs in a clinical trial participant (3) whether or not it is considered related to the study drug, (4) even if the patient never receives the study drug (intent-to-treat). As a result we can define an AE as any illness, sign or symptom that has appeared or worsened during the course of the clinical study regardless of causal relationship to the medicine(s) under study. The situation of serious adverse events (SAE) is somewhat different. The FDA has defined that with respect to human clinical experience, a serious adverse drug experience includes any experience that is fatal or life-threatening, is permanently disabling, requires inpatient hospitalization, or is a congenital anomaly, cancer, or overdose. On the other hand, the ICH defines an SAE as any untoward medical occurrence that at any dose results in death, is life-threatening, requires inpatient hospitalization or prolongation of existing hospitalization, causes persistent or significant disability/incapacity, or is manifest in congenital anomaly/birth defect. Unlike the definitions of an adverse event, these definitions are more precise, which makes for uniformity in collecting and summarizing SAEs. Note that the ICH also defines other *significant* adverse events as marked hematological and other laboratory abnormalities and any adverse events that lead to an intervention, including withdrawal of drug treatment, dose reduction, or significant additional concomitant therapy.

With respect to adverse events, in the 1988 *Guideline for the Format and Content of the Clinical and Statistical Section of an NDA*, the FDA states that an adverse event tabulation of particular interest that should be produced for all studies would include all new adverse events (i.e., those not seen at the baseline or that worsened during treatment). Adverse events of this kind is known as treatment-emergent adverse events (TEAE). The definition, however, is vague and open to a number of different interpretations. Northington (1996) provides a more explicit definition of TEAE as follows: An adverse event that occurs during the active phase of the study will be considered a TEAE if (1) it was not present at the time the active phase of the study began and it is not a chronic condition that is part of the patient's medical history, or (2) it was present at the start of the active phase of the study or as part of the patient's medical history but the severity or frequency increased during therapy. Note that when the active phase of study begins depends on whether there is a run-in period. For example, for studies without lead-in periods, the active phase of the study begins at the time of the first dose of the study drug. The active phase of the study begins at the time of the first dose of double-blind therapy for studies with single-blind placebo run-in periods.

Note that adverse events that occur while patients are taking study drug typically receive much attention. The treatment-emergent concept described is generally applied only to events that occur while the patient is on therapy. In most clinical trials there are usually some adverse events that are reported after the last day of study drug. The adverse events of this kind are usually referred to as post-therapy adverse events.

Basically an adverse event can be generated either (1) during a clinical trial or (2) spontaneously from reports on drugs already on the market. The purpose of collecting adverse events from clinical trials is to enable a complete and accurate summarization of adverse events which can be expected in the target patient population. The information can also be used to guide the practicing physician in the use of the drug, and the events reported by a patient as to whether or not they are likely to be related to the drug. The purpose of obtaining reports of spontaneous events is to detect marked changes in frequency and seriousness of events from what was observed during the clinical trials. The collection and reporting of adverse events in clinical trials help to identify and evaluate adverse events and can lead to improvements in the problem areas.

Scherer and Wiltse (1996) pose some important questions on the definition of adverse events: (1) Should expected clinical outcomes be distinguished from serious adverse events? (2) Are surgical and diagnostic procedures adverse events? (3) What changes in laboratory variables are adverse events? On the first question, Scherer and Wiltse point out that the FDA permits the separation of clinical outcomes from SAEs except for the deaths if the expected clinical outcomes of the disease are efficacy endpoints of the study. It should be noted that the underlying assumption of this approach is that there is no significant increase in the frequency of observed clinical outcomes due to the active treatment. Clinical outcomes that are not classified as SAEs are not subject to expedited reporting. On the second question, Scherer and Wilse propose that only the illness leading to the surgical or diagnostic procedures be reported, including medical therapy, radiation, or nuclear medicine therapy, although the procedure could be captured as part of the action taken in response to the events. On the third question, the FDA guideline states that laboratory findings that constitute an adverse event (ECG abnormality suggesting infarction, serious arrhythmia, etc.) should be included. However, the interpretation of this FDA guideline seems to vary from individual to individual. For example, in practice, abnormal laboratory values are almost always entered as adverse events if they fall outside a reference range specified by some pharmaceutical companies. The difficulty is that there is always a large number of abnormal laboratory findings in patients without any obvious clinical abnormalities in clinical trials. As an alternative, Scherer and Wiltse suggests considering abnormal laboratory values as adverse events only when they are secondary (e.g., an oliguric patient with a creatinine of 6 mg/dL) to a primary clinical event (e.g., renal failure).

Coding System

The FDA recommends that a dictionary for grouping similar events be used in reporting the adverse events. The purpose of grouping the observed adverse events is to assess the safety profile of the drug under investigation. There are many dictionaries currently available for this purpose. These dictionaries include the *Coding Symbols for a Thesaurus of a Adverse Reaction Terms* (COSTART) developed by the FDA, the *World Health Organization Adverse Reaction Terminology* (WHOART) recommended by the World Health Organization, the *International Classification of Diseases* (ICD) adopted codes, HARTS, a dictionary developed by Hoechst, and the *Medical Dictionary for Drug Regulatory Affairs* (MEDDRA) adopted by the International Conference on Harmonization (ICH). Among these dictionaries, the COSTART and WHOART are probably the most commonly used dictionaries for grouping similar events in the summarization and reporting of adverse events. Since dictionaries designed for different purposes have their relative advantages and disadvantages, the FDA pointed out that experience at this time is too limited to recommend a particular one for grouping similar events. The FDA, however, encourages the sponsor to also give the corresponding COSTART terms when reporting adverse events regardless of which dictionary is used. Therefore in this section we will only focus on the COSTART.

The COSTART is the terminology developed and used by the FDA for the coding, filing, and retrieving of adverse reaction reports. It provides a method to deal with the variation in vocabulary used by those who submit adverse events to the FDA. The COSTART dictionary was derived in the 1960s from the dictionary of adverse reaction terms (DART). As of the fifth edition, the COSTART has more than 6000 glossary terms collapsing to approximately 1200 unique COSTART terms. Basically the COSTART is divided into seven indexes that provide different information on adverse events' coding, filing, and reporting. The use of these indexes is briefly summarized below.

Index A of the COSTART dictionary is composed of a hierarchical *Body System Classification*, *General Search Categories*, and *Special Search Categories*. Hierarchical Body System Classification is the primary category, and it contains 12 subcategories which are summarized in Table 12.3.1. For each primary category there may be a number of subcategories associated with it. The coding symbols are chosen to reflect both the primary category and its subcategories. For example, for *Body as a Whole*, the primary category is coded as *BODY* and its subcategories Head, Neck, and Thorax are coded as *BODY/HEAD*, *BODY/NECK*, and *BODY/THOR*, respectively. The purpose of the General Search Categories is for search and retrieval strategies. The Body System Classification sometimes serves as the basis of the search strategy. For the most part the need has been superseded by the use of the pathophysiologic classification of COSTART as presented in Index F. The Special Search Categories are useful for the assessment of possible fetal and neonatal disorders associated with drug or biologic product used *in utero* or in early life. The Special Search Categories include codes for recording whether a suspected drug was given to (1) either parent before conception, (2) the mother during gestation, (3) the mother at a particular time during delivery or while nursing, (4) the fetus directly *in utero*, (5) the infant during the first two years of life. These symbols signify that a particular reaction occurred during a particular time, fetal life, or in the new born.

Table 12.3.1 Body System Classification of COSTART Dictionary

Classification	Coding Symbol
Body as a whole	BODY
Cardiovascular system	CV
Digestive system	DIG
Endocrine system	ENDO
Hemic and lymphatic system	HAL
Metabolic and nutritional disorders	MAN
Musculoskeletal system	MS
Nervous system	NER
Respiratory system	RES
Skin and appendages	SKIN
Special sense	SS
Urogenital system	UG

Table 12.3.2 General Categories of Drug-Induced Diseases

Category	Description
1	Automatic nervous
2	Cardiovascular
3	Endocrine
4	Gastrointestinal
5	Genitourinary
6	Gynecdogic
7	Hematologic
8	Maternal-fetal
9	Metabolic
10	Nervous
11	General/nonspecific
12	Opthalmic
13	Pathological
14	Pulmonary
15	Renal
16	Reticuloendothelial
17	Skin

Index B of the COSTART dictionary gives a comprehensive alphabetical listing of the COSTART symbols and obligatory Body System categories and subcategories, and Index C lists COSTART symbols stratified by Body System categories. Index C is useful in reviewing and/or determining the selection of a COSTART term. Index D of the COSTART dictionary is a glossary that contains nearly 6000 reported terms with appropriate COSTART symbols. Index E of the COSTART dictionary reflects the FDA's attempt for a pathophysiologic classification of COSTART terminology. The arrangement of terms was developed by identifying 17 categories of drug-induced diseases and subdividing these into sometimes overlapping subcategories of more specific types of dysfunction or disease (see Table 12.3.2). Note that some of the definitions in Index E are broad. Therefore, although there is a high level of sensitivity, the false positive rate is high.

Since many companies use the WHOART and the FDA encourages the sponsors to also give the corresponding COSTART terms when reporting adverse drug reactions, there is a need to translate WHOART to COSTART. Indexes F and G describe the relationship between the COSTART dictionary and the dictionary adopted by the World Health Organization Adverse Reaction Terminology (WHOART). Index F of the COSTART dictionary lists the translation of each and every COSTART term to an acceptable WHOART term, while Index G of the COSTART dictionary provides translations of each and every preferred WHOART term to an acceptable COSTART term.

The COSTART dictionary is widely used for the purpose of grouping similar events. Any potential trend in adverse events can be detected in this efficient way. However, there are potential problems with any dictionary used for grouping. First, we may map synonymous clinical events to different class terms. This is known as the problem of *one-to-many* problems. For example, *Heart attack* could be classified to either *Cardiovascular disorder, Asystole, Myocardial infarction,* or *Coronary artery disorder.* On the other hand, we may map clinically different events to a single class term, which is usually referred to as *many-to-one.* For example, the terms such as *Angiography, Coronary, Left ventricular aneurysm and trombus,* and *Heart murmur* may be classified to the category of *Surgical procedure.* In practice, it is difficult to avoid these mapping problems.

Some Concerns

How should syndromes be recorded on the case report form? as the syndrome, as symptom separately, or both? A syndrome is a collection of signs, symptoms, or laboratory findings that together describes a distinct clinical entity. For example, congestive heart failure is described by one or more of shortness of breath, peripheral edema, easy fatiguability, pulmonary rales, elevated venous pressures, low plasma sodium, and elevated BUN. The commonly encountered problem with the summarizing syndrome and its components is that symptoms that occur alone are not distinguished from those that occur as part of the syndrome. Moreover, increased incidence of true adverse events related to the drug may be lost in a high incidence of the same event that occurs as a component of a syndrome in both groups. These problems may misrepresent the effect of the drug.

How should primary versus secondary adverse event be recorded? The problems include (1) multiple counting of same events and (2) over-reporting of components of syndrome or secondary events, which are not in and of themselves related to drug. These concerns may lead to inaccurate conclusions about undesirable effects of the drug. Thus analysis of primary events leads to correct understanding of causal relationship between the drug's administration and adverse events. Both primary and secondary adverse events can be recorded on CRF, with the secondary event clearly linked to the primary event or syndrome.

12.4 ANALYSIS OF ADVERSE EVENTS

As indicated earlier, the evaluation of safety-related data can be considered at three levels. The next levels involve the more common adverse events and serious adverse events. The analysis of adverse events is not only to identify factors that may affect the frequency of adverse reactions/events such as time dependence, relation to demographic characteristics, relation to dose or drug concentration but also to determine whether the identified adverse events are drug related. For this purpose we may consider the approaches of data listing, summary tables, and statistical analysis.

Table 12.4.1 Information to be Included in the List of Adverse Events

Patient identifier
Age, race, sex, weight
Location of case report forms if provided
Adverse event
Duration of the adverse event
Severity
Seriousness
Action taken
Outcome
Causality assessment
Date of onset or date of clinic visit at which the event was discovered
Timing of onset of the adverse event in relation to the last dose of the last drug
Study treatment at the time of event or the most recent study taken
Test drug/investigational product dose in absolute amount, mg/kg or mg/m^2, at time of event
Drug concentration (if known)
Duration of test drug/investigational product treatment
Concomitant treatment during study

Adverse Event Data Listing

For evaluation of adverse events, it is helpful to list all adverse events including the same event on several occasions by the patient and by the treatment group. The listing should include both the preferred term and the original term used by the investigators. For a complete listing of adverse events, the ICH guidelines suggest that the information listed in Table 12.4.1 be included in the data listing.

In Table 12.4.1 the patient characteristics may include height if relevant. The adverse event should be displayed both as the preferred term and the reported term. The degree of severity may be classified as mild, moderate, or severe, while the seriousness can be expressed as either serious or nonserious. For action taken, it could be none, dose reduced, treatment stopped, specific treatment instituted, and so forth. The causality assessment may include two categories of related and not related. However, how it was determined should be clearly described in the protocol or table.

Note that also required is that all deaths during the study, including the posttreatment follow-up period and deaths that resulted from a process that began during the study should be listed by patient. All serious adverse events other than death but including the serious adverse events temporally associated with or preceding the deaths should be listed. In addition other significant adverse events and any events that led to an intervention, including withdrawal of test drug/investigational product treatment, dose reduction, or significant additional concomitant therapy, other than those adverse events reported as serious should be listed.

Summary Tables of Adverse Events

The ICH guidelines suggest that all adverse events occurring after the initiation of study treatments be displayed in summary tables. The adverse events include events that are likely to be related to the underlying disease or events that are likely to represent a concomitant illness. The summary tables list each adverse event, the number of patients in each treatment group in whom the event occurred, and the rate of occurrence. In most cases it is helpful to identify and summarize events not seen at the baseline and events that worsened even if present at the baseline. Moreover, the summary tables should include changes in vital signs and any laboratory changes that are considered serious adverse events or other significant adverse events.

In practice, adverse events are grouped by body system and then divided into defined severity categories. However, in presenting adverse events, it is important to display both the original terms used by the investigators and to attempt to group related events so that the true occurrence rate is not obscured. The ICH guidelines recommend that a standard adverse reaction/events dictionary be used for grouping. As indicated in the previous section, the COSTART is probably the most commonly used dictionary for this purpose.

In most cases the summary tables help divide the adverse events into those considered at least possibly related to drug use and those considered unrelated, or by another causality scheme such as unrelated or possibly, probably, or definitely related. However, when a causality scheme is used, it is suggested that the summary tables include all adverse events regardless of whether they are drug related or represent intercurrent illnesses. It is also important to identify each patient having the adverse event for safety evaluation. An example of such a tabular presentation as specified in the ICH guidelines is reproduced in Table 12.4.2. In addition to Table 12.4.2, the summary tables should include a comparison of common adverse events between the treatment and control groups. The observed safety profiles from the summary tables can then be confirmed by some formal statistical tests which will be discussed in the next subsection.

Graphical Presentation

To provide a preliminary examination of adverse events between treatment groups, a graphical presentation is usually helpful. A creative graphical presentation enables clinical scientists/researchers to discover trends/patterns about the data that are unanticipated. Levine (1996) recommends that a study design and patient flow display, raw data display, and inferential displays of clinical trial adverse events be provided. Levine indicates that these graphical presentations provide not only rapid answers to prespecified questions but also the insight into the structure of the raw data. In addition a creative graphical pre-

Table 12.4.2 Adverse Events: Number Observed and Rate with Patient Identifications

	Treatment Group X						$N = 50$	
	Mild		Moderate		Severe		Total	Total
	Related[a]	NR[a]	Related	NR	Related	NR	Related	NR	R + NR
Body System A								
Event 1	6(12%)	2(4%)	3(6%)	1(2%)	3(6%)	1(2%)	12(24%)	4(8%)
	N11[b]	N21	N31	N41	N51	N61		
	N12	N22	N32		N52			
	N13		N33		N53			
	N14							
	N15							
	N16							
Event 2								

[a] NR = not related; related could be expanded (e.g., as definite, probable, possible).
[b] Patient identification number.

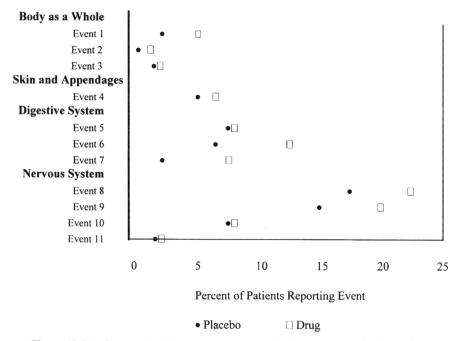

Figure 12.4.1 Scatter plot of adverse events versus incidence rates: ● placebo, ☐ drug.

sentation can generate new questions regarding safety and provide rapid assessment to these new questions. For example, Figure 12.4.1 gives a scatter plot showing adverse events by the body system against the percentage of patients reporting events from a placebo-controlled clinical trial. This scatter plot not only provides a preliminary examination of adverse events rates within treatments but also compares the adverse event rates between treatments.

Analysis of Adverse Events

Although adverse event data listing, summary tables of adverse events, and graphical presentations of adverse events provide useful information for the safety evaluation of a study drug, they do not provide any statistical inference for the safety assessment. As was indicated earlier, most clinical studies are not designed to examine safety. As a result the analysis of adverse events is less standardized than that for efficacy data. Silliman (1996) classifies adverse events into rare adverse events and common adverse events. For rare adverse events the primary interest is to estimate the occurrence. For common adverse events, in addition to estimation of occurrence, it is also of interest to make comparisons between treatments and/or among subgroups. Basically the analysis of adverse events depends on the type of data, which can be classified as nominal (binary) or ordinal, counts or rates, or time to occurrence. These types

Table 12.4.3 The 2 × 2 Contingency Table for Adverse Events

	Response		Total
Treatment	a	b	n_1
Control	c	d	n_2
Total	m_1	m_2	N

of data are among those described in Section 12.2 and are usually analyzed by Fisher's exact test or the Mantel-Haenszel test, logistic regression, and survival analysis. In this section, for illustration purposes, we will only consider the commonly used approaches for comparison of adverse event rates between treatments (e.g., treatment vs. control). For other types of adverse events data such as absorbing events and recurring events with and/or without duration as described in Section 12.2, statistical methods including the Kaplan-Meier and the Cox proportional hazards models, as described in Chapter 9, can be directly applied.

For simplicity, consider the case where the event severity categories and/or the causality categories of a given response (adverse event) are combined into a 2 × 2 contingency table as shown in Table 12.4.3. For a 2 × 2 contingency table, a typical approach is to use the Mantel-Haenszel test. For the 2 × 2 table given in Table 12.4.3, the Mantel-Haenszel test is given by

$$\chi^2_{MH} = \frac{N(ad-bc)^2}{n_1 n_2 m_1 m_2} \qquad (12.4.1)$$

or with a continuity correction as

$$\chi^2_{MH} = \frac{N(|ad-bc| - n/2)^2}{n_1 n_2 m_1 m_2}.$$

Note that χ^2_{MH} can be written as

$$\chi^2_{MH} = \frac{N}{N-1} \left[\frac{a - E(a)}{\sqrt{\text{var}(a)}} \right]^2$$

$$= \frac{N}{N-1} \chi^2_P,$$

where

ANALYSIS OF ADVERSE EVENTS

$$E(a) = \frac{n_1 m_1}{N}$$

and

$$\operatorname{var}(a) = \frac{n_1 n_2 m_1 m_2}{N^2(N-1)}.$$

When the sample size is sufficiently large, the Mantel-Haenszel test is approximately distributed as a chi-square random variable with one degree of freedom. Therefore we can reject the null hypothesis of no difference between treatment groups at the α level of significance if

$$\chi^2_{MH} > \chi^2(\alpha, 1).$$

When the sample size is small, Fisher's exact test as described in (8.3.16) is often used as an alternative to the Mantel-Haenszel test. For a 2×2 table, Fisher's exact test yields the probability of observing a table that gives at least as much evidence of association as the one actually observed, given that the null hypothesis is true. In other words, with row and column margins, namely n_1, n_2, m_1, m_2 considered fixed, the distribution of a has the hypergeometric form

$$P(a|n_1, n_2, m_1, m_2) = \frac{\binom{n_1}{a}\binom{n_2}{m_1 - a}}{\binom{N}{m_1}},$$

where a takes on values between $\max(0, n_1 + m_1 - M)$ and $M = \min(n_1, m_1)$. Thus we can reject the null hypothesis of no difference if

$$p\text{-value} = \sum_{i=1}^{M} P(i|n_1, n_2, m_1, m_2) \qquad (12.4.2)$$

is less than the α level of significance.

In clinical trials, although an adverse event can be classified into dichotomous groups such as presence versus absence or serious versus nonserious (see Table 12.4.3), it is often of interest to examine the intensity or severity of an adverse event that may be evaluated according to c categories such as *mild*, *moderate*, and *severe* (where $c = 3$). In this case the 2×2 table becomes $2 \times c$ table. On the other hand, the intended clinical trial may be designed to com-parc r treatments. In this case the 2×2 table becomes a $r \times c$ table. Both the Mantel-Haenszel test and Fisher's exact test can be extended to analyze $r \times c$ tables. Note that the Mantel-Haenszel test for $r \times c$ tables is also known as the Cochran-Mantel-Haenszel test. Agresti and Wackerly (1977) consider generations of Fisher's exact test for $r \times c$ tables. Pagano and Halvorsen (1981) propose an algorithm, which does not require the total enumeration of all tables consistent with the given marginals, for calculating the exact permutation significance value for $r \times c$ tables.

In practice, since most clinical trials are conducted at more than one study site, it is necessary to adjust for the effects that may be due to the study site (or center) when comparing adverse event rates between treatment groups. For this purpose the Cochran-Mantel-Haenszel (CMH) test is useful. The concept behind the CMH test is to consider the study site (or center) as a stratum and assume that the strata are independent and that the marginal totals of each stratum (center) are fixed. Under these assumptions the null hypothesis is that there is no association between treatment and response (a given adverse event say) in any of the strata. Consequently the multiple hypergeometric model can be used to derive the CMH test. Let $\mathbf{n}_{hi} = (n_{hi1}, n_{hi2}, \ldots, n_{hic})'$, where n_{hij} is the cell frequency of the response at the jth intensity or severity category for the ith treatment group at the hth center, $i = 1, \ldots, r$ (the number of treatments), $j = 1, \ldots, c$ (the number of categories for the response), and $h = 1, \ldots, H$ (the number of centers). Also let

$$\mathbf{n}_h = (\mathbf{n}_{h1}, \mathbf{n}_{h2}, \ldots, \mathbf{n}_{hr})'$$

and

$$P_{hi.} = \frac{n_{hi.}}{N_h}, \quad P_{h.j} = \frac{n_{h.j}}{N_h},$$

where N_h is the total at the hth center. Thus we have

$$\mathbf{P}_{h*.} = (P_{h1.}, P_{h2.}, \ldots, P_{hr.})'$$

and

$$\mathbf{P}_{h.*} = (P_{h.1}, P_{h.2}, \ldots, P_{h.c})'.$$

Under the null hypothesis the expected value and covariance matrix of the frequencies are then given by

$$\mathbf{m}_h = E(\mathbf{n}_h) = N_h(\mathbf{P}_{h.*} \otimes \mathbf{P}_{h*.})$$

ANALYSIS OF ADVERSE EVENTS

and

$$\text{var}(\mathbf{n}_h) = \frac{N_h^2}{N_h - 1} [(\mathbf{D}_{P_{h.*}} - \mathbf{P}_{h.*}\mathbf{P}'_{h.*}) \otimes (\mathbf{D}_{P_{h*.}} - \mathbf{P}_{h*.}\mathbf{P}'_{h*.}),$$

where $\otimes$ denotes Kronecker product multiplication and $\mathbf{D}_A$ is a diagonal matrix with elements of A on the main diagonal. The generalized CMH test is then defined as

$$\chi^2_{\text{CMH}} = \mathbf{G}'\mathbf{V}_G^{-1}\mathbf{G}, \tag{12.4.3}$$

where

$$\mathbf{G} = \sum_h \mathbf{B}_h(\mathbf{n}_h - \mathbf{m}_h),$$

$$\mathbf{V}_G = \sum_h \mathbf{B}_h \, \text{var}(\mathbf{n}_h)\mathbf{B}'_h,$$

and where

$$\mathbf{B}_h = \mathbf{C}_h \otimes \mathbf{R}_h$$

is a matrix of fixed constants based on column scores $\mathbf{C}_h$ and row scores $\mathbf{R}_h$. Under the null hypothesis, χ^2_{CMH} is approximately distributed as a chi-square random variable with degrees of freedom equal to the rank of $\mathbf{B}_h$. Note that the CMH test given in (12.4.3) is for the general case of $r \times c$ tables stratifying for the study sites. When $r = c = 2$, the CMH test can be expressed as

$$\chi^2_{\text{CMH}} = \frac{(\sum_{h=1}^{H} [a_h - (n_{1h}m_{1h}/N_h)])^2}{\sum_{h=1}^{H} [n_{1h}n_{2h}m_{1h}m_{2h}/N_h^2(N_h - 1)]},$$

where a_h, n_{1h}, n_{2h}, m_{1h}, m_{2h}, and N_h are defined as in Table 12.4.3 for the hth center.

As was indicated earlier, for a premarketing safety assessment, the FDA requires that an integrated summary of all available information about the safety of the drug product be submitted (Section 314 of CFR [314.50 (d)(5)(ii)]). The purpose of such an integrated summary is to allow examination of differences among population subsets not possible with the relatively small numbers of patients in individual studies. Therefore the integrated summary is, in part, simply a summarization of data from individual studies and, in part, a new analysis that goes beyond what can be done with individual studies. Silliman (1996) suggests that combining data across studies in the integrated summary of safety be done by either subgroup analysis (pooling the data) or by performing a meta-analysis. For analysis of subgroup safety data by pooling, the Fisher exact test, the CMH test (or Breslow-Day test), logistic regression, and survival analysis are commonly employed. The objectives of subgroups analysis are to address the following questions in an integrated summary safety report:

1. Are adverse event rates the same across a subgroup for patients taking experimental drug?
2. Within subgroup levels, are adverse event rates the same across treatment groups?
3. Is there a consistent association between the treatment group and the adverse event response across levels of a subgroup?
4. Does the subgroup predict an adverse event response?
5. Is the time until the adverse event the same across levels of a subgroup?

For an analysis of subgroup safety data using meta-analysis, the commonly used statistical methods include the CMH test (or Breslow-Day test), logistic regression such as proportional odds model, and survival analysis such as the Kaplan-Meier (log-rank), piecewise exponential, and incidence density tests. The objective of a meta-analysis is to address the above questions by controlling for the study and/or time interval.

12.5 ANALYSIS OF LABORATORY DATA

In clinical trials the analysis plan for laboratory data is usually vague. A typical approach is to compare the change from baseline to the last patient visit both between and within each treatment group. The comparison can be made either based on the absolute mean change from baseline in laboratory values or based on the frequencies of the values outside the reference or normal ranges. Either way provides a different analysis and interpretation of the laboratory data. In this section we will focus on the analysis of frequencies of the values according to the reference or normal ranges. For analysis of the absolute mean changes from baseline, the statistical methods described in Chapter 7 can be used.

Reference Ranges

In practice, most clinical trials are conducted at more than one study site in order to enroll enough patients within a desired time frame. In this case a concern may be whether the laboratory tests should be performed by local laboratories or a central laboratory. The relative advantages and drawbacks between the use of a central laboratory and local laboratories include (1) the combinability of data,

(2) timely access to laboratory data, (3) laboratory data management, and (4) cost. For example, a central laboratory provides combinable data with unique normal ranges, while local laboratories may produce uncombinable data due to different equipment, analysts, and normal ranges. As a result laboratory data obtained from central laboratories are more accurate and reliable compared with those obtained from local laboratories.

Combinability of Laboratory Data

Since different local laboratories may have different normal ranges, it is necessary to transfer laboratory results according to the investigators' normal ranges or local laboratories' normal ranges to a *standard* range before analysis. Typically a laboratory result can be transformed according to the following formula:

$$R_t = S_L + \frac{R_u - I_L}{I_H - I_L}(S_H - S_L), \qquad (12.5.1)$$

where R_u and R_t denote the untransformed result and transformed result, respectively, and (I_L, I_H) and (S_L, S_H) are the investigators' and *standard* lower and upper limits of normality, respectively. Note that if both lower limits are equal to 0, then the transformation is based only on the ratio of upper limits. Moreover, if an untransformed result is within the investigator's limits, then the transformed result will be within the *standard* limits. However, data cannot be transformed when lower and upper limits are equal. An example of such a case is hemoglobinuria whose normal ranges are known to be (0, 0); that is, complete absence is the only normal condition. As can be seen from (12.5.1), if results for a specific test have different investigators' ranges, the transformation procedure effectively removes data variations due to the different sources to be examined under equivalent conditions and displays comparable values.

Chuang-Stein (1996) propose similar methods to combine data from different laboratories. Let I_W be the width of the investigators' normal range,

$$I_W = I_H - I_L.$$

Then

$$I_M = \tfrac{1}{2}(I_L + I_H)$$

and

$$I_S = \tfrac{1}{2}(I_H - I_M)$$

are the midpoint and the $\tfrac{1}{4}$ of the investigators' normal range. Chuang-Stein

(1996) suggests using the following two methods for combining data from different laboratories: The first method is to combine the data by means of the transformation

$$R_1 = \frac{R_u - I_L}{I_W}. \qquad (12.5.2)$$

As can be seen from the above, $R_1 \in (0, 1)$ when $R_u \in (I_L, I_H)$; otherwise, R_1 is outside of the interval. The second method is to consider

$$R_2 = \frac{R_u - I_M}{I_S}. \qquad (12.5.3)$$

When $R_u \in (I_L, I_H)$, R_2 is within the interval of (-2, 2); otherwise, it is outside of this interval.

For safety assessment based on laboratory data, correct normal ranges are important for the following reasons: First, notable outlying values and/or clinically significant changes from the baseline must be identified in order to protect the patients during the clinical trial. Second, patients who exhibit unusual laboratory results may be excluded from the trial. In addition, patients with notable test results must be identified, studied, and discussed in the final study report. As a result the determination of normal ranges will have an impact on the assessment of safety of the drug under study. In practice, the commonly encountered difficulties with normal ranges are (1) normal ranges are usually based on healthy volunteers who may not be representative of the patient population of the intended clinical trial, (2) some normal ranges may be inaccurate due to a small-sample estimate, and (3) the underlying distribution of many laboratory analyses is not normally distributed. To overcome these difficulties, it is suggested that the so-called Lilly reference limits be used. The concept behind Lilly reference limits is to use a large number of laboratory data obtained from the baseline visits. The laboratory data are first examined for the effect of various covariates such as age, gender, origin, smoking, and alcohol use. Lilly reference limits are then obtained using actual percentiles without any distributional assumptions.

Note that before the data from different laboratories can be combined for analysis, it may be of interest to evaluate the repeatability and reproducibility of the results. The repeatability of the results is referred to as within-laboratory variability, while the reproducibility of the results is the assessment of between-site variability. Typically the repeatability and reproducibility can be assessed by sending to each laboratory identical samples that represent a wide range of possible values. The results can then be analyzed using the method analysis of variance.

Clinically Significant Changes

Before one can determine whether there is a clinically significant change in laboratory data, the definition of clinically significant change is essential. The definition of clinically significant change depends on the objectives of the laboratory evaluation. For example, if the objective is to determine whether a patient's change in laboratory value from the baseline is within the normal range, then we can claim that a clinically significant change has occurred provided that the patient's change in laboratory value from the baseline results in moving the patient from the status of normality to abnormality or from abnormality to normality. For assessing normality change, the most straightforward approach is to transform the result to the categories of either *normal* and *abnormal* according to some established reference or normal range.

The classification of patients' laboratory results to either *normal* or *abnormal* categories, however, cannot assess the magnitude of change. To assess the magnitude of change in terms of absolute (or actual) change or relative (percent) change, we need to establish an equivalence range so that any changes in laboratory values from the baseline within the range is not of clinical significance. This equivalence range can then be used as a reference range for determination of a clinically significant change. It should be noted that a clinically significant change may not result in a change in the status of normality or abnormality. To provide a consistent assessment, some pharmaceutical companies have suggested not to classify a subject into the category of *abnormal* unless (1) the result is outside of the normal range in a direction potentially adverse and (2) the change is of clinical significance. In practice, one may want to divide the magnitude of change from the baseline into a number of ranges to distinguish between different grades of change. For example, for a laboratory test of hematocrit, the normal range for a male is between 38% and 53% (see Table 12.2.2). We may consider three grades of change in either direction (i.e., increasing or decreasing direction) as 5% to 10% (grade 1), 10% to 15% (grade 2), and greater than 15% (grade 3). In this case clinically significant changes depend on the magnitude of change in the grades of interest.

In some clinical trials a patient's laboratory results are classified into several categories to assess the degree of concern depending on the frequency with which the abnormalities, singly or in succession, were found. Table 12.5.1 provides some examples of categories used in assessing the degree of concern. From Table 12.5.1 it can be seen that the categories range from all results normal (category I), to two are more successive abnormalities, and to the last observed value was abnormal (category V). That is to say, the definition of a clinically significant change is a change from one category to another category.

In many clinical trials, laboratory tests are performed at the visits (or time points). The change is assessed as a trend over time or at some specific time points. For patients without a value at a specified time, but with a value prior to and another following the specified time, an estimate at the specified time can be made by linear interpolation between given times as follows:

Table 12.5.1 An Example of Categories for Assessing the Degree of Concern

Category	Definition
I	All results normal
II	1. No two successive abnormalities 2. Last result normal
III	1. Two or more successive abnormalities 2. Last result normal
IV	1. No two successive abnormalities 2. Last result abnormal
V	1. Two or more successive abnormalities 2. Last result abnormal

$$R_k = R_{i-1}\left(\frac{t_i - t_k}{t_i - t_{i-1}}\right) + R_i\left(\frac{t_k - t_{i-1}}{t_i - t_{i-1}}\right),$$

where R_{i-1} is the nearest result preceding the desired result (R_k), R_i is the nearest result following R_k, and t_k is the desired observation time. The definition of a clinically significant change for detection of a trend over time is based on the slope of the fitted linear regression if the relationship between the laboratory results and time is linear. For specific time points, a clinically significant change may depend on the disease status or duration of the treatment at that time point.

Shift Analysis

For a given laboratory test, subjects are usually classified into categories such as below the normal range, within the normal range, and above the normal range at pre- and post-treatment evaluation. Let n_{ij} be the number of subjects allocated to category i before treatment but to category j after treatment. Then the allocation of the subjects to the categories pre- and post-treatment can be set out in a 3 × 3 table (see Table 12.5.2). In Table 12.5.2 the entries on the main diagonal, namely n_{ij} for $i = j$, give the numbers of subjects for whom pre-treatment and post-treatment are in agreement. The off-diagonal cells show the disagreements. For example, n_{12} is the number of subjects whose laboratory values are below the normal range before treatment but are within the normal range after treatment, while n_{21} is the number of subjects whose laboratory values are within the normal range before treatment but fall below the normal range after the treatment.

A typical approach for the analysis of laboratory data as presented in Table 12.5.2 is to perform a shift analysis between pre- and post-treatment. The shift

ANALYSIS OF LABORATORY DATA

Table 12.5.2 Allocation of Subjects to Categories of Pre-Treatment and Post-Treatment

	Post-treatment			
	Below	Within	Above	Total
Pre-treatment				
Below	n_{11}	n_{12}	n_{13}	$n_{1.}$
Within	n_{21}	n_{22}	n_{23}	$n_{2.}$
Above	n_{31}	n_{32}	n_{33}	$n_{3.}$
Total	$n_{.1}$	$n_{.2}$	$n_{.3}$	n

Note: Below, within, and above mean below, within, and above normal range: $n = \sum_{j=1}^{3} \sum_{i=1}^{3} n_{ij}$.

analysis for safety assessment is to test whether the number of subjects about which before and after treatment disagreed was similarly distributed across categories. The first analysis of interest is then to compare the distributions of row and column totals. In other words, it is of interest to examine whether the overall distribution of subjects to categories, irrespective of which subjects are allocated to which categories, is the same between pre- and post-treatment. For this purpose consider

$$d_i = n_{i.} - n_{.i},$$

where $i = 1, \ldots, K$, the number of categories at pre- and post-treatment and

$$n_{i.} = \sum_{j=1}^{K} n_{ij}, \quad n_{.j} = \sum_{i=1}^{K} n_{ij}.$$

It can be easily verified that

$$\widehat{\text{var}}(d_i) = n_{i.} + n_{.j} - 2n_{ij}$$

for $i = j$ and

$$\widehat{\text{cov}}(d_i, d_j) = -(n_{ij} + n_{ji})$$

for $i \neq j$. Since the covariance matrix for the $d_1, d_2, \ldots, d_K$ is singular, we may consider omitting one of the d's; that is let

$$\mathbf{d} = (d_1, \ldots, d_{K-1})'.$$

Then, under the null hypothesis that the distributions of row and column totals are the same, statistic

$$\mathbf{d}'\hat{\Sigma}^{-1}\mathbf{d},$$

is asymptotically distributed as a chi-square random variable with $K-1$ degrees of freedom, where $\hat{\Sigma}$ is an estimate of the covariance matrix of $\mathbf{d}$. Therefore we reject the null hypothesis that the distributions of row and column totals are the same at the αth level of significance if

$$\mathbf{d}'\hat{\Sigma}^{-1}\mathbf{d} > \chi^2(\alpha, K-1), \qquad (12.5.4)$$

where $\chi^2(\alpha, K-1)$ is the upper αth quantile of a chi-square distribution with $K-1$ degrees of freedom. Note that the rejection of the null hypothesis would imply that for at least one of the categories of the pre- and post-treatment differs in the proportions of the total sample allocated to that category. However, failure to reject the null hypothesis would not imply that the proportions of the subjects allocated to the different categories at pre- and post-treatment are necessarily the correct proportions, much less would it imply that the subjects are correctly classified.

In practice, it is also of interest to investigate whether the number of subjects at which pre- and post-treatment disagreed was distributed by them in a similar manner among other categories. In other words, it is of interest to compare frequencies in corresponding cells about the main diagonal in Table 12.5.2. Thus for $i < j$ the hypotheses of interest can be expressed as

$$\begin{aligned} &\text{H}_0\text{: } \mu_{ij} = \mu_{ji}, \\ \text{vs.} \quad &\text{H}_a\text{: } \mu_{ij} \neq \mu_{ji}, \end{aligned} \qquad (12.5.5)$$

where

$$\mu_{ij} = E(n_{ij}).$$

Under the null hypothesis of (12.5.5), the test statistic

$$X_{\text{SM}} = \sum_{i<j} \frac{(n_{ij} - n_{ji})^2}{(n_{ij} + n_{ji})} \qquad (12.5.6)$$

follows a chi-square distribution with $\frac{1}{2}K(k-1)$ degrees of freedom. Therefore, at the αth level of significance, we can reject the null hypothesis of (12.5.5) if

ANALYSIS OF LABORATORY DATA

Table 12.5.3 Allocation of Subjects for Hemoglobin Laboratory Test

	Post-treatment			
	Below	Within	Above	Total
Pre-treatment				
Below	10	2	0	12
Within	15	32	2	49
Above	0	13	41	54
Total	25	47	43	115

Note: Normal ranges for hemoglobin test are 12.9–17.9 g/dL for male and 11.0–15.6 g/dL for female.

$$X_{SM} > \chi^2[\alpha, \tfrac{1}{2}K(K-1)],$$

where $\chi^2[\alpha, \tfrac{1}{2}K(K-1)]$ is the upper αth quantile of a chi-square distribution with $\tfrac{1}{2}K(K-1)$ degrees of freedom. The test statistic given in (12.5.6) is known as the Stuart-Maxwell test.

Note that when $K = 3$, the Stuart-Maxwell statistic can be rewritten as

$$X_{SM} = \frac{\bar{n}_{23}d_1^2 + \bar{n}_{13}d_2^2 + \bar{n}_{12}d_3^2}{2(\bar{n}_{12}\bar{n}_{13} + \bar{n}_{12}\bar{n}_{23} + \bar{n}_{13}\bar{n}_{23})}, \quad (12.5.7)$$

where

$$\bar{n}_{ij} = \frac{n_{ij} + n_{ji}}{2}.$$

Example 12.5.1

To illustrate the statistical methods for comparing the distributions of row and column totals and frequencies in corresponding cells about the main diagonal as described above, consider the laboratory test results of a given parameter for a sample of 115 subjects in a clinical trial. The results were summarized in Table 12.5.3. From Table 12.5.3 we have

$$d_1 = 12 - 25 = -13$$
$$d_2 = 49 - 47 = 2,$$
$$d_3 = 44 - 43 = 11;$$

$$\widehat{\text{var}}(d_1) = 12 + 25 - 2(10) = 17,$$
$$\widehat{\text{var}}(d_2) = 49 + 47 - 2(32) = 32,$$
$$\widehat{\text{var}}(d_3) = 54 + 43 - 2(41) = 15;$$

and

$$\widehat{\text{cov}}(d_1, d_2) = -(15 + 2) = -17,$$
$$\widehat{\text{cov}}(d_1, d_3) = -(0 + 0) = 0,$$
$$\widehat{\text{cov}}(d_2, d_3) = -(13 + 2) = -15.$$

Thus the matrix of variance and covariance of the d's is given by

$$\begin{bmatrix} 17 & -17 & 0 \\ -17 & 32 & -15 \\ 0 & -15 & 15 \end{bmatrix}.$$

If we delete the last row and the last column of the matrix, we then have

$$\hat{\Sigma} = \begin{bmatrix} 17 & -17 \\ -17 & 32 \end{bmatrix}.$$

As a result the inverse matrix is given by

$$\hat{\Sigma}^{-1} = \frac{1}{255} \begin{bmatrix} 32 & 17 \\ 17 & 17 \end{bmatrix} = \begin{bmatrix} 0.125 & 0.067 \\ 0.067 & 0.067 \end{bmatrix}.$$

Thus

$$\mathbf{d}' \hat{\Sigma}^{-1} \mathbf{d} = 17.909 > \chi^2(0.05, 2) = 7.378.$$

Therefore, at the 5% level of significance, we can conclude that there is a significant difference in the distributions of row and column totals.

To compare frequencies in corresponding cells about the main diagonal, we apply the test statistic given in (12.5.6). The test result and the contributions of each cell to the overall chi-square test from the three pairs of cells are summarized in Table 12.5.4. The test result indicates that there is a shift in classifications between pre- and post-treatment. Table 12.5.4 also indicates that the greatest discrepancy occurs between cells n_{12} and n_{21}. Therefore an attempt to improve agreement could be directed to eliminate the discrepancy.

As described in Chapter 8, another approach is to collapse Table 12.5.2 into a 2 × 2 table as illustrated in Table 12.5.5. Under Table 12.5.5 we can then apply the following McNemar's statistic to test the difference in changes:

ANALYSIS OF LABORATORY DATA

Table 12.5.4 Individual Contributions to Overall Chi-Square Test

Cells	Contribution to χ^2
n_{12} and n_{21}	9.941
n_{13} and n_{31}	0.000
n_{23} and n_{32}	8.067
Total	$\chi^2 = 18.008$

$$X_M = \left\{ \frac{|p_2 - p_1| - 1/n}{\text{s.e.}(p_2 - p_1)} \right\}^2$$

$$= \frac{(|b - c| - 1)^2}{b + c}.$$

Note that

$$a = n_{22},$$
$$b = n_{21} + n_{23},$$
$$c = n_{12} + n_{32},$$
$$d = n_{11} + n_{13} + n_{31} + n_{33}.$$

If it is not significant, we can conclude that the change is due to chance. In other words, the treatment does not have any influence on the change.

Other Analyses

In addition to shift analysis, the ICH guidelines also suggest that the number or fraction of patients who had a parameter change of a predetermined size

Table 12.5.5 Summary Table for Pre- and Post-Study Laboratory Test Results

	Post-study		
	Normal	Abnormal	Total
Pre-study			
Normal	a	b	$a + b$
Abnormal	c	d	$c + d$
Total	$a + c$	$b + d$	n

at selected time intervals be summarized in tables. For example, for BUN, it might be decided that a change of more than a predetermined value of BUN should be noted. For this parameter the number of patients having a similar or greater change would be shown for one or more visits, usually grouping patients separately depending on the baseline BUN. The advantage of this table is that changes of a certain size are noted even if the final value is not abnormal.

Another commonly used approach is the scatter plot with a 45° line. This approach is a graph combining the initial value and the on-treatment values of a laboratory measurement for each patient by locating the point defined by the initial value on the ordinate. If no changes occur, the point representing each patient will be located on the 45° line. A general shift to higher values will show a clustering of points above the 45° line. The scatter plot with a 45° line not only shows the baseline and most extreme on-treatment values but also helps identify potential outliers. Since the scatter plot with a 45° line usually shows a single time point for a single treatment, a time series approach for the analysis and interpretation of these plots is employed in comparing adverse events between treatment groups.

12.6 DISCUSSION

Unlike hypotheses for efficacy assessment, safety hypotheses cannot be specified a priori due to the following reasons. First, clinical trials are not designed to detect differences in safety outcomes. Second, clinical trials are usually not prepared to test hypotheses regarding rare safety events. However, as indicated by Levine (1996), failure to achieve statistical significance does not mean that a safety finding can be ignored.

In clinical trials, as was shown in this chapter, an adequate protocol must clearly define the responsibilities of the investigator in reporting and documenting all adverse events. The person(s) responsible for notifying appropriate authorities (regulatory, hospital ethics committee, etc) must be identified. When clinical and laboratory tests are used to monitor safety, parameters to be measured, timing and frequency, normal values for laboratory parameters, and definition of test abnormalities should be provided. For safety assessment, it is suggested that all patients entered into treatment who receive just one dose of the treatment should be included in the safety analysis. If a patient is not included, an explanation must be provided.

For measurement of risk of various adverse events type and data available, Tremmel (1996) suggests that the crude rate or cumulative rate be used for short-term clinical trials and all types of adverse events. For long-term clinical trials, the hazard function or median survival time can be used as a meaningful measure of risk for absorbing adverse events. For recurring events with short duration, hazard functions or expected counts can be employed to have meaningful measures of risk. For recurring events with long duration, however, it is suggested that alternative approaches such as prevalence functions or expected proportions of time affected be used. For analysis and interpretation of adverse events, the FDA guidelines suggest that attention be given to (1) frequency of treatment-emergent events; (2) relevant Body System categories; (3) severity categories (if used); (4) relationship/causality (if used); (5) original terms used by the investigator (individual study report) and group related reactions, such as defined in one of the dictionaries (integrated safety summary). If a dictionary is not used in the study report, any synonymous reactions should be grouped together. If the study size permits, more common adverse events that seem to be drug related should be examined for their relationship to the levels of drug exposure and to the baseline characteristics. Laboratory findings can reflect an adverse event (ECG abnormality suggesting infarction, serious arrhythmia, etc).

For safety assessments in clinical trials, it is helpful to examine the correlation between the drug concentration data such as concentration at the time of an event, maximum plasma concentration, or area under the curve (if available) and adverse events or changes in laboratory variables.

Note that for the safety assessment of laboratory data, since a majority of the laboratory measurements are surrogates and reference ranges can vary from method to method, many different types of errors can occur during laboratory testing. The sources of errors can be the technician, an instrument, the environment, or reagents. As a result multiple laboratory methods within a statistical analysis should be avoided. In practice, it is strongly recommended that standard procedures for good laboratory practice (GLP) such as quality assurance/control, well-defined procedures, well-trained staff, and well-defined data management system be employed to minimize these potential errors.

In summary, there are still problems encountered in defining, capturing and evaluating safety-related data including adverse events and laboratory data in clinical trials. These problems may be best resolved by the ICH in a standardized approach to defining, evaluating, recording, and summarizing safety-related data for a complete safety assessment of the drug product under investigation. It should be noted that both FDA and ICH Guidelines pointed out that it is not intended that every adverse event be subjected to rigorous statistical evaluation. As a result, the analysis of adverse events is basically descriptive in nature.

Bibliography

Agresti, A. (1983). Testing marginal homogeneity for ordinal categorical variables. *Biomet.*, **39,** 505–510.

Agresti, A. (1984). *Analysis of Ordered Categorical Data.* Wiley, New York.

Agresti, A. (1989). A survey of models for repeated ordered categorical response data. *Stat. Med.*, **8,** 1209–1224.

Agresti, A. (1990). *Categorical Data Analysis.* Wiley, New York.

Agresti, A., and Wackerly, D. (1977). Some exact conditional tests of independence for R × C cross-classification tables. *Psychometrika*, **42,** 111–125.

Albanse, M. A., Clarke, W. R., Adams, H. P., Woolson, R. F., and TOAST Investigators (1994). Ensuring reliability of outcome measures in multicenter clinical trials of treatments for acute ischemic stroke. *Stroke*, **25,** 1746–1751.

Amberson, J. B., McMahon, B. T., and Pinner, M. A. (1931). A clinical trial of sanocrysin in pulmonary tuberculosis. *Amer. Rev. Tuber.*, **24,** 401–435.

Andersen, P. K., Borgan, O., Gill, R. D., and Keiding, N. (1993). *Statistical Models Based on Counting Process.* Springer, New York.

Andersen, P. K., and Gill, R. D. (1982). Cox's regression model for counting processes: A large sample study. *Ann. Stat.*, **10,** 1100–1120.

APA (1987). *Diagnostic and Statistical Manual of Mental Disorders*, 3d ed. American Psychiatric Association, Washington, DC.

Armitage, P. (1975). *Sequential Medical Trials*, 2d ed. Blackwell Scientific, Oxford.

Armitage, P., and Berry, G. (1987). *Statistical Methods in Medical Research.* Blackwell Scientific Publications, Oxford.

Armitage, P., McPherson, C. K., and Rowe, B. C. (1969). Repeated significance tests on accumulating data. *J. Roy. Stat. Soc.*, A **132,** 235–244.

Attal, M., and the Intergroupe Francis Du Myelome (1996). A prospective, randomized trial of autologous bone marrow transplantation and chemotherapy in multiple myeloma. *New Eng. J. Med.*, **335,** 91–97.

Bailar, J. C. (1992). Some uses of statistical thinking. In *Medical Uses of Statistics*, ed. by Bailar, J. C., and Mosteller, F. New England Journal of Medicine Books, Boston, 5–26.

Bailar, J. C., and Mosteller, F. (1986). *Medical Uses of Statistics.* Massachusetts Medical Society, Waltham, MA.

Bailey, C. (1992). Biguanides and NIDDM. *Diabet. Care*, **15,** 755–772.

Balaam, L. N. (1968). A two-period design with t^2 experimental units. *Biometrics*, **24,** 61–73.

Barry, M. J., Fowler, F. J., Jr., O'Leary, M. P., Bruskewitz, R. C., Holtgrewe, H. L., Mebust, W. K., and Cockett, A. T. (1992). The American Urological Association Symptom Index for Benign Prostatic Hyperplasia. *J. Urol.*, **148,** 1549–1557.

Begg, C. B. (1990). On inference for Wei's based coin design on the randomized play-by-winner rule. *Biometrika*, **76,** 467–484.

Berger, R. L. (1982). Multiparameter hypothesis testing in acceptance sampling. *Technomet.* **24,** 295–300.

Berger, R. L., and Hsu, J. C. (1996). Bioequivalence trials, intersection-union tests, and equivalence confidence sets. *Stat. Sci.*, **11,** 283–319.

Berry, D. A. (1990). *Statistical Methodology in the Pharmaceutical Science.* Dekker, New York.

Beta-Blocker Heart Attack Trial Research Group (1982). A randomized trial of propranolol in patients with acute myocardial infarction I. Mortality results. Beta-blocker Heart Attack Trial—Final Report. *J. Am. Med. Assoc.*, **247,** 1707–1714.

Beta-Blocker Heart Attack Trial Research Group (1983). A randomized trial of propranolol in patients with acute myocardial infarction: Morbidity. *J. Am. Med. Assoc.*, **250,** 2814–2819.

Bhapkar, V. P. (1966). A note on the equivalence of two criteria for hypothesis in categorical data. *J. Am. Stat. Assoc.*, **61,** 256–264.

Blackwelder, W. C. (1982). "Proving the null hypothesis" in clinical trials. *Controlled Clin. Trials*, **3,** 345–353.

Blackwell, D., and Hodges, J. L., Jr., (1957) Design for the control of selection bias. *Ann. Math. Stat.*, **28,** 449–460.

Bloomfield, S. S. (1969). Conducting the clinical drug study. In *Proceedings of the Institute on Drug Literature Evaluation.* American Society of Hospital Pharmacists, Washington, DC, 147–154.

Blyth, C. R., and Still, H. A. (1983). Binomial confidence intervals. *J. Amer. Stat. Assoc.*, **78,** 108–116.

BMS (1994). Dose-response study of various dose levels of metformin hydrochloride compared to placebo in patients with non-insulin dependent diabetes mellitus. Bristol-Myers Squibb, CV138-001, Princeton, NJ.

Brown, L. D., Hwang, J. T. G., and Munk, A. (1998). An unbiased test for bioequivalence problem. In press.

Box, G. E. P., and Draper, N. R. (1987). *Empirical Model-Building and Response Surface.* Wiley, New York.

Box, G. E. P., Hunter, J. S., and Hunter, W. G. (1978). *Statistics for Experimenter: An Introduction to Design, Data Analysis, and Model Building.* Wiley, New York.

Boyarsky, S. J., and Paulson, D. F. (1977). A new look at bladder neck obstruction by the Food and Drug Administration regulators: Guideline for investigation of benign prostatic hypertrophy. *Trans. Am. Assoc. Genito. Surg.*, **68,** 29–32.

Breslow, N. E., and Day, N. E. (1980). *Statistical Methods in Cancer Research. Vol. 1: The Analysis of Case-Control Studies.* Oxford University Press, New York.

Breslow, N. E., and Day, N. E. (1987). *Statistical Methods in Cancer Research. Vol. 2: The Analysis of Cohort Studies.* Oxford University Press, New York.

Brown, B. W. (1980). The crossover experiment for clinical trials. *Biomet.*, **36,** 69–79.

Brownell, K. D., and Stunkard, A. J. (1982). The double-blind in danger: Untoward consequences of informed consent. *Am. J. Psychiat.*, **139,** 1487–1489.

Brody, H. (1981). *Placebos and the Philosophy of Medicine.* University of Chicago Press, Chicago.

Brody, H. (1982). The lie that heals: The ethics of giving placebos. *Ann. Int. Med.*, **97,** 112–118.

Brunelle, R., and Wilson, M. (1996). Analysis and interpretation of standard laboratory values. Presented at Biopharmaceutical Section Workshop on Adverse Events, Bethesda, MD, October 28, 1996.

Buncher, C. R., and Tsay, J. Y. (1994). *Statistics in Pharmaceutical Industry*, 2d ed. Dekker, New York.

Buyse, M. E., Staquet, M. J., and Sylvester, R. J. (1984). *Cancer Clinical Trials: Methods and Practice.* Oxford Medical Publications, New York.

Byar, D. P., and Piantadosi, S. (1985). Factorial designs for randomized clinical trials. *Cancer Treat. Rep.*, **69,** 1055–1063.

Byar, D. P., Simon, R. M., Friedwald, W. T., Schlesselman, J. J., DeMets, D. L., Ellenberg, J. H., Gail, M. L. and Ware, J. H. (1976). Randomized clinical trials—Perspectives on some recent ideas. *New Engl. J. Med.*, **295,** 74–80.

Byington, R. P., Curb, J. D., and Mattson, M. E. (1985). Assessment of double-blindness at the conclusion of the beta-blocker heart attack trial. *J. Am. Med. Assoc.*, **263,** 1733–1783.

Calimlim, J. F., Wardell, W. M., Lasagna, L., and Gillies, A. J. (1977). Analgesic efficacy of an orally administered combination of pentazocine and aspirin, with observations on the use and statistical efficiency of "global" subjective efficacy ratings. *Clin. Pharmacol. Ther.*, **21,** 34–43.

Canner, P. L. (1981). Practical aspects of decision-making in clinical trials: The coronary drug project as a case study. *Controlled Clin. Trials*, **1,** 363–376.

Capizzi, T., and Zhang, J. (1996). Testing the hypothesis that matters for multiple primary endpoints. *Drug Info. J.*, **30,** 949–956.

Carey, R. A., Eby, R. Z., Beg, M. A., McNally, C. F., and Fox, M. J. (1984). Patient selection of blood pressure in clinical trials. *Clin. Ther.*, **7,** 121–126.

CAST (1989). Cardiac Arrhythmia Supression Trial. Preliminary report: Effect of encainide and flecainide on mortality in a randomized trial of arrhythmia supression after myocardial infarction. *New Eng. J. Med.*, **321,** 406–412.

Chakravorti, S. R., and Grizzle, J. E. (1975). Analysis of data from multiclinics experiments. *Biometrics*, **31,** 325–338.

Chan, J. C., Tomlinson, B., Critchley, J. A., Cockram, C. S., and Walden, R. (1993). Metabolic and hemodynamic effects of Metformin and Glibenclamide in normotensive NIDDM patients. *Diabet. Care*, **16,** 1035–1038.

Chang, M. N. (1989). Confidence intervals for a normal mean following group sequential test. *Biometrics*, **45,** 247–254.

Chang, M. N., and O'Brien, P. C. (1986). Confidence intervals following group sequential test. *Controlled Clin. Trials*, **7,** 18–26.

Chang, M. N., Therneau, T. M., Wieand, H. S., and Cha, S. S. (1987). Designs for group sequential phase II clinical trials. *Biometrics*, **43,** 865–874.

Chang, M. N., Wieand, H. S., and Chang, V. T. (1989). The bias of the sample proportion following a group sequential phase II trial. *Stat. Med.*, **8,** 563–570.

Chevret, S. (1993). The continual reassessment method in cancer phase I clinical trials: a simulation study. *Stat. Med.*, **12,** 1093–1108.

Chinchilli, V. M., and Bortey, E. B. (1991). Testing for consistency in a single multicenter trial. *J. Biopharm. Stat.*, **1,** 67–80.

Cholesterol and Recurrent Events Trial Investigators (1996). The effect of Pravastatin on coronary events after myocardial infarction in patients with average cholesterol levels. *New Eng. J. Med.*, **335,** 1001–1009.

Chow, S. C. (1996). Statistical considerations for replicated design. In *Bioavailability, Bioequivalence and Pharmacokinetics Studies, Proceedings of FIP Bio-International '96*, ed. by Midha, K. K. and Nagai, T. Business Center for Academic Societies, Tokyo, Japan, 107–112.

Chow, S. C. (1997). Good statistics practice in drug development and regulatory approval process. *Drug Info. J.*, **31,** 1157–1166.

Chow, S. C., and Liu, J. P. (1992a). *Design and Analysis of Bioavailability and Bioequivalence Studies*. Dekker, New York.

Chow, S. C., and Liu, J. P. (1992b). On assessment of bioequivalence under a higher-order crossover design. *J. Biopharm. Stat.*, **2,** 239–256.

Chow, S. C., and Liu, J. P. (1995a). *Statistical Design and Analysis in Pharmaceutical Science*. Dekker, New York.

Chow, S. C., and Liu, J. P. (1995b). Current issues in bioequivalence trials. *Drug Info. J.*, **29,** 795–804.

Chow, S. C., and Liu, J. P. (1997). Meta-analysis for bioequivalence review. *J. Biopharm. Stat.*, **7,** 97–111.

Chuang, C. (1987) The analysis of a titration study. *Stat. Med.*, **6,** 583–590.

Chuang-Stein, C. (1993). Personal communications. Upjohn Company, Kalamazoo, MI.

Chuang-Stein, C., and Shih, W. J. (1991). A note of the analysis of titration studies. *Stat. Med.*, **10,** 323–328.

Chuang-Stein, C. (1996). Summarizing laboratory data with different reference ranges in multi-center clinical trials. *Drug Info. J.*, **26,** 77–84.

Clopper, C. J. and Pearson, E. S. (1934). The use of confidence or fiducial limits illustrated in the case of the binomial. *Biometrika*, **26,** 404–413.

Cochran, W. G. (1954). The combination of estimates from different experiments. *Biometrics*, **10**, 101–129.

Cochran, W. G. (1977). *Sampling Techniques.* Wiley, New York.

Cochran, W. G., and Cox, G. M. (1957). *Experimental Designs*, 2d ed. Wiley, New York, p. 18.

Cohen, J. (1977). *Statistical Power Analysis for the Behavioral Sciences.* Academic Press, New York.

Colton, T. (1974). *Statistics in Medicine.* Little Brown, Boston.

Congress (1993). Section 492B, National Institutes of Health Reauthorization Bill, the 103rd U.S. Congress.

Coniff, R. F., Shapiro, J. A., Seaton, T. B., and Bray, G. A. (1995). Multicenter placebo-controlled trial comparing acarbose (Bay g 5421) with placebo, tolbutamide, and tolbutamide-plus-acarbose in non-insulin-dependent diabetes mellitus. *Am. J. Med.*, **98**, 443.

Conover, W. J. (1980). *Practical Nonparametric Statistics.* Wiley, New York.

Cooppan, R. (1994). Pathophysiology and current treatment of NIDDM. Presented at Diabetes Education Updates. Harvard Medical School, Boston, MA.

Cornell, R. G. (1990). Handling dropouts and related issues. In *Statistical Methodology in the Pharmaceutical Sciences*, ed. by Berry, D. A. Dekker, New York.

Cornell, R. G., Landenberger, B. D., and Bartlett, R. H. (1986). Randomized play-by-winner clinical trials, *Comm. Stat. Theory Meth.*, **15**, 159–178.

Coronary Drug Project Research Group (1973). The coronary drug project: Design, method, and baseline results. *Circul.*, **47** (suppl. I), I-1–I-50.

Coronary Drug Project Research Group (1980). Influence of adherence to treatment and response of cholesterol on mortality in the Coronary Project. *New Eng. J. Med.*, **303**, 1038–1041.

Cox, D. R. (1952). A note of the sequential estimation of means. *Proc. Camb. Phil. Soc.*, **48**, 447–450.

Cox, D. R. (1972). Regression models and life tables. *J. Royal Stat. Soc., Series B*, **34**, 187–220.

Cox, D. R. (1990). Discussion of paper by C. B. Begg, *Biometrika*, **77**, 483–484.

Cox, D. R., and Snell, E. J. (1989). *Analysis of Binary Data.* 2nd Ed. Chapman and Hall, New York.

CPMP Working Party on Efficacy on Medicinal Products (1990). Good clinical practice for trials on medicinal products in the European Community. *Pharmacol. Toxicol.*, **67**, 361–372.

CPMP Working Party on Efficacy of Medicinal Products (1994). Biostatistical methodology in clinical trials in applications for marketing authorisations for medicinal products, European Commission, Brussels, Belgium.

CPMP Working Party on Efficacy on Medicinal Products (1995). Biostatistical methodology in clinical trials in applications for marketing authorisations for medicinal products: Note for guidance. *Stat. Med.*, **14**, 1659–1682.

Cramer, J. A., Mattson, R. H., Prevey, M. L., Scheyer, R. D., and Ouellette, V. L. (1989).

How often is medication taken as prescribed. *J. Am. Med. Assoc.*, **261**, 3273–3277.

Crawford, E. D., Eisenberg, M. A., Mcleod, D. G., Spaulding, J. T., Benson, R., Dorr, A., Blumenstein, B. A., Davis, M. A., and Goodman, P. J. (1989). A controlled trial of Leuprolide with and without flutamide in prostatic carcinoma. *New Eng. J. Med.*, **321**, 419–424.

Crépeau, H., Koziol, J., Reid, N., and Yuh, Y. S. (1985). Analysis of incomplete multivariate data from repeated measurement experiments. *Biometrics*, **41**, 505–514.

Crowder, M. J., and Hand, D. J. (1990). *Analysis of Repeated Measures*, Chapman and Hall, London.

Dannenberg, O., Dette, H., and Munk, A. (1994). An extension of Welch's approximation t-solution to comparative bioequivalence trials. *Biometrika*, **81**, 91–101.

Davis, C. S., and Chung, Y. (1995). Randomization model methods for evaluating treatment efficacy in multicenter clinical trials. *Biometrics*, **51**, 1163–1174.

Davis, R., Ribner, H. S., Keung, E., Sonnenblick, E. H., and LeJemtel, T. H. (1979). Treatment of chronic congestive heart failure with catopril, an oral inhibitor of angiotension-converting enzyme. *New Eng. J. Med.*, **301**, 117–121.

Davis, K. L., Thal, L. J., Gamzu, E. R., Davis, C. S., Woolson, R. F., Gracon, S. I., Drachman, D. A., Schneider, L. S., Whitehouse, P. J., Hoover, T. M., Morris, J. C., Kawas, G. H., Knopman, D. S., Earl, N. L., Jumar, V., Doody, R. S., and the Tacrine Collaborative Study Group (1992). A double-blind, placebo-controlled multicenter study for Alzheimer's disease. *New Eng. J. Med.*, **327**, 1253–1259.

DeMets, D., and Lan, K. K. G. (1994). Interim analysis: The alpha spending function approach. *Stat. Med.*, **13**, 1341–1352.

Deyo, R. A., Walsh, N. E., Schoenfeld, L. S., and Ramamurthy, S. (1990). Can trials of physical treatments be blinded: The example of transcutaneous electrical nerve stimulation for chronic pain. *Am. J. Phys. Med. Rehab.*, **69**, 6–10.

Dietrich, F. H., and Kearns, T. J. (1986). *Basic Statistics: An Inferential Approach.* Dellen, San Francisco.

Diggle, P. J. (1988). An approach to the analysis of repeated measurements. *Biometrics*, **44**, 959–971.

Diggle, P. J. (1989). Test for random dropouts in repeated measurement data. *Biometrics*, **45**, 1255–1258.

Diggle, P. J., and Kenward, M. G. (1994). Informative dropout in longitudinal data analysis (with discussion). *Appl. Stat.*, **43**, 49–93.

Diggle, P. J., Liang, K. Y., and Zeger, S. L. (1994). *Analysis of Longitudinal Data.* Oxford Science, New York.

Dornan, T. L., Heller, S. R., Peck, G. M., and Tattersol, R. B. (1991). Double-blind evaluation of efficacy and tolerability of Metformin in NIDDM. *Diabet. Care*, **14**, 342–344.

Dubey, S. D. (1991). Some thoughts on the one-sided and two-sided tests. *J. Biopharm. Stat.*, **1**, 139–150.

Dunnett, C. W. (1955). A multiple comparison procedure for comparing several treatments with a control. *J. Am. Stat. Assoc.*, **50**, 1096–1121.

Dunnett, C. W., and Gent, M. (1977). Significance testing to establish equivalence between treatments with special reference to data in the form of 2 × 2 table. *Biometrics*, **33**, 593–602.

Dunnett, C. W., and Goldsmith, C. H. (1995). When and how to do multiple comparisons. In *Statistics in the Pharmaceutical Industry*, ed. by Buncher, C. R., and Tsay, J. Y. Dekker, New York.

Dunnett, C. W., and Tamhane, A. J. (1991). Step-down multiple tests for comparing treatments with a control in unbalanced one-way layouts. *Stat. Med.*, **10**, 939–947.

Dunnett, C. W., and Tamhane, A. J. (1992). Comparisons between a new drug and active drug and placebo controls in an efficacy clinical trial. *Stat. Med.*, **11**, 1057–1063.

Durrleman, S., and Simon (1990). Planning and Monitoring of equivalence studies. *Biometrics*, **46**, 329–336.

Ebbeling, C. B., and Clarkson, P. M. (1989). Exercise-induced muscle damage and adaptation. *Sport Med.*, **7**, 207–234.

Echt, D. S., Liebson, P. R., Mitchell, L. B., Peters, R. W., Obias-Manno, D., Barker, A. H., Arensberg, D., Baker, A., Friedman, L., Greene, H. L., Huther, M. L. Richardson, D. W., and the CAST Investigators (1991). Mortality and morbidity in patients receiving encainide and flecainide or placebo. *New Eng. J. Med.*, **324**, 781–788.

Efron, B. (1971). Forcing a sequential experiment to be balanced. *Biometrika*, **58**, 403–417.

Elkeles, R. S. (1991). The effects of oral hypoglycemic drugs on serum lipids and lipoproteins in NIDDM. *Diabete Metab.*, **17**, 197–200.

Ellenburg, J. H. (1990). Biostatistical collaboration in medical research. *Biometrics*, **46**, 1–32.

van Elteren, P. H. (1960). On the combinations of independent two-sample tests of Wilcoxon. *Bull. Int. Stat. Inst.*, **37**, 351–361.

Ensign, L. G., Gehan, E. A., Kamen, D. S. and Thall, P. F. (1994). An optimal three-stage design for phase II clinical trials. *Stat. Med.*, **13**, 1727–1736.

Espland, M. A., Bush, T. L., Mebane-Sims, I., Stefanick, M. L., Johnson, S., Sherwin, R., and Waclawiw, M. (1995). Rationale, design, and conduct of the PEPI trial. *Controlled Clin. Trials*, **16**, 3S–19S.

ESVEM (1989). The ESVEM trial: Electrophysiologic study versus electrocardiographic monitoring for selection of antiarrhythmic therapy of ventricular tachyarrhythmias. *Circul.*, **79**, 1354–1360.

ESVEM (1993). Determinants of predicted efficacy of antiarrhythmic drugs in the Electrophysiologic study versus electrocardiographic monitoring trials. *Circul.*, **87**, 323–329.

Fairweather, W. R. (1994). Statisticians, the FDA and a time of transition. Presented at Pharmaceutical Manufacturers Associated Education and Research Institute Training Course in Non-clinical Statistics, Georgetown University Conference Center, February 6–8, 1994, Washington, DC.

Farlow, M., Gracon, S. I., Hershey, L. A., Lewis, K. W., Sadowsky, C. H., and Dolan-Ureno, J. (1992). A controlled trials of tacrine in Alzheimer's disease. *J. Am. Med. Assoc.*, **268**, 2523–2529.

Farrington, C. P., and Manning, G. (1990). Test statistics and sample size formulae for comparing binomial trials with null hypothesis of no-zero risk difference of non-unity relative risk. *Stat. Med.*, **9,** 1447–1454.

FDA (1988). *Guideline for the Format and Content of the Clinical and Statistical Sections of New Drug Applications*, U.S. Food and Drug Administration, Rockville, MD.

Feigl, P., Blumestein, B., Thompson, I., Crowley, J., Wolf, M., Kramer, B. S., Coltman, C. A., Jr., Brawley, O. W., and Ford, L. G. (1995). Design of the Prostate Cancer Prevention Trial (PCPT). *Controlled Clin. Trial*, **16,** 150–163.

Feingold, M., and Gillespie, B. W. (1996). Cross-over trials with censored data. *Stat. Med.*, **15,** 953–967.

Feinstein, A. R. (1977). *Clinical Biostatistics.* C.V. Mosby, St. Louis, MO.

Feinstein, A. R. (1989). Models, methods, and goals. *J. Clin. Epidemiol.*, **42,** 301–308.

Feldt, L. S., and Mahmound, M. W. (1958). Power function charts for specifying numbers of observations in analyses of variance of fixed effects. *Ann. Math. Stat.*, **29,** 871–877.

Feuer, E. J., and Kessler, L. G. (1989). Test statistic and sample size for a two-sample McNemar test. *Biometrics*, **45,** 629–636.

Fisher, L. D. (1990). *Biostatistics: Methodology for the Health Sciences.* Wiley, New York.

Fisher, L. D. (1991). The use of one-sided tests in drug trials: An FDA advisory committee member's perspective. *J. Biopharm. Stat.*, **1,** 151–156.

Fisher, R. A. (1925). *Statistical Methods for Research Workers.* Oliver and Boyd, Edinburgh.

Fisher, R. A. (1947). *The Design of Experiments*, 4th ed. Oliver and Boyd, Edinburgh.

Fisher, R. A., and Mackenzie, W. A. (1923). Studies in crop variation II: The manurial response of different potato varieties. *J. Agri. Sci.*, **13,** 311–320.

Fitzmaurice, G. M., Molenberghs, G., and Lipsitz, S. R. (1995). Regression models for longitudinal binary responses with informative drop-outs. *J. Roy. Stat. Soc.*, **57,** 691–704.

Fleiss, J. L. (1986a). *The Design and Analysis of Clinical Experiments.* Wiley, New York, NY.

Fleiss, J. L. (1986b). Analysis of data from multiclinic trials. *Controlled Clin. Trials*, **7,** 267–275.

Fleiss, J. (1987). Some thoughts on two-tailed tests (Letter to the Editor). *Controlled Clin. Trials*, **8,** 394.

Fleming, T. R. (1982). One sample multiple testing procedure for phase II clinical trials. *Biometrics*, **38,** 143–151.

Fleming, T. R., O'Fallon, J. P., O'Brien, P. C., and Harrington, D. P. (1980). Modified Kolmogorov–Smirnov test procedures with application to arbitrary right-censored data. *Biometrics*, **36,** 607–626.

Folstein, M. F., Folstein, S. E., and McHugh, P. R. (1975). "Mini-mental state": A practical method for grading the cognitive state of patients for the clinician. *J. Psychiat. Rev.*, **12**, 189–198.

Foulds, G. A. (1958). Clinical research in psychiatry. *J. Mental Sci.*, **104**, 259–265.

France, L. A., Lewis, J. A., and Kay, R. (1991). The analysis of failure time data in crossover studies. *Stat. Med.*, **10**, 1099–1113.

Freedman, L. S. (1982). Tables of the number of patients required in clinical trials using logrank test. *Stat. Med.*, **1**, 121–129.

Freedman, L. S., and White, S. J. (1976). On the use of Pocock and Simon's method for balancing treatment numbers over prognostic factors in the controlled clinical trials. *Biometrics*, **32**, 691–694.

Frick, M. H., Elo, O., Haapa, K., Heinonen, O. P., Heinsalmi, P., Helo, P., Huttunen, J. K., Kaitaniemi, P., Koskinen, P., Manninen, V., Maenpaa, H., Malkonen, M., Manttari, M., Norola, S., Pasternack, A., Pikkarainen, J., Romo, M., Sjoblom, T., and Nikkila, E. A. (1987). Helsikini heart study: Primary-prevention trial with gemfibrozil in middle-aged men with dyslipidemia. *New Eng. J. Med.*, **317**, 1237–1245.

Friedman, L. M., Furberg, C. D., and DeMets, D. L. (1981). *Fundamentals of Clinical Trials.* Wiley, New York.

Gail, M. H. (1985). Applicability of sample size calculations based on a comparison of proportions for use with the logrank test. *Controlled Clin. Trials*, **6**, 112-119.

Gail, M. H., and Simon, R. (1985). Testing for qualitative interactions between treatment effects and patient subsets. *Biometrics*, **41**, 361–372.

Gail, M. H., Wieand, S., and Paintadosi, S. (1984). Biased estimates of treatment effect in randomized experiments with nonlinear regressions and omitting covariates. *Biometrika*, **71**, 431–444.

Gehan, E. A. (1965a). A generalized Wilcoxon test for comparing arbitrarily singly-censored samples. *Biometrika*, **52**, 203–223.

Gehan, E. A. (1965b). A generalized two-sample Wilcoxon test for doubly censored data. *Biometrika*, **52**, 1965.

Geller, N. L. (1994). Discussion of interim analysis: The alpha spending approach. *Stat. Med.*, **13**, 1353–1356.

George, S. L., and Desu, M. M. (1973). Planning the size and duration of a clinical trial studying the time to some critical event. *J. Chron. Dis.*, **27**, 15–24.

Gilbert, G. S. (1992). *Drug Safety Assessment in Clinical Trials.* Dekker, New York.

Giugliano, D., Quatraro, A., Consoli, G., Minei, A., Ceriello, A., De Rosa, N., and D'Onofrio, F. (1993). Metformin for obese, insulin-treated diabetic patients: Improvement in glycemic control and reduction of metabolic risk factors. *Eur. J. Clin. Pharmacol.*, **44**, 107–112.

Glantz, S. A. (1987). *Primer of Biostatistics*, 2d ed. McGraw-Hill, New York.

Glanz, K., Fiel, S. B., Swartz, M. A., and Francis, M. E. (1984). Compliance with an experimental drug regimen for treatment of asthma: Its magnitude, importance, and correlates. *J. Chron. Dis.*, **37**, 815–824.

The Global Use of Strategies to Open Occluded Coronary Arteries (GUSTO) IIb Investigators (1996). A comparison of recombinant hirubin with heparin for the treatment of acute coronary syndromes. *New Engl. J. Med.*, **335**, 775–782.

Goldberg, J. D., and Koury, K. J. (1990). Design and Analysis of Multicenter Trials. In *Statistical Methodology in the Pharmaceutical Industry*, ed. by Berry, D. Dekker, New York, 201–237.

Gormley, G. J., Stoner, E., Bruskewitz, R. C., Imperato-McGinley, J., Walsh, P. C., McConnell, J. D., Andriole, G. L., Geller, J., Bracken, B. R., Tenover, J. S., Vaughan, E. D., Pappas, F., Taylor, A., Binkowitz, B., Ng, J., for the Finasteride Study Group (1992). The effect of finasteride in men with benign prostatic hyperplasia. *New Eng. J. Med.*, **327**, 1185–1191.

Gould, A. L. (1992). Interim analyses for monitoring clinical trials that do not maternally affect the type I error rate. *Stat. Med.*, **11**, 55–66.

Gould, A. L. (1995a). Planning and revising the sample size for a trial. *Stat. Med.*, **14**, 1039–1051.

Gould, L. A. (1995b). Group sequential extensions of standard bioequivalence testing procedure. *J. Pharmacokinetics Biopharmaceutics*, **23**, 57–85.

Gould, A. L., and Shih, W. J. (1992). Sample size re-estimation without unblinding for normally distributed outcomes with unknown variance. *Comm. Stat. Theory Meth.*, **21**, 2833–2853.

Greenberg, R. P. and Fisher, S. (1994). Seeing through the double-masked design: a commentary. *Controlled Clinical Trials*, **15**, 244–246.

Greene, J. G., and Hart, D. M. (1987). Evaluation of a psychological treatment programme for climacteric women. *Maturitas*, **9**, 41–48.

Grizzle, J. E., and Allen, D. M. (1969). Analysis of growth and dose response curves. *Biometrics*, **25**, 357–381.

GUSTO (1993). An investigational randomized trial comparing four thrombolytic strategies for acute myocardial function. The GUSTO investigators. *New Eng. J. Med.*, **329**, 673–682.

Guyatt, G., Sackett, M. D., Taylor, D. W., Chong, M. D., Roberts, M. S., and Pugsley, M. D. (1986). Determining optimal therapy-randomized trials in individual patients. *New Eng. J. Med.*, **314**, 889–892.

Haaland, P. D. (1991). *Experimental Design in Biotechnology*. Dekker, New York.

Hand, D. J., and Crowder, M. J. (1996). *Practical Longitudinal Data Analysis*. Chapman and Hall, London.

Harris, E. K., and Albert, A. (1991). *Survivorship Analysis for Clinical Studies*. Dekker, New York.

Harville, D. A. (1977). Maximum likelihood approaches to variance components estimation and to related problems. *J. Am. Stat. Assoc.*, **72**, 320–338.

Haseman, J. K. (1978). Exact sample sizes for use with the Fisher-Irwin test for 2×2 tables. *Biometrics*, **34**, 106–109.

Haybittle, J. L. (1971). Repeated assessment of results in clinical trials of cancer treatment. *Br. J. Radiol.*, **44**, 793–797.

Heart Special Project Committee (1988). Organization, review and administration of cooperative studies (Greenberg report): A report from the Heart Special Project Committee to the National Advisory Council, May 1967. *Controlled Clin. Trials*, **9**, 137–148.

Henderson, J. D. (1993). Bioequivalence and bioavailability. Invited presentation at Generic Drug Approvals Workshop. Regulatory Affairs Professionals Society, May 4, 1993, Reston, VA.

Hennekens, C. H., Buring, J. E., Manson, J. E., Stampfer, M., Rosner, B., Cook, N. R., Belanger, C., LaMotte F., Gaziano, J. M., Ridker, P. M., Willett, W., and Peto, R. (1996). Lack of effect of long-term supplementation with beta carotene on the incidence of malignant neoplasms and cardiovascular disease. *New Eng. J. Med.*, **334,** 1145–1149.

Hill, A. B. (1962). *Statistical Methods in Clinical and Preventive Medicine.* Oxford University Press, New York.

Hines, L. K., Laird, N. M., Hewitt, P., and Chalmers, T. C. (1989). Meta-analysis of empirical long-term antiarrhythmic therapy after myocardial infarction. *J. Am. Med. Assoc.*, **262,** 3037–3040.

Hochberg, Y. (1988). A sharper Bonferroni's procedure for multiple tests of significance. *Biometrika*, **75,** 800–803.

Hochberg, Y., and Tamhane, A. C. (1987). *Multiple Comparison Procedures.* Wiley, New York.

Hollander, M., and Wolfe, D. A. (1973). *Nonparametric Statistical Methods.* Wiley, New York.

Holm, S. (1979). A sharper Bonferroni procedure for multiple tests of significance. *Scand. J. Stat.*, **6,** 65–70.

Holt, J. D., and Prentice, R. L. (1974). Survival analysis in twin studies and matched-pair experiments. *Biometrika*, **61,** 17–30.

Hoover, D. R. (1996). Extension of life table to repeating and changing events. *Am. J. Epidemiol.*, 1266–1276.

Hosmer, D. W., and Lemeshow, S. (1989). *Applied Logistic Regression.* Wiley, New York.

Hsu, J. C., Hwang, J. T. G., Liu, H. K., and Ruberg, S. J. (1994). Confidence intervals associated with tests for bioequivalence. *Biometrika*, **81,** 103–114.

Hsu, J. P. (1983). The assessment of statistical evidence from active control clinical trials. *Proc. of Biopharmaceutical Section of the American Statistical Association*, 12–17.

Hughes, M. D. (1993). Stopping guidelines for clinical trials with multiple treatments. *Stat. Med.*, **12,** 901–913.

Hughes, M. D., and Pocock, S. J. (1988). Stopping rules and estimation problems in clinical trials. *Stat. Med.*, **7,** 1231–1242.

Hui, S. L. (1984). Curve fitting for repeated measurements made at urregular time points. *Biometrics*, **40,** 691–697.

Huitson, A., Poloniecki, J., Hews, R., and Barker, N. (1982). A review of crossover trials. *Statistician*, **31,** 71–80.

Hung, J. H. M. (1992). On identifying a positive dose-response surface for combination agents. *Stat. Med.*, **11,** 703–711.

Hung, J. H. M. (1994). Correspondence: Test for the existence of a desirable dose combination. *Biometrics*, **50,** 307–308.

Hung, J. H. M. (1996). Global tests for combination drug studies in factorial trials. *Stat. Med.*, **15,** 233–247.

Hung, J. H. M., Chi, G. Y. H., and Lipicky, R. J. (1993). Testing for the existence of a desirable dose combination. *Biometrics*, **49,** 85–94.

Hung, J. H. M., Ng, T. H., Chi, G. Y. H., and Lipicky, R. J. (1990). Response surface and factorial designs for combination antihypertensive drugs. *Drug Info. J.*, **24,** 371–378.

Hung, J. H. M., Ng, T. H., Chi, G. Y. H., and Lipicky, R. J. (1989). Testing for the existence of a dose combination beating its components. *Proc. of Biopharmaceutical Section of the American Statistical Association*, Alexandria, VA, 53–59.

Hung, J. H. M., Chi, G. Y. H., and O'Neill, R. T. (1995). Efficacy evaluation for monotherapies in two-by-two factorial trials. *Biometrics*, **51,** 1483–1493.

Huque, M. F., Dubey, S., and Fredd, S. (1989). Establishing therapeutic equivalence with clinical endpoints. *Proc. of Biopharmaceutical Section of the American Statistical Association*, Alexandria, VA, 46–52.

Huque, M. F., and Dubey, S. (1990). Design and analysis for therapeutic equivalence clinical trials with binary clinical endpoints. *Proc. of Biopharmaceutical Section of the American Statistical Association*, Alexandria, VA, 91–97.

Hurwitz, N. (1969). Admission to hospital due to drugs. *Br. Med. J.*, **1,** 536–539.

Huster, W. J., and Louv, W. C. (1992). Demonstration of the reproducibility of treatment efficacy from a single multicenter trial. *J. Biopharmaceut. Stat.*, **2,** 219–238.

Huster, W. J., Brookmeyer, R., and Self, S. G. (1989). Model paired survival data with covariates. *Biometrics*, **45,** 145–156.

ICH (1994). International Conference on Harmonisation Guideline for Clinical Safety Data Management: Definitions and Standards for Expedited Reporting. A Step 4 Paper, ICH-2 EWG E2, February 22, 1994.

ICH (1995). Third International Conference on Harmonisation Structure and Content of Clinical Study Reports, Yokohama, Japan.

ICH (1996). International Conference on Harmonisation Tripartite Guideline for Good Clinical Practice. Recommended for Adoption at Step 4 of the ICH Process. ICH Steering Committee, May 1, 1996.

IDSA (1990). Infectious Diseases Society of America: Guidelines for the use of antimicrobial agents in neutropenic patients with unexplained fever. *J. Infect. Dis.*, **161,** 381–396.

IHS (1990). The design, analysis, and reporting of clinical trials on the empirical antibiotic management of the neutropenic patient: Report of a consensus panel for the Immunocompromised Host Society. *J. Infect. Dis.*, **161,** 397–401.

Iman, R. L., Quade, D., and Alexander, D. A. (1975). Exact probability levels for the Kruskal–Wallis test, *Selected Tables in Mathematical Statistics*, **3,** 329–384.

ISIS-2 Group (1988). Randomized trial of intravenous streptokinase, oral aspirin, both, or neither among 17,187 cases of suspected acute myocardial infarction. *Lancet*, **13,** 349–360.

Jennison, C., and Turnbull, B. (1989). Interim analysis: the repeated confidence interval approach (with discussion). *J. Roy. Stat. Soc.*, **B51,** 305–361.

Jennison, C., and Turnbull, B. W. (1990). Statistical approaches to interim monitoring of medical trials: A review and commentary. *Stat. Sci.*, **5**, 299–317.

Jennison, C., and Turnbull, B. (1991). Exact calculation for sequential, *t*, chi-square, and *F* tests. *Biometrika*, **78**, 133–141.

Jennison, C., and Turnbull, B. (1993). Sequential equivalence testing and repeated confidence intervals, with application to normal and binary responses. *Biometrics*, 31–44.

John, P. W. M. (1971). *Statistical Design and Analysis of Experiments*. Macmillan, New York.

Jones, B., and Kenward, M. G. (1989). *Design and Analysis of Crossover Trials*. Chapman-Hall, London.

Jung, S. H., and Su, J. Q. (1995). Non-parametric estimation for the difference of ratio of median failure times for paired observations. *Stat. Med.*, **14**, 275–281.

Kalbfleisch, J. D., and Prentice, R. L. (1980). *The Statistical Analysis of Failure Time Data*. Wiley, New York.

Kalbfleisch, J. D., and Street, J. O. (1990). Survival analysis. In *Statistical Methodology in Pharmaceutical Science*, ed. by Berry, D. A. Dekker, New York.

Kaplan, E. L. and Meier, P. (1958). Nonparametric estimation from incomplete observations. *J. Amer. Stat. Assoc.*, **53**, 457–481.

Karlowski, T. R., Chalmers, T. C., Frenkel, L. D., Kapikian, A. Z., Lewis, T. L., and Lynch, J. M. (1975). Ascorbic acid for the common cold: A prophylactic and therapeutic trial. *J. Am. Med. Assoc.*, **231**, 1038–1042.

Kastenbaum, M. A., Hoel, D. G., and Bowman, K. O. (1970). Sample size requirements: one way analysis of variance. *Biometrika*, **57**, 421–430.

Kenward, M. G. (1987). A method for comparing profiles of repeated measurements. *Appl. Stat.*, **36**, 296–308.

Kershner, R. P., and Federer, W. T. (1981). Two-treatment crossover design for estimating a variety of effects. *J. Am. Stat. Assoc.*, **76**, 612–618.

Kessler, D. A. (1989). The regulation of investigational drugs. *New Eng. J. Med.*, **320**, 281–288.

Kessler, D. A., and Feiden, K. L. (1995). Faster evaluation of vital drugs. *Sci. Am.*, **272**, 48–54.

Keuls, M. (1952). The use of the studenized range in connection with an analysis of variance. *Euphytica*, **1**, 112–122.

Khatri, C. G., and Patel, H. I. (1992). Analysis of a multicenter trial using a multivariate approach to a mixed linear model. *Comm. Stat. Theory Meth.*, **21**, 21–39.

Kim, K. (1989). Point estimation following group sequential tests. *Biometrics*, **45**, 613–617.

Kim, K., and DeMets, D. L. (1987). Confidence intervals following group sequential tests in clinical trials. *Biometrics*, **43**, 857–864.

Kleinbaum, D. G. (1996). *Survival Analysis, A Self-learning Text*. Springer-Verlag, New York.

Knapp, M. J., Knopman, D. S., Solomon, P. R., Pendlebury, W. W., Davis, C. S., and

Gracon, S. I. (1994). A 30-week randomized controlled trial of high-dose tacrine in patients with Alzheimer's disease. *J. Am. Med. Assoc.*, **271,** 985–991.

Koch, G. C. (1991). One-sided and two-sided tests and *p*-values. *J. Biopharm. Stat.*, **1,** 161–170.

Koch, G. C., and Bhapkar, V. P. (1982). Chi-square test. *Encyclopedia of Statistical Sciences*, Vol. 1, ed. by Johnson, N. L. and Kotz, S. Wiley, New York.

Koch, G. G., and Edwards, S. (1988). Clinical efficacy trials with categorical data. In *Biopharmaceutical Statistics for Drug Development*, ed. by Peace, K. Dekker, New York.

Koch, G. G., Carr, G. J., Amara, I. A., Stokes, M. E., and Uryniak, T. J. (1990). Categorical data analysis. In *Statistical Methodology in the Pharmaceutical Sciences*, ed. by Berry, D. A. Dekker, New York.

Koch, G. G., Amara, I. A., Davis, G. W., and Gillings, D. B. (1982). A review of some statistical methods for covariance analysis of categorical data. *Biometrics*, **38,** 563–595.

Korn, E. L., Midthune, D., Chen, T. T., Rubinstein, L. V., Christian, M. C., and Simon, R. M. (1994). A comparison of two phase I designs. *Stat. Med.*, **13,** 1799–1806.

Lachin, J. M. (1981). Introduction to sample size determination and power analysis for clinical trials. *Controlled Clin. Trials*, **2,** 93–113.

Lachin, J. M. (1988a). Statistical properties of randomization in clinical trials. *Controlled Clin. Trials*, **9,** 289–311.

Lachin, J. M. (1988b). Properties of simple randomization in clinical trials. *Controlled Clin. Trials*, **9,** 312–326.

Lachin, J. M., and Foulkes, M. A. (1986). Evaluation of sample size and power for analyses of survival with allowance for nonuniform patient entry, losses to follow-up, noncompliance, and stratification. *Biometrics*, **42,** 507–519.

Lachin, J. M., Matts, J. P., and Wei, L. J. (1988). Randomization in clinical trials: Conclusions and recommendations. *Controlled Clin. Trials*, **9,** 365–374.

Laird, N. M. (1988). Missing data in longitudinal studies. *Stat. Med.*, **7,** 305–315.

Lamborn, K. R. (1983). Some practical issues and concerns in active control clinical trials. *Proc. of Biopharmaceutical Section of the American Statistical Association*, Alexandria, VA, 8–12.

Lakatos, E. (1986). Sample size determination in clinical trials with time-dependent rates of losses and noncompliance. *Controlled Clin. Trials*, **7,** 189–199.

Lakatos, E. (1988). Sample sizes based on the log-rank statistic in complex clinical trials. *Biometrics*, **44,** 229–241.

Lan, K. K. G. and DeMets, D. L. (1983). Discrete sequential boundaries for clinical trials. *Biometrika*, **70,** 659–663.

Lan, K. K. G., and DeMets, D. L. (1989). Group sequential procedures: Calendar versus information time. *Stat. Med.*, **8,** 1191–1198.

Lan, K. K. G., and DeMets, D. L. (1994). Interim analysis: the alpha spending function approach. *Stat. Med.*, **13,** 1341–1352.

Lan, K. K. G., Simon, R., and Halperin, M. (1982). Stochastically curtailed testing in long-term clinical trials. *Comm. Stat.*, **C1,** 207–219.

Lan, K. K. G., and Wittes, J. (1988). The *B*-value: A tool for monitoring data. *Biometrics*, **44,** 579–585.

Lan, K. K. G., and Zucker, D. M. (1993). Sequential monitoring of clinical trials: The role of information and Brownian motion. *Stat. Med.*, **12,** 753–765.

Lasagna, L. (1975). *Combination Drugs: Their Use and Regulations.* Stratton Intercontinental Medical Book, New York.

Lasagna, L., Laties, V. G., and Dohan, J. L. (1958). Further studies on the "Pharmacology of placebo administration." *J. Clin. Invest.*, **37,** 533–537.

Laska, E. M., and Meisner, M. (1985). A variational approach to optimal two-treatment crossover designs: Applications to carryover effects methods. *J. Am. Stat. Assoc.*, **80,** 704–710.

Laska, E. M., and Meisner, M. (1989). Testing whether an identified treatment is best. *Biometrics*, **45,** 1139–1151.

Laska, E. M., and Meisner, M. (1990). Hypothesis testing for combination treatments. In *Statistical Issues in Drug Research and Development*, ed. by Peace, K. L. Dekker, New York, 276–284.

Laska, E. M., Meisner, M., and Kushner, H. B. (1983). Optimal crossover designs in the presence of carryover effects. *Biometrics*, **39,** 1089–1091.

Laubscher, N. F. (1960). Normalizing the concentral t and F distributions. *Ann. Math. Stat.*, **31,** 1105–1112.

Lee, E. T. (1992). *Statistical Methods for Survival Data Analysis.* Wiley, New York.

Lee, E. T., Desu, M. M., and Gehan, E. A. (1975). A Monte Carlo study of the power of some tw—sample tests. *Biometrika*, **62,** 423–425.

Lee, K. L., Califf, R. M., Simers, J., Werf, F. V., and Topol, E. J. (1994). Holding GUSTO up to the light. *Ann. Internal Med.*, **120,** 876–881.

Lehmann, E. L. (1953). The power of rank tests. *Annals of Math. Stat.*, **24,** 23–43.

Lehmann, E. L. (1975). *Nonparametrics: Statistical Methods Based on Ranks.* Holden-Day, San Francisco, 132–334.

Lemeshow, S., Hosmer, D. W., and Stewart, J. P. (1981). A comparison of sample size determination methods in the two-group trials where the underlying diseases is rare. *Comm. Stat. Simu. Computa*, **B10,** 437–449.

Lepor, H., Williford, W. O., Barry, M. J., Brawer, M. K., Dixon, C. M., Gormley, G., Haakenson, C., Machi., M., Narayan, P., and Padley, R. J. (1996). The efficacy of terazosin, finasteride, or both in benign prostatic hyperplasia. *New Engl. J. Med.*, **335,** 533–539.

Levine, R. J. (1987). The apparent incompatibility between informed consent and placebo-controlled clinical trials. *Clin. Pharmacol. Ther.*, **42,** 247–249.

Levine, J. G. (1996). Analysis and presentation of clinical trial adverse events data. Presented at the Biopharmaceutical Section Workshop on Adverse Events, October 28–29, 1996, Bethesda, MD.

Levine, J. G., and Szarfman, A. (1996). Standardized data structures and visualization tool: a way to accelerate the regulatory review of the integrated summary of safety of new drug applications. *Biopharmaceut. Rep.*, **4,** 12–17.

Lewis, J. A. (1995). Statistical issues in the regulation of medicine. *Stat. Med.*, **14,** 127–136.

Lewis, J. A., Jones, D. R., and Röhmel, J. (1995). Biostatistical methodology in clinical trials—A European guideline. *Stat. Med.*, **14,** 1655–1657.

Liang, K. Y., and Zeger, S. L. (1986). Longitudinal data analysis using generalized linear models. *Biometrika*, **73,** 13–22.

Lindsey, J. K. (1993). *Models for Repeated Measurements.* Oxford Science, New York.

Lipid Research Clinics Program (1984). The lipid research clinics coronary primary prevention trials results. I: Reduction in incidence of coronary heart disease. *J. Am. Med. Assoc.*, **251,** 351–364.

Lipsitz, S. R., Fitzmaurice, G. M., Orav, E. J., and Laird, N. M. (1944). Performance of generalized estimating equations in practical situations. *Biometrics*, **50,** 270–279.

Little, R. J. A. (1995). Modeling the drop-out mechanism in repeated-measures studies. *J. Am. Stat. Assoc.*, **90,** 1112–1121.

Little, R. J. A., and Rubin, D. B. (1987). *Statistical Analysis with Missing Data.* Wiley, New York.

Liu, G., and Liang, K. Y. (1995). Sample size calculations for studies with correlated observations. Unpublished manuscript.

Liu, J. P. (1994). Invited discussion of "Individual bioequivalence: A problem for switchability" by S. Anderson. *Biopharmaceut. Rep.*, **2,** 7–9.

Liu, J. P. (1995a). Letter to the editor on "Sample size for therapeutic equivalence based on confidence interval" by S. C. Lin, *Drug Info. J.*, **29,** 45–50.

Liu, J. P. (1995b). Use of the replicated crossover designs in assessing bioequivalence. *Stat. Med.*, **14,** 1067–1078.

Liu, J. P. and Chow, S. C. (1992). Sample size determination for the two one-sided tests procedure in bioequivalence. *J. Pharmacokinetics & Biopharmaceutics*, **20,** 101–104.

Liu, J. P., and Chow, S. C. (1993). Assessment of bioequivalence for drugs with negligible plasma levels. *Biomet. J.*, **36,** 109–123.

Liu, J. P., and Chow, S. C. (1995). Replicated crossover designs in bioavailability and bioequivalence studies. *Drug Info. J.*, **29,** 871–884.

Liu, J. P., and Weng, C. S. (1995). Bias of two one-sided tests procedures in assessment of bioequivalence. *Stat. Med.*, **14,** 853–862.

Lundh, L. G. (1987). Placebo, relief, and health: A cognitive-emotional model. *Scand. J. Psycho.*, **28,** 128–143.

Lyden, P., Brott, T., Tilley, B. Welch, K. M. A., Mascha, E. J., Levine, S., Haley, E. C., Grotta, J., Marler, J., and the NINDS TPA Stroke Study Group (1994). Improved reliability of the NIH stroke scale using video training. *Stroke*, **25,** 2220–2226.

Makuch, R. and Johnson, M. (1989). Issues in planning and interpretating active control equivalence studies. *J. Clin. Epidemiol.*, **42,** 503–511.

Makuch, R. W., and Johnson, M. (1990). Active control equivalence studies: Planning and interpretation. In *Statistical Issues in Drug Research and Development*, ed. by Peace, K. E. Dekker, New York, 238–246.

Makuch, R. W., and Simon, R. (1978). Sample size requirements for evaluating a conservative therapy. *Cancer Treat. Rep.*, **6**, 1037–1040.

Manninnen, V., Elo, O., and Frick, M. H. (1988). Lipid alterations and decline in the incidence of coronary heart disease in the Helsinki Heart Study. *J. Am. Med. Assoc.*, **260**, 641–651.

Mantel, N. (1963). Chi-square tests with one degree of freedom: extensions of the Mantel-Haenszel procedure. *J. Am. Stat. Assoc.*, **58**, 690–700.

Mantel, N. (1967). Ranking procedure for arbitrarily restricted observations. *Biometrics*, **23**, 65–78.

Mantel, N., and Haenzsel, W. (1959). Statistical aspects of the analysis of data from retrospective studies of disease. *J. Nat. Cancer Inst.*, **22**, 719–748.

Margolies, M. E. (1994). Regulations of combination products. *Appl. Clin. Trials*, **3**, 50–65.

Marubini, E., and Valsecchi, M. G. (1995). *Analyzing Survival Data from Clinical Trials and Observational Studies*. Wiley, New York.

Mason, J. W., for the ESVEM Investigators (1993a). A comparison of seven antiarrhythmic drugs in patients with ventricular tachyarrhythmias. *New Engl. J. Med.*, **329**, 452–458.

Mason, J. W., for the ESVEM Investigators (1993b). A comparison of electrophysiologic testing versus Holter monitoring to predict antiarrhythmic–drug efficacy for ventricular tachyarrhythmias. *New Eng. J. Med.*, **329**, 445–451.

Matts, J. P., and Lachin, J. M. (1988). Properties of permutated-block randomization in clinical trials. *Controlled Clin. Trials*, **9**, 327–344.

McCullagh, P., and Nelder, J. A. (1983). Quasi-likelihood functions. *Ann. Stat.*, **11**, 59–67.

McCullagh, P., and Nelder, J. A. (1989). *Generalized Linear Models*. 2nd Ed. Chapman and Hall, New York.

McGuire, W. P., Hoskins, W. J., Brady, M. F., Kucera, P. R., Patridge, E. E., Look, K. Y., Clarke-Pearson, D. L., and Davidson, M. (1996). Cyclophosphamide and cisplatin compared with paclitaxal and cisplatin in patients with stage III and stage IV ovarian cancer. *New Eng. J. Med.*, **334**, 1–6.

McHugh, R., and Matts, J. (1983). Post-stratification in the randomized clinical trial, *Biometrics*, **39**, 217–225.

Medical Research Council (1948). Streptomycin treatment of pulmonary: A Medical Research Council investigation. *Br. Med. J.*, **2**, 769–782.

Meier, P. (1989). The biggest public health experiment ever, the 1954 field trial of the Salk poliomyeitis vaccine. In *Statistics: A Guide to the Unknown*, ed. by Tanur, J. M., Mosteller, F., and Kruskal, W. H., 3d ed. Wadsworth, Belmont, CA, 3–14.

Meinert, C. L. (1986). *Clinical Trials: Design, Conduct, and Analysis*. Oxford University Press, New York.

Metzler, C. M. (1974). Bioavailability: A problem in equivalence. *Biometrics*, **30**, 309–317.

Mezey, K. C. (1980). *Fixed Drug Combination—Rationale and Limitations*. International Congress and Symposium Series, No. 22. Royal Society of Medicine, London; Academic Press, London.

Miao, L. L. (1977). Gastric freezing: An example of the evaluation of medical therapy for randomized trials. In *The Costs, Risks, and Benefits of Surgery*, ed. by Bunker, J. P., Barnes, B. A., Mosteller, F. Oxford University Press, New York.

Mielke, P., and McHugh, R. B. (1965). Two-way analysis of variance for the mixed model with disproportionate sub-class frequencies. *Biometrics*, **21**, 308–323.

Mike, V., and Stanley, K. E. (1982). *Statistics in Medical Research: Methods and Issues, with Applications in Cancer Research*. Wiley, New York.

Miller, R. G., Jr. (1981). *Survival Analysis*. Wiley, New York.

Miller, M. E., Davis, C. S., and Landis, J. R. (1993). The analysis of longitudinal polytomous data: GEE and connections with weighted least squares. *Biometrics*, **49**, 1033–1044.

Miller, R. G., Efron, B., and Brown, B. W. (1980). *Biostatistics Case Book*. Wiley, New York.

Møller, S. (1995). An extension of the continual reassessment methods using a preliminary up-and-down design in a dose finding study in cancer patients, in order to investigate a greater range of doses, *Stat. Med.*, **14**, 911–922.

Moore, T. J. (1989). The cholesterol myth. *Atlantic Monthly*, Sept. 37–70.

Morgan, P. P. (1985). Randomized clinical trials need to be more clinical. *J. Am. Med. Assoc.*, **253**, 1782–1783.

Morrision, D. F. (1976). *Multivariate Statistical Methods*, 2d ed. McGraw-Hill, New York.

Moses, L. E. (1992). Statistical concepts fundamental to investigations. In *Medical Uses of Statistics*, ed. by Bailar, J. C., and Mosteller, F. New England Journal of Medicine Books, Boston, 5–26.

Myers, R. H., and Montgomery, D. C. (1995). *Response Surface Methodology: Product and Process and Optimization with Designed Experiments*. Wiley, New York.

Nagi, D. K., and Yudkin, J. S. (1993). Effects of Metformin on insulin resistance, risk factors for cardiovascular disease and plasminogen activator inhibitor in NIDDM subjects: A study of 2 ethnic groups. *Diabet. Care*, **16**, 621–629.

National Institute of Neurological Disorders and Stroke rt-PA Stroke Group (1995). Tissue plasminogen activator for acute ischemic stroke. *New Eng. J. Med.*, **333**, 1581–1587.

National Institute of Health: NIH Almanac (1981). Publication No. 81-5, Division of Public Information, Bethesda, MD.

Nevius, S. E. (1988). Assessment of evidence from a single multicenter trial. *Proc. of Biopharmaceutical Section of the American Statistical Association*, Alexandria, VA, 43–45.

NIH Consensus Development Panel (1994). Helicobacter plyori in peptic ulcer disease. *J. Am. Med. Assoc.*, **272**, 65–69.

Nickas, J. (1995). Adverse event data collection and reporting: A discussion of two grey areas. *Drug Info. J.*, **29**, 1247–1251.

Northington, B. (1996). A review of issues in the collection and reporting of adverse events. *Biopharmaceut. Rep.*, **4,** 1–5.

O'Brien, P. C. (1984). Procedures for comparing samples with multiple endpoints. *Biometrics*, **40,** 1079–1087.

O'Brien, P. C., and Fleming, T. R. (1979). A multiple testing procedure for clinical trials. *Biometrics*, **35,** 549–556.

O'Brien, W. M. (1968). Indomethacin: A survey of clinical trials. *Clin. Pharmacol. Ther.*, **9,** 94–107.

O'Dell, J. R., Haire, C. E., Erikson, N., Drymalski, W., Palmer, W., Eckhoff, P. J., Garwood, V., Maloley, P., Klassen, L. W., Wees, S., Klein, H., Moore, G. F. (1996). Treatment of rheumatoid arthritis with methotrexate alone, sulfasalazine, and hydroxychloroquine, or a combination of all three treatments. *New Eng. J. Med.*, **334,** 1287–1291.

Olkin, I. (1995). Meta-analysis: Reconciling the results of independent studies. *Stat. Med.*, **14,** 457–472.

O'Neill, R. T. (1988). Assessment of safety. In *Biopharmaceutical Statistics for Drug Development*, ed. by Peace, K. Dekker, New York.

O'Neill, R. T. (1993). Some FDA perspectives on data monitoring in clinical trials in drug development. *Stat. Med.*, **12,** 601–608.

O'Quigley, J., and Chervet, S. (1991). Methods for dosing finding studies in cancer trials: a review and results of a Monte Carlo study. *Stat. Med.*, **10,** 1647–1664.

O'Quigley, J., Pepe, M., and Fisher, L. (1990). Continual reassessment method: A practical design for phase I clinical trials in cancer. *Biometrics*, **46,** 33–48.

Pagano, M., and Halvorsen, K. T. (1981). An algorithm for finding the exact significance levels of $r \times c$ contingency tables. *J. Am. Stat. Assoc.*, **76,** 931–934.

Pan, H. (1996). Research and development process. *Bio/Pharma Quarterly*, 2(2), 34–35.

Park, T., and Davis, C. S. (1993). A test of the missing mechanism for repeated categorical data. *Biometrics*, **49,** 631–638.

Patel, J. A., Reisner, B., Vizirnia, N., Owen, M., Chonmaitree, T., and Howie, V. (1995). Bacteriologic failure of amoxicillin-clavulanate in treatment of acute oititis media caused by nontypeable haemophilus influenzae. *J. Pediat.*, **126,** 799–806.

Patulin Clinical Trials Committee (of the Medical Research Council) (1944). Clinical trial of Patulin in the common cold. *Lancet*, **2,** 373–375.

Pawitan, Y. and Hallstrom, A. (1990). Statistical interim monitoring of the cardiac arriythmia suppression trial. *Stat. Med.*, **9,** 1081–1090.

PDR (1992). *Physicians Desk Reference.* p. 1089.

Peace, K. E., ed. (1987). *Biopharmaceutical Statistics for Drug Development*. Dekker, New York.

Peace, K. E., ed. (1990). *Statistical Issues in Drug Research and Development*. Dekker, New York.

Peace, K. E. (1990). Response surface methodology in the development of antianginal drugs. In *Statistical Issues in Drug Research and Development*, ed. by Peace, K. Dekker, New York, 285–301.

Peace, K. E. (1991). One-sided or two-sided p values: Which most appropriately address the question of drug efficacy. *J. Biopharm. Stat.*, **1**, 133–138.

Peace, K. E., ed. (1992). *Biopharmaceutical Sequential Statistical Applications.* Dekker, New York.

Pearson, E. S., and Hartley, H. O. (1951). Charts of the power function of the analysis of variance tests, derived from the non-central F distribution. *Biometrika*, **38**, 112–130.

Peterson, W. L. (1991). Drug therapy: *Helicobacter plyori* and peptic ulcer disease. *New Eng. J. Med.*, **324**, 1043–1048.

Peto, R., Pike, M. C., Armitage, P., Breslow, N. E., Cox, D. R., Howard, S. V., Mantel, N., McPherson, K., Peto, J., and Smith, P. G. (1976). Design and analysis of randomized clinical trials requiring prolonged observation of each patient. *Int. Br. J. Cancer*, **34**, 585–612.

Petricciani, J. C. (1981). An overview of FDA, IRBs and regulations. *IRB*, **3**, 1.

PHSRG (1989). Steering Committee of the Physician's Health Study Research Group. Final report of the aspirin component of the ongoing physicians' health study. *New Eng. J. Med.*, **321**, 129–135.

Piantadosi, S. (1997). *Clinical Trials: A Methodologic Perspective.* Wiley, New York.

Pierce, M., Crampton, S., Henry, D., Heifets, L., LaMarca, A., Montecalvo, M., Wormser, G. P., Jablonowski, H., Jemsek, J., Cynamon, M., Yangco, B. G., Notario, G., Craft, J. C. (1996). A randomized trial of clarithromycin as prophylaxis against disseminated Mycobacterium avium complex infection in patients with advanced acquired immunodeficiency syndrome. *New Eng. J. Med.*, **335**, 384–391.

Phillips, K. F. (1990). Power of the two one-sided tests procedure in bioequivalence. *J. Pharmacokinetics & Biopharmaceutics*, **18**, 137–144.

Pledger, G. W., and Hall, D. (1986). Active control trials: Do they address the efficacy issue? *Proc. of Biopharmaceutical Section of the American Statistical Association*, Alexandria, VA, 1–7.

Pledger, G., and Hall, D. (1990). Active control equivalence studies: Do they address the efficacy issue. In *Statistical Issues in Drug Research and Development*, ed. by Peace, K. E. Dekker, New York, 226–238.

PMA (1989). Issues in data monitoring and interim analysis in the pharmaceutical industry. The PMA Biostatistics and Medical Ad hoc Committee on Interim Analysis. Pharmaceutical Manufacturing Association.

PMA Biostatistics and Medical Ad hoc Committee on Interim Analysis (1993). Interim analysis in the pharmaceutical industry. *Controlled Clin. Trials*, **14**, 160–173.

Pocock, S. J. (1977). Group sequential methods in the design and analysis of clinical trials. *Biometrika*, **64**, 191–199.

Pocock, S. J. (1983). *Clinical Trials: A Practical Approach.* Wiley, New York.

Pocock, S. J. (1990). Discussion of paper by C. B. Begg. *Biometrika*, **77**, 480–481.

Pocock, S. J., and Hughes, M. D. (1989). Practical problems in interim analyses with particular regard to estimation. *Controlled Clin. Trials*, **10**, 209S–221S.

Pocock, S. J., Geller, N. L., and Tsiatis, A. A. (1987). The analysis of multiple endpoints in clinical trials. *Biometrics*, **43**, 487–498.

Pocock, S. J., O'Brien, P. C., and Fleming, T. R. (1987). A paired Prentice-Wilcoxon test for censored paired data. *Biometrics*, **43,** 169–180.

Pocock, S. J., and Simon, R. (1975). Sequential treatment assignment with balancing for prognostic factors in the controlled clinical trials. *Biometrics*, **31,** 103–115.

Powderly, W. G., Finkelstein, D. M., Feinberg, J., Frame, P., He, W., Van Der Horst, C., Koletar, S. L., Eyster, M. E., Carey, J., Waskin, H., Hooton, T. M., Hyslop, N., Spector, S., and Bozzette, S. A. (1995). A randomized trial comparing fluconazole with clotrimazole troches for the prevention of fungal infections in patients with advanced human immunodeficiency virus infection. *New Eng. J. Med.*, **332,** 700–705.

Prentice, R. L. (1978). Linear rank tests with right censored data. *Biometrika*, **65,** 167–179.

Proschan, M. A., Follmann, D. A., and Geller, N. L. (1994). Monitoring multiarmed trials. *Stat. Med.*, **13,** 1441–1452.

Randles, R. H., and Wofle, D. A. (1979). *Introduction to the Theory of Nonparametric Statistics.* Wiley, New York.

Rao, C. R. (1973). *Linear Statistical Inference and Its Applications.* Wiley, New York.

Rapaport, E. (1993). GUSTO: Assessment of the preliminary results. *J. Myocardial Ischemia*, **5,** 15–24.

Recommendations for the treatment of hyperlipidemia in adults (1984). A joint statement of the Nutrition Committee and the Council on arteriosclerosis of the American Heart Association. *Arteriosclerosis*, **4,** 445A–468A.

Rider, P. M., O'Donnell, C., Marder, V. J., Hennekens, C. H. (1993). Large-scale trials of thrombolytic therapy for acute myocardial infarction: GISSI-2, ISIS-3, and GUSTO-1 (Editorial). *Ann. Int. Med.*, **119,** 530–532.

Rider, P. M., O'Donnell, C., Marder, V. J., and Hennekens, C. H. (1994). A response to "Holding GUSTO up to the light." *Ann. Int. Med.*, **120,** 882–885.

Ridout, M. (1991). Testing for random dropouts in repeated measurement data. *Biometrics*, **47,** 1617–1721.

Rodda, B. E., Tsianco, M. C., Bolognese, J. A., and Kersten, M. K. (1988). Clinical development. In *Biopharmaceutical Statistics for Drug Development*, ed. by Peace, K. E. Dekker, New York.

Rosenberger, W. F. (1993). Asymptotic inference with response-adaptive treatment allocation designs. *Annals of Statistics*, **210,** 2098–2107.

Rosenberger, W. F., and Lachin, J. M. (1993). The use of responsive-adaptive designs in clinical trials. *Controlled Clin. Trials*, **14,** 471–484.

Rosner, B., and Muñoz, A. (1988). Autoregressive modelling for analysis of longitudinal data with unequally space examinations. *Stat. Med.*, **7,** 59–71.

Rothman, K. S. (1986). *Modern Epidemiology.* Little Brown, Boston.

Rowinsky, E. K., and Donehower, R. C. (1995). Drug therapy: Paclitaxal (taxol). *New Eng. J. Med.*, **332,** 1004–1014.

Ruberg, S. J. (1995a). Dose response studies: I. Some design considerations. *J. Biopharmaceut. Stat.*, **5,** 1–14.

Ruberg, S. J. (1995b). Dose response studies: II. Analysis and interpretation. *J. Biopharmaceut. Stat.*, **5**, 15–42.

Rubin, D. B. (1976). Inference and missing data. *Biometrika*, **63**, 581–592.

Rubins, H. R. (1994). From clinical trials to clinical practice: Generation from participant to patient. *Controlled Clin. Trials*, **17**, 7–10.

Rubinstein, L. V., Gail, M. H., and Santner, T. J. (1981). Planning the duration of a comparative clinical trial with loss to follow-up and a period of continued observation. *J. Chronic Dis.*, **34**, 469–479.

Ruskin, J. N. (1989). The Cardiac Arrhythmia Suppression Trial (CAST). *New Eng. J. Med.*, **321**, 386–388.

Sackett, D. L. (1979). Bias in analytical research. *J. Chronic Disease*, **32**, 51–63.

Sackett, D. L. (1989). Inference and decision at the bedside. *J. Clin. Epidemiol.*, **42**, 309–316.

Sackett, D. L., Haynes, R. B., and Tugnell, P. (1991). *Clinical Epidemiology, A Basic Science for Clinical Medicine.* 2nd Ed., Little, Brown and Company, Boston, MA.

Sahai, H., and Khurshid, A. (1996). Formulae and tables for the determination of sample sizes and power in clinical trials for testing difference in proportions for the two-sample design: a review. *Stat. Med.*, **15**, 1–21.

Salsburg, D. S. (1993). The use of hazard functions in safety analysis. In *Drug Safety Assessment in Clinical Trials.* ed. by Sogliero-Gibert, G. Dekker, New York.

Sanford, R. L. (1994). The wonders of placebo. In *Statistics in the Pharmaceutical Industry*, ed. by Buncher, C. R., and Tsay, J. Y. Dekker, New York.

Scherer, J. C., and Wiltse, C. G. (1996). Adverse events: after 58 years, do we have it right yet? *Biopharmaceut. Rep.*, **4**, 1–5.

Schlesselman, J. J. (1982). *Case-Control Studies: Design, Conduct, Analysis.* Oxford University Press, New York.

Schneiderman, M. A. (1967). Mouse to man: Statistical problems in bringing a drug to clinical trial. *Proc. of Fifth Berkeley Symposium on Mathematical Statistics and Probability*, vol. 4. University of California Press. Berkeley, CA, 855–866.

Schoenfeld, D. (1981). Table, life; test, logmark; test, Wilcoxon: The asymptotic properties of nonparametric tests for comparing survival distributions. *Biometrika*, **68**, 316–319.

Schuirmann, D. J. (1987). A comparison of the two one-sided tests procedure and the power approach for assessing the equivalence of average bioquivalence. *J. Pharmacokinetics and Biopharmaceutics*, **15**, 657–680.

Searle, S. R. (1971). *Linear Models.* Wiley, New York.

Seidl, L. G., Thornton, G. F., Smith, J. W., and Gluff, L. E. (1966). Studies on epidemiology of adverse drug reactions. *Bull. Hopkins Hosp.*, **119**, 299–315.

Self, S. and Mauritsen, R. (1988). Power/sample size calculations for generalized linear models. *Biometrics*, **44**, 79–86.

Self, S., Mauritsen, R., and Ohara, J. (1992). Power calculations for likelihood ratio tests in generalized linear models. *Biometrics*, **48**, 31–39.

Self, S., Prentice, R., Iverson, D., Henderson, M., Thompson, D., Byar, D., Insull, W., Gorbach, S. L., Clifford, C., Goldman, S., Urban, N., Sheppard, L., and Greenwald, P. (1988). Statistical design of the women's health trial. *Controlled Clin. Trials*, **9**, 119–136.

Segraves, R. T. (1988). Sexual side-effects of psychiatric drugs. *Int. J. Psychiat. Med.*, **18**, 243–251.

Segraves, R. T. (1992). Sexual dysfunction complicating the treatment of depression. *J. Clin. Psychiat. Monograph*, **10**, 75–79.

Senn, S. (1993). Inherent difficulties which active control equivalence studies. *Stat. Med.*, **12**, 2367–2375.

SERC (1993). EGRET SIZ: sample size and power for nonlinear regression models. Reference Manual, Version 1. Statistics and Epidemiology Research Corporation.

Shao, J., and Chow, S. C. (1993). Two-stage sampling with pharmaceutical applications. *Stat. Med.*, **12**, 1999–2008.

Shapiro, S. H., and Louis, T. A. (1983). *Clinical Trials, Issues and Approaches*. Dekker, New York.

Shih, J. H. (1995). Sample size calculation for complex clinical trials with survival endpoints. *Controlled Clin. Trials*, **16**, 395–407.

Shih, W. J., Gould, A. L., and Hwang, I. K. (1989). The analysis of titration studies in phase III clinical trials. *Stat. Med.*, **8**, 583–591.

Silliman, N. P. (1996). Analysis of subgroups in safety data. Presented at the Biopharmaceutical Section Workshop on Adverse Events, October 28–29, 1996, Bethesda, MD.

Simon, R. (1989). Optimal two-stage designs for phase II clinical trials. *Controlled Clin. Trials*, **10**, 1–10.

Simon, R. (1991). A decade of progress in statistical methodology for clinical trials. *Stat. Med.*, **10**, 1789–1817.

Simon, R., and Korn, E. L. (1991). Selection combinations of chemotherapeutic drugs to maximize dose intensity. *J. Biopharmaceut. Stat.*, **1**, 247–259.

Sinclair, J. C. (1966). Prevention and treatment of the respiratory distress syndromes. *Pediatric. Clin. North Am.*, **13**, 711–730.

Sleight, P. (1993). Thrombolysis after GUSTO: A European perspective. *J. Myocardial Ischemia*, **5**, 25–30.

Slud, E. V., and Wei, L. J. (1982). Two-sample repeated significance tests based on the modified Wilcoxon statistic. *J. Am. Stat. Assoc.*, **77**, 862–868.

Smith, A., Traganza, E., and Harrison, G. (1969). Studies on the effectiveness of antidepressant drugs. *Psychopharmacol. Bull.*, **5**(suppl. 1), 1–53.

Smoking and Health (1964). Report of Advisory Committee to the Surgeon General of the Public Health Services. Government Printing Office, Washington, DC 235–257 (Public Health Service Publication no. 1103).

Snedecor, G. W., and Cochran, W. G. (1980). *Statistical Methods*, 7th ed. Iowa State University, Ames.

Spilker, B. (1991). *Guide to Clinical Trials*. Raven Press, New York.

Spriet, A. and Dupin-Spriet, T. (1992). *Good Practice of Clinical Drug Trials.* Karger, S. Karger AG, Medical & Scientific Publication, Basel.

Stampfer, M. J., Willett, W. C., and Colditz, G. A. (1985). A prospective study of postmenopause estrogen therapy and coronary heart disease. *New Eng. J. Med.*, **313**, 1044–1049.

Stanley, B. (1988). An integration of ethical and clinical considerations in the use of placebos. *Psychopharmacol. Bull.*, **24**, 18–20.

Steward, R. B., and Cluff, L. E. (1972). A review of medication errors and compliance in ambulant patients. *Clin. Pharmacol. Ther.*, **13**, 463–468.

Stolley, P. D. and Strom, B. L. (1986). Sample size calculations for clinical pharmology studies. *Clin. Pharmcol. Ther.*, **27**, 489–490.

Storer, B. E. (1989). Design and analysis of phase I trials. *Biometrics*, **45**, 925–937.

Storer, B. E. (1993). Small-sample confidence sets for the MTD in a phase I clinical trial. *Biometrics*, **49**, 1117–1125.

Stuart, A. (1955). A test for homogeneity of the marginal distributions in a two-way classification. *Biometrika*, **42**, 412–416.

Suissa, S., and Shuster, J. J. (1991). The 2×2 matched-pairs trial: Exact unconditional design and analysis. *Biometrics*, **47**, 361–372.

Tamura, R. N., Faries, D. E., Andersen, J. S., and Heiligenstein (1994). A case study of an adaptive clinical trial in the treatment of out-patients with depression disorder. *J. Am. Stat. Assoc.*, **89**, 768–776.

Tang, D.-I., Geller, N. L., and Pocock. S. J. (1993). On the design and analysis of randomized clinical trials with multiple endpoints. *Biometrics*, **49**, 23–30.

Tarone, R. E., and Ware, J. (1977). On distribution-free tests for equality of survival distributions. *Biometrika*, **64**, 156–160.

Taves, D. R. (1974). Minimization: A new method of assigning patients to treatment and control groups. *Clin. Pharmacol. Ther.*, **15**, 443–453.

Temple, R. (1982). Government viewpoint of clinical trials. *Drug Info. J.*, **16**, 10–17.

Temple, R. (1983). Difficulties in evaluating positive control trials. *Proc. of Biopharmaceutical Section of the American Statistical Association*, Alexandria, VA, 1–7.

Temple, R. (1993). Trends in pharmaceutical development. *Drug Info. J.*, **27**, 355–366.

Tessman, D. K., Gipson, B., and Levins, M. (1994). Cooperative fast-track development: The fludara story. *Appl. Clin. Trials*, **3**, 55–62.

Testa, M. A., Anderson, R. B., Nackley, J. F., Hollenberg, N. K., and the Quality-of-Life Hypertension Study Group (1993). Quality of life and antihypertensive therapy in men—A comparison of Captopril with Enalapril. *New Eng. J. Med.*, **328**, 907–913.

The Expert Panel (1988). Report of the National Cholesterol Education Program Expert Panel on Detection, Evaluation and Treatment of High Blood Cholesterol in Adults. *Archiv. Int. Med.*, **148**, 36–69.

Therneau, T. M., Wieand, H. S., and Chang, M. N. (1990). Optimal designs for a grouped sequential binomial trial. *Biometrics*, **46**, 771–781.

Tremmel, L. (1996). Describing risk in long-term clinical trials. *Biopharmaceut. Rep.*, **4**, 5–8.

Tsai, K. T., and Patel, H. I. (1992). Exploratory data analysis of a multicenter trial. *Proc. of Biopharmaceutical Section of the American Statistical Association*, Alexandria, VA, 56–61.

Tsiatis, A. A., Rosner, G. L., and Mehta, C. R. (1984). Exact confidence interval following a group sequential test. *Biometrics*, **40**, 797–803.

Tugwell, P., Pincus, T., Yocum, D., Stein, M., Gluck, O. Kraag, G., McKendry, R., Tesser, J., Baker, P., and Wells, G., for the Methotrexate-Cyclosporine Combination Study Group (1995). *New Eng. J. Med.*, **333**, 137–141.

Tygstrup, N., Lachin, J. M., and Juhl, E. (1982). *The Randomized Clinical Trials and Therapeutic Decisions*. Dekker, New York.

The United States Food and Drug Administration (1988). *Guideline for the Format and Content of the Clinical and Statistical Sections of New Drug Applications*, FDA, US Department of Health and Human Services, Rockville, MD.

USP/NF (1995). The United States Pharmacopeia XXIII and the National Formulatry XVIII. The United States Pharmacopeial Convention, Rockville, MD.

Veteran's Administration Cooperative Urological Research Group (1967). Treatment and survival of patients with cancer of the prostate. *Surg. Gynecol. Obstet.*, **124**, 1011–1017.

Vonesh and Chinchilli, V. (1997). *Linear and Nonlinear Models for the Analysis of Repeated Measures*. Dekker, New York.

Walsh, J. H., and Peterson, W. L. (1995). Drug therapy: The treatment of *Helicobacter plyori* infection in the management of peptic ulcer disease. *New Eng. J. Med.*, **333**, 984–991.

Wang, S. G. and Chow, S. C. (1995). *Advanced Linear Models*. Marcel Dekker, Inc., New York, New York.

Wang, W., Hsuan, F., and Chow, S. C. (1996). Patient compliance and the fluctuation of the serum drug concentration. *Stat. Med.*, **15**, 659–669.

Ward, D. E., and Camm, A. J. (1993). Dangerous ventricular arrhythmia: Can we predict drug efficacy. *New Eng. J. Med.*, **329**, 498–499.

Ware, J. H. (1989). Investigating therapies of potentially great benefit: ECMO. *Stat. Sci.*, **4**, 298–340.

Ware, J., Lipsitz, S., and Speizer, F. E. (1988). Issue in the analysis of repeated categorical outcomes. *Stat. Med.*, **7**, 95–107.

Warren, J. R. (1982). Unidentified curved bacilli on gastric epithelium in active chronic gastritis. *Lancet*, 1273.

Wedderburn, R. W. M. (1974). Quasi-likelihood functions, generalized linear models, and the Gauss-Newton method. *Biometrika*, **61**, 439–447.

Wei, L. J. (1977). A class of designs for sequential clinical trials. *J. Am. Stat. Assoc.*, **72**, 382–386.

Wei. L. J. (1978). The adaptive biased-coin design for sequential experiments. *Ann. Stat.*, **9**, 92–100.

Wei, L. J. (1988). Exact two-sample permutation tests based on the randomized play-the-winner rule. *Biometrika*, **75**, 603–606.

Wei, L. J. (1990). Discussion of paper by C. B. Begg, *Biometrika*, **77**, 476–477.

Wei, L. J., and Durham, S. (1978). The randomized play-the-winner rule in medical trials. *J. Am. Stat. Assoc.*, **73**, 840–843.

Wei, L. J., and Lachin, J. M. (1988). Properties of the urn randomization in clinical trials. *Controlled Clin. Trials*, **9**, 345–364.

Wei, L. J., Smythe, R. T., Lin, D. Y., and Park, T. S. (1990). Statistical inference with data-dependent treatment allocation rules. *J. Am. Stat. Assoc.*, **73**, 840–843.

Weintraub, M., and Calimlim, J. F. (1994). Selecting patients for a clinical trial. In *Statistics in the Pharmaceutical Industry*, ed. by Buncher, C. R., and Tsay, J. Y. Dekker, New York.

Wellek, S. (1993). A log-rank test for equivalence of two survivor function. *Biometrics*, **49**, 877–881.

Weschler, H., Grosser, G. H., and Greenblatt, M. (1965). Research evaluating antidepressant medications on hospitalized mental patients: A survey of published reports during a five-year period. *J. Nervous Mental Dis.*, **141**, 21–239.

Westlake, W. J. (1972). Use of confidence intervals in analysis of comparative bioavailability trials. *J. Pharmaceut. Sci.*, **61**, 1340–1341.

The West of Scotland Coronary Prevention Study Group (1995). Prevention of coronary heart disease with pravastatin in men with hypercholesterolemia, *New Eng. J. Med.*, **333**, 1301–1307.

Westlake, W. J. (1976). Symmetrical confidence intervals for bioequivalence trials. *Biometrics*, **32**, 741–744.

White, S. J., and Freedman, L. S. (1978). Allocation of patients to treatment groups in a controlled clinical study. *British J. of Cancer*, **37**, 849–857.

Whitehead, J. (1991). *The Design and Analysis of Sequential Clinical Trials*, 2d ed. Ellis Horwood, Chichester, England.

Wilcoxon, F. (1945). Individual comparisons by ranking methods. *Biometrics*, **1**, 80–83.

Williams, D. A. (1971). A test for differences between treatment means when several dose levels are compared with a zero dose control. *Biometrics*, **27**, 103–118.

Williams, D. A. (1972). The comparison of several dose levels with a zero dose control. *Biometrics*, **28**, 519–531.

Williams, E. J. (1949). Experimental designs balanced for the residual effects of treatment. *Austral. J. Sci. Res.*, **2**, 149–168.

Williams, G. W., Davis, R. L., Geston, A. J., Gould, L., Hwang, I. K., Mathews, Shih, W. J., Snapinn, S. M., and Walton-Bowen, K. L. (1993). Monitoring of clinical trials and interim analyses for a drug sponsor's point of view. *Stat. Med.*, **12**, 481–492.

Wilson, P. W. F., Garrison, R. J., and Castelli, W. P. (1985). Post-menopause estrogen use, cigarette smoking and cardiovascular morbidity in women over 50: The Framingham study. *New Eng. J. Med.*, **313**, 1038–1043.

Wittes, J. (1994). Introduction: From clinical trials to clinical practice—four papers from a plenary session. *Controlled Clin. Trials*, **15**, 5–6.

Wittes, J. (1996). A statistical perspective on adverse event reporting in clinical trials. *Biopharmaceut. Rep.*, **4**, 5–10.

Wolf, S. (1950). Affects of suggestion and conditioning on the action of chemical agents in human subjects—The pharmacology of placebos. *J. Clin. Invest.*, **29,** 100–109.

Wooding, W. M. (1994). *Planning Pharmaceutical Clinical Trials: Basic Statistical Principles.* Wiley, New York.

Wu, M., Fisher, M., and DeMets, D. (1980). Sample sizes of long-term medical trials with time-dependent noncompliance and event rates. *Controlled Clin. Trials,* **1,** 109–121.

Yateman, N. A. and Skene, A. M. (1992). Sample sizes for proportional hazards survival studies with arbitrary patient entry and loss to follow-up distributions. *Stat. Med.,* **11,** 1103–1113.

Yates, F. (1934). The analysis of multiple classifications with unequal numbers in the different classes. *J. Am. Stat. Assoc.,* **29,** 51–66.

Yates, F., and Cochran, W. G. (1938). The analysis of groups of experiments. *J. Agri. Sci.,* **28,** 556–580.

Yusuf, S., Peto, R., Lewis, J., Collins, R., and Sleight, P. (1985). Beta blocker during and after myocardial infarction: an overview of the randomized trials. *Prog. Cardiovas. Dis.,* **27,** 335–371.

Zeger, S. L., and Liang, K. Y. (1986). Longitudinal data analysis for discrete and continuous outcomes. *Biometrics,* **44,** 1825–1829.

Zeger, S. L., Liang, K. Y., and Albert, P. S. (1988). Models for longitudinal data: A generalized estimating equation approach. *Biometrics,* **44,** 1049–1960.

Zeger, S. L., and Liang, K. Y. (1992). An overview of methods for the analysis of longitudinal data. *Stat. Med.,* **11,** 1825–1839.

Zeger, S. L., and Qagish, B. (1988). Markov regression models for time series: a quasi-likelihood approach. *Biometrics,* **44,** 1019–1031.

Zelen, M. (1969). Play the winner rule and the controlled clinical trial. *J. Am. Stat. Assoc.,* **64,** 131–146.

Zoloft (1992). Zoloft (sertraline hydrochloride) tablets, package insert, Pfizer Inc. Groton, CT.

APPENDIX A

Tables

Table A.1 Areas of Upper Tail of the Standard Normal Distribution
Table A.2 Upper Quantiles of a χ^2 Distribution
Table A.3 Upper Quantiles of a Central t Distribution
Table A.4 Upper Quantiles of an F Distribution $\alpha = 0.05$
Table A.5 Quantiles of the Distribution of Wilcoxon-Mann-Whitney Statistic
Table A.6 Quantiles of the Wilcoxon Signed Rank Test Statistic
Table A.7 Tolerance Factor for the Degrees of Confidence $1 - \alpha$
Table A.8 Upper Quantiles of the Studentized Range Distribution
Table A.9 Upper Quantiles of the Dennett's Distribution: One-Sided Comparisons with Control

Table A.1 Areas of Upper Tail of the Standard Normal Distribution

z	0.00	0.01	0.02	0.03	0.04	0.05	0.06	0.07	0.08	0.09
0.0	0.5000	0.4960	0.4920	0.4880	0.4840	0.4801	0.4761	0.4721	0.4681	0.4641
0.1	0.4602	0.4562	0.4522	0.4483	0.4443	0.4404	0.4364	0.4325	0.4286	0.4247
0.2	0.4207	0.4168	0.4129	0.4090	0.4052	0.4013	0.3974	0.3936	0.3897	0.3859
0.3	0.3821	0.3783	0.3745	0.3707	0.3669	0.3632	0.3594	0.3557	0.3520	0.3483
0.4	0.3446	0.3409	0.3372	0.3336	0.3300	0.3264	0.3228	0.3192	0.3156	0.3121
0.5	0.3085	0.3050	0.3015	0.2981	0.2946	0.2912	0.2877	0.2843	0.2810	0.2776
0.6	0.2743	0.2709	0.2676	0.2643	0.2611	0.2578	0.2546	0.2514	0.2483	0.2451
0.7	0.2420	0.2389	0.2358	0.2327	0.2296	0.2266	0.2236	0.2206	0.2177	0.2148
0.8	0.2119	0.2090	0.2061	0.2033	0.2005	0.1977	0.1949	0.1922	0.1894	0.1867
0.9	0.1841	0.1814	0.1788	0.1762	0.1736	0.1711	0.1685	0.1660	0.1635	0.1611
1.0	0.1587	0.1562	0.1539	0.1515	0.1492	0.1469	0.1446	0.1423	0.1401	0.1379
1.1	0.1357	0.1335	0.1314	0.1292	0.1271	0.1251	0.1230	0.1210	0.1190	0.1170
1.2	0.1151	0.1131	0.1112	0.1093	0.1075	0.1056	0.1038	0.1020	0.1003	0.0985
1.3	0.0968	0.0951	0.0934	0.0918	0.0901	0.0885	0.0869	0.0853	0.0838	0.0823
1.4	0.0808	0.0793	0.0778	0.0764	0.0749	0.0735	0.0721	0.0708	0.0694	0.0681
1.5	0.0668	0.0655	0.0643	0.0630	0.0618	0.0606	0.0594	0.0582	0.0571	0.0559
1.6	0.0548	0.0537	0.0526	0.0516	0.0505	0.0495	0.0485	0.0475	0.0465	0.0455
1.7	0.0446	0.0436	0.0427	0.0418	0.0409	0.0401	0.0392	0.0384	0.0375	0.0367
1.8	0.0359	0.0351	0.0344	0.0336	0.0329	0.0322	0.0314	0.0307	0.0301	0.0294
1.9	0.0287	0.0281	0.0274	0.0268	0.0262	0.0256	0.0250	0.0244	0.0239	0.0233
2.0	0.02275	0.02222	0.02169	0.02118	0.02068	0.02018	0.01970	0.01923	0.01876	0.01831
2.1	0.01786	0.01743	0.01700	0.01659	0.01618	0.01578	0.01539	0.01500	0.01463	0.01426
2.2	0.01390	0.01355	0.01321	0.01287	0.01255	0.01222	0.01191	0.01160	0.01130	0.01101
2.3	0.01072	0.01044	0.01017	0.00990	0.00964	0.00939	0.00914	0.00889	0.00866	0.00842
2.4	0.00820	0.00798	0.00776	0.00755	0.00734	0.00714	0.00695	0.00676	0.00657	0.00639
2.5	0.00621	0.00604	0.00587	0.00570	0.00554	0.00539	0.00523	0.00508	0.00494	0.00480
2.6	0.00466	0.00453	0.00440	0.00427	0.00415	0.00402	0.00391	0.00379	0.00368	0.00357
2.7	0.00347	0.00336	0.00326	0.00317	0.00307	0.00298	0.00289	0.00280	0.00272	0.00264
2.8	0.00256	0.00248	0.00240	0.00233	0.00226	0.00219	0.00212	0.00205	0.00199	0.00193
2.9	0.00187	0.00181	0.00175	0.00169	0.00164	0.00159	0.00154	0.00149	0.00144	0.00139

Source: Table 3 of *Statistical Tables for Science. Engineering and Management*, J. Murdock and J. A. Barnes, Macmillian, London, 1968.

Table A.2 Upper Quantiles of a χ^2 Distribution

ν/α	0.995	0.990	0.975	0.950	0.900	0.100	0.050	0.025	0.010	0.005
1	$392704 \cdot 10^{-10}$	$157088 \cdot 10^{-9}$	$982069 \cdot 10^{-9}$	$393214 \cdot 10^{-8}$	0.0157908	2.70554	3.84146	5.02389	6.63490	7.87944
2	0.0100251	0.0201007	0.0506356	0.102587	0.210720	4.60517	5.99147	7.37776	9.21034	10.5966
3	0.0717212	0.114832	0.215795	0.351846	0.584375	6.25139	7.81473	9.34840	11.3449	12.8381
4	0.206990	0.297110	0.484419	0.710721	1.063623	7.77944	9.48773	11.1433	13.2767	14.8602
5	0.411740	0.554300	0.831211	1.145476	1.61031	9.23635	11.0705	12.8325	15.0863	16.7496
6	0.675727	0.872085	1.237347	1.63539	2.20413	10.6446	12.5916	14.4494	16.8119	18.5476
7	0.989265	1.239043	1.68987	2.16735	2.83311	12.0170	14.0671	16.0128	18.4753	20.2777
8	1.344419	1.646482	2.17973	2.73264	3.48954	13.3616	15.5073	17.5346	20.0902	21.9550
9	1.734926	2.087912	2.70039	3.32511	4.16816	14.6837	16.9190	19.0228	21.6660	23.5893
10	2.15585	2.55821	3.24697	3.94030	4.86518	15.9871	18.3070	20.4831	23.2093	25.1882
11	2.60321	3.05347	3.81575	4.57481	5.57779	17.2750	19.6751	21.9200	24.7250	26.7569
12	3.07382	3.57056	4.40379	5.22603	6.30380	18.5494	21.0261	23.3367	26.2170	28.2995
13	3.56503	4.10691	5.00874	5.89186	7.04150	19.8119	22.3621	24.7356	27.6883	29.8194
14	4.07468	4.66043	5.62872	6.57063	7.78953	21.0642	23.6848	26.1190	29.1413	31.3193
15	4.60094	5.22935	6.26214	7.26094	8.54675	22.3072	24.9958	27.4884	30.5779	32.8013
16	5.14224	5.81221	6.90766	7.96164	9.31223	23.5418	26.2962	28.8454	31.9999	34.2672
17	5.69724	6.40776	7.56418	8.67176	10.0852	24.7690	27.5871	30.1910	33.4087	35.7185
18	6.26481	7.01491	8.23075	9.39046	10.8649	25.9894	28.8693	31.5261	34.8053	37.1564
19	6.84398	7.63273	8.90655	10.1170	11.6509	27.2036	30.1435	32.8523	36.1908	38.5822
20	7.43386	8.26040	9.59083	10.8508	12.4426	28.4120	31.4104	34.1696	37.5662	39.9968
21	8.03366	8.89720	10.28293	11.5913	13.2396	29.6151	32.6705	35.4789	38.9321	41.4010
22	8.64272	9.54249	10.9823	12.3380	14.0415	30.8133	33.9244	36.7807	40.2894	42.7956
23	9.26042	10.19567	11.6885	13.0905	14.8479	32.0069	35.1725	38.0757	41.6384	44.1813
24	9.88623	10.8564	12.4011	13.8484	15.6587	33.1963	36.4151	39.3641	42.9798	45.5585
25	10.5197	11.5240	13.1197	14.6114	16.4734	34.3816	37.6525	40.6465	44.3141	46.9278
26	11.1603	12.1981	13.8439	15.3791	17.2919	35.5631	38.8852	41.9232	45.6417	48.2899
27	11.8076	12.8786	14.5733	16.1513	18.1138	36.7412	40.1133	43.1944	46.9630	49.6449
28	12.4613	13.5648	15.3079	16.9279	18.9392	37.9159	41.3372	44.4607	48.2782	50.9933
29	13.1211	14.2565	16.0471	17.7083	19.7677	39.0875	42.5569	45.7222	49.5879	52.3356
30	13.7867	14.9535	16.7908	18.4926	20.5992	40.2560	43.7729	46.9792	50.8922	53.6720
40	20.7065	22.1643	24.4331	26.5093	29.0505	51.8050	55.7585	59.3417	63.6907	66.7659
50	27.9907	29.7067	32.3574	34.7642	37.6886	63.1671	67.5048	71.4202	76.1539	79.4900
60	35.5346	37.4848	40.4817	43.1879	46.4589	74.3970	79.0819	83.2976	88.3794	91.9517
70	43.2752	45.4418	48.7576	51.7393	55.3290	85.5271	90.5312	95.0231	100.425	104.215
80	51.1720	53.5400	57.1532	60.3915	64.2778	96.5782	101.879	106.629	112.329	116.321
90	59.1963	61.7541	65.6466	69.1260	73.2912	107.565	113.145	118.136	124.116	128.299
100	67.3276	70.0648	74.2219	77.9295	82.3581	118.498	124.342	129.561	135.807	140.169

Source: Tables of Percentage Points of the χ^2-Distribution by C. M. Thompson. *Biometrika* (1941). Vol. 32, pp. 188–189.

Table A.3 Upper Quantiles of a Central t Distribution

ν/α	0.050	0.025	0.010	0.005
1	6.3138	12.706	25.452	63.647
2	2.9200	4.3027	6.2053	9.9248
3	2.3534	3.1825	4.1765	5.8409
4	2.1318	2.7764	3.4954	4.6041
5	2.0150	2.5706	3.1634	4.0321
6	1.9432	2.4469	2.9687	3.7074
7	1.8946	2.3646	2.8412	3.4995
8	1.8595	2.3060	2.7515	3.3554
9	1.8331	2.2622	2.6850	3.2498
10	1.8125	2.2281	2.6338	3.1693
11	1.7959	2.2010	2.5931	3.1058
12	1.7823	2.1788	2.5600	3.0545
13	1.7709	2.1604	2.5326	3.0123
14	1.7613	2.1448	2.5096	2.9768
15	1.7530	2.1315	2.4899	2.9467
16	1.7459	2.1199	2.4729	2.9208
17	1.7396	2.1098	2.4581	2.8982
18	1.7341	2.1009	2.4450	2.8784
19	1.7291	2.0930	2.4334	2.8609
20	1.7247	2.0860	2.4231	2.8453
21	1.7207	2.0796	2.4138	2.8314
22	1.7171	2.0739	2.4055	2.8188
23	1.7139	2.0687	2.3979	2.8073
24	1.7109	2.0639	2.3910	2.7969
25	1.7081	2.0595	2.3846	2.7874
26	1.7056	2.0555	2.3788	2.7787
27	1.7033	2.0518	2.3734	2.7707
28	1.7011	2.0484	2.3685	2.7633
29	1.6991	2.0452	2.3638	2.7564
30	1.6973	2.0423	2.3596	2.7500
40	1.6839	2.0211	2.3289	2.7045
60	1.6707	2.0003	2.2991	2.6603
120	1.6577	1.9799	2.2699	2.6174
∞	1.6449	1.9600	2.2414	2.5758

Source: Tables of Percentage Points of the t-Distribution by M. Merrington, *Biometrika* (1941), Vol. 32, p. 300.

Table A.4 Upper Quintiles of an *F* Distribution

$\alpha = 0.05$

ν_2/ν_1	1	2	3	4	5	6	7	8	9
1	161.45	199.50	215.71	224.58	230.16	233.99	236.77	238.88	240.54
2	18.513	19.000	19.164	19.247	19.296	19.330	19.353	19.371	19.385
3	10.128	9.5521	9.2766	9.1172	9.0135	8.9406	8.8868	8.8452	8.8123
4	7.7086	6.9443	6.5914	6.3883	6.2560	6.1631	6.0942	6.0410	5.9988
5	6.6079	5.7861	5.4095	5.1922	5.0503	4.9503	4.8759	4.8183	4.7725
6	5.9874	5.1433	4.7571	4.5337	4.3874	4.2839	4.2066	4.1468	4.0990
7	5.5914	4.7374	4.3468	4.1203	3.9715	3.8660	3.7870	3.7257	3.6767
8	5.3177	4.4590	4.0662	3.8378	3.6875	3.5806	3.5005	3.4381	3.3881
9	5.1174	4.2565	3.8626	3.6331	3.4817	3.3738	3.2927	3.2296	3.1789
10	4.9646	4.1028	3.7083	3.4780	3.3258	3.2172	3.1355	3.0717	3.0204
11	4.8443	3.9823	3.5874	3.3567	3.2039	3.0946	3.0123	2.9480	2.8962
12	4.7472	3.8853	3.4903	3.2592	3.1059	2.9961	2.9134	2.8486	2.7964
13	4.6672	3.8056	3.4105	3.1791	3.0254	2.9153	2.8321	2.7669	2.7144
14	4.6001	3.7389	3.3439	3.1122	2.9582	2.8477	2.7642	2.6987	2.6458
15	4.5431	3.6823	3.2874	3.0556	2.9013	2.7905	2.7066	2.6408	2.5876
16	4.4940	3.6337	3.2389	3.0069	2.8524	2.7413	2.6572	2.5911	2.5377
17	4.4513	3.5915	3.1968	2.9647	2.8100	2.6987	2.6143	2.5480	2.4943
18	4.4139	3.5546	3.1599	2.9277	2.7729	2.6613	2.5767	2.5102	2.4563
19	4.3808	3.5219	3.1274	2.8951	2.7401	2.6283	2.5435	2.4768	2.4227
20	4.3513	3.4928	3.0984	2.8661	2.7109	2.5990	2.5140	2.4471	2.3928
21	4.3248	3.4668	3.0725	2.8401	2.6848	2.5727	2.4876	2.4205	2.3661
22	4.3009	3.4434	3.0491	2.8167	2.6613	2.5491	2.4638	2.3965	2.3419
23	4.2793	3.4221	3.0280	2.7955	2.6400	2.5277	2.4422	2.3748	2.3201
24	4.2597	3.4028	3.0088	2.7763	2.6207	2.5082	2.4226	2.3551	2.3002
25	4.2417	3.3852	2.9912	2.7587	2.6030	2.4904	2.4047	2.3371	3.2821
26	4.2252	2.3690	2.9751	2.7426	2.5868	2.4741	2.3883	2.3205	2.2655
27	4.2100	3.3541	2.9604	2.7278	2.5719	2.4591	2.3732	2.3053	2.2501
28	4.1960	3.3404	2.9467	2.7141	2.5581	2.4453	2.3593	2.2913	2.2360
29	4.1830	3.3277	2.9340	2.7014	2.5454	2.4324	2.3463	2.2782	2.2229
30	4.1709	3.3158	2.9223	2.6896	2.5336	2.4205	2.3343	2.2662	2.2107
40	4.0848	3.2317	2.8387	2.6060	2.4495	2.3359	2.2490	2.1802	2.1240
60	4.0012	3.1504	2.7581	2.5252	2.3683	2.2540	2.1665	2.0970	2.0401
120	3.9201	3.0718	2.6802	2.4472	2.2900	2.1750	2.0867	2.0164	1.9588
∞	3.8415	2.9957	2.6049	2.3719	2.2141	2.0986	2.0096	1.9384	1.8799

Table A.4 (*Continued*)

					$\alpha = 0.05$					
$v2/v1$	10	12	15	20	24	30	40	60	120	∞
1	241.88	243.91	245.95	248.01	249.05	250.09	251.14	252.20	253.25	254.32
2	19.396	19.413	19.429	19.446	19.454	19.462	19.471	19.479	19.487	19.496
3	8.7855	8.7446	8.7029	8.6602	8.6385	8.6166	8.5944	8.5720	8.5494	8.5265
4	5.9644	5.9117	5.8578	5.8025	5.7744	5.7459	5.7170	5.6878	5.6581	5.6281
5	4.7351	4.6777	4.6188	4.5581	4.5272	4.4957	4.4638	4.4314	4.3984	4.3650
6	4.0600	3.9999	3.9381	3.8742	3.8415	3.8082	3.7743	3.7398	3.7047	3.6688
7	3.6365	3.5747	3.5108	3.4445	3.4105	3.3758	3.3404	3.3043	3.2674	3.2298
8	3.3472	3.2840	3.2184	3.1503	3.1152	3.0794	3.0428	3.0053	2.9669	2.9276
9	3.1373	3.0729	3.0061	2.9365	2.9005	2.8637	2.8259	2.7872	2.7475	2.7067
10	2.9782	2.9130	2.8450	2.7740	2.7372	2.6996	2.6609	2.6211	2.5801	2.5379
11	2.8536	2.7876	2.7186	2.6464	2.6090	2.5705	2.5309	2.4901	2.4480	2.4045
12	2.7534	2.6866	2.6169	2.5436	2.5055	2.4663	2.4259	2.3842	2.3410	2.2962
13	2.6710	2.6037	2.5331	2.4589	2.4202	2.3803	2.3392	2.2966	2.2524	2.2064
14	2.6021	2.5342	2.4630	2.3879	2.3487	2.3082	2.2664	2.2230	2.1778	2.1307
15	2.5437	2.4753	2.4035	2.3275	2.2878	2.2468	2.2043	2.1601	2.1141	2.0658
16	2.4935	2.4247	2.3522	2.2756	2.2354	2.1938	2.1507	2.1058	2.0589	2.0096
17	2.4499	2.3807	2.3077	2.2304	2.1898	2.1477	2.1040	2.0584	2.0107	1.9604
18	2.4117	2.3421	2.2686	2.1906	2.1497	2.1071	2.0629	2.0166	1.9681	1.9168
19	2.3779	2.3080	2.2341	2.1555	2.1141	2.0712	2.0264	1.9796	1.9302	1.8780
20	2.3479	2.2776	2.2033	2.1242	2.0825	2.0391	1.9938	1.9464	1.8963	1.8432
21	2.3210	2.2504	2.1757	2.0960	2.0540	2.0102	1.9645	1.9165	1.8657	1.8117
22	2.2967	2.2258	2.1508	2.0707	2.0283	1.9842	1.9380	1.8895	1.8380	1.7831
23	2.2747	2.2036	2.1282	2.0476	2.0050	1.9605	1.9139	1.8649	1.8128	1.7570
24	2.2547	2.1834	2.1077	2.0267	1.9838	1.9390	1.8920	1.8424	1.7897	1.7331
25	2.2365	2.1649	2.0889	2.0075	1.9643	1.9192	1.8718	1.8217	1.7684	1.7110
26	2.2197	2.1479	2.0716	1.9898	1.9464	1.9010	1.8533	1.8027	1.7488	1.6906
27	2.2043	2.1323	2.0558	1.9736	1.9299	1.8842	1.8361	1.7851	1.7307	1.6717
28	2.1900	2.1179	2.0411	1.9586	1.9147	1.8687	1.8203	1.7689	1.7138	1.6541
29	2.1768	2.1045	2.0275	1.9446	1.9005	1.8543	1.8055	1.7537	1.6981	1.6377
30	2.1646	2.0921	2.1048	1.9317	1.8874	1.8409	1.7918	1.7396	1.6835	1.6223
40	2.0772	2.0032	1.9245	1.8389	1.7929	1.7444	1.6928	1.6373	1.5766	1.5089
60	1.9926	1.9174	1.8364	1.7480	1.7001	1.6491	1.5943	1.5343	1.4673	1.3893
120	1.9105	1.8337	1.7505	1.6587	1.6084	1.5543	1.4952	1.4290	1.3519	1.2539
∞	1.8307	1.7522	1.6664	1.5705	1.5173	1.4591	1.3940	1.3180	1.2214	1.0000

Table A.4 (*Continued*)

$\alpha = 0.025$

ν_2/ν_1	1	2	3	4	5	6	7	8	9
1	647.79	799.50	864.16	899.58	921.85	937.11	948.22	956.66	963.28
2	38.506	39.000	39.165	39.248	39.298	39.331	39.355	39.373	39.387
3	17.443	16.044	15.439	15.101	14.885	14.735	14.624	14.540	14.473
4	12.218	10.649	9.9792	9.6045	9.3645	9.1973	9.0741	8.9796	8.9047
5	10.007	8.4336	7.7636	7.3879	7.1464	6.9777	6.8531	6.7572	6.6810
6	8.8131	7.2598	6.5988	6.2272	5.9876	5.8197	5.6955	5.5996	5.5234
7	8.0727	6.5415	5.8898	5.5226	5.2852	5.1186	4.9949	4.8994	4.8232
8	7.5709	6.0595	5.4160	5.0526	4.8173	4.6517	4.5286	4.4332	4.3572
9	7.2093	5.7147	5.0781	4.7181	4.4844	4.3197	4.1971	4.1020	4.0260
10	6.9367	5.4564	4.8256	4.4683	4.2361	4.0721	3.9498	3.8549	3.7790
11	6.7241	5.2559	4.6300	4.2751	4.0440	3.8807	3.7586	3.6638	3.5879
12	6.5538	5.0959	4.4742	4.1212	3.8911	3.7283	3.6065	3.5118	3.4358
13	6.4143	4.9653	4.3472	3.9959	3.7667	3.6043	3.4827	3.3880	3.3120
14	6.2979	4.8567	4.2417	3.8919	3.6634	3.5014	3.3799	3.2853	3.2093
15	6.1995	4.7650	4.1528	3.8043	3.5764	3.4147	3.2934	3.1987	3.1227
16	6.1151	4.6867	4.0768	3.7294	3.5021	3.3406	3.2194	3.1248	3.0488
17	6.0420	4.6189	4.0112	3.6648	3.4379	3.2767	3.1556	3.0610	2.9849
18	5.9781	4.5597	3.9539	3.6083	3.3820	3.2209	3.0999	3.0053	2.9291
19	5.9216	4.5075	3.9034	3.5587	3.3327	3.1718	3.0509	2.9563	2.8800
20	5.8715	4.4613	3.8587	3.5147	3.2891	3.1283	3.0074	2.9128	2.8365
21	5.8266	4.4199	3.8188	3.4754	3.2501	3.0895	2.9686	2.8740	2.7977
22	5.7863	4.3828	3.7829	3.4401	3.2151	3.0546	2.9338	2.8392	2.7628
23	5.7498	4.3492	3.7505	3.4083	3.1835	3.0232	2.9024	2.8077	2.7313
24	5.7167	4.3187	3.7211	3.3794	3.1548	2.9946	2.8738	2.7791	2.7027
25	5.6864	4.2909	3.6943	3.3530	3.1287	2.9685	2.8478	2.7531	2.6766
26	5.6586	4.2655	3.6697	3.3289	3.1048	2.9447	2.8240	2.7293	2.6528
27	5.6331	4.2421	3.6472	3.3067	3.0828	2.9228	2.8021	2.7074	2.6309
28	5.6096	4.2205	3.6264	3.2863	3.0625	2.9027	2.7820	2.6872	2.6106
29	5.5878	4.2006	3.6072	3.2674	3.0438	2.8840	2.7633	2.6686	2.5919
30	5.5675	4.1821	3.5894	3.2499	3.0265	2.8667	2.7460	2.6513	2.5746
40	5.4239	4.0510	3.4633	3.1261	2.9037	2.7444	2.6238	2.5289	2.4519
60	5.2857	3.9253	3.3425	3.0077	2.7863	2.6274	2.5068	2.4117	2.3344
120	5.1524	3.8046	3.2270	2.8943	2.6740	2.5154	2.3948	2.2994	2.2217
∞	5.0239	3.6889	3.1161	2.7858	2.5665	2.4082	2.2875	2.1918	2.1136

Table A.4 (*Continued*)

$\alpha = 0.025$

ν_2/ν_1	10	12	15	20	24	30	40	60	120	∞
1	968.63	976.71	984.87	993.10	997.25	1001.4	1005.6	1009.8	1014.0	1018.3
2	39.398	39.415	39.431	39.448	39.456	39.465	39.473	39.481	39.490	39.498
3	14.419	14.337	14.253	14.167	14.124	14.081	14.037	13.992	13.947	13.902
4	8.8439	8.7512	8.6565	8.5599	8.5109	8.4613	8.4111	8.3604	8.3092	8.2573
5	6.6192	6.5246	6.4277	6.3285	6.2780	6.2269	6.1751	6.1225	6.0693	6.0153
6	5.4613	5.3662	5.2687	5.1684	5.1172	5.0652	5.0125	4.9589	4.9045	4.8491
7	4.7611	4.6658	4.5678	4.4667	4.4150	4.3624	4.3089	4.2544	4.1989	4.1423
8	4.2951	4.1997	4.1012	3.9995	3.9472	3.8940	3.8398	3.7844	3.7279	3.6702
9	3.9639	3.8682	3.7694	3.6669	3.6142	3.5604	3.5055	3.4493	3.3918	3.3329
10	3.7168	3.6209	3.5217	3.4186	3.3654	3.3110	3.2554	3.1984	3.1399	3.0798
11	3.5257	3.4296	3.3299	3.2261	3.1725	3.1176	3.0613	3.0035	2.9441	2.8828
12	3.3736	3.2773	3.1772	3.0728	3.0187	2.9633	2.9063	2.8478	2.7874	2.7249
13	3.2497	3.1532	3.0527	2.9477	2.8932	2.8373	2.7797	2.7204	2.6590	2.5955
14	3.1469	3.0501	2.9493	2.8437	2.7888	2.7324	2.6742	2.6142	2.5519	2.4872
15	3.0602	2.9633	2.8621	2.7559	2.7006	2.6437	2.5850	2.5242	2.4611	2.3953
16	2.9862	2.8890	2.7875	2.6808	2.6252	2.5678	2.5085	2.4471	2.3831	2.3163
17	2.9222	2.8249	2.7230	2.6158	2.5598	2.5021	2.4422	2.3801	2.3153	2.2474
18	2.8664	2.7689	2.6667	2.5590	2.5027	2.4445	2.3842	2.3214	2.2558	2.1869
19	2.8173	2.7196	2.6171	2.5089	2.4523	2.3937	2.3329	2.2695	2.2032	2.1333
20	2.7737	2.6758	2.5731	2.4645	2.4076	2.3486	2.2873	2.2234	2.1562	2.0853
21	2.7348	2.6368	2.5338	2.4247	2.3675	2.3082	2.2465	2.1819	2.1141	2.0422
22	2.6998	2.6017	2.4984	2.3890	2.3315	2.2718	2.2097	2.1446	2.0760	2.0032
23	2.6682	2.5699	2.4665	2.3567	2.2989	2.2389	2.1763	2.1107	2.0415	1.9677
24	2.6396	2.5412	2.4374	2.3273	2.2693	2.2090	2.1460	2.0799	2.0099	1.9353
25	2.6135	2.5149	2.4110	2.3005	2.2422	2.1816	2.1183	2.0517	1.9811	1.9055
26	2.5895	2.4909	2.3867	2.2759	2.2174	2.1565	2.0928	2.0257	1.9545	1.8781
27	2.5676	2.4688	2.3644	2.2533	2.1946	2.1334	2.0693	2.0018	1.9299	1.8527
28	2.5473	2.4484	2.3438	2.2324	2.1735	2.1121	2.0477	1.9796	1.9072	1.8291
29	2.5286	2.4295	2.3248	2.2131	2.1540	2.0923	2.0276	1.9591	1.8861	1.8072
30	2.5112	2.4210	2.3072	2.1952	2.1359	2.0739	2.0089	1.9400	1.8664	1.7867
40	2.3882	2.2882	2.1819	2.0677	2.0069	1.9429	1.8752	1.8028	1.7242	1.6371
60	2.2702	2.1692	2.0613	1.9445	1.8817	1.8152	1.7440	1.6668	1.5810	1.4822
120	2.1570	2.0548	1.9450	1.8249	1.7597	1.6899	1.6141	1.5299	1.4327	1.3104
∞	2.0483	1.9447	1.8326	1.7085	1.6402	1.5660	1.4835	1.3883	1.2684	1.0000

Table A.4 (*Continued*)

$\alpha = 0.010$

ν_2/ν_1	1	2	3	4	5	6	7	8	9
1	4052.2	4999.5	5403.3	5624.6	5763.7	5859.0	5928.3	5981.6	6022.5
2	98.503	99.000	99.166	99.249	99.299	99.332	99.356	99.374	99.388
3	34.116	30.817	29.457	28.710	28.237	27.911	27.672	27.489	27.345
4	21.198	18.000	16.694	15.977	15.522	15.207	14.976	14.799	14.659
5	16.258	13.274	12.060	11.392	10.967	10.672	10.456	10.289	10.158
6	13.745	10.925	9.7795	9.1483	8.7459	8.4661	8.2600	8.1016	7.9761
7	12.246	9.5466	8.4513	7.8467	7.4604	7.1914	6.9928	6.8401	6.7188
8	11.259	8.6491	7.5910	7.0060	6.6318	6.3707	6.1776	6.0289	5.9106
9	10.561	8.0215	6.9919	6.4221	6.0569	5.8018	5.6129	5.4671	5.3511
10	10.044	7.5594	6.5523	5.9943	5.6363	5.3858	5.2001	5.0567	4.9424
11	9.6460	7.2057	6.2167	5.6683	5.3160	5.0692	4.8861	4.7445	4.6315
12	9.3302	6.9266	5.9526	5.4119	5.0643	4.8206	4.6395	4.4994	4.3875
13	9.0738	6.7010	5.7394	5.2053	4.8616	4.6204	4.4410	4.3021	4.1911
14	8.8616	6.5149	5.5639	5.0354	4.6950	4.4558	4.2779	4.1399	4.0297
15	8.6831	6.3589	5.4170	4.8932	4.5556	4.3183	4.1415	4.0045	3.8948
16	8.5310	6.2262	5.2922	4.7726	4.4374	4.2016	4.0259	3.8896	3.7804
17	8.3997	6.1121	5.1850	4.6690	4.3359	4.1015	3.9267	3.7910	3.6822
18	8.2854	6.0129	5.0919	4.5790	4.2479	4.0146	3.8406	3.7054	3.5971
19	8.1850	5.9259	5.0103	4.5003	4.1708	3.9386	3.7653	3.6305	3.5225
20	8.0960	5.8489	4.9382	4.4307	4.1027	3.8714	3.6987	3.5644	3.4567
21	8.0166	5.7804	4.8740	4.3688	4.0421	3.8117	3.6396	3.5056	3.3981
22	7.9454	5.7190	4.8166	4.3134	3.9880	3.7583	3.5867	3.4530	3.3458
23	7.8811	5.6637	4.7649	4.2635	3.9392	3.7102	3.5390	3.4057	3.2986
24	7.8229	5.6136	4.7181	4.2184	3.8951	3.6667	3.4959	3.3629	3.2560
25	7.7698	5.5680	4.6755	4.1774	3.8550	3.6272	3.4568	3.3239	3.2172
26	7.7213	5.5263	4.6366	4.1400	3.8183	3.5911	3.4210	3.2884	3.1818
27	7.6767	5.4881	4.6009	4.1056	3.7848	3.5580	3.3882	3.2558	3.1494
28	7.6356	5.4529	4.5681	4.0740	3.7539	3.5276	3.3581	3.2259	3.1195
29	7.5976	5.4205	4.5378	4.0449	3.7254	3.4995	3.3302	3.1982	3.0920
30	7.5625	5.3904	4.5097	4.0179	3.6990	3.4735	3.3045	3.1726	3.0665
40	7.3141	5.1785	4.3126	3.8283	3.5138	3.2910	3.1238	2.9930	2.8876
60	7.0771	4.9774	4.1259	3.6491	3.3389	3.1187	2.9530	2.8233	2.7185
120	6.8510	4.7865	3.9493	3.4796	3.1735	2.9559	2.7918	2.6629	2.5586
∞	6.6349	4.6052	3.7816	3.3192	3.0173	2.8020	2.6393	2.5113	2.4073

Table A.4 (*Continued*)

ν_2/ν_1	10	12	15	20	24	30	40	60	120	∞
					$\alpha = 0.010$					
1	6055.8	6106.3	6157.3	6208.7	6234.6	6260.7	6286.8	6313.0	6339.4	6366.0
2	99.399	99.416	99.432	99.449	99.458	99.466	99.474	99.483	99.491	99.501
3	27.229	27.052	26.872	26.690	26.598	26.505	26.411	26.316	26.221	26.125
4	14.546	14.374	14.198	14.020	13.929	13.838	13.745	13.652	13.558	13.463
5	10.051	9.8883	9.7222	9.5527	9.4665	9.3793	9.2912	9.2020	9.1118	9.0204
6	7.8741	7.7183	7.5590	7.3958	7.3127	7.2285	7.1432	7.0568	6.9690	6.8801
7	6.6201	6.6491	6.3143	6.1554	6.0743	5.9921	5.9084	5.8236	5.7372	5.6495
8	5.8143	5.6668	5.5151	5.3591	5.2793	5.1981	5.1156	5.0316	4.9460	4.8588
9	5.2565	5.1114	4.9621	4.8080	4.7290	4.6486	4.5667	4.4831	4.3978	4.3105
10	4.8492	4.7059	4.5582	4.4054	4.3269	4.2469	4.1653	4.0819	3.9965	3.9090
11	4.5393	4.3974	4.2509	4.0990	4.0209	3.9411	3.8596	3.7761	3.6904	3.6025
12	4.2961	4.1553	4.0096	3.8584	3.7805	3.7008	3.6192	3.5355	3.4494	3.3608
13	4.1003	3.9603	3.8154	3.6646	3.5868	3.5070	3.4253	3.3413	3.2548	3.1654
14	3.9394	3.8001	3.6557	3.5052	3.4274	3.3476	3.2556	3.1813	3.0942	3.0040
15	3.8049	3.6662	3.5222	3.3719	3.2940	3.2141	3.1319	3.0471	2.9595	2.8684
16	3.6909	3.5527	3.4089	3.2588	3.1808	3.1007	3.0182	2.9330	2.8447	2.7528
17	3.5931	3.4552	3.3117	3.1615	3.0835	3.0032	2.9205	2.8348	2.7459	2.6530
18	3.5082	3.3706	3.2273	3.0771	2.9990	2.9185	2.8354	2.7493	2.6597	2.5660
19	3.4338	3.2965	3.1533	3.0031	2.9249	2.8442	2.7608	2.6742	2.5839	2.4893
20	3.3682	3.2311	3.0880	2.9377	2.8594	2.7785	2.6947	2.6077	2.5168	2.4212
21	3.3098	3.1729	3.0299	2.8796	2.8011	2.7200	2.6359	2.5484	2.4568	2.3603
22	3.2576	3.1209	2.9780	2.8274	2.7488	2.6675	2.5831	2.4951	2.4029	2.3055
23	3.2106	3.0740	2.9311	2.7805	2.7017	2.6202	2.5355	2.4471	2.3542	2.2559
24	3.1681	3.0316	2.8887	2.7380	2.6591	2.5773	2.4923	2.4035	2.3099	2.2107
25	3.1294	3.9931	2.8502	2.6993	2.6203	2.5383	2.4530	2.3637	2.2695	2.1694
26	3.0941	2.9579	2.8150	2.6640	2.5848	2.5026	2.4170	2.3273	2.2325	2.1315
27	3.0618	2.9256	2.7827	2.6316	2.5522	2.4699	2.3840	2.2938	2.1984	2.0965
28	3.0320	2.8959	2.7530	2.6017	2.5223	2.4397	2.3535	2.2629	2.1670	2.0642
29	3.0045	2.8685	2.7256	2.5742	2.4946	2.4118	2.3253	2.2344	2.1378	2.0342
30	2.9791	2.8431	2.7002	2.5487	2.4689	2.3860	2.2992	2.2079	2.1107	2.0062
40	2.8005	2.6648	2.5216	2.3689	2.2880	2.2034	2.1142	2.0194	1.9172	1.8047
60	2.6318	2.4961	2.3523	2.1978	2.1154	2.0285	1.9360	1.8363	1.7263	1.6006
120	2.4721	2.3363	2.1915	2.0346	1.9500	1.8600	1.7628	1.6557	1.5330	1.3805
∞	2.3209	2.1848	2.0385	1.8783	1.7908	1.6964	1.5923	1.4730	1.3246	1.0000

Table A.4 (*Continued*)

$\alpha = 0.005$

$\nu2/\nu1$	1	2	3	4	5	6	7	8	9
1	16211	20000	21615	22500	23056	23437	23715	23925	24091
2	198.50	199.00	199.17	199.25	199.30	199.33	199.36	199.37	199.39
3	55.552	49.799	47.467	46.195	45.392	44.838	41.434	44.126	43.882
4	31.333	26.284	24.259	23.155	22.456	21.975	21.622	21.352	21.139
5	22.785	18.314	16.530	15.556	14.940	14.513	14.200	13.961	13.722
6	18.635	14.544	12.917	12.028	11.464	11.073	10.786	10.566	10.391
7	16.236	12.404	10.882	10.050	9.5221	9.1554	8.8854	8.6781	8.5138
8	14.688	11.042	9.5965	8.051	8.3018	7.9520	7.6942	7.4960	7.3386
9	13.614	10.107	8.7171	7.9559	7.4711	7.1338	6.8849	6.6933	6.5411
10	12.826	9.4270	8.0807	7.3428	6.8723	6.5446	6.3025	6.1159	5.9676
11	12.226	8.9122	7.6004	6.8809	6.4217	6.1015	5.8648	5.6821	5.5368
12	11.754	8.5096	7.2258	6.5211	6.0711	5.7570	5.5245	5.3451	5.2021
13	11.374	8.1805	6.9257	6.2335	5.7910	5.4819	5.2529	5.0761	4.9351
14	11.060	7.9217	6.6803	5.9984	5.5623	5.2574	5.0313	4.8566	4.7173
15	10.798	7.7008	6.4760	5.8029	5.3721	5.0708	4.8473	4.6743	4.5464
16	10.575	7.5138	6.3034	5.6378	5.2117	4.9134	4.6920	4.5207	4.3838
17	10.384	7.3536	6.1556	5.4967	5.0746	4.7789	4.5594	4.3893	4.2535
18	10.218	7.2148	6.0277	5.3746	4.9560	4.6627	4.4448	4.2759	4.1410
19	10.073	7.0935	5.9161	5.2681	4.8526	4.5614	4.3448	4.1770	4.0428
20	19.9439	6.9865	5.8177	5.1743	4.7616	4.4721	4.2569	4.0900	3.9564
21	19.8295	6.8914	5.7304	5.0911	4.6808	4.3931	4.1789	4.0128	3.8799
22	9.7271	6.8064	5.6524	5.0168	4.6088	4.3225	4.1094	3.9440	3.8116
23	9.6348	6.7300	5.5823	4.9500	4.5441	4.2591	4.0469	3.8822	3.7502
24	9.5513	6.6610	5.5190	4.8898	4.4857	4.2019	3.9905	3.8264	3.6949
25	9.4753	6.5982	5.4615	4.8351	4.4327	4.1500	3.9394	3.7758	3.6447
26	9.4059	6.5409	5.4091	4.7852	4.3844	4.1027	3.8928	3.7297	3.5989
27	9.3423	6.4885	5.3611	4.7396	4.3402	4.0594	3.8501	3.6875	3.5571
28	9.2838	6.4403	5.3170	4.6977	4.2996	4.0197	3.8110	3.6487	3.5186
29	9.2297	6.3958	5.2764	4.6591	4.2622	3.9830	3.7749	3.6130	3.4832
30	9.1797	6.3547	5.2388	4.6233	4.2276	3.9492	3.7416	3.5801	3.4505
40	8.8278	6.0664	4.9759	4.3738	3.9860	3.7129	3.5088	3.3498	3.2220
60	8.4946	5.7950	4.7290	4.1399	3.7600	3.4918	3.2911	3.1344	3.0083
120	8.1790	5.5393	4.4973	3.9207	3.5482	3.2849	3.0874	2.9330	2.8083
∞	7.8794	5.2983	4.2794	3.7151	3.3499	3.0913	2.8968	2.7444	2.6210

Table A.4 *(Continued)*

					$\alpha = 0.005$					
ν_2/ν_1	10	12	15	20	24	30	40	60	120	∞
1	24224	24426	24630	24836	24940	25044	25148	25253	25359	25465
2	199.40	199.42	199.43	199.45	199.46	199.47	199.47	199.48	199.49	199.51
3	43.686	43.387	43.085	42.778	42.622	42.466	42.308	42.149	41.989	41.829
4	20.967	20.705	20.438	20.167	20.030	19.892	19.752	19.611	19.468	19.325
5	13.618	13.384	13.146	12.903	12.780	12.656	12.530	12.402	12.274	12.161
6	10.250	10.034	9.8140	9.5888	9.4741	9.3583	9.2408	9.1219	9.0015	8.8793
7	8.3803	8.1764	7.9678	7.7540	7.6450	7.5345	7.4225	7.3088	7.1933	7.0760
8	7.2107	7.0149	6.8143	6.6082	6.5029	6.3961	6.2875	6.1772	6.0648	5.9505
9	6.4171	6.2274	6.0325	5.8318	5.7292	5.6248	5.5186	5.4104	5.3001	5.1875
10	5.8467	5.6613	5.4707	5.2740	5.1732	5.0705	4.9659	4.8592	4.7501	4.6385
11	5.4182	5.2363	5.0489	4.8552	4.7557	4.6543	4.5508	4.4450	4.3367	4.2256
12	5.0855	4.9063	4.7214	4.5299	4.4315	4.3309	4.2282	4.1229	4.0149	3.9039
13	4.8199	4.6429	4.4600	4.2703	4.1726	4.0727	3.9704	3.8655	3.7577	3.6465
14	4.6034	4.4281	4.2468	4.0585	3.9614	3.8619	3.7600	3.6553	3.5473	3.4359
15	4.4236	4.2498	4.0698	3.8826	3.7859	3.6867	3.5850	3.4803	3.3722	3.2602
16	4.2719	4.0994	3.9205	3.7342	3.6378	3.5388	3.4372	3.3324	3.2240	3.1115
17	4.1423	3.9709	3.7929	3.6073	3.5112	3.4124	3.3107	3.2058	3.0971	2.9839
18	4.0305	3.8599	3.6827	3.4977	3.4017	3.3030	3.2014	3.0962	2.9871	2.8732
19	3.9329	3.7631	3.5866	3.4020	3.3062	3.2075	3.1058	3.0004	2.8908	2.7762
20	3.8470	3.6779	3.5020	3.3178	3.2220	3.1234	3.0215	2.9159	2.8058	2.6904
21	3.7709	3.6024	3.4270	3.2431	3.1474	3.0488	2.9467	2.8408	2.7302	2.6140
22	3.7030	3.5350	3.3600	3.1764	3.0807	2.9821	2.8799	2.7736	2.6625	2.5455
23	3.6420	3.4745	3.2999	3.1165	3.0208	2.9221	2.8198	2.7132	2.6016	2.4837
24	3.5870	3.4199	3.2456	3.0624	2.9667	2.8679	2.7654	2.6585	2.5463	2.4276
25	3.5370	3.3704	3.1963	3.0133	2.9176	2.8187	2.7160	2.6088	2.4960	2.3765
26	3.4916	3.3252	3.1515	2.9685	2.8728	2.7738	2.6709	2.5633	2.4501	2.3297
27	3.4499	3.2839	3.1104	2.9275	2.8318	2.7327	2.6296	2.5217	2.4078	2.2867
28	3.4117	3.2460	3.0727	2.8899	2.7941	2.6949	2.5916	2.4834	2.3689	2.2469
29	3.3765	3.2111	3.0379	2.8551	2.7594	2.6601	2.5565	2.4479	2.3330	2.2102
30	3.3440	3.1787	3.0057	2.8230	2.7272	2.6278	2.5241	2.4151	2.2997	2.1760
40	3.1167	2.9531	2.7811	2.5984	2.5020	2.4015	2.2958	2.1838	2.0635	1.9318
60	2.9042	2.7419	2.5705	2.3872	2.2898	2.1874	2.0789	1.9622	1.8341	1.6885
120	2.7052	2.5439	2.3727	2.1881	2.0890	1.9839	1.8709	1.7469	1.6055	1.4311
∞	2.5188	2.3583	2.1868	1.9998	1.8983	1.7891	1.6691	1.5325	1.3637	1.0000

Source: Tables of Percentage Points of the Inverted beta (F)-Distribution by M. Merrington and C. M. Thompson, *Biometrika* (1942), Vol. 33, pp. 73–88.

Table A.5 Quantiles of the Distribution of Wilcoxon–Mann–Whitney Statistic

n_1	α	$n_2=2$	3	4	5	6	7	8	9	10	11	12	13	14	15	16	17	18	19	20
2	0.001	0	0	0	0	0	0	0	0	0	0	0	0	0	0	0	0	0	0	0
	0.005	0	0	0	0	0	0	0	0	0	0	0	0	0	0	0	0	0	1	1
	0.01	0	0	0	0	0	0	0	0	0	0	0	0	1	1	1	1	1	2	2
	0.025	0	0	0	0	0	0	0	0	1	1	1	1	1	2	2	3	3	3	3
	0.05	0	0	0	1	1	1	2	2	2	2	3	3	4	4	4	4	5	5	5
	0.10	0	1	1	2	2	3	3	4	4	5	5	6	6	7	7	8	8	8	8
3	0.001	0	0	0	0	0	0	0	0	0	0	0	0	0	0	0	1	1	1	1
	0.005	0	0	0	0	0	0	0	1	1	1	2	2	2	3	3	3	3	4	4
	0.01	0	0	0	0	0	1	1	2	2	2	3	3	3	4	4	5	5	5	6
	0.025	0	0	0	1	2	2	3	3	4	4	5	5	6	6	7	7	8	8	9
	0.05	0	1	1	2	3	3	4	5	5	6	6	7	8	8	9	10	10	11	12
	0.10	1	2	2	3	4	5	6	6	7	8	9	10	11	11	12	13	14	15	16
4	0.001	0	0	0	0	0	0	0	0	1	1	1	2	2	2	3	3	4	4	4
	0.005	0	0	0	0	1	1	2	2	3	3	4	4	5	6	6	7	7	8	9
	0.01	0	0	0	1	2	2	3	4	4	5	6	6	7	8	8	9	10	10	11
	0.025	0	0	1	2	3	4	5	5	6	7	8	9	10	11	12	12	13	14	15
	0.05	0	1	2	3	4	5	6	7	8	9	10	11	12	13	15	16	17	18	19
	0.010	1	2	4	5	6	7	8	10	11	12	13	14	16	17	18	19	21	22	23
5	0.001	0	0	0	0	0	0	1	2	2	3	3	4	4	5	6	6	7	8	8
	0.005	0	0	0	1	2	2	3	4	5	6	7	8	8	9	10	11	12	13	14
	0.01	0	0	1	2	3	4	5	6	7	8	9	10	11	12	13	14	15	16	17
	0.025	0	1	2	3	4	6	7	8	9	10	12	13	14	15	16	18	19	20	21
	0.05	1	2	3	5	6	7	9	10	12	13	14	16	17	19	20	21	23	24	26
	0.10	2	3	5	6	8	9	11	13	14	16	18	19	21	23	24	26	28	29	31

Table A.5 (Continued)

n_1	α	$n_2 = 2$	3	4	5	6	7	8	9	10	11	12	13	14	15	16	17	18	19	20
6	0.001	0	0	0	0	0	0	2	3	4	5	5	6	7	8	9	10	11	12	13
	0.005	0	0	1	2	3	4	5	6	7	8	10	11	12	13	14	16	17	18	19
	0.01	0	0	2	3	4	5	7	8	9	10	12	13	14	16	17	19	20	21	23
	0.025	0	2	3	4	6	7	9	11	12	14	15	17	18	20	22	23	25	26	28
	0.05	1	3	4	6	8	9	11	13	15	17	18	20	22	24	26	27	29	31	33
	0.10	2	4	6	8	10	12	14	16	18	20	22	24	26	28	30	32	35	37	39
7	0.001	0	0	0	0	1	2	3	4	6	7	8	9	10	11	12	14	15	16	17
	0.005	0	0	1	2	4	5	7	8	10	11	13	14	16	17	19	20	22	23	25
	0.01	0	1	2	4	5	7	8	10	12	13	15	17	18	20	22	24	25	27	29
	0.025	0	2	4	6	7	9	11	13	15	17	19	21	23	25	27	29	31	33	35
	0.05	1	3	5	7	9	12	14	16	18	20	22	25	27	29	31	34	36	38	40
	0.10	2	5	7	9	12	14	17	19	22	24	27	29	32	34	37	39	42	44	47
8	0.001	0	0	0	1	2	3	5	6	7	9	10	12	13	15	16	18	19	21	22
	0.005	0	0	2	3	5	7	8	10	12	14	16	18	19	21	23	25	27	29	31
	0.01	0	1	3	5	7	8	10	12	14	16	18	21	23	25	27	29	31	33	35
	0.025	1	3	5	7	9	11	14	16	18	20	23	25	27	30	32	35	37	39	42
	0.05	2	4	6	9	11	14	16	19	21	24	27	29	32	34	37	40	42	45	48
	0.10	3	6	8	11	14	17	20	23	25	28	31	34	37	40	43	46	49	52	55
9	0.001	0	0	0	2	3	4	6	8	9	11	13	15	16	18	20	22	24	26	27
	0.005	0	1	2	4	6	8	10	12	14	17	19	21	23	25	28	30	32	34	37
	0.01	0	2	4	6	8	10	12	15	17	19	22	24	27	29	32	34	37	39	41
	0.025	1	3	5	8	11	13	16	18	21	24	27	29	32	35	38	40	43	46	49
	0.05	2	5	7	10	13	16	19	22	25	28	31	34	37	40	43	46	49	52	55
	0.10	3	6	10	13	16	19	23	26	29	32	36	39	42	46	49	53	56	59	63

n	α																				
10	0.001	0	0	0	1	2	4	6	7	9	11	13	15	18	20	22	24	26	28	30	33
	0.005	0	0	1	3	5	7	10	12	14	17	19	22	25	27	30	32	35	38	40	43
	0.01	0	0	2	4	7	9	12	14	17	20	23	25	28	31	34	37	39	42	45	48
	0.025	1	1	4	6	9	12	15	18	21	24	27	30	34	37	40	43	46	49	53	56
	0.05	2	2	5	8	12	15	18	21	25	28	32	35	38	42	45	49	52	56	59	63
	0.10	4	4	7	11	14	18	22	25	29	33	37	40	44	48	52	55	59	63	67	71
11	0.001	0	0	0	1	3	5	7	9	11	13	16	18	21	23	25	28	30	33	35	38
	0.005	0	0	1	3	6	8	11	14	17	19	22	25	28	31	34	37	40	43	46	49
	0.01	0	0	2	5	8	10	13	16	19	23	26	29	32	35	38	42	45	48	51	54
	0.025	1	1	4	7	10	13	17	19	24	27	30	34	38	41	45	48	52	56	59	63
	0.05	2	2	6	9	13	17	20	24	28	32	35	39	43	47	51	55	58	62	66	70
	0.10	4	4	8	12	16	20	24	28	32	37	41	45	49	54	58	62	66	70	74	79
12	0.001	0	0	0	1	3	5	8	10	13	15	18	21	24	26	29	32	35	38	41	43
	0.005	0	0	2	4	7	10	13	16	19	22	25	28	32	35	38	42	45	48	52	55
	0.01	0	0	3	6	9	12	15	18	22	25	29	32	36	39	43	47	50	54	57	61
	0.025	2	2	5	8	12	15	19	23	27	30	34	38	42	46	50	54	58	62	66	70
	0.05	3	3	6	10	14	18	22	27	31	35	39	43	48	52	56	61	65	69	73	78
	0.10	5	5	9	13	18	22	27	31	36	40	45	50	54	59	64	68	73	78	82	87
13	0.001	0	0	0	2	4	6	9	12	15	18	21	24	27	30	33	36	39	43	46	49
	0.005	0	0	2	4	8	11	14	18	21	25	28	32	35	39	43	46	50	54	58	61
	0.01	1	1	3	6	10	13	17	21	24	28	32	36	40	44	48	52	56	60	64	68
	0.025	2	2	5	9	13	17	21	25	29	34	38	42	46	51	55	60	64	68	73	77
	0.05	3	3	7	11	16	20	25	29	34	38	43	48	52	57	62	66	71	76	81	85
	0.10	5	5	10	14	19	24	29	34	39	44	49	54	59	64	69	75	80	85	90	95

Table A.5 (Continued)

n_1	α	$n_2=2$	3	4	5	6	7	8	9	10	11	12	13	14	15	16	17	18	19	20
14	0.001	0	0	2	4	7	10	13	16	20	23	26	30	33	37	40	44	47	51	55
	0.005	0	2	5	8	12	16	19	23	27	31	35	39	43	47	51	55	59	64	68
	0.01	1	3	7	11	14	18	23	27	31	35	39	44	48	52	57	61	66	70	74
	0.025	2	6	10	14	18	23	27	32	37	41	46	51	56	60	65	70	75	79	84
	0.05	4	8	12	17	22	27	32	37	42	47	52	57	62	67	72	78	83	88	93
	0.10	5	11	16	21	26	32	37	42	48	53	59	64	70	75	81	86	92	98	103
15	0.001	0	0	2	5	8	11	15	18	22	25	29	33	37	41	44	48	52	56	60
	0.005	0	3	6	9	13	17	21	25	30	34	38	43	47	52	56	61	65	70	74
	0.01	1	4	8	12	16	20	25	29	34	38	43	48	52	57	62	67	71	76	81
	0.025	2	6	11	15	20	25	30	35	40	45	50	55	60	65	71	76	81	86	91
	0.05	4	8	13	19	24	29	34	40	45	51	56	62	67	73	78	84	89	95	101
	0.10	6	11	17	23	28	34	40	46	52	58	64	69	75	81	87	93	99	105	111
16	0.001	0	0	3	6	9	12	16	20	24	28	32	36	40	44	49	53	57	61	66
	0.005	0	3	6	10	14	19	23	28	32	37	42	46	51	56	61	66	71	75	80
	0.01	1	4	8	13	17	22	27	32	37	42	47	52	57	62	67	72	77	83	88
	0.025	2	7	12	16	22	27	32	38	43	48	54	60	65	71	76	82	87	93	99
	0.05	4	9	15	20	26	31	37	43	49	55	61	66	72	78	84	90	96	102	108
	0.10	6	12	18	24	30	37	43	49	55	62	68	75	81	87	94	100	107	113	120
17	0.001	0	1	3	6	10	14	18	22	26	30	35	39	44	48	53	58	62	67	71
	0.005	0	3	7	11	16	20	25	30	35	40	45	50	55	61	66	71	76	82	87
	0.01	1	5	9	14	19	24	29	34	40	45	50	56	61	67	72	78	83	89	94
	0.025	3	7	12	18	23	29	35	40	46	52	58	64	70	76	82	88	94	100	106
	0.05	4	10	14	21	27	34	40	46	52	58	65	71	78	84	90	97	103	110	116
	0.10	7	13	19	26	32	39	46	53	59	66	73	80	86	93	100	107	114	121	128

n_1	α																			
18	0.001	0	1	4	7	11	15	19	24	28	33	38	43	47	52	57	62	67	72	77
	0.005	0	3	7	12	17	22	27	32	38	43	48	54	59	65	71	76	82	88	93
	0.01	1	5	10	15	20	25	31	37	42	48	54	60	66	71	77	83	89	95	101
	0.025	3	8	13	19	25	31	37	43	49	56	62	68	75	81	87	94	100	107	113
	0.05	5	10	17	23	29	36	42	49	56	62	69	76	83	89	96	103	110	117	124
	0.10	7	14	21	28	35	42	49	56	63	70	78	85	92	99	107	114	121	129	136
19	0.001	0	1	4	8	12	16	21	26	30	35	41	46	51	56	61	67	72	78	83
	0.005	1	4	8	13	18	23	29	34	40	46	52	58	64	70	75	82	88	94	100
	0.01	2	5	10	16	21	27	33	39	45	51	57	64	70	76	83	89	95	102	108
	0.025	3	8	14	20	26	33	39	46	53	59	66	73	79	86	93	100	107	114	120
	0.05	5	11	18	24	31	38	45	52	59	66	73	81	88	95	102	110	117	124	131
	0.10	8	15	22	29	37	44	52	59	67	74	82	90	98	105	113	121	129	136	144
20	0.001	0	1	4	8	13	17	22	27	33	38	43	49	55	60	66	71	77	83	89
	0.005	1	4	9	14	19	25	31	37	43	49	55	61	68	74	80	87	93	100	106
	0.01	2	6	11	17	23	29	35	41	48	54	61	68	74	81	88	94	101	108	115
	0.025	3	9	15	21	28	35	42	49	56	63	70	77	84	91	99	106	113	120	128
	0.05	5	12	19	26	33	40	48	55	63	70	78	85	93	101	108	116	124	131	139
	0.10	8	16	23	31	39	47	55	63	71	79	87	95	103	111	120	128	136	144	152

Source: Table 1 of Extended Tables of Critical Values for Wilcoxon's Test Statistic by L. R. Verdooren, *Biometrika* (1963), Vol. 50, pp. 177–186.

Table A.6 Quantiles of the Wilcoxon Signed-Rank Test Statistic[a]

n	$W_{0.005}$	$W_{0.01}$	$W_{0.025}$	$W_{0.05}$	$W_{0.10}$	$W_{0.20}$	$W_{0.30}$	$W_{0.40}$	$W_{0.50}$	$\frac{n(n+1)}{2}$
4	0	0	0	0	1	3	3	4	5	10
5	0	0	0	1	3	4	5	6	7.5	15
6	0	0	1	3	4	6	8	9	10.5	21
7	0	1	3	4	6	9	11	12	14	28
8	1	2	4	6	9	12	14	16	18	36
9	2	4	6	9	11	15	18	20	22.5	45
10	4	6	9	11	15	19	22	25	27.5	55
11	6	8	11	14	18	23	27	30	33	66
12	8	10	14	18	22	28	32	36	39	78
13	10	13	18	22	27	33	38	42	45.5	91
14	13	16	22	26	32	39	44	48	52.5	105
15	16	20	26	31	37	45	51	55	60	120
16	20	24	30	36	43	51	58	63	68	136
17	24	28	35	42	49	58	65	71	76.5	153
18	28	33	41	48	56	66	73	80	85.5	171
19	33	38	47	54	63	74	82	89	95	190
20	38	44	53	61	70	83	91	98	105	210
21	44	50	59	68	78	91	100	108	115.5	131
22	49	56	67	76	87	100	110	119	126.5	153
23	55	63	74	84	95	110	120	130	138	176
24	62	70	82	92	105	120	131	141	150	300
25	69	77	90	101	114	131	143	153	162.5	325
26	76	85	99	111	125	142	155	165	175.5	351
27	84	94	108	120	135	154	167	178	189	378
28	92	102	117	131	146	166	180	192	203	406
29	101	111	127	141	158	178	193	206	217.5	435
30	110	121	138	152	170	191	207	220	232.5	465
31	119	131	148	164	182	205	221	235	248	496
32	129	141	160	176	195	219	236	250	264	528
33	139	152	171	188	208	233	251	266	280.5	561
34	149	163	183	201	222	248	266	282	297.5	595
35	160	175	196	214	236	263	283	299	315	630
36	172	187	209	228	251	279	299	317	333	666
37	184	199	222	242	266	295	316	335	351.5	703
38	196	212	236	257	282	312	334	353	370.5	741
39	208	225	250	272	298	329	352	372	390	780
40	221	239	265	287	314	347	371	391	410	820
41	235	253	280	303	331	365	390	411	430.5	861
42	248	267	295	320	349	384	409	431	451.5	903
43	263	282	311	337	366	403	429	452	473	946
44	277	297	328	354	385	422	450	473	495	990
45	292	313	344	372	403	442	471	495	517.5	1035
46	308	329	362	390	423	463	492	517	540.5	1081
47	324	346	379	408	442	484	514	540	564	1128
48	340	363	397	428	463	505	536	563	588	1176
49	357	381	416	447	483	527	559	587	612.5	1225
50	374	398	435	467	504	550	583	611	637.5	1275

Source: Adapted from Harder and Owen (1970), with permission from the Institute of Mathematical Statistics.

[a]For n larger than 50, the pth quantile ω_p, of the Wilcoxon signed-rank test statistic may be approximated by $\omega_p = [n(n+1)/4] + \chi_p \sqrt{n(n+1)(2n+1)/24}$, where χ_p is the pth quantile of a standard normal random variable, obtained from Appendix A.1.

Table A.7 Tolerance Factor For the Degrees Of Confidence $1 - \alpha$

	$1 - \alpha = 0.95$			$1 - \alpha = 0.99$		
n	0.90	0.95	0.99	0.90	0.95	0.99
2	32.019	37.674	48.430	160.193	188.491	242.300
3	8.380	9.916	12.861	18.930	22.401	29.055
4	5.369	6.370	8.299	9.398	11.150	14.527
5	4.275	5.079	6.634	6.612	7.855	10.260
6	3.712	4.414	5.775	5.337	6.345	8.301
7	3.369	4.007	5.248	4.613	5.488	7.187
8	3.136	3.732	4.891	4.147	4.936	6.468
9	2.967	3.532	4.631	3.822	4.550	5.966
10	2.839	3.379	4.433	3.582	4.265	5.594
11	2.737	3.259	4.277	3.397	4.045	5.308
12	2.655	3.162	4.150	3.250	3.870	5.079
13	2.587	3.081	4.044	3.130	3.727	4.893
14	2.529	3.012	3.955	3.029	3.608	4.737
15	2.480	2.954	3.878	2.945	3.507	4.605
16	2.437	2.903	3.812	2.872	3.421	4.492
17	2.400	2.858	3.754	2.808	3.345	4.393
18	2.366	2.819	3.702	2.753	3.279	4.307
19	2.337	2.784	3.656	2.703	3.221	4.230
20	2.310	2.752	3.615	2.659	3.168	4.161
25	2.208	2.631	3.457	2.494	2.972	3.904
30	2.140	2.549	3.350	2.385	2.841	3.733
35	2.090	2.490	3.272	2.306	2.748	3.611
40	2.052	2.445	3.213	2.247	2.677	3.518
45	2.021	2.408	3.165	2.200	2.621	3.444
50	1.996	2.379	3.126	2.162	2.576	3.385
55	1.976	2.354	3.094	2.130	2.538	3.335
60	1.958	2.333	3.066	2.103	2.506	3.293
65	1.943	2.315	3.042	2.080	2.478	3.257
70	1.929	2.299	3.021	2.060	2.454	3.225
75	1.917	2.285	3.002	2.042	2.433	3.197
80	1.907	2.272	2.986	2.026	2.414	3.173
85	1.897	2.261	2.971	2.012	2.397	3.150
90	1.889	2.251	2.958	1.999	2.382	3.130
95	1.881	2.241	2.945	1.967	2.368	3.112
100	1.874	2.233	2.934	1.977	2.355	3.096
150	1.825	2.175	2.859	1.905	2.270	2.983
200	1.798	2.143	2.816	1.865	2.222	2.921
250	1.780	2.121	2.788	1.839	2.191	2.880
300	1.767	2.106	2.767	1.820	2.169	2.850
400	1.749	2.084	2.739	1.794	2.138	2.809
500	1.737	2.070	2.721	1.777	2.117	2.783
600	1.729	2.060	2.707	1.764	2.102	2.763
700	1.722	2.052	2.697	1.755	2.091	2.748
800	1.717	2.046	2.688	1.747	2.082	2.736
900	1.712	2.040	2.682	1.741	2.075	2.726
1000	1.709	2.036	2.676	1.736	2.068	2.718
∞	1.645	1.960	2.576	1.645	1.960	2.576

Source: Adapted by permission from *Techniques of Statistical Analysis* by C. Eisenhart, M. W. Hastay, and W. A. Wallis. Copyright 1947, McGraw-Hill Book Company, Inc.

Table A.8 Upper Quantiles of the Studentized Range Distribution

ν	α	2	3	4	5	6	7	8	9	10
					(a) $t = 2$–10					
1	.20	4.353	6.615	8.075	9.138	9.966	10.64	11.21	11.70	12.12
	.10	8.929	13.44	16.36	18.49	20.15	21.51	22.64	23.62	24.48
	.05	17.97	26.98	32.82	37.08	40.41	43.12	45.4	47.36	49.07
	.01	90.03	135.0	164.3	185.6	202.2	215.8	227.2	237.0	245.6
2	.20	2.667	3.820	4.559	5.098	5.521	5.867	6.158	6.409	6.630
	.10	4.130	5.733	6.773	7.538	8.139	8.633	9.049	9.409	9.725
	.05	6.085	8.331	9.798	10.88	11.74	12.44	13.03	13.54	13.99
	.01	14.04	19.02	22.29	24.72	26.63	28.20	29.53	30.68	31.69
3	.20	2.316	3.245	3.833	4.261	4.597	4.872	5.104	5.305	5.481
	.10	3.328	4.467	5.199	5.738	6.162	6.511	6.806	7.062	7.287
	.05	4.501	5.910	6.825	7.502	8.037	8.478	8.853	9.177	9.462
	.01	8.261	10.62	12.17	13.33	14.24	15.00	15.64	16.20	16.69
4	.20	2.168	3.004	3.527	3.907	4.205	4.449	4.655	4.832	4.989
	.10	3.015	3.976	4.586	5.035	5.388	5.679	5.926	6.139	6.327
	.05	3.927	5.040	5.757	6.287	6.707	7.053	7.347	7.60	7.826
	.01	6.512	8.120	9.173	9.958	10.58	11.10	11.55	11.93	12.27
5	.20	2.087	2.872	3.358	3.712	3.988	4.214	4.405	4.57	4.715
	.10	2.850	3.717	4.264	4.664	4.979	5.238	5.458	5.648	5.816
	.05	3.635	4.602	5.218	5.673	6.033	6.330	6.582	6.802	6.995
	.01	5.702	6.976	7.804	8.421	8.913	9.321	9.669	9.972	10.24
6	.20	2.036	2.788	3.252	3.588	3.850	4.065	4.246	4.403	4.540
	.10	2.748	3.559	4.065	4.435	4.726	4.966	5.168	5.344	5.499
	.05	3.461	4.339	4.896	5.305	5.628	5.895	6.122	6.319	6.493
	.01	5.243	6.331	7.033	7.556	7.973	8.318	8.613	8.869	9.097
7	.20	2.001	2.731	3.179	3.503	3.756	3.962	4.136	4.287	4.419
	.10	2.680	3.451	3.931	4.28	4.555	4.780	4.972	5.137	5.283
	.05	3.344	4.165	4.681	5.06	5.359	5.606	5.815	5.998	6.158
	.01	4.949	5.919	6.543	7.005	7.373	7.679	7.939	8.166	8.368
8	.20	1.976	2.689	3.126	3.440	3.686	3.886	4.055	4.201	4.330
	.10	2.630	3.374	3.834	4.169	4.431	4.646	4.829	4.987	5.126
	.05	3.261	4.041	4.529	4.886	5.167	5.399	5.597	5.767	5.918
	.01	4.746	5.635	6.204	6.625	6.960	7.237	7.474	7.681	7.863
9	.20	1.956	2.658	3.085	3.393	3.633	3.828	3.994	4.136	4.261
	.10	2.592	3.316	3.761	4.084	4.337	4.545	4.721	4.873	5.007
	.05	3.199	3.949	4.415	4.756	5.024	5.244	5.432	5.595	5.739
	.01	4.596	5.428	5.957	6.348	6.658	6.915	7.134	7.325	7.495
10	.20	1.941	2.632	3.053	3.355	3.590	3.782	3.944	4.084	4.206
	.10	2.563	3.270	3.704	4.018	4.264	4.465	4.636	4.783	4.913
	.05	3.151	3.877	4.327	4.654	4.912	5.124	5.305	5.461	5.599
	.01	4.482	5.270	5.769	6.136	6.428	6.669	6.875	7.055	7.213

Table A.8 (*Continued*)

ν	α	\multicolumn{9}{c}{t}
		2	3	4	5	6	7	8	9	10
		\multicolumn{9}{c}{(a) t = 2–10}
11	.20	1.928	2.612	3.027	3.325	3.557	3.745	3.905	4.042	4.162
	.10	2.540	3.234	3.658	3.965	4.205	4.401	4.568	4.711	4.838
	.05	3.113	3.820	4.256	4.574	4.823	5.028	5.202	5.353	5.487
	.01	4.392	5.146	5.621	5.970	6.247	6.476	6.672	6.842	6.992
12	.20	1.918	2.596	3.006	3.300	3.529	3.715	3.872	4.007	4.126
	.10	2.521	3.204	3.621	3.922	4.156	4.349	4.511	4.652	4.776
	.05	3.082	3.773	4.199	4.508	4.751	4.950	5.119	5.265	5.395
	.01	4.320	5.046	5.502	5.836	6.101	6.321	6.507	6.670	6.814
13	.20	1.910	2.582	2.988	3.279	3.505	3.689	3.844	3.978	4.095
	.10	2.505	3.179	3.589	3.885	4.116	4.305	4.464	4.602	4.724
	.05	3.055	3.735	4.151	4.453	4.690	4.885	5.049	5.192	5.318
	.01	4.260	4.964	5.404	5.727	5.981	6.192	6.372	6.528	6.667
14	.20	1.902	2.570	2.973	3.261	3.485	3.667	3.820	3.953	4.069
	.10	2.491	3.158	3.563	3.854	4.081	4.267	4.424	4.560	4.680
	.05	3.033	3.702	4.111	4.407	4.639	4.829	4.990	5.131	5.254
	.01	4.210	4.895	5.322	5.634	5.881	6.085	6.258	6.409	6.543
15	.20	1.896	2.560	2.960	3.246	3.467	3.648	3.800	3.931	4.046
	.10	2.479	3.140	3.540	3.828	4.052	4.235	4.390	4.524	4.641
	.05	3.014	3.674	4.076	4.367	4.595	4.782	4.940	5.077	5.198
	.01	4.168	4.836	5.252	5.556	5.796	5.994	6.162	6.309	6.439
16	.20	1.891	2.551	2.948	3.232	3.452	3.631	3.782	3.912	4.026
	.10	2.469	3.124	3.520	3.804	4.026	4.207	4.360	4.492	4.608
	.05	2.998	3.649	4.046	4.333	4.557	4.741	4.897	5.031	5.150
	.01	4.131	4.786	5.192	5.489	5.722	5.915	6.079	6.222	6.349
17	.20	1.886	2.543	2.938	3.220	3.439	3.617	3.766	3.895	4.008
	.10	2.460	3.110	3.503	3.784	4.004	4.183	4.334	4.464	4.579
	.05	2.984	3.628	4.020	4.303	4.524	4.705	4.858	4.991	5.108
	.01	4.099	4.742	5.140	5.430	5.659	5.847	6.007	6.147	6.270
18	.20	1.882	2.536	2.930	3.210	3.427	3.604	3.753	3.881	3.993
	.10	2.452	3.098	3.488	3.767	3.984	4.161	4.311	4.440	4.554
	.05	2.971	3.609	3.997	4.277	4.495	4.673	4.824	4.956	5.071
	.01	4.071	4.703	5.094	5.379	5.603	5.788	5.944	8.081	6.201
19	.20	1.878	2.530	2.922	3.200	3.416	3.592	3.740	3.867	3.979
	.10	2.445	3.087	3.474	3.751	3.966	4.142	4.290	4.418	4.531
	.05	2.960	3.593	3.977	4.253	4.469	4.645	4.794	4.924	5.038
	.01	4.046	4.670	5.054	5.334	5.554	5.735	5.889	6.022	6.141
20	.20	1.874	2.524	2.914	3.192	3.407	3.582	3.729	3.855	3.966
	.10	2.439	3.078	3.462	3.736	3.950	4.124	4.271	4.398	4.510
	.05	2.950	3.578	3.958	4.232	4.445	4.620	4.768	4.896	5.008
	.01	4.024	4.639	5.018	5.294	5.510	5.688	5.839	5.970	6.087

Table A.8 (*Continued*)

ν	α	\multicolumn{9}{c}{t}
		2	3	4	5	6	7	8	9	10
		\multicolumn{9}{c}{(a) $t = 2\text{--}10$}
24	.20	1.864	2.507	2.892	3.166	3.377	3.549	3.694	3.818	3.927
	.10	2.420	3.047	3.423	3.692	3.900	4.070	4.213	4.336	4.445
	.05	2.919	3.532	3.901	4.166	4.373	4.541	4.684	4.807	4.915
	.01	3.956	4.546	5.907	5.168	5.374	5.542	5.685	5.809	5.919
30	.20	1.853	2.490	2.870	3.140	3.348	3.517	3.659	3.781	3.887
	.10	2.400	3.017	3.386	3.648	3.851	4.016	4.155	4.275	4.381
	.05	2.888	3.486	3.845	4.102	4.302	4.464	4.602	4.720	4.824
	.01	3.889	4.455	4.799	5.048	5.242	5.401	5.536	5.653	5.756
40	.20	1.843	2.473	2.848	3.114	3.318	3.484	3.624	3.743	3.848
	.10	2.381	2.988	3.349	3.605	3.803	3.963	4.099	4.215	4.317
	.05	2.858	3.442	3.791	4.039	4.232	4.389	4.521	4.635	4.735
	.01	3.825	4.367	4.696	4.931	5.114	5.265	5.392	5.502	5.599
60	.20	1.833	2.456	2.826	3.089	3.290	3.452	3.589	3.707	3.809
	.10	2.363	2.959	3.312	3.562	3.755	3.911	4.042	4.155	4.254
	.05	2.829	3.399	3.737	3.977	4.163	4.314	4.441	4.550	4.646
	.01	3.762	4.282	4.595	4.818	4.991	5.133	5.253	5.356	5.447
120	.20	1.822	2.440	2.805	3.063	3.260	3.420	3.554	3.669	3.770
	.10	2.344	2.930	3.276	3.520	3.707	3.859	4.987	4.096	4.191
	.05	2.800	3.356	3.685	3.917	4.096	4.241	4.363	4.468	4.560
	.01	3.702	4.200	4.497	4.709	4.872	5.005	5.118	5.214	5.299
∞	.20	1.812	2.424	2.784	3.037	3.232	3.389	3.520	3.632	3.730
	.10	2.326	2.902	3.240	3.478	3.661	3.808	3.931	4.037	4.129
	.05	2.772	3.314	3.633	3.858	4.030	4.170	4.286	4.387	4.474
	.01	3.643	4.120	4.403	4.603	4.757	4.882	4.987	5.078	5.157

Table A.8 (*Continued*)

ν	α	\multicolumn{9}{c}{t}								
		11	12	13	14	15	16	17	18	19
		\multicolumn{9}{c}{(b) t = 11–19}								
10	.20	4.316	4.414	4.503	4.585	4.660	4.730	4.795	4.856	4.913
	.10 !	5.029	5.134	5.229	5.317	5.397	5.472	5.542	5.607	5.668
	.05	5.722	5.833	5.935	6.028	6.114	6.194	6.269	6.339	6.405
	.01	7.356	7.485	7.603	7.712	7.812	7.906	7.993	8.076	8.153
11	.20	4.270	4.366	4.454	4.534	4.608	4.677	4.741	4.801	4.857
	.10	4.951	5.053	5.146	5.231	5.309	5.382	5.450	5.514	5.573
	.05	5.605	5.713	5.811	5.901	5.984	6.062	6.134	6.202	6.265
	.01	7.128	7.250	7.362	7.465	7.560	7.649	7.732	7.809	7.883
12	.20	4.231	4.327	4.413	4.492	4.565	4.633	6.696	4.755	4.810
	.10	4.886	4.986	5.077	5.160	5.236	5.308	5.374	5.436	5.495
	.05	5.511	5.615	5.710	5.798	5.878	5.953	6.023	6.089	6.151
	.01	6.943	7.060	7.167	7.265	7.356	7.441	7.520	7.594	7.665
13	.20	4.199	4.293	4.379	4.457	4.529	4.596	4.658	4.716	4.770
	.10	4.832	4.930	5.019	5.100	5.176	5.245	5.311	5.372	5.429
	.05	5.431	5.533	5.625	5.711	5.789	5.862	5.931	5.995	6.055
	.01	6.791	6.903	7.006	7.101	7.188	7.269	7.345	7.417	7.485
14	.20	4.172	4.265	4.349	4.426	4.498	4.564	4.625	4.683	4.737
	.10	4.786	4.882	4.970	5.050	5.124	5.192	5.256	5.316	5.373
	.05	5.364	5.463	5.554	5.637	5.714	5.786	5.852	5.915	5.974
	.01	6.664	6.772	6.871	6.962	7.047	7.126	7.199	7.268	7.333
15	.20	4.148	4.240	4.324	4.400	4.471	4.536	4.597	4.654	4.707
	.10	4.746	4.841	4.927	5.006	5.079	5.147	5.209	5.269	5.324
	.05	5.306	5.404	5.493	5.574	5.649	5.720	5.785	5.846	5.904
	.01	6.555	6.660	6.757	6.845	6.927	7.003	7.074	7.142	7.204
16	.20	4.127	4.218	4.301	4.377	4.447	4.512	4.572	4.628	4.681
	.10	4.712	4.805	4.890	4.968	5.040	5.107	5.169	5.227	5.282
	.05	5.256	5.352	5.439	5.520	5.593	5.662	5.727	5.786	5.843
	.01	6.562	6.564	8.658	6.744	6.823	6.898	6.967	7.032	7.093
17	.20	4.109	4.199	4.282	4.357	4.426	4.490	4.550	4.606	4.659
	.10	4.682	4.774	4.858	4.935	5.005	5.071	5.133	5.190	5.244
	.05	5.212	5.307	5.392	5.471	5.544	5.612	5.675	5.734	5.790
	.01	6.381	6.480	6.572	6.656	6.734	6.806	6.873	6.937	6.997
18	.20	4.093	4.182	4.264	4.339	4.407	4.471	4.531	4.586	4.638
	.10	4.655	4.746	4.829	4.905	4.975	5.040	5.101	5.158	5.211
	.05	5.174	5.267	5.352	5.429	5.501	5.568	5.630	5.688	5.743
	.01	6.310	6.407	6.497	6.579	6.655	6.725	6.792	6.854	6.912
19	.20	4.078	4.167	4.248	4.323	4.391	4.454	4.513	4.569	4.620
	.10	4.631	4.721	4.803	4.879	4.948	5.012	5.073	5.129	5.182
	.05	5.140	5.231	5.315	5.391	5.462	5.528	5.589	5.647	5.701
	.01	6.247	6.342	6.430	6.510	6.585	6.654	6.719	6.780	6.837

Table A.8 (*Continued*)

ν	α	\multicolumn{9}{c}{t}
		11	12	13	14	15	16	17	18	19

(b) t = 11–19

ν	α	11	12	13	14	15	16	17	18	19
20	.20	4.065	4.154	4.234	4.308	4.376	4.439	4.498	4.552	4.604
	.10	4.609	4.699	4.780	4.855	4.924	4.987	5.047	5.103	5.155
	.05	5.108	5.199	5.282	5.357	5.427	5.493	5.553	5.610	5.663
	.01	6.191	6.285	6.371	6.450	6.523	6.591	6.654	6.714	6.771
24	.20	4.024	4.111	4.190	4.262	4.329	4.391	4.448	4.502	4.552
	.10	4.541	4.628	4.708	4.780	4.847	4.909	4.966	5.021	5.071
	.05	5.012	5.099	5.179	5.251	5.319	5.381	5.439	5.494	5.545
	.01	6.017	6.106	6.186	6.261	6.330	6.394	6.453	6.510	6.563
30	.20	3.982	4.068	4.145	4.216	4.281	4.342	4.398	4.451	4.500
	.10	4.474	4.559	4.635	4.706	4.770	4.830	4.886	4.939	4.988
	.05	4.917	5.001	5.077	5.147	5.211	5.271	5.327	5.379	5.429
	.01	5.849	5.932	6.008	6.078	6.143	6.203	6.259	6.311	6.361
40	.20	3.941	4.025	4.101	4.170	4.234	4.293	4.348	4.399	4.447
	.10	4.408	4.490	4.564	4.632	4.695	4.752	4.807	4.857	4.905
	.05	4.824	4.904	4.977	5.044	5.106	5.163	5.216	5.266	5.313
	.01	5.686	5.764	5.835	5.900	5.961	6.017	6.069	6.119	6.165
60	.20	3.900	3.982	4.056	4.124	4.186	4.244	4.297	4.347	4.395
	.10	4.342	4.421	4.493	4.558	4.619	4.675	4.727	4.775	4.821
	.05	4.732	4.808	4.878	4.942	5.001	5.056	5.107	5.154	5.199
	.01	5.528	5.601	5.667	5.728	5.785	5.837	5.886	5.931	5.974
120	.20	3.859	3.938	4.011	4.077	4.138	4.194	4.246	4.295	4.341
	.10	4.276	4.353	4.422	4.485	4.543	4.597	4.647	4.694	4.738
	.05	4.641	4.714	4.781	4.842	4.898	4.950	4.998	5.044	5.086
	.01	5.375	5.443	5.505	5.562	5.614	5.662	5.708	5.750	5.790
∞	.20	3.817	3.895	3.966	4.030	4.089	4.144	4.195	4.242	4.287
	.10	4.211	4.285	4.351	4.412	4.468	4.519	4.568	4.612	4.654
	.05	4.552	4.622	4.685	4.743	4.796	4.845	4.891	4.934	4.974
	.01	5.227	5.290	5.348	5.400	5.448	5.493	5.535	5.574	5.611

Table A.8 (Continued)

ν	α	\multicolumn{9}{c}{t}
		20	22	24	26	28	30	32	34	36
		\multicolumn{9}{c}{(c) t = 20–36}
19	.20	4.669	4.759	4.840	4.914	4.981	5.044	5.102	5.156	5.206
	.10	5.232	5.324	5.407	5.483	5.552	5.616	5.676	5.732	5.784
	.05	5.752	5.846	5.932	6.009	6.081	6.147	6.209	6.267	6.321
	.01	6.891	6.992	7.082	7.166	7.242	7.313	7.379	7.440	7.498
20	.20	4.652	4.742	4.822	4.895	4.963	5.025	5.082	5.136	5.186
	.10	5.205	5.296	5.378	5.453	5.522	5.586	5.645	5.700	5.752
	.05	5.714	5.807	5.891	5.968	6.039	6.104	6.165	6.222	6.275
	.01	6.823	6.922	7.011	7.092	7.168	7.237	7.302	7.362	7.419
24	.20	4.599	4.687	4.766	4.838	4.904	4.964	5.021	5.073	5.122
	.10	5.119	5.208	5.287	5.360	5.427	5.489	5.546	5.600	5.650
	.05	5.594	5.683	5.764	5.838	5.906	5.968	6.027	6.081	6.132
	.01	6.612	6.705	6.789	6.865	6.936	7.001	7.062	7.119	7.173
30	.20	4.546	4.632	4.710	4.779	4.844	4.903	4.958	5.010	5.058
	.10	5.034	5.120	5.197	5.267	5.332	5.392	5.447	5.499	5.547
	.05	5.475	5.561	5.638	5.709	5.774	5.833	5.889	5.941	5.990
	.01	6.407	6.494	6.572	6.644	6.710	6.772	6.828	6.881	6.932
40	.20	4.493	4.576	4.652	4.720	4.783	4.841	4.895	4.945	4.993
	.10	4.949	5.032	5.107	5.174	5.236	5.294	5.347	5.397	5.444
	.05	5.358	5.439	5.513	5.581	5.642	5.700	5.753	5.803	5.849
	.01	6.209	6.289	6.362	6.429	6.490	6.547	6.600	6.650	6.697
60	.20	4.439	4.520	4.594	4.661	4.722	4.778	4.831	4.880	4.925
	.10	4.864	4.944	5.015	5.081	5.141	5.196	5.247	5.295	5.340
	.05	5.241	5.319	5.389	5.453	5.512	5.566	5.617	5.664	5.708
	.01	6.015	6.090	6.158	6.220	6.277	6.330	6.378	6.424	6.467
120	.20	4.384	4.463	4.535	4.600	4.659	4.714	4.765	4.812	4.857
	.10	4.779	4.856	4.924	4.987	5.044	5.097	5.146	5.192	5.235
	.05	5.126	5.200	5.266	5.327	5.382	5.434	5.481	5.526	5.568
	.01	5.827	5.897	5.959	6.016	6.069	6.117	6.162	6.204	6.244
∞	.20	4.329	4.405	4.475	4.537	4.595	4.648	4.697	4.743	4.786
	.10	4.694	4.767	4.832	4.892	4.947	4.997	5.044	5.087	5.128
	.05	5.012	5.081	5.144	5.201	5.253	5.301	5.346	5.388	5.427
	.01	5.645	5.709	5.766	5.818	5.866	5.911	5.952	5.990	6.026

Table A.8 (*Continued*)

ν	α	\multicolumn{8}{c}{t}
		38	40	50	60	70	80	90	100

(d) t = 38–100

ν	α	38	40	50	60	70	80	90	100
30	.20	5.103	5.146	5.329	5.475	5.597	5.701	5.791	5.871
	.10	5.593	5.636	5.821	5.969	6.093	6.198	6.291	6.372
	.05	6.037	6.080	6.267	6.417	6.543	6.650	6.744	6.827
	.01	6.978	7.023	7.215	7.370	7.500	7.611	7.709	7.796
40	.20	5.037	5.078	5.257	5.399	5.518	5.619	5.708	5.786
	.10	5.488	5.529	5.708	5.850	5.969	6.071	6.160	6.238
	.05	5.893	5.934	6.112	6.255	6.375	6.477	6.566	6.645
	.01	6.740	6.782	6.960	7.104	7.225	7.328	7.419	7.500
60	.20	4.969	5.009	5.183	5.321	5.437	5.535	5.621	5.697
	.10	5.382	5.422	5.593	5.730	5.844	5.941	6.026	6.102
	.05	5.750	5.789	5.958	6.093	6.206	6.303	6.387	6.462
	.01	6.507	6.546	6.710	6.843	6.954	7.050	7.133	7.207
120	.20	4.899	4.938	5.106	5.240	5.352	5.447	5.530	5.603
	.10	5.275	5.313	5.476	5.606	5.715	5.808	5.888	5.960
	.05	5.607	5.644	5.802	5.929	6.035	6.126	6.205	6.275
	.01	6.281	6.316	6.467	6.588	6.689	6.776	6.852	6.919
∞	.20	4.826	4.864	5.026	5.155	5.262	5.353	5.433	5.503
	.10	5.166	5.202	5.357	5.480	5.582	5.669	5.745	5.812
	.05	5.463	5.498	5.646	5.764	5.863	5.947	6.020	6.085
	.01	6.060	6.092	6.228	6.338	6.429	6.507	6.575	6.636

Source: Values were extracted by permission from H. L. Harter, *1969, Order Statistics and Their Use in Testing and Estimation*, Vol. 1, Aerospace Research Laboratory, USAF, U.S. Government Printing Office, Washington, D.C., pp. 648–657.

Table A.9 Upper Quantiles of the Dunnett's t Distribution: One-Sided Comparisons with Control

		\multicolumn{9}{c}{m}
ν	α	1	2	3	4	5	6	7	8	9
5	.05	2.02	2.44	2.68	2.85	2.98	3.08	3.16	3.24	3.30
	.01	3.37	3.90	4.21	4.43	4.60	4.73	4.85	4.94	5.03
6	.05	1.94	2.34	2.56	2.71	2.83	2.92	3.00	3.07	3.12
	.01	3.14	3.61	3.88	4.07	4.21	4.33	4.43	4.51	4.59
7	.05	1.89	2.27	2.48	2.62	2.73	2.82	2.89	2.95	3.01
	.01	3.00	3.42	3.66	3.83	3.96	4.07	4.15	4.23	4.30
8	.05	1.86	2.22	2.42	2.55	2.66	2.74	2.81	2.87	2.92
	.01	2.90	3.29	3.51	3.67	3.79	3.88	3.96	4.03	4.09
9	.05	1.83	2.18	2.37	2.50	2.60	2.68	2.75	2.81	2.86
	.01	2.82	3.19	3.40	3.55	3.66	3.75	3.82	3.89	3.94
10	.05	1.81	2.15	2.34	2.47	2.56	2.64	2.70	2.76	2.81
	.01	2.76	3.11	3.31	3.45	3.56	3.64	3.71	3.78	3.83
11	.05	1.80	2.13	2.31	2.44	2.53	2.60	2.67	2.72	2.77
	.01	2.72	3.06	3.25	3.38	3.48	3.56	3.63	3.69	3.74
12	.05	1.78	2.11	2.29	2.41	2.50	2.58	2.64	2.69	2.74
	.01	2.68	3.01	3.19	3.32	3.42	3.50	3.56	3.62	3.67
13	.05	1.77	2.09	2.27	2.39	2.48	2.55	2.61	2.66	2.71
	.01	2.65	2.97	3.15	3.27	3.37	3.44	3.51	3.56	3.61
14	.05	1.76	2.08	2.25	2.37	2.46	2.53	2.59	2.64	2.69
	.01	2.62	2.94	3.11	3.23	3.32	3.40	3.46	3.51	3.56
15	.05	1.75	2.07	2.24	2.36	2.44	2.51	2.57	2.62	2.67
	.01	2.60	2.91	3.08	3.20	3.29	3.36	3.42	3.47	3.52
16	.05	1.75	2.06	2.23	2.34	2.43	2.50	2.56	2.61	2.65
	.01	2.58	2.88	3.05	3.17	3.26	3.33	3.39	3.44	3.48
17	.05	1.74	2.05	2.22	2.33	2.42	2.49	2.54	2.59	2.64
	.01	2.57	2.86	3.03	3.14	3.23	3.30	3.36	3.41	3.45
18	.05	1.73	2.04	2.21	2.32	2.41	2.48	2.53	2.58	2.62
	.01	2.55	2.84	3.01	3.12	3.21	3.27	3.33	3.38	3.42
19	.05	1.73	2.03	2.20	2.31	2.40	2.47	2.52	2.57	2.61
	.01	2.54	2.83	2.99	3.10	3.18	3.25	3.31	3.36	3.40
20	.05	1.72	2.03	2.19	2.30	2.39	2.46	2.51	2.56	2.60
	.01	2.53	2.81	2.97	3.08	3.17	3.23	3.29	3.34	3.38
24	.05	1.71	2.01	2.17	2.28	2.36	2.43	2.48	2.53	2.57
	.01	2.49	2.77	2.92	3.03	3.11	3.17	3.22	3.27	3.31

Table A.9 (*Continued*)

ν	α	\multicolumn{9}{c}{m}
		1	2	3	4	5	6	7	8	9
30	.05	1.70	1.99	2.15	2.25	2.33	2.40	2.45	2.50	2.54
	.01	2.46	2.72	2.87	2.97	3.05	3.11	3.16	3.21	3.24
40	.05	1.68	1.97	2.13	2.23	2.31	2.37	2.42	2.47	2.51
	.01	2.42	2.68	2.82	2.92	2.99	3.05	3.10	3.14	3.18
60	.05	1.67	1.95	2.10	2.21	2.28	2.35	2.39	2.44	2.48
	.01	2.39	2.64	2.78	2.87	2.94	3.00	3.04	3.08	3.12
120	.05	1.66	1.93	2.08	2.18	2.26	2.32	2.37	2.41	2.45
	.01	2.36	2.60	2.73	2.82	2.89	2.94	2.99	3.03	3.06
∞	.05	1.64	1.92	2.06	2.16	2.23	2.29	2.34	2.38	2.42
	.01	2.33	2.56	2.68	2.77	2.84	2.89	2.93	2.97	3.00

Source: Values taken by permission from C. W. Dunnett, *J. Am. Stat. Assoc.*, **50**(1955):1115–1116.

APPENDIX B

SAS Programs

Table B.1 SAS Program Used for Generation of Random Codes by Simple Randomization for Two Parallel Groups
Table B.2 SAS Program Used for Generation of Random Codes by Simple Randomization for Three Parallel Groups
Table B.3 SAS Program Used for Generation of Random Codes by Permuted-Block Randomization and Random Allocation for Two Parallel Groups
Table B.4 SAS Program Used for Generation of Random Codes by Permuted-Block by Random Selection for Parallel Groups

Table B.1 SAS Program Used For Generation of Random Codes by Simple Randomization for Two Parallel Groups

```
options linesize=70 pagesize=40 nodate;
data one;
seed=4576891;
n=1;
p=0.5;
do center=1 to 4;
do sub=1 to 24;
tmt=ranbin(seed,n,p)+1;
subject=(14*100000)+(center*1000)+sub;
output;
end;
end;
run;
proc format;
value tmtf 1='Active Drug'
           2='Placebo       ';
value centerf 1='J. Smith, MD'
              2='M. Dole, MD '
              3='A. Hope, MD '
              4='C. Price, MD';
proc sort data=one;   by center;
proc print data=one split='*' ; pageby center; by center;
id  subject;   var tmt;
label
     subject='Subject*Number *____'
     tmt='Treatment *Assignment*_____';
format tmt tmtf. center centerf.;
title1 '                   Table 4.3.3                          ';
title2 ' Example of Simple Randomization for Four Centers       ';
title3 ' Program:[DRUGXXX.PXXX014.SAS.NCKUJPL]SIMPRAN.SAS';
title4 ' Random Codes for Drug XXX, Protocol XXX-014            ';
title5 ' Double-blind, Randomized, Placebo Control, Two Parallel Groups';
run;
proc freq data=one;
tables center*tmt/nopercent nocol norow;
format tmt tmtf.;
run;
```

Table B.2 SAS Program Used for Generation of Random Codes by Simple Randomization for Three Parallel Groups

```
options linesize=70 pagesize=40 nodate;
data one;
seed=716891;
n=1;
p=1/3;
do center=1 to 4;
do sub=1 to 24;
seed=seed+(sub*(center-1));
tmt=rantbl(seed,p,p,p);
subject=(14*100000)+(center*1000)+sub;
output;
end;
end;
run;
proc print; run;
proc format;
value tmtf 1='Placebo'
           2='100 mg '
           3='200 mg ';
value centerf 1='J. Smith, MD'
              2='M. Dole, MD '
              3='A. Hope, MD '
              4='C. Price, MD';
proc sort data=one;   by center;
proc print data=one split='*' ; pageby center; by center;
id   subject;    var tmt;
label
      subject='Subject*Number *____'
      tmt='Treatment *Assignment*_____';
format tmt tmtf. center centerf.;
title1 '                   Table 4.3.4                        ';
title2 ' Example of Simple Randomization for Four Centers     ';
title3 ' Program:[DRUGXXX.PXXX016.SAS.NCKUJPL]SIMPRAN.SAS    ';
title4 ' Random Codes for Drug XXX, Protocol XXX-014          ';
title5 ' Double-blind, Randomized, Placebo Control, Three Parallel Groups';
run;
proc freq data=one;
tables center*tmt/nopercent nocol norow;
format tmt tmtf.;
run;
```

Table B.3 SAS Program Used for Generation of Random Codes by Permuted-Block Randomization and Random Allocation for Two Parallel Groups

```
options linesize=70 pagesize=40 nodate;
proc plan ordered;
factors center=4 blocks=6 cell=4;
treatments t=4 random;
output out=perblock;
run;
data perblock; set perblock;
tmt=1;
if t gt 2 then tmt=2;
sub=_n_;
subject=(14*100000)+center*1000+(sub-(center-1)*24);
proc format;
value tmtf 1='Active Drug'
           2='Placebo       ';
      centerf 1='J. Smith, MD'
              2='M. Dole, MD '
              3='A. Hope, MD '
              4='C. Price, MD';
proc sort data=perblk;   by center;
proc print data=perblock split='*' ; pageby center; by center;
id subject;
var   t tmt;
label
      subject='Subject*Number *____'
      t='   Random   *Permutation*_____'
      tmt='Treatment *Assignment*_____';
format tmt tmtf. center centerf.;
title1 '                Table 4.3.7                      ';
title2 ' Example of Permutated-Block Randomization       ';
title3 ' for Four Centers and a Block Size of Four       ';
title4 ' Program:[DRUGXXX.PXXX014.SAS.NCKUJPL]PERBLK.SAS ';
title5 ' Random Codes for Drug XXX, Protocol XXX-014     ';
title6 ' Double-blind, Randomized, Placebo-Control, Two Parallel Groups';
run;
proc sort data=perblock;   by center;
proc freq data=perblock;
tables center*tmt/nopercent nocol norow;
format tmt tmtf.; run;
proc freq data=perblock; by center;
tables blocks*tmt/nopercent nocol norow;
format tmt tmtf.; run;
proc plan ordered;
factors center=4   cell=24;
treatments t=24 random;
```

Table B.3 (*Continued*)

```
output out=perblock;
run;
data perblock;   set perblock;
tmt=1;
if t gt 12 then tmt=2;
sub=_n_;
subject=(14*100000)+center*1000+(sub-(center-1)*24);
proc format;
value tmtf 1='Active Drug'
           2='Placebo     ';
proc sort data=perblk;   by center;
proc print data=perblock split='*'; pageby center; by center;
id subject;
var   t tmt;
label
      subject='Subject*Number *____'
      t='  Random   *Permutation*_____'
      tmt='Treatment *Assignment*_____';
format tmt tmtf. center centerf.;
title1 '                    Table 4.3.6                              ';
title2 ' Example of Random Allocation for Four Centers               ';
title3 ' Program:[DRUGXXX.PXXX014.SAS.NCKUJPL]RANALC.SAS ';
title4 ' Random Codes for Drug XXX, Protocol XXX-014                 ';
title5 ' Double-blind, Randomized, Placebo-Control, Two Parallel Groups';
run;
proc freq data=perblock;
tables center*tmt/nopercent nocol norow;
format tmt tmtf.;
run;
```

Table B.4 SAS Program Used for Generation of Random Codes by Permuted-Block by Random Selection for Parallel Groups

```
options linesize=70 pagesize=40 nodate;
data one;
do bloc=1 to 6;
do sub=1 to 4;
input tmt @;
output;
end; end;
cards;
1 1 2 2 2 2 1 1 1 2 1 2 1 2 2 1 2 1 2 1 2 1 1 2
data one;  set one;
do center=1 to 4;
output;
end;
proc sort data=one;   by center block sub;
proc plan ordered;
factors center=4 nblock=6 ;
treatments block=6 random;
output out=perblock;
run;
proc sort data=perblock;   by center block ;
data perblock; merge perblock one;   by center block ;
proc sort data=perblock;   by center nblock sub;
proc print;
data perblock;   set perblock;
sub=_n_;
subject=(14*100000)+center*1000+(sub-(center-1)*24);
proc format;
value tmtf 1='Active Drug'
           2='Placebo        ';
value centerf 1='J. Smith, MD'
              2='M. Dole, MD '
              3='A. Hope, MD '
              4='C. Price, MD';
proc sort data=perblk;   by center;
proc print data=perblock split='*' ; pageby center; by center;
id    subject;
var   tmt;
label subject='Subject*Number *____'
      tmt='Treatment *Assignment*____';
format tmt tmtf. center centerf.;
```

Table B.4 (*Continued*)

```
title1 '                  Table 4.3.8                              ';
title2 ' Example of Permutated-Block Randomization by Random Selection';
title3 ' of Blocks for Four Centers and a Block Size of Four        ';
title4 ' Program:[DRUGXXX.PXXX014.SAS.NCKUJPL]PERBLK.SAS            ';
title5 ' Random Codes for Drug XXX, Protocol XXX-014                ';
title6 ' Double-blind, Randomized, Placebo-Control, Two Parallel Groups ';
run;
proc sort data=perblock;   by center;
proc freq data=perblock;
tables center*tmt/nopercent nocol norow;
format tmt tmtf.; run;
proc freq data=perblock; by center;
tables nblock*tmt/nopercent nocol norow;
format tmt tmtf.; run;
```

Index

Absorbing event, 538, 556
Accelerated approval, 6, 32
Accuracy, 52, 100, 425
Additivity, 345
Adverse drug experience, 536
Adverse drug reaction (ADR), 118, 536, 542, 543
Adverse event (AE), 22, 37, 103, 110, 130, 173, 228, 363, 536, 537, 542, 543, 546–548, 551–553, 555, 557, 558, 571, 586, 589, 593, 594, 597
  absorbing, 571
  common, 551, 555
  post-therapy, 547
  primary, 551
  rare, 555
  secondary, 551
  serious (SAE), 22, 484, 536, 537, 546, 547, 551–553
  significant, 547, 552, 553
  treatment-emergent (TEAE), 547
Adverse experience, 23, 364
  serious, 23
  unexpected, 23
Adverse reaction, 16, 118, 536
  serious, 536
Advisory committee, 33
Alternative
  fixed local, 455
  Lehmann's, 386
  one-sided, 418, 463
  two-sided, 510
Alzheimer's disease assessment scale (ADAS), 204
Analysis
  change from baseline, 322
  cross-sectional, 300

  efficacy, 484
  evaluable, 492–494
  final, 529
  intention-to-treat (ITT), 67, 110, 403, 491–493
  interim, 21, 22, 94, 112, 114, 156, 168, 212, 256, 363–366, 407–410, 413, 414, 418, 422, 484, 491, 515, 525, 526, 530, 577, 580, 582, 583, 585, 591, 597
    administrative, 528
    formal, 528
    unplanned, 526
    unreported, 526
  meta-, 115, 117, 221, 560, 575, 582, 590
  ordinary regression, 390
  power, 424, 430, 482, 585
  precision, 424, 425, 428, 429
  preferred, 494
  prestudy power, 425
  primary, 491, 504
  regression, 501
  shift, 564
  stratified, 418, 502, 509, 515
  subgroup, 21, 484, 502, 515, 523, 524
  survival, 366, 453, 556, 560, 584
Analysis of covariance (ANCOVA), 281, 283, 284, 501, 514, 524
Analysis of variance (ANOVA), 146, 217, 272, 273, 275, 277, 279, 284, 293, 295, 297, 305, 562, 584, 590
  one-way, 337, 441, 501
  two-way, 507, 589
Analysis of variance table, 518
Angiotensin converting enzyme (ACE) inhibitor, 200
Approach
  alpha spending, 580

635

Approach (*Continued*)
  confidence interval, 267
  population-average, 305
  quasi-likelihood, 300, 598
Approximation, 52
  normal, 454
Arrhythmia, 97
  ventricular, 200
Aspirin, 67
Assumption
  fundamental bioequivalence, 116, 241
  normality, 284, 295, 345, 520
  proportional hazard, 389–391, 393, 394
Average, 163, 487
  adjusted treatment, 500
  unadjusted treatment, 500
AUA-7 symptom score, 307
Audit trail, 156

Baseline, 19, 22, 115, 125, 281, 485
Baseline characteristic, 119
Baseline hazard, 385
Baseline value, 497, 500
Beta-block, 200
Beta-blocker Heart Attack Trial (BHAT), 368, 402, 573
Bias, 4, 19, 26, 47, 52, 53, 94, 99, 103, 113, 114, 120, 121, 144, 165, 170, 171, 209, 256, 423, 491, 526, 532, 593
  accidental, 131, 134, 145, 497
  investigator's, 130
  judgmental, 166
  selection, 19, 53, 129, 130, 134, 145, 172, 179, 359
  subjective, 129
Binary response, 306, 307
Bioavailability, 4
Bioequivalence, 4, 91, 165, 252, 421, 423, 582, 587, 591
  individual, 587
Biological product, 3, 8, 10
Biostatistician, 44, 120, 163
Biostatistics, 35, 43, 529
Blind
  Double, 26, 107, 156, 166, 167, 182, 183, 191, 574
  Single, 166, 167
  Triple, 26, 113, 155, 166, 167
Blinded, triple, 527
Blinding, 19, 21, 26, 94, 120, 122, 165, 166
  double, 122
  single, 122
  triple, 122, 175

Blindness, 113, 114, 156, 159, 167, 171, 173
Blocking, 60, 122
Body System Classification, 549
Bone mineral density (BMD), 61
Bonferroni adjustment, 401, 481, 519
Bonferroni technique, 516
Boundary
  group sequential, 408, 410, 411, 413
  O'Brien–Fleming, 411, 413
  O'Brien–Fleming group sequential, 413
B-value, 365, 419, 585

Cardiac Arrhythmia Suppression Trial (CAST), 49, 206, 402, 419, 574, 590, 593
Case report form, 38, 120, 155
Case report tabulation, 484
Center for Biologics Evaluation and Research (CBER), 8
Center for Devices and Radiological Health (CDRH), 8
Center for Drug Evaluation and Research (CDER), 8
Central limit theorem, 52, 262, 263, 269, 287
Central tendency, 50
Change from the baseline, 408, 486, 541
  percent, 486
Chebyshev inequality, 429
Cholesterol, 50
Cholesterol and Recurrent Events Trial Investigator, 575
Classification
  one-way, 273, 283
  two-way, 277
Clinical chemistry, 542
Clinical Development Plan, 34, 43
Clinical Development Program, 36
Clinical response, 147
Clinical significant change, 563, 564
Clinician, 44, 163
Code of Federal Regulations (CFR), 1
Coding Symbols for a Thesaurus of Adverse Reaction Terms (COSTART), 537, 548–550, 586
Coding system, 548
Combination
  fixed-drug, 588
  optimal, 235
  optimal dose, 240, 241
Combination product, 226, 588
Combination therapy, 181, 226, 228, 229, 231
Combined product, 14
Combined therapy, 14

INDEX   **637**

Committee for Proprietary Material Products
    (CPMP) Working Party, 484, 576
Comparison
    between-group, 327
    marginal, 323
    multiple, 21, 275, 485, 515
    pairwise, 401
    within-patient, 183
    within-treatment, 268
Compliance, 66, 120, 180
    patient, 118
Composite clinical index, 307
Composite index, 520
Confidence interval, 75, 79, 249, 250, 263,
        269, 275, 313, 327, 342, 372, 383,
        422, 423, 428, 573, 597
    Bonferroni simultaneous, 275–277
    exact, 313, 326
    large-sample, 371, 388, 398
    repeated, 365, 413, 583
    Satterwaite's, 271
    simultaneous, 274, 517
Confidence limit, 250
Confounding, 47, 63
Consistency, 61
Contingency table, 331
    2 × 2, 332, 556
Continuity correction, 325, 334
Contract Research Organization (CRO), 528
Contrast
    linear, 459
    orthogonal, 487
    within-subject, 186
Control, 1, 6, 19, 26, 94, 177, 183
    active, 102, 178, 221, 222, 242, 256, 269,
        276, 425, 426, 440
    active concurrent, 26, 104
    active treatment concurrent, 107, 200
    concurrent, 104, 158, 167, 169
    dose-comparison concurrent, 26, 104, 106,
        107
    historical, 26, 104, 108, 221, 231
    matched, 169
    no treatment, 26, 104
    no treatment current, 108, 221
    parallel placebo current, 200, 201
    placebo, 120, 178, 191, 211, 221, 223, 224,
        269, 272, 276, 481, 519, 578
    placebo concurrent, 26, 102, 104–107, 192
    positive, 221
    quality, 529
Control group, 434
Cooperative North Scandinavian Enalapril
    Survival Study II (CONSENSUS II),
        530
Correlation, 252, 501, 516, 518, 571
    intrasubject, 299, 300
    pairwise, 521
    within-subject, 301
Correlation coefficient, 519
    intrasubject, 146
Correlation matrix, within-group, 520
Coronary Drug Project Research, 576
Cost-benefit analysis, 2
Cost-effectiveness, 2, 36
Cost-minimization, 2
Covariance, 301
Covariance matrix, 358, 422, 487, 489, 520,
        558, 565
    sample, 488
    within-group sample, 489
Covariance structure, 301
Covariate, 21, 67, 124, 128, 129, 147, 179,
        300, 312, 340, 341, 351, 354, 383,
        384, 390–392, 400, 453, 484, 485,
        497, 524
    stratified, 155
    subject-specific, 502
    time-dependent, 394, 540
    time-independent, 384
    two time-dependent, 397
Criterion, 95
    eligibility, 95, 101
    entry, 485
    evaluable, 109
    exclusion, 19, 95, 160, 485, 491
    inclusion, 19, 85, 160, 485, 491
Critical value, 84, 413

Data
    baseline, 485
    binary, 312, 429, 576
    categorical, 259, 306, 507
        ordered, 259, 307, 308, 310, 312, 334,
            572
        ordinal, 351
        repeated, 312
    censored, 363, 364, 369, 373, 382, 420,
        421, 454, 507
    correlated, 421, 422
    continuous, 260
    cross-section, 202
    demographic, 485
    efficacy, 484
    longitudinal, 202, 299, 305, 359, 577, 586,
        592, 598

638 INDEX

Data (*Continued*)
  measurement, 259
  missing, 21, 310, 484, 587
  numerical, 259
  paired survival, 583
  ranked, 259
  repeated, 354, 359
  repeated measurement, 592
  repeated ordered categorical response, 572
Database, 75, 156
Database management system, 530
Data coordination and statistical analysis center, 529
Data coordinator, 494
Data management, 529
Data management system, 530
Data monitoring, 94, 112, 364, 484, 485, 525–528, 591
Data monitoring board, 114
  external, 114
  internal, 114
Data Monitoring Committee (DMC), 21, 406, 529
  independent, 527
Data monitoring process, 407
Dataset
  evaluable, 484, 485
  intention-to-treat, 484, 485, 497
Death per patient year (DPPY), 539
Deescalation, 193, 195
Degree of freedom, 263, 266, 274, 276, 283, 291, 293, 295, 297, 332, 337, 341, 356, 358, 359, 377, 380, 388, 391, 392, 396, 468, 473, 488, 512, 520, 557, 566, 567, 588
Design
  adaptive biased-coin, 596
  Balaam, 473, 476
  complete crossover, 183
  completely binomial, 131
  complete randomized, 178
  crossover, 60, 65, 129, 178, 190–211, 420, 425, 467, 586
    higher-order, 65, 184, 186, 473, 575
    replicated, 587
  dose-ranging, 192
  dual, 187
  enrichment, 203, 210
  experimental, 19, 21
  extended-period, 187
  factorial, 6, 233, 254, 574, 582
    2 × 2, 67, 255
    fractional, 255

    full, 256
    incomplete, 255
    multilevel, 232, 235
    partial, 235
  forced dose-escalation, 201
  four-period, 473
  group comparison, 178
  matched pairs parallel, 177, 178
  multiple-stage, 475
  optimal, 177, 595
  optimal three-stage, 476, 480
  parallel, 178, 181, 192, 210, 211
    two-group, 178
  parallel-group, 233, 269, 420
  phase I, 585
  randomized block, 183, 189
  randomized incomplete block, 189
  replicated, 187
  responsive-adaptive treatment allocation, 592
  single-stage, 189
  titration, 191, 192, 197, 200, 201, 203, 210
  two-period, two-sequence (2 × 2)
    crossover, 185, 252, 420, 422, 467, 473
    double standard, 187
  two-sequence dual, 473, 477
  two-stage, 194, 594
  two-stage Simon, 475
  up-and-down, 192, 195, 196, 589
  variance-balance, 187
  Williams, 187, 188, 233
Design matrix, 453
Diagonal matrix, 559
Difference
  clinical, 48
  clinically
    important, 90
    meaningful, 110, 250, 426, 427, 430, 435, 463, 468, 532, 536
    scientific meaningful, 424
  significant, 89
  period, 467
  significance, 89
  statistical, 48
  statistical significant, 89, 90, 110, 409, 426, 514
Disease, 2
  Alzheimer, 102
Dispersion, 51
Distribution
  binomial, 315
  Cauchy, 486

INDEX

chi-square, 354, 356, 358, 359, 377, 380, 388, 391, 566, 567. *See also* Distribution, χ^2
extreme-value, 349
F-, 274, 283, 293, 295, 297, 488, 512
   noncentral, 442, 591
hypergeometric, 332, 376
   product, 343
independent and identical, 123
lognormal, 251
normal, 51, 251, 432, 454, 507, 508
   multivariate, 348
permutation, 151
   joint, 418
Poisson, 360
population, 50
probability, 123
sampling, 51. 84
standard normal, 262, 288, 313, 314, 331, 378, 408, 428, 461, 462
Student t-, 263, 266
t-, 268, 276, 468, 473, 520
   central, 470, 471
   noncentral, 434
   tolerance, 192
χ^2, 291, 331, 335
Dose
   maximum effective, 183
   maximum tolerable (MTD), 106, 167, 192, 227, 234, 517, 595
   minimum effective (MED), 106, 162, 197, 459, 461, 466
Dose combination, 582, 583
Dose-response, 34
Dose-response curve, 420
Dose-response surface, 582
Dosing range, 15
Double-dummy, 169
Dropout, 66, 113, 119, 184, 305, 401, 484
   informative, 577, 579
   random, 577, 592
Drug, 2, 3, 8, 10
   combination, 88, 235, 254, 582, 585
   fixed-combination, 229
Drug Price Competition and Patient Term Restoration Act, 4, 23
Duration, maximum, 404, 406, 414

Effect
   adverse, 97, 419, 525, 531, 535
   blocking, 146
   carryover, 64, 65, 184, 211, 407, 473, 586
   center, 510
   common treatment, 502
   confounding, 63, 185
   direct drug, 184, 185, 187
   drift, 61
   drug, 426
   estimated treatment, 400, 497
   fixed, 292, 467, 579
   interaction, 63
   main, 67
   optimal therapeutic, 118, 176, 292, 299
   overall average, 292, 299
   overall treatment, 509, 510, 512–514, 524
   pharmacokinetic, 15
   pharmacological, 15, 173
   placebo, 104, 108, 180, 201, 487, 488, 490, 528
   prophylactic, 363
   random, 297, 467
   residual, 178, 184
   sequence, 65, 185
   side, 15, 22, 177, 204, 527, 528, 593
   subject, 540
   synergistic, 227, 231
   therapeutic, 212, 227, 514
   time, 293, 299, 487
   treatment, 21, 48, 64, 65, 71, 128, 130, 134, 146, 166, 171, 217, 300, 361, 392, 399, 422, 423, 426, 485, 486, 492, 494, 500, 501, 508, 511, 526
   treatment-by-time, 292
   trend, 488
Effectiveness, 1, 2, 10, 15, 25, 35, 36, 47, 66, 84, 86, 102, 103, 119, 177, 403, 425, 426, 430, 486, 514, 528
Efficacy, 1, 3, 6, 7, 19, 28, 36, 47, 52, 75, 90, 94, 106, 181, 224, 364, 382, 403, 490, 491, 497, 519, 525, 531, 536
   average, 161, 165
   clinically meaningful, 177
   false negative, 66
   false positive, 66
   individual, 161, 164, 165
   population, 161, 165
   superior, 519, 523
   synergistic, 235
Electrophysiologic Study versus Electrocardiographic Monitoring (ESVEM) Trial, 206, 578
Endpoint, 6, 15, 153, 487, 501
   binary, 323, 327
   binary clinical, 331, 534
   categorical, 363, 504

640　　　　　　　　　　　　　　　　　　　　　　　　　　　　　　　　　　　　INDEX

Endpoint (*Continued*)
  censored, 363
  censored clinical, 420
  clinical, 109, 129, 239, 421, 483, 501, 518
  clinical surrogate, 49
  continuous, 363, 504
  efficacy, 22, 59, 161, 206, 514, 516, 532
    hard, 167
    objective, 167
    ordinal categorical, 310
  exploratory, 518
  multiple, 484, 485, 515, 518, 589, 591, 595
  multiple binary clinical, 521
  multiple primary, 574
  normal continuous, 507
  primary, 155, 160, 207, 383, 394, 400, 430, 518, 519, 522, 523, 530, 532, 533
  primary clinical, 59, 128, 165, 197, 242, 423, 486, 497, 514
  primary efficacy, 19, 21, 364
  repeated categorical, 308
  safety, 19, 22
  secondary, 518
  surrogate, 30, 32, 106, 206, 306
  survival, 425, 594
  tertiary, 518
Entry, staggering, 404
Equality, 91
Equation
  Fieldewald, 542
  generalized estimating (GEE), 198, 292, 299, 300, 359, 504, 587, 589, 598
  normal, 347
Equivalence, 91, 222, 224, 256, 469, 577, 588, 596
  average, 244
  clinical, 48, 89–91
  individual, 244
  one-sided, 250
  population, 242
  therapeutic, 108, 242, 361, 427, 494, 583, 587
  two-sided, 250
Equivalence confidence set, 573
Error
  maximum, 428
  measurement, 60, 61
  random, 53, 59, 273, 284, 292, 487
    within-subject, 467
  type I, 82, 112, 146, 248, 400, 403, 414, 419, 424, 427–430, 526
  type II, 82, 424, 427, 428, 480
  within-subject, 404

Establishment License Application (ELA), 14
Estimate, 51, 402
  best linear unbiased, 488
  consistent, 348
  inconsistent, 299
  inefficient, 299
  interval, 260, 262
  Kaplan–Meier, 370, 372, 389
  Kaplan–Meier survival, 394
  least squares (LSE), 500
  maximum likelihood (MLE), 301, 347, 387, 462, 534
  point, 79, 262, 382
  unbiased, 64, 75, 262, 462, 501
Estimation, 26, 47, 260, 403
  interval, 311
    large sample, 382
  Kaplan–Meier nonparametric, 370
  point, 584
  sample size, 110
  sequential, 576
  unbiased, 422
Estimator
  consistent, 511
  estimated generalized least square (EGLS), 489, 491
  least square, 281
  maximum likelihood, 347
    modified, 423
  median unbiased, 423
  minimum variance unbiased (MVUE), 511, 524
  point, 423
  quasi-likelihood, 301
  unbiased, 488, 501, 508, 511, 512
  unbiased consistent, 489
European Community (EC), 8
Evaluation, 1, 19
  safety, 110
Evidence
  qualitative, 422
  quantitative, 422
  substantial, 25
Expanded access, 28
Expectation, 534
Expected value, 558
Experimental unit, 1, 2, 19

Factor, 232
  classification, 510, 524
  confounding, 64, 104, 497
  demographic, 119
  designed, 510

INDEX 641

expected bias, 130, 148, 172
finite population correction, 337
fixed, 509
prognostic, 149, 179, 383, 394, 396, 497, 523, 580
random, 509
risk, 400, 497
stratified, 340
Federal Food, Drug, and Cosmetic Act, 3, 535
Fibonacci sequence, 195
Fieller theorem, 197
Food and Drug Administration (FDA), 3, 4, 6, 8, 24, 483, 531, 548
Function
  baseline hazard, 383, 384, 386
  cumulative distribution, 257, 368
  hazard, 368, 369, 376, 383, 384, 387, 539, 570, 571, 593
  Kaplan–Meier survival, 371
  identity, 396
  likelihood, 386, 391
  link, 301, 349
  log-likelihood, 390, 392, 534
  power, 425, 462, 579
  probability density, 368
  quasi-likelihood, 596
  quasi-score, 300
  spending, 365, 418
    alpha, 407, 414, 417, 423, 577, 585
    continuous, 418
  step, 370, 390
  survival, 368–370, 373, 381, 384–386

General search category, 549
Generic drug product, 33, 116
Good clinical practice (GCP), 4, 37, 38, 95, 108, 154, 259, 292, 525
Good laboratory practice (GLP), 571
Global Utilization of Streptokinase and Tissue Plasminogen Activator for Occluded Coronary Arteries Trial (GUSTO or Global Use of Strategies to Open Occluded Coronary Arteries), 7, 49, 58, 159, 170, 531, 580, 581, 592
Greenwood formula, 371
Growth curve, 305

Hamilton depression scale, 152
Helmert matrix, 488
Helmert transformation, 488
Hematology, 542

Heterogeneity, 512
Holter monitor, 49, 208
Holter recording, 198
Homogeneity, 507
  marginal, 572
Homoscedasticity, 345
Hotelling T^2, 487
Human, 1, 2
Human immunodeficiency virus (HIV), 30
Hypercholesterolemia, 50
Hypothesis, 26, 269, 425, 426, 454, 523
  alternative, 21, 81, 84, 86, 402, 433, 459, 461
    specific, 431
  interval, 90, 246, 469, 473
  null, 21, 81, 82, 84, 88, 266, 267, 270, 272, 274, 283, 286, 290, 293, 334, 377, 392, 408, 409, 420, 426, 427, 441, 462, 468, 470, 509, 510, 557, 566
  one-sided, 86, 247, 249, 469
  point, 90, 244, 246, 467, 473
  statistical, 19
  two one-sided, 230, 252
  two-sided, 86, 432, 433
  two-sided alternative, 532
Hypothesis of equality, 250
Hypothesis testing, 47, 75, 81, 239, 311, 346

Imbalance
  covariate, 131, 136, 496, 504
  treatment, 132, 134, 135
Inference, 593
  biased, 492
  clinical, 94, 115
  statistical, 19, 47, 50, 51, 75, 94, 115, 123, 128, 166, 302, 366, 386, 402, 485, 486, 491, 514, 515
  unbiased, 492
Information, 403, 404
  clinical, 403
  individual, 404, 405
  maximum, 405, 406, 413, 419
  statistical, 403
  total, 404–406, 416
Information matrix, 348, 388
Informed consent form, 103, 124, 155
Institutional review board (IRB), 22, 24, 103
Interaction, 63, 231, 390, 392, 394, 492, 501, 502, 508
  crossover, 71
  drug-to-drug, 212, 235, 239, 240, 535
  qualitative, 71, 72, 219, 512, 580
  quantitative, 71, 72, 217, 219, 512

Interaction (*Continued*)
  time-covariate, 391
  treatment-by-age, 399, 400
  treatment-by-center, 68, 117, 214, 219, 221, 278, 507, 511, 512, 514
    qualitative, 214, 215, 507, 513
    quantitative, 214, 215
  treatment-by-covariate, 524
  treatment-by-period, 65
  treatment-by-study, 221
  treatment-by-time, 295, 297, 299, 396, 398–400
  treatment-by-visit, 297, 298
  two-factor, 350
Interaction between treatment and time, 390
Interaction mean square, 509
International Classification of Diseases (ICD), 548
International Conference on Harmonisation (ICH), 38, 484, 531, 548, 583
Intervention, 2
Investigational Device Exemptions (IDE), 14
Investigational New Drug Application (IND), 14, 16, 24, 36
  commercial, 17
  noncommercial, 17
  treatment, 6, 23, 30
Investigator, 166, 529
In Vitro, 177
In Vivo, 177
Iterative reweighted least squares, 347

Kefauver–Harris Amendment, 3, 353
Kronecker product, 559

Last observation carried forward (LOCF), 492
Latin squares, orthogonal, 188
Laboratory
  central, 529, 560
  local, 560
Level, 232
  confidence, 80
  nominal, 314, 325, 326, 424, 428
  significance, 82, 110, 267, 315, 402, 420, 426, 430, 532, 534, 590
    nominal, 409, 519, 522, 524
    overall, 409, 410, 414, 418
Level of significance, 81, 83, 266, 267, 269, 270, 274, 285, 287, 290, 326, 341, 358, 361, 381, 427, 428, 475, 510, 516, 517, 533
  nominal, 22, 430

Limit, equivalence, 247, 250–252
Lilly reference, 562
Link
  complementary log, 349
  identity, 349
  log, 349
  logit, 349
  profit, 349
Lipid Research Clinics Coronary Primary Prevention Trial (CPPT), 160
Location, 50
Location shift, 336–339, 343–345
Logit, 347
  cumulative, 353, 354, 359
Lost to follow-up, 363, 401

Masking, 113
Mean, 75, 260, 269, 270, 273, 428, 432, 454, 537
  arithmetic, 50
  overall sample, 273, 292, 533
  population, 260, 261, 402, 403, 441, 534
  sample, 261, 270, 313, 408, 423, 519
Mean square error (MSE), 275, 295, 423, 461
Measure, repeated, 291
Measurement, repeated, 299, 305, 514, 576, 577, 582, 584, 587
Median, 50, 51, 537
Medical device, 3, 8, 10
Medical Dictionary for Drug Regulatory Affairs (MEDDRA), 537, 548
Medical monitor, 494
Medication event monitor system (MEMS), 118
Medicine, 1
Method
  continual reassessment, 196, 590
  delta, 197
  estimated generalized least squares (EGLS), 520
  GEE, 301, 305
  group sequential, 365, 407, 591
  Kaplan–Meier, 364
  least squares, 240
  log-likelihood, 351
  maximum likelihood, 240, 422
  minimization, 149
  model-based, 312, 346, 360
  nonparametric, 370, 520
  randomization-based, 312, 346
  response surface, 241
  sequential, 256
  stochastic curtailment, 419

INDEX

Missing value, 119, 531
  completely random, 531
  informative, 531
  intermittent, 362, 531
  random, 31
Missing at random (MR), 119, 362
Missing completely at random (MCAR), 362
Mode, 50, 51
Model
  analysis of covariance, 500
  analysis of variance, 461
  autoregressive, 305
  cell-means, 511
  Cox's proportional hazard, 382, 383, 385, 388, 390, 504, 524, 540, 556
    stratified, 391, 392
  Cox's proportional hazard regression, 384, 386, 421
  fixed effect, 443
  fixed linear, 584
  frailty, 540
  generalized linear (GLM), 299, 452, 586, 596
  interaction, 390
  invoked population, 124
  linear, 345, 422, 596
  linear random effect, 403
  linear regression, 453
  logistic regression, 312
  loglinear, 361
  main-effect, 511
  marginal, 299, 300, 305
  mixed effect, 509
  nested, 292
  permutation, 146
  population, 123
  proportional hazard, 364
  proportional hazard regression, 383
  proportional odds, 351, 354
  random effect, 300, 305, 540
  randomization, 122, 123, 125, 577
  randomized complete block, 443
  reduced, 390
  subject-specific, 305
  transition, 300
  two-way classification fixed, 279
Monotherapy, 181, 237
Mortality, 115, 167, 208, 402, 406, 525, 536
Multiple baseline measurement, 486
Multiple-evaluator, 171
Multiple-placebo, 169
Multiplicity, 514, 515

Myocardial infarction, 49, 67, 97, 115, 159, 178, 206, 306, 382, 402, 573, 583

N-of-1 randomized trial, 164
National Cancer Institute (NCI), 4
National formulary, 10, 32
National Heart, Lung, and Blood Institute, 402
National Institute of Allergy and Infectious Disease, 531
National Institutes of Health (NIH), 3, 400, 408, 529
National Institutes of Health Reauthorization Bill, 576
National Institutes of Health Stroke Scale (NIHSS), 61, 307, 587
Negative event, 545
New Drug Application (NDA), 6, 14, 17, 25
  abbreviated (ANDA), 14, 32, 241
  supplemental (SNDA), 33
Newton–Raphson algorithm, 347
Normality, 345
Numerical integration, 414

Objective, primary study, 425
Open-label, 122, 158, 170
Outcome, repeated categorical, 596
Outcomes research, 2

Parallel track regulation, 6, 30
Parameter, 51
  noncentrality, 434, 442
  nuisance, 300
  overdispersion, 301
  primary efficacy, 435
Partial likelihood, 386, 387, 392
Patient, 1, 166
  evaluable, 481
  qualified, 481
Patient characteristic, 485
Patient data listing, 484
Patient identification number, 114
Peak urinary flow rate, 125
Period
  active treatment, 118, 180
  baseline, 487
  follow-up, 404
  lead-in, 547
  placebo run-in, 420, 487, 490, 514
  recruiting, 404
  run-in, 19, 180, 181, 183, 547
  single-blind placebo run-in, 547
  washout, 180, 486

Permutation, 141
Pharmaceutical entity, 1, 35, 43, 364
Pharmaceutical identity, 2
Pharmacoeconomics, 2, 7
Phase
  active, 547
  dose-titration enrichment, 203
  double-blind active, 182
  enrichment, 203
  maintenance, 118
  placebo baseline, 203
  placebo-blind, placebo-controlled, 203
  placebo run-in, 118
  placebo washout, 197
  run-in single-blind, 182
  sustained active, 203
Phase I clinical investigation, 15
Placebo, 2, 58, 102, 104, 121, 224, 231, 307, 373, 383, 385, 386, 408, 426, 522, 586, 593, 597
  active, 173
Placebo concurrent group, 106
Placebo responder, 205
Play-the-winner role, 150
  modified, 150
  randomized, 150
Population, 1
  efficacy patient, 493
  geriatric, 103
  homogeneous, 124
  intention-to-treat, 514
  patient, 34, 94, 115, 123, 493
  preprotocol patient, 493
  targeted, 1, 19, 75, 426, 485, 505, 510
  targeted patient, 47, 48, 50, 100, 115, 128, 277, 278, 547
  time-heterogeneous, 125, 143
Postmenopausal Estrogen/Progestin Interventions Trial (PEPI), 400, 578
Power, 4, 82, 99, 110, 133, 146, 401–403, 424, 427, 430, 433, 441, 453, 461, 471, 473, 475, 505, 585, 586, 590, 591, 593, 594
  conditional, 419
Power approach, 245, 246
Precision, 58, 59, 100, 403, 424, 427
  optimal, 128
Premarket Approval of Medical Devices (PMA), 14
Premarketing clinical investigation, 536
Premarketing safety assessment, 535
Premature ventricular beat (PVB), 49
Principle, intention-to-treat, 532

Probability, 47, 48
  conditional, 368
  coverage, 314
Problem
  one-sided equivalence, 249
  two-sided equivalence, 249
Procedure
  blinding, 483
  Bonferroni, 582
  four one-sided tests, 254
  generalized least square (GLS), 488
  group sequential, 408, 416, 422, 423, 585
    K-stage, 409
  group sequential bioequivalence testing, 423
  Holm's Bonferroni, 517
  intersection union test (IUT), 523
  Mantel–Haenszel, 588
  semiparametric statistical, 384
  two one-sided tests, 247, 252, 470, 471, 587, 591, 593
  unconditional large-sample, 326
Process
  random, 510
  titration, 201, 240
Product License Application (PLA), 6, 14
Proportion, 364
  binomial, 313, 314
  marginal, 358, 359
  population, 312, 313, 327
  sample, 312, 313
Prostate Cancer Prevention Trial, 579
Protocol, 18, 19, 21, 47, 48, 93, 113, 115, 123, 155, 183, 401, 529
Protocol amendment, 22
Protocol review committee, 38
  internal, 155
Pure Food and Drug Act, 3
P-value, 47, 81, 83, 89, 113, 322, 383, 490, 492, 514, 516–519, 524, 527, 590
  conditional exact, 326
  observed, 85

Quality assurance, 571
Quality of life, 2, 7, 36, 75, 177, 403

Random allocation, 131, 133, 134, 146, 150
Random assignment, 124, 141
Random censoring, 370
Randomization, 4, 21, 26, 94, 118, 121, 174, 219, 364, 366, 368, 369, 378, 483, 485, 491, 492, 542, 585
  adaptive, 122, 131, 146

INDEX 645

covariate, 149, 150
   response, 150, 153, 154
   treatment, 147
bias coin, 147
complete, 122, 131, 133, 134, 137, 148, 151
minimization, 159
permuted-block, 122, 131, 135, 140, 146, 152, 255
restricted, 133, 148
simple, 131, 137, 585
stratified, 128, 146, 281
urn, 147, 158, 596
Randomization code, 113, 114, 132, 146, 155, 157, 171, 494, 527
Randomization method, 129, 131
Random number, 132
Random selection, 124, 131, 142
Range, 51
   equivalence, 563
   normal, 561–563
   reference, 560, 575
   therapeutic dose, 240, 242
Rank, 285
Rate, 536
   crude, 570
   crude incidence, 538, 539
   cumulative, 570
   eradication, 228
   experimentwise false positive, 516, 522, 524
   experimentwise type I error, 516
   false positive, 519, 526
   hazard, 454, 455
   incidence, 540, 541
   instantaneous death, 368
   mortality, 207, 231, 368, 373, 402
   prevalence, 540, 541
   relative hazard, 377, 382
   type I error, 401
Ratio
   benefit-to-risk, 237
   cumulative, 354
   hazard, 384, 386, 390, 392, 398, 421
   odds, 310, 328
   relative hazard, 381
Recurring event, 538, 539, 540, 556, 571
Region
   acceptance, 84
   rejection, 81, 84, 245
Regression
   binary, 421
   linear, 353

   logistic, 346, 347, 349, 386, 504, 524, 556, 560, 582
Regression coefficient, 299, 300, 350, 383, 385–387, 396, 399, 500
   estimated, 301
Regulation, 14
Relationship, dose-response, 64, 65, 106, 192, 197, 234, 235, 239, 457, 459, 515, 517
Reliability, 52, 103, 425
Repeatability, 562
Report
   clinical, 483
   full integrated, 483
   integrated clinical, 484
   integrated statistical, 484
   statistical, 483
Reproducibility, 59, 61, 562
Residual plot, 241
Response, longitudinal binary, 579
Response surface, 239, 582, 590
   estimated, 240
Risk
   consumer's, 81
   exposure, 538
   false positive, 414
   producer's, 81
   relative, 310, 328, 329, 384–386, 435, 536
Risk set, 370, 371, 387

Safety, 2, 3, 6, 7, 10, 15, 19, 25, 28, 35, 36, 47, 52, 59, 66, 75, 86, 90, 94, 103, 106, 120, 177, 181, 364, 382, 403, 425, 430, 486, 490, 492, 497, 535, 536
Safety assessment, 536
Safety data, 536
Sample
   random, 81, 115, 269, 426
   representative, 51, 94, 100, 115, 123, 218, 509
Sample size, 155, 326, 400, 401, 424, 427, 429, 455, 481, 532, 536, 581, 585, 593, 594, 597
   adjustment, 534
   determination, 21, 110, 424, 585
   justification, 110, 424
   planned, 532–534
   reestimation, 534, 588
Scale
   nominal, 306
   ordinal, 306, 307, 336
Score
   integer, 338
   logrank, 338

Score (*Continued*)
  standardized midrank, 338
Second International Study of Infarct Survival (ISIS2), 67, 159, 170, 583
Sequential equivalence testing, 583
Session
  closed, 530
  executive, 530
  open, 530
Sherley Amendment, 3
Significance
  clinical, 563
  prognostic, 486
Similarity circle, 163
Size, 522
Skewness, 50
Small-sample confidence set, 595
Special search category, 549
Sponsor, 166, 529
Spontaneous event, 547
Spread, 50
Standard deviation, 51, 75, 261, 269, 270, 273, 429, 430
  population, 261
  within-group sample, 520
Standard operating procedure (SOP), 38, 122, 154, 506, 526–528
Standard error, 51, 75, 79, 248, 350, 392, 408
  large-sample, 388, 399
Statistics, 51
  Bhapkar Q, 358, 359
  descriptive, 47, 74, 75, 311, 507, 536
  Gehan test, 382
  inferential, 47, 74, 75
  likelihood ratio, 452
  linear rank, 127, 146
  logrank, 364, 377–379, 521
  logrank test, 406
  Mantel–Haenszel, 340, 376
    extended, 343
  McNemar, 569
  paired Prentice–Wilcoxon, 421
  sample, 75
  score, 452
  summary, 528
  t-, 263, 511
    two-sample, 467
    two-sample unpaired, 524
  test, 81
  Wald, 388
  Wilcoxon–Mann–Whitney test, 287
  Wilcoxon rank sum, 344, 364
    blocked, 344

  Z-, 332, 408
Steering committee, 529, 530
Stochastically curtailed testing, 585
Stratification, 26, 60, 122, 127, 128, 179
Stratum, 391
Streptokinase, 50, 67
Study
  active control, 484
  adequate and well-controlled, 25, 58, 159, 201
  adequate and well-controlled clinical, 34
  bioequivalence, 575
  blind reader, 159
  case-control, 128, 174, 593
  controlled clinical, 597
  dose-escalating, 167, 197
  dose proportionality, 106
  dose response (ranging), 17, 425, 457, 573, 592
  dose titration, 60
  double-blind, placebo-controlled multicenter, 577
  Helsinki Heart (Health), 2, 580, 588
  multicenter, 21, 214, 485
  observational, 174
  parallel-group, 183
  phase I, 15, 17, 192
  phase I and II, 17
  phase I safety and tolerance, 191, 192
  phase II, 15
  phase II and III, 535
  phase III, 15, 17
  phase IV, 16
  phase V, 16
  Physicians' Health, 2, 7, 159, 591
  placebo-controlled, double-blind, 363
  positive control, 112
  postmarketing surveillance, 167
  premarketing surveillance, 167
  prospective cohort, 435
  randomized, 363
  retrospective case-control, 435
  titration, 575, 594
  two-group parallel, 363
Study center, 100, 166
Study site, 100, 166
Subgroup, 537
Subject, 1, 2, 3
Subpopulation
  heterogeneous, 124
  homogeneous, 124
Sum of squares, 295, 297
  correct, 404

INDEX **647**

due to error, 273
due to treatment, 273
error, 293
residual, 501
total, 273, 279
Summary statistical table, 484
Superiority
   global, 235
      strict, 236–239, 241
      wide, 236, 237
Survival, 167
   event-free, 383
   overall, 383
Symmetry, compound, 422

Term
   preferred, 552
   repeated, 552
Test
   blocked Wilcoxon rank sum, 504
   Breslow–Day, 560
   chi-square, 331, 332, 588
   Cochran–Mantel–Haenszel, 558
   distribution-free, 284, 422
   F-, 304
   Fisher exact, 333, 334, 492, 556–558, 560
   Fisher–Irwin, 581
   goodness-of-fit, 241
   group sequential, 584
   intersection-union, 522, 573
   Kolmogorov–Smirnov, 258
   Kruskal–Wallis, 284, 290, 291, 304, 583
   lack-of-fit, 241
   logrank, 148, 376, 386, 388, 392, 420, 455, 596
   log-likelihood chi-square, 293
   Mantel–Haenszel, 146, 556, 557, 558
   McNemar, 323, 325
   Min, 238
   Newman–Keuls range, 293
   one-sample, 431, 432
   one-sided, 48, 86, 88, 126, 434, 579
   paired Prentice–Wilcoxon, 591
   paired t-, 267, 268
   Pearson, 348
   Pearson chi-square, 331, 335, 338
   permutation, 125, 127, 153, 418
      conditional, 126
      unconditional, 126
   Peto–Peto–Prentice–Wilcoxon randomization, 338
   score, 388
   sign, 421
   statistical, 390
   Stuart–Maxwell, 567
   Student t-, 258
   t-, 434
   treadmill, 421
   two-sample, 433
   two-sample McNemar, 355–357
   two-sample t-, 269–272, 304
   two-sided, 48, 84, 86, 88, 126, 420, 434, 577
   Wilcoxon rank sum, 126, 256, 286, 288, 289, 304, 520
   Wilcoxon signed rank, 284, 285
   Wilcoxon two-sample rank sum, 378
   Williams, 465
Therapeutic range, 177, 197
Therapeutic window, 197
Time
   calendar, 364, 400, 402, 405, 406, 413, 410
   censoring, 369, 402
   discrete information, 405
   information, 364, 400, 405, 406, 409, 413, 414, 416, 420, 585
   median failure, 422, 584
   median survival, 363, 371, 453, 454, 570
   survival, 368
Titration
   crossover, 19
   dose, 292, 299
   forced, 19, 183
   parallel, 19
T-PA, 50, 68
Tolerability, 167
Toxicity, 119, 193, 528, 537, 542
   clinical, 192
   systematic, 536
Transformation, logarithmic, 251
Treatment, 1, 2, 19, 147
   active, 102
Treatment-emergent events, 571
Treatment group, 232
Trial
   active control, 88, 108, 212, 221, 591
   active control clinical, 582, 585
   active control equivalence, 246, 250, 253
   adaptive clinical, 595
   adequate, well-controlled, 423
   adequate, well-controlled clinical, 10, 34, 47, 86, 103, 167, 483
   bioequivalence, 79, 116, 213, 469, 573, 575, 597

648 INDEX

Trial (*Continued*)
  clinical, 1, 3, 4, 19, 580
  combination, 212, 226
  comparative, 167
  comparative clinical, 102, 219
  confirmatory clinical, 112
  controlled randomized, 115
  crossover, 579
  dose-ranging, 58
  double-blind, placebo-controlled, 181
  double-blind, randomized multicenter clinical, 514
  equivalence, 212, 241, 484
  factorial, 583
  maximum duration, 406, 416
  maximum information, 406
  multicenter, 100, 115, 212, 213, 219, 256, 484, 505, 506, 580, 585, 595
  multicenter clinical, 339, 572
  phase I, 6, 594
  phase I clinical, 590
  phase II, 30, 575
  phase II clinical, 169, 178, 578, 594
  phase III, 17, 30, 178, 530
  phase III clinical, 594
  phase IV, 16
  phase I, II, and III, 28, 32
  placebo-controlled clinical, 555
  positive control, 226
  primary prevention, 2
  randomized clinical, 574
  randomized, double-blind, parallel-group clinical, 524
  randomized, double-blind, placebo-controlled, 402
  randomized, triple-blind, two-parallel-group, 408
  sequential, 212
  sequential clinical, 597
  single-center, 115
  telescoping, 30
  well-controlled randomized clinical, 160
  women-health, 2

Unbiasedness, 47, 52
Uncertainty, 47, 48, 51
Unit, experimental, 100
Urinalysis, 542
U.S. National Cancer Institute, *see* National Cancer Institute
U.S. National Heart, Lung, and Blood Institute, *see* National Heart, Lung, and Blood Institute
U.S. National Institute of Allergy and Infectious Disease, *see* National Institute of Allergy and Infectious Disease
U.S. National Institutes of Health (NIH), *see* National Institutes of Health
U.S. Pharmacopeia, 10, 33

Validation, external, 231
Validity
  internal, 104, 115, 252
  external, 115
Variability, 47, 58, 59, 94, 99, 103, 110, 122, 128, 163, 179, 252, 260, 269, 486, 532
  between-patient, 297
  between-site, 562
  interpatient, 180, 183
  intrapatient, 179
  intrasubject, 165, 186, 473
  within-laboratory, 562
Variable
  baseline, 486
  continuous, 260
  discrete, 259
  explanatory, 300, 349, 354
  latent, 334
  ordinal categorical, 572
  Poisson, 349
  primary efficacy, 109, 440, 482, 536, 542
  primary response, 109
  primary safety, 482
  prognostic, 486
  random, 261, 402
    Bernoulli, 432
    binary, 313
    chi-square, 332, 341, 557, 566. *See also* Variable, random, χ^2
    normal, 428, 456
    χ^2-, 325, 337
  secondary response, 109
  standard normal, 431, 433
  time-dependent, 384
Variance, 51, 52, 402, 423, 428, 432, 454
  estimated, 519
  estimated large sample, 372
  large-sample, 310, 327, 329, 356
  pooled, 509
  population, 260, 403
  sample, 270, 533
  within-group, 532, 533, 534

Variation, 51, 59, 128, 506
  biological, 50, 59, 128
  intersubject, 59
  intrasubject, 60
  random, 273
  temporal, 60
Ventricular ectopy, 198
Ventricular tachycardia, 207, 208

Weighted least squares, 359, 589
West of Scotland Coronary Prevention Study Group, 382, 589
World Health Organization (WHO), 39, 548
World Health Organization Adverse Reaction Terminology (WHOART), 537, 548, 550

WILEY SERIES IN PROBABILITY AND STATISTICS
ESTABLISHED BY WALTER A. SHEWHART AND SAMUEL S. WILKS

Editors
*Vic Barnett, Ralph A. Bradley, Noel A. C. Cressie, Nicholas I. Fisher,
Iain M. Johnstone, J. B. Kadane, David G. Kendall, David W. Scott,
Bernard W. Silverman, Adrian F. M. Smith, Jozef L. Teugels;
J. Stuart Hunter, Emeritus*

Probability and Statistics Section

*ANDERSON · The Statistical Analysis of Time Series
ARNOLD, BALAKRISHNAN, and NAGARAJA · A First Course in Order Statistics
ARNOLD, BALAKRISHNAN, and NAGARAJA · Records
BACCELLI, COHEN, OLSDER, and QUADRAT · Synchronization and Linearity:
  An Algebra for Discrete Event Systems
BASILEVSKY · Statistical Factor Analysis and Related Methods: Theory and
  Applications
BERNARDO and SMITH · Bayesian Statistical Concepts and Theory
BILLINGSLEY · Convergence of Probability Measures
BOROVKOV · Asymptotic Methods in Queuing Theory
BRANDT, FRANKEN, and LISEK · Stationary Stochastic Models
CAINES · Linear Stochastic Systems
CAIROLI and DALANG · Sequential Stochastic Optimization
CONSTANTINE · Combinatorial Theory and Statistical Design
COVER and THOMAS · Elements of Information Theory
CSÖRGŐ and HORVÁTH · Weighted Approximations in Probability Statistics
CSÖRGŐ and HORVÁTH · Limit Theorems in Change Point Analysis
DETTE and STUDDEN · The Theory of Canonical Moments with Applications in
  Statistics, Probability, and Analysis
*DOOB · Stochastic Processes
DRYDEN and MARDIA · Statistical Analysis of Shape
DUPUIS and ELLIS · A Weak Convergence Approach to the Theory of Large Deviations
ETHIER and KURTZ · Markov Processes: Characterization and Convergence
FELLER · An Introduction to Probability Theory and Its Applications, Volume 1,
  Third Edition, Revised; Volume II, *Second Edition*
FULLER · Introduction to Statistical Time Series, *Second Edition*
FULLER · Measurement Error Models
GELFAND and SMITH · Bayesian Computation
GHOSH, MUKHOPADHYAY, and SEN · Sequential Estimation
GIFI · Nonlinear Multivariate Analysis
GUTTORP · Statistical Inference for Branching Processes
HALL · Introduction to the Theory of Coverage Processes
HAMPEL · Robust Statistics: The Approach Based on Influence Functions
HANNAN and DEISTLER · The Statistical Theory of Linear Systems
HUBER · Robust Statistics
IMAN and CONOVER · A Modern Approach to Statistics
JUREK and MASON · Operator-Limit Distributions in Probability Theory
KASS and VOS · Geometrical Foundations of Asymptotic Inference

*Now available in a lower priced paperback edition in the Wiley Classics Library.

Probability and Statistics (Continued)

KAUFMAN and ROUSSEEUW · Finding Groups in Data: An Introduction to Cluster Analysis
KELLY · Probability, Statistics, and Optimization
LINDVALL · Lectures on the Coupling Method
McFADDEN · Management of Data in Clinical Trials
MANTON, WOODBURY, and TOLLEY · Statistical Applications Using Fuzzy Sets
MARDIA and JUPP · Statistics of Directional Data, *Second Edition*
MORGENTHALER and TUKEY · Configural Polysampling: A Route to Practical Robustness
MUIRHEAD · Aspects of Multivariate Statistical Theory
OLIVER and SMITH · Influence Diagrams, Belief Nets and Decision Analysis
*PARZEN · Modern Probability Theory and Its Applications
PRESS · Bayesian Statistics: Principles, Models, and Applications
PUKELSHEIM · Optimal Experimental Design
RAO · Asymptotic Theory of Statistical Inference
RAO · Linear Statistical Inference and Its Applications, *Second Edition*
*RAO and SHANBHAG · Choquet-Deny Type Functional Equations with Applications to Stochastic Models
ROBERTSON, WRIGHT, and DYKSTRA · Order Restricted Statistical Inference
ROGERS and WILLIAMS · Diffusions, Markov Processes, and Martingales, Volume I: Foundations, *Second Edition;* Volume II: Îto Calculus
RUBINSTEIN and SHAPIRO · Discrete Event Systems: Sensitivity Analysis and Stochastic Optimization by the Score Function Method
RUZSA and SZEKELY · Algebraic Probability Theory
SCHEFFE · The Analysis of Variance
SEBER · Linear Regression Analysis
SEBER · Multivariate Observations
SEBER and WILD · Nonlinear Regression
SERFLING · Approximation Theorems of Mathematical Statistics
SHORACK and WELLNER · Empirical Processes with Applications to Statistics
SMALL and McLEISH · Hilbert Space Methods in Probability and Statistical Inference
STAPLETON · Linear Statistical Models
STAUDTE and SHEATHER · Robust Estimation and Testing
STOYANOV · Counterexamples in Probability
TANAKA · Time Series Analysis: Nonstationary and Noninvertible Distribution Theory
THOMPSON and SEBER · Adaptive Sampling
WELSH · Aspects of Statistical Inference
WHITTAKER · Graphical Models in Applied Multivariate Statistics
YANG · The Construction Theory of Denumerable Markov Processes

Applied Probability and Statistics Section

ABRAHAM and LEDOLTER · Statistical Methods for Forecasting
AGRESTI · Analysis of Ordinal Categorical Data
AGRESTI · Categorical Data Analysis
ANDERSON, AUQUIER, HAUCK, OAKES, VANDAELE, and WEISBERG · Statistical Methods for Comparative Studies
ARMITAGE and DAVID (editors) · Advances in Biometry
*ARTHANARI and DODGE · Mathematical Programming in Statistics
ASMUSSEN · Applied Probability and Queues
*BAILEY · The Elements of Stochastic Processes with Applications to the Natural Sciences

*Now available in a lower priced paperback edition in the Wiley Classics Library.

Applied Probability and Statistics (Continued)
  BARNETT and LEWIS · Outliers in Statistical Data, *Third Edition*
  BARTHOLOMEW, FORBES, and McLEAN · Statistical Techniques for Manpower Planning, *Second Edition*
  BATES and WATTS · Nonlinear Regression Analysis and Its Applications
  BECHHOFER, SANTNER, and GOLDSMAN · Design and Analysis of Experiments for Statistical Selection, Screening, and Multiple Comparisons
  BELSLEY · Conditioning Diagnostics: Collinearity and Weak Data in Regression
  BELSLEY, KUH, and WELSCH · Regression Diagnostics: Identifying Influential Data and Sources of Collinearity
  BHAT · Elements of Applied Stochastic Processes, *Second Edition*
  BHATTACHARYA and WAYMIRE · Stochastic Processes with Applications
  BIRKES and DODGE · Alternative Methods of Regression
  BLOOMFIELD · Fourier Analysis of Time Series: An Introduction
  BOLLEN · Structural Equations with Latent Variables
  BOULEAU · Numerical Methods for Stochastic Processes
  BOX · Bayesian Inference in Statistical Analysis
  BOX and DRAPER · Empirical Model-Building and Response Surfaces
  BOX and DRAPER · Evolutionary Operation: A Statistical Method for Process Improvement
  BUCKLEW · Large Deviation Techniques in Decision, Simulation, and Estimation
  BUNKE and BUNKE · Nonlinear Regression, Functional Relations and Robust Methods: Statistical Methods of Model Building
  CHATTERJEE and HADI · Sensitivity Analysis in Linear Regression
  CHOW and LIU · Design and Analysis of Clinical Trials: Concepts and Methodologies
  CLARKE and DISNEY · Probability and Random Processes: A First Course with Applications, *Second Edition*
  *COCHRAN and COX · Experimental Designs, *Second Edition*
  CONOVER · Practical Nonparametric Statistics, *Second Edition*
  CORNELL · Experiments with Mixtures, Designs, Models, and the Analysis of Mixture Data, *Second Edition*
  *COX · Planning of Experiments
  CRESSIE · Statistics for Spatial Data, *Revised Edition*
  DANIEL · Applications of Statistics to Industrial Experimentation
  DANIEL · Biostatistics: A Foundation for Analysis in the Health Sciences, *Sixth Edition*
  DAVID · Order Statistics, *Second Edition*
  *DEGROOT, FIENBERG, and KADANE · Statistics and the Law
  DODGE · Alternative Methods of Regression
  DOWDY and WEARDEN · Statistics for Research, *Second Edition*
  DUNN and CLARK · Applied Statistics: Analysis of Variance and Regression, *Second Edition*
  ELANDT-JOHNSON and JOHNSON · Survival Models and Data Analysis
  EVANS, PEACOCK, and HASTINGS · Statistical Distributions, *Second Edition*
  FLEISS · The Design and Analysis of Clinical Experiments
  FLEISS · Statistical Methods for Rates and Proportions, *Second Edition*
  FLEMING and HARRINGTON · Counting Processes and Survival Analysis
  GALLANT · Nonlinear Statistical Models
  GLASSERMAN and YAO · Monotone Structure in Discrete-Event Systems
  GNANADESIKAN · Methods for Statistical Data Analysis of Multivariate Observations, *Second Edition*
  GOLDSTEIN and LEWIS · Assessment: Problems, Development, and Statistical Issues
  GREENWOOD and NIKULIN · A Guide to Chi-Squared Testing
  *HAHN · Statistical Models in Engineering
  HAHN and MEEKER · Statistical Intervals: A Guide for Practitioners

*Now available in a lower priced paperback edition in the Wiley Classics Library.

Applied Probability and Statistics (Continued)

HAND · Construction and Assessment of Classification Rules
HAND · Discrimination and Classification
HEIBERGER · Computation for the Analysis of Designed Experiments
HINKELMAN and KEMPTHORNE: · Design and Analysis of Experiments, Volume 1: Introduction to Experimental Design
HOAGLIN, MOSTELLER, and TUKEY · Exploratory Approach to Analysis of Variance
HOAGLIN, MOSTELLER, and TUKEY · Exploring Data Tables, Trends and Shapes
HOAGLIN, MOSTELLER, and TUKEY · Understanding Robust and Exploratory Data Analysis
HOCHBERG and TAMHANE · Multiple Comparison Procedures
HOCKING · Methods and Applications of Linear Models: Regression and the Analysis of Variables
HOGG and KLUGMAN · Loss Distributions
HOLLANDER and WOLFE · Nonparametric Statistical Methods
HOSMER and LEMESHOW · Applied Logistic Regression
HØYLAND and RAUSAND · System Reliability Theory: Models and Statistical Methods
HUBERTY · Applied Discriminant Analysis
JACKSON · A User's Guide to Principle Components
JOHN · Statistical Methods in Engineering and Quality Assurance
JOHNSON · Multivariate Statistical Simulation
JOHNSON and KOTZ · Distributions in Statistics
    Continuous Multivariate Distributions
JOHNSON, KOTZ, and BALAKRISHNAN · Continuous Univariate Distributions, Volume 1, *Second Edition*
JOHNSON, KOTZ, and BALAKRISHNAN · Continuous Univariate Distributions, Volume 2, *Second Edition*
JOHNSON, KOTZ, and BALAKRISHNAN · Discrete Multivariate Distributions
JOHNSON, KOTZ, and KEMP · Univariate Discrete Distributions, *Second Edition*
JUREČKOVÁ and SEN · Robust Statistical Procedures: Aymptotics and Interrelations
KADANE · Bayesian Methods and Ethics in a Clinical Trial Design
KADANE AND SCHUM · A Probabilistic Analysis of the Sacco and Vanzetti Evidence
KALBFLEISCH and PRENTICE · The Statistical Analysis of Failure Time Data
KELLY · Reversability and Stochastic Networks
KHURI, MATHEW, and SINHA · Statistical Tests for Mixed Linear Models
KLUGMAN, PANJER, and WILLMOT · Loss Models: From Data to Decisions
KLUGMAN, PANJER, and WILLMOT · Solutions Manual to Accompany Loss Models: From Data to Decisions
KOVALENKO, KUZNETZOV, and PEGG · Mathematical Theory of Reliability of Time-Dependent Systems with Practical Applications
LAD · Operational Subjective Statistical Methods: A Mathematical, Philosophical, and Historical Introduction
LANGE, RYAN, BILLARD, BRILLINGER, CONQUEST, and GREENHOUSE · Case Studies in Biometry
LAWLESS · Statistical Models and Methods for Lifetime Data
LEE · Statistical Methods for Survival Data Analysis, *Second Edition*
LePAGE and BILLARD · Exploring the Limits of Bootstrap
LINHART and ZUCCHINI · Model Selection
LITTLE and RUBIN · Statistical Analysis with Missing Data
MAGNUS and NEUDECKER · Matrix Differential Calculus with Applications in Statistics and Econometrics
MALLER and ZHOU · Survival Analysis with Long Term Survivors
MANN, SCHAFER, and SINGPURWALLA · Methods for Statistical Analysis of Reliability and Life Data

*Now available in a lower priced paperback edition in the Wiley Classics Library.

Applied Probability and Statistics (Continued)

McLACHLAN and KRISHNAN · The EM Algorithm and Extensions
McLACHLAN · Discriminant Analysis and Statistical Pattern Recognition
McNEIL · Epidemiological Research Methods
MEEKER and ESCOBAR · Statistical Methods for Reliability Data
MILLER · Survival Analysis
MONTGOMERY and PECK · Introduction to Linear Regression Analysis, *Second Edition*
MYERS and MONTGOMERY · Response Surface Methodology: Process and Product in Optimization Using Designed Experiments
NELSON · Accelerated Testing, Statistical Models, Test Plans, and Data Analyses
NELSON · Applied Life Data Analysis
OCHI · Applied Probability and Stochastic Processes in Engineering and Physical Sciences
OKABE, BOOTS, and SUGIHARA · Spatial Tesselations: Concepts and Applications of Voronoi Diagrams
PANKRATZ · Forecasting with Dynamic Regression Models
PANKRATZ · Forecasting with Univariate Box-Jenkins Models: Concepts and Cases
PIANTADOSI · Clinical Trials: A Methodologic Perspective
PORT · Theoretical Probability for Applications
PUTERMAN · Markov Decision Processes: Discrete Stochastic Dynamic Programming
RACHEV · Probability Metrics and the Stability of Stochastic Models
RÉNYI · A Diary on Information Theory
RIPLEY · Spatial Statistics
RIPLEY · Stochastic Simulation
ROUSSEEUW and LEROY · Robust Regression and Outlier Detection
RUBIN · Multiple Imputation for Nonresponse in Surveys
RUBINSTEIN · Simulation and the Monte Carlo Method
RUBINSTEIN and MELAMED · Modern Simulation and Modeling
RYAN · Statistical Methods for Quality Improvement
SCHUSS · Theory and Applications of Stochastic Differential Equations
SCOTT · Multivariate Density Estimation: Theory, Practice, and Visualization
*SEARLE · Linear Models
SEARLE · Linear Models for Unbalanced Data
SEARLE, CASELLA, and McCULLOCH · Variance Components
STOYAN, KENDALL, and MECKE · Stochastic Geometry and Its Applications, *Second Edition*
STOYAN and STOYAN · Fractals, Random Shapes and Point Fields: Methods of Geometrical Statistics
THOMPSON · Empirical Model Building
THOMPSON · Sampling
TIJMS · Stochastic Modeling and Analysis: A Computational Approach
TIJMS · Stochastic Models: An Algorithmic Approach
TITTERINGTON, SMITH, and MAKOV · Statistical Analysis of Finite Mixture Distributions
UPTON and FINGLETON · Spatial Data Analysis by Example, Volume 1: Point Pattern and Quantitative Data
UPTON and FINGLETON · Spatial Data Analysis by Example, Volume II: Categorical and Directional Data
VAN RIJCKEVORSEL and DE LEEUW · Component and Correspondence Analysis
WEISBERG · Applied Linear Regression, *Second Edition*
WESTFALL and YOUNG · Resampling-Based Multiple Testing: Examples and Methods for p-Value Adjustment
WHITTLE · Systems in Stochastic Equilibrium
WOODING · Planning Pharmaceutical Clinical Trials: Basic Statistical Principles

*Now available in a lower priced paperback edition in the Wiley Classics Library.

WOOLSON · Statistical Methods for the Analysis of Biomedical Data
*ZELLNER · An Introduction to Bayesian Inference in Econometrics

Texts and References Section

AGRESTI · An Introduction to Categorical Data Analysis
ANDERSON · An Introduction to Multivariate Statistical Analysis, *Second Edition*
ANDERSON and LOYNES · The Teaching of Practical Statistics
ARMITAGE and COLTON · Encyclopedia of Biostatistics: Volumes 1 to 6 with Index
BARTOSZYNSKI and NIEWIADOMSKA-BUGAJ · Probability and Statistical Inference
BERRY, CHALONER, and GEWEKE · Bayesian Analysis in Statistics and Econometrics: Essays in Honor of Arnold Zellner
BHATTACHARYA and JOHNSON · Statistical Concepts and Methods
BILLINGSLEY · Probability and Measure, *Second Edition*
BOX · R. A. Fisher, the Life of a Scientist
BOX, HUNTER, and HUNTER · Statistics for Experimenters: An Introduction to Design, Data Analysis, and Model Building
BOX and LUCEÑO · Statistical Control by Monitoring and Feedback Adjustment
BROWN and HOLLANDER · Statistics: A Biomedical Introduction
CHATTERJEE and PRICE · Regression Analysis by Example, *Second Edition*
COOK and WEISBERG · An Introduction to Regression Graphics
COX · A Handbook of Introductory Statistical Methods
DILLON and GOLDSTEIN · Multivariate Analysis: Methods and Applications
DODGE and ROMIG · Sampling Inspection Tables, *Second Edition*
DRAPER and SMITH · Applied Regression Analysis, *Third Edition*
DUDEWICZ and MISHRA · Modern Mathematical Statistics
DUNN · Basic Statistics: A Primer for the Biomedical Sciences, *Second Edition*
FISHER and VAN BELLE · Biostatistics: A Methodology for the Health Sciences
FREEMAN and SMITH · Aspects of Uncertainty: A Tribute to D. V. Lindley
GROSS and HARRIS · Fundamentals of Queueing Theory, *Third Edition*
HALD · A History of Probability and Statistics and their Applications Before 1750
HALD · A History of Mathematical Statistics from 1750 to 1930
HELLER · MACSYMA for Statisticians
HOEL · Introduction to Mathematical Statistics, *Fifth Edition*
JOHNSON and BALAKRISHNAN · Advances in the Theory and Practice of Statistics: A Volume in Honor of Samuel Kotz
JOHNSON and KOTZ (editors) · Leading Personalities in Statistical Sciences: From the Seventeenth Century to the Present
JUDGE, GRIFFITHS, HILL, LÜTKEPOHL, and LEE · The Theory and Practice of Econometrics, *Second Edition*
KHURI · Advanced Calculus with Applications in Statistics
KOTZ and JOHNSON (editors) · Encyclopedia of Statistical Sciences: Volumes 1 to 9 wtih Index
KOTZ and JOHNSON (editors) · Encyclopedia of Statistical Sciences: Supplement Volume
KOTZ, REED, and BANKS (editors) · Encyclopedia of Statistical Sciences: Update Volume 1
KOTZ, REED, and BANKS (editors) · Encyclopedia of Statistical Sciences: Update Volume 2
LAMPERTI · Probability: A Survey of the Mathematical Theory, *Second Edition*
LARSON · Introduction to Probability Theory and Statistical Inference, *Third Edition*
LE · Applied Survival Analysis
MALLOWS · Design, Data, and Analysis by Some Friends of Cuthbert Daniel
MARDIA · The Art of Statistical Science: A Tribute to G. S. Watson

*Now available in a lower priced paperback edition in the Wiley Classics Library.

Texts and References (Continued)

    MASON, GUNST, and HESS · Statistical Design and Analysis of Experiments with Applications to Engineering and Science
    MURRAY · X-STAT 2.0 Statistical Experimentation, Design Data Analysis, and Nonlinear Optimization
    PURI, VILAPLANA, and WERTZ · New Perspectives in Theoretical and Applied Statistics
    RENCHER · Methods of Multivariate Analysis
    RENCHER · Multivariate Statistical Inference with Applications
    ROSS · Introduction to Probability and Statistics for Engineers and Scientists
    ROHATGI · An Introduction to Probability Theory and Mathematical Statistics
    RYAN · Modern Regression Methods
    SCHOTT · Matrix Analysis for Statistics
    SEARLE · Matrix Algebra Useful for Statistics
    STYAN · The Collected Papers of T. W. Anderson: 1943–1985
    TIERNEY · LISP-STAT: An Object-Oriented Environment for Statistical Computing and Dynamic Graphics
    WONNACOTT and WONNACOTT · Econometrics, *Second Edition*

WILEY SERIES IN PROBABILITY AND STATISTICS
ESTABLISHED BY WALTER A. SHEWHART AND SAMUEL S. WILKS

Editors
Robert M. Groves, Graham Kalton, J. N. K. Rao, Norbert Schwarz, Christopher Skinner

Survey Methodology Section

    BIEMER, GROVES, LYBERG, MATHIOWETZ, and SUDMAN · Measurement Errors in Surveys
    COCHRAN · Sampling Techniques, *Third Edition*
    COX, BINDER, CHINNAPPA, CHRISTIANSON, COLLEDGE, and KOTT (editors) · Business Survey Methods
  *DEMING · Sample Design in Business Research
    DILLMAN · Mail and Telephone Surveys: The Total Design Method
    GROVES and COUPER · Nonresponse in Household Interview Surveys
    GROVES · Survey Errors and Survey Costs
    GROVES, BIEMER, LYBERG, MASSEY, NICHOLLS, and WAKSBERG · Telephone Survey Methodology
  *HANSEN, HURWITZ, and MADOW · Sample Survey Methods and Theory, Volume 1: Methods and Applications
  *HANSEN, HURWITZ, and MADOW · Sample Survey Methods and Theory, Volume II: Theory
    KASPRZYK, DUNCAN, KALTON, and SINGH · Panel Surveys
    KISH · Statistical Design for Research
  *KISH · Survey Sampling
    LESSLER and KALSBEEK · Nonsampling Error in Surveys
    LEVY and LEMESHOW · Sampling of Populations: Methods and Applications
    LYBERG, BIEMER, COLLINS, de LEEUW, DIPPO, SCHWARZ, TREWIN (editors) · Survey Measurement and Process Quality
    SKINNER, HOLT, and SMITH · Analysis of Complex Surveys

*Now available in a lower priced paperback edition in the Wiley Classics Library.